AF687397

# Plant Physiology and Development

# Plant Physiology and Development

Edited by Edgar Crombie

MURPHY & MOORE
www.murphy-moorepublishing.com

Published by Murphy & Moore Publishing,
1 Rockefeller Plaza,
New York City, NY 10020, USA
www.murphy-moorepublishing.com

**Plant Physiology and Development**
Edited by Edgar Crombie

© 2022 Murphy & Moore Publishing

International Standard Book Number: 978-1-63987-434-7 (Hardback)

This book contains information obtained from authentic and highly regarded sources. Copyright for all individual chapters remain with the respective authors as indicated. All chapters are published with permission under the Creative Commons Attribution License or equivalent. A wide variety of references are listed. Permission and sources are indicated; for detailed attributions, please refer to the permissions page and list of contributors. Reasonable efforts have been made to publish reliable data and information, but the authors, editors and publisher cannot assume any responsibility for the validity of all materials or the consequences of their use.

**Trademark Notice:** Registered trademark of products or corporate names are used only for explanation and identification without intent to infringe.

**Cataloging-in-Publication Data**

Plant physiology and development / edited by Edgar Crombie.
p. cm.
Includes bibliographical references and index.
ISBN 978-1-63987-434-7
1. Plant physiology. 2. Plants--Development. 3. Growth (Plants).
I. Crombie, Edgar.
QK711.2 .P53 2022
571.2--dc23

# TABLE OF CONTENTS

# PREFACE

I am honored to present to you this unique book which encompasses the most up-to-date data in the field. I was extremely pleased to get this opportunity of editing the work of experts from across the globe. I have also written papers in this field and researched the various aspects revolving around the progress of the discipline. I have tried to unify my knowledge along with that of stalwarts from every corner of the world, to produce a text which not only benefits the readers but also facilitates the growth of the field.

The subdiscipline of botany which studies the functioning of plants is known as plant physiology. It aims to understand the fundamental processes which occur in plants, such as respiration, photosynthesis, photoperiodism, transpiration, nastic movements, hormone functions, etc. It studies all the chemical and physical processes of plants at a small as well as large scale. Phytochemistry is a broad area of study within plant physiology. It refers to the study of the chemicals derived from plants. These chemicals are known as phytochemicals. Plant development is another area which is focused in this discipline. It is concerned with the growth and development of plant structures like roots, buds, flowers, shoots and leaves. This book aims to shed light on some of the unexplored aspects of plant physiology and the recent researches in this field. Some of the diverse topics covered herein address the varied branches that fall under this category. For all readers who are interested in plant physiology and plant development, the case studies included in this book will serve as an excellent guide to develop a comprehensive understanding.

Finally, I would like to thank all the contributing authors for their valuable time and contributions. This book would not have been possible without their efforts. I would also like to thank my friends and family for their constant support.

**Editor**

# 1

# Multiple regulatory roles of AP2/ERF transcription factor in angiosperm

Chao Gu*, Zhi-Hua Guo, Ping-Ping Hao, Guo-Ming Wang, Zi-Ming Jin and Shao-Ling Zhang*

**Abstract**

APETALA2/ethylene response factor (AP2/ERF) transcription factor (TF) is a superfamily in plant kingdom, which has been reported to be involved in regulation of plant growth and development, fruit ripening, defense response, and metabolism. As the final response gene in ethylene signaling pathway, AP2/ERF TF could feedback modulate phytohormone biosynthesis, including ethylene, cytokinin, gibberellin, and abscisic acid. Moreover, AP2/ERF TF also participates in response to the signals of auxin, cytokinin, abscisic acid, and jasmonate. Thus, this superfamily is key regulator for connecting the phytohormonal signals. In this review, based on the evidence of structural and functional studies, we discussed the multiple regulator roles of AP2/ERF TF in angiosperm, and then constructed the network model of AP2/ERF TF in response to various phytohormonal signals and regulatory mechanism of the cross-talk.

**Keywords:** AP2/ERF TF, Stress, Plant growth and development, Fruit ripening, Phytohormones

## Background

The surperfamily APELATA 2/ethylene response factor (AP2/ERF) has been studies in many plants, which have a range of 119–200 members (Du et al. 2014; Nakano et al. 2006; Rao et al. 2015; Zhuang et al. 2008), and have been reported in responses to ethylene, stress, metabolic, fruit ripening and senescence (Han et al. 2016; Koyama et al. 2013; Lee et al. 2012; Li et al. 2007; Fits and Memelink 2000; Trujillo et al. 2008; Zhu et al. 2014). All the time, regulatory mechanism of AP2/ERF TF in these fields were wide-spread studies by many scientists and their research teams, and increasing experimental evidence was exploited to elucidate the detailed roles in each field (Guo and Ecker 2004; Liu et al. 2014; Pré et al. 2008; Taketa et al. 2008; Tang et al. 2016; Xiao et al. 2013; Yin et al. 2016). Herein, research advance of AP2/ERF TF was reviewed in plant, and the doubtful viewpoints were also discussed.

## Classification and DNA-binding elements

According to previous reports, the superfamily AP2/ERF members contain a common DNA binding domain, AP2 domain. Based on the difference of this domain in copy numbers, AP2/ERF TF could usually be divided into four families, AP2, ERF, RAV, and Soloist (Nakano et al. 2006; Licausi et al. 2010a). AP2 members constitute by one or additionally taking a tandem repeated AP2 domain (Kagaya et al. 1999; Licausi et al. 2013). ERF members characterize by a single AP2 domain (Nakano et al. 2006; Licausi et al. 2013). RAV members comprise by a consensus sequence elements for both AP2 domain and B3 domains (Kagaya et al. 1999; Swaminathan et al. 2008). Soloist family have little members (one or two) that also contain a single AP2 domain in all sequenced plant genome, but they strongly diverged in gene sequence from other AP2/ERF members (Du et al. 2013; Licausi et al. 2010a; Rao et al. 2015; Zhuang et al. 2008). Because of ERF family members could bind to two mainly DNA-binding elements (Hao et al. 2002; Sakuma et al. 2002), resulting in a novel DREB family is separated from ERF family (Du et al. 2013; Rao et al. 2015; Sakuma et al. 2002; Zhuang et al. 2008). Of the DREB and ERF families, all the members are further classified into six groups, A1 to A6 and B1 to B6, respectively (Sakuma et al. 2002). However, these twelve groups are re-designated with group I

*Correspondence: guchao@njau.edu.cn; zhangsl@njau.edu.cn
State Key Laboratory of Crop Genetics and Germplasm Enhancement, Center of Pear Engineering Technology Research, Nanjing Agricultural University, Nanjing 210095, China

to X, VI-L, and Xb-L or group A to J (Nakano et al. 2006). The re-designated classification is employed in horticultural plants, such as *Vitis vinifera*, *Prunus mume*, and *Solanum lycopersicon* (Licausi et al. 2010a; Du et al. 2013; Pirrello et al. 2012), whereas the traditionally classification is used in other plant species, including *Salix arbutifolia*, *Nicotiana tabacum*, and *Populus trichocarpa* (Rao et al. 2015; Sasaki et al. 2007; Zhuang et al. 2008).

AP2/ERF proteins have strongly capacity to bind a wide range of cis-regulatory elements in promoter of target genes (Sasaki et al. 2007). Of these cis-regulatory elements, GCC-box (AGCCGCC element) and DRE/CRT (dehydrationresponsive element/C-repeat, RCCGCC element) are the mainly two DNA-binding elements (De Boer et al. 2011; Fujimoto et al. 2000; Hao et al. 1998, 2002; Oñate-Sánchez et al. 2007; Wang et al. 2012). Noteworthy, most AP2/ERF proteins can bind GCC-box containing promoter, but the activation degree is different among members in various groups. For instance, the members are weak activators in group A, B and E, neutral in class G and H, and strong in group C, whereas that are as repressor in group F (Pirrello et al. 2012). Besides GCC-box and DRE/CRT, the elements diverged from these two also belong to cis-regulatory elements, which may be in response to different stimuli underlying various stresses (Mizoi et al. 2012; Shaikhali et al. 2008; Welsch et al. 2007). Moreover, ERF protein can also bind to VWRE (vascular wounding responsive element, GAAAAGAAAATTTC) and CE1 (coupling element, CACCG) in tobacco (Sasaki et al. 2007; Wu et al. 2008). In addition, few reports reveal that ERF proteins could interact directly with a non-GCC element containing promoters (Chakravarthy et al. 2003).

## Ethylene response

Ethylene is an important phytohorome for plant growth, development, senescence, and stress tolerance. Ethylene is synthesized by ACS (1-aminocyclopropane-1- carboxyla synthase) catalyzing substrate of SAM (S-adenosyl methionine) to form ACC (1-aminocyclopropane-1-carboxyla acid), and then impel by ACO (1-aminocyclopropane-1-carboxyla oxidase). Sequentially, how much ethylene produced in plant tissues are positive correlated to ACS and ACO activities. The produced ethylene in plant tissues is combined with ETR (Ethylene receptor) to activate constitutive triple response (CTR), and then induce expression of a set of ethylene insensitive (EIN) and Ethylene insensitive-like (EIL). The EIN/EIL proteins bind to upstream regions of ERF TFs to promote it expressed in tissues (Alexander and Grierson 2002; Guo and Ecker 2003; Solano et al. 1998). However, due to GCC-box usually presented in the promoter of ACS and ACO in many plants, the expressed *ERF* genes will enhance the activities of the two genes, thereby accelerate ethylene biosynthesis and signal transduction, such as *LeERF1*, *AtERF73/HRE1*, *TERF2/LeERF2*, and *MaERF9* (Li et al. 2007; Xiao et al. 2013; Yang et al. 2011; Zhang et al. 2009). Besides the positive feedback genes, few ERF TFs also represent as repressor of ACS and ACO activities to prevent ethylene biosynthesis, including *AtERF4*, *AtERF11*, *SlERF6*, and *MaERF11* (Lee et al. 2012; Li et al. 2011; Xiao et al. 2013; Yang et al. 2005). In addition, ERF. B3 has the ability to modulate the transcription levels of a subset of other ERF TFs (Liu et al. 2013). Noteworthy, this subset contains the aforementioned activators and repressors of ethylene biosynthesis and signal pathway genes. Thus, AP2/ERF TF is not only in response to ethylene signal transduction, but also can feedback regulate ethylene synthesis in plant tissues.

## Stress tolerance

Stresses are the negative environment factors around plant growth and development. Both abiotic and biotic stresses are mediated by multiple transcriptional factors, such as NAC, WRKY, MYB, bHLH, bZIP, and ERF (Abe et al. 2003; Li et al. 2013; Puranik et al. 2012; Rushton et al. 2010; Singh et al. 2002; Zhang et al. 2012a). Most studies have found the importance of AP2/ERF TF in defense of various stresses. In general, the AP2/ERF TFs in response to abiotic stresses are the members of DREB family (Licausi et al. 2013; Sakuma et al. 2002). Such as *AtERF98*, *MsERF8*, *JcERF011*, and *CaERFLP1* that enhance tolerance to salt (Chen et al. 2012; Lee et al. 2004; Tang et al. 2016; Zhang et al. 2004, 2012b). *TERF2/LeERF2*, *CBF1*, and *CBF3* exalt cold and freezing tolerances (Novillo et al. 2007; Tian et al. 2011; Zhang et al. 2010b). *Sub1A*, *SNORKEL1* and *SNORKEL2* allow rice to adapt to deep water (Fukao et al. 2006, 2011; Hattori et al. 2009; Xu et al. 2006). *HRE1* and *HRE2* improve the tolerance of the plant to the hypoxia stress (Licausi et al. 2010b). *OsWR1*, *JERF1*, *TERF1*, and *SHINE* are positive regulators of resistance to drought (Aharoni et al. 2004; Wang et al. 2012; Zhang et al. 2005, 2010a). Moreover, few of AP2/ERF TFs are involved to modulate at least two different abiotic stresses in defense response. For example, over-expression of *SlERF5* in transgenic tomato plants result in high tolerance to drought and salt stress (Pan et al. 2012). Over-expression of *JERF3* and *SodERF3* improve resistance to drought, osmotic, salt, and freezing stresses in transgenic rice and tobacco (Trujillo et al. 2008; Wu et al. 2008; Zhang et al. 2010c). Ectopic expression of *DREB2A* in *Arabidopsis* increase endurance to drought, stress, and heat stresses (Sakuma et al. 2006a, b).

Unless enhanced tolerance to abiotic stresses, AP2/ERF TF also are reported to be concerned in raising resistance to biotic stresses. Over-expression of *NtERF5* contributes to high tolerance to *Tobacco mosaic virus* in *Nicotiana tabacum* (Fischer and Droge-Laser 2004). Silence-expression of *ORA59* or *RAP2.2* results in low tolerance to *Botrytis cinerea* in *Arabidopsis thaliana* (Pré et al. 2008; Zhao et al. 2012). Loss-of-function mutants of *AtERF2* or *AtERF14* are more susceptible against *Fusarium oxysporum* in *Arabidopsis thaliana* (McGrath et al. 2005; Oñate-Sánchez et al. 2007). Exceptionally, *AtERF4* is the negatively genes in regulating *Fusarium oxysporum* resistance (McGrath et al. 2005). Similar to that in abiotic stress defense, few of AP2/ERF TFs have the ability to coordinate two or more biotic stresses in defense response. For instance, Over-expression of *MtERF1-1* improves tolerance to *Rhizoctonia solani* and *Phytophthora medicaginis* in *Medicago* roots (Anderson et al. 2010). Over-expression of *ERF1* in *Arabidopsis* conferred resistance to necrotrophic fungi including *B. cinerea* and *Plectosphaerella cucumerina* (Berrocal-Lobo et al. 2002). The tomato Transcription Factor *Pti4* Regulates Defense-Related Gene Expression for *Pseudomonas syringae* and *Erysiphe orontii* by combined to GCC Box and Non-GCC Box cis Elements (Chakravarthy et al. 2003).

In addition, few AP2/ERF TFs had been reported responsible for biotic and abiotic stress, simultaneously. A typical example is the positively regulator *TaPIE1* that raise the defense responses to *R. cerealis* and freezing stresses by activating defense- and stress-related genes (Zhu et al. 2014). Taken together, AP2/ERF TF plays very important roles in regulating defense response to all kinds of biotic and abiotic stresses.

## Plant growth, development, and senescence

The life of plant is cycled through seed germination, seedling growth, organ development, and senescence. In this cycle, AP2/ERF TF also displays their regulatory roles for shaping many architectural traits. In the process of seed germination, *SlERF2* positively improve transcription level of marker gene, mannanase 2, resulting in a stimulation of premature germination, and enhance hook formation of darkgrown (Pirrello et al. 2006). In the progression of plant growth and development, *AINTEGUMENTA* and *AINTEGUMENTA-LIKE6* are related to flower organ growth and ovule development in *Arabidopsis* (Elliott et al. 1996; Jofuku et al. 1994; Klucher et al. 1996; Krizek 2009; Mizukami and Fischer 2000). Rice ethylene-response AP2/ERF factor *OsEATB* restricts internode elongation by down-regulating ent-kaurene synthase A, leading to a reduction of rice plant height and panicle length at maturity (Qi et al. 2011). In contrast, *AtERF1*, *AtDREB1*, and *TINYT* present their ability in dwarfing plant height (Liu et al. 1998; Solano et al. 1998; Wilson et al. 1996). Moreover, *NtERF3*, *AtERF4* and *AtERF8* had been found to be associated with plant aging (Koyama et al. 2013). Of these three genes, *AtERF4* and *AtERF8* belonged to class II ERFs in *Arabidopsis*, which can accelerate precocious leaf senescence by targeting the *EPITHIOSPECIFIER PROTEIN/EPITHIOSPECIFYING SENESCENCE REGULATOR* gene and regulating the expression of many genes related to senescence (Koyama et al. 2013). In addition, AP2/ERF TF is involved in regulating metabolite productions, such as chlorophyll, wax and cutin. The present evidences show that *CitERF13* is negative regulator for chlorophyll degradation during *Citrus* fruit degreening by directly binding to the *CitPPH* promoter and enhancing the activity of a metabolite of pheophorbide hydrolase (Yin et al. 2016). *AtWIN1*, *AtSHN*, and *HvNUD* could increase an accumulation of wax and cutin on the epidermis by regulating a lipid biosynthesis pathway (Aharoni et al. 2004; Broun et al. 2004; Taketa et al. 2008). Obviously, the functions of these AP2/ERF TFs are distinctly elucidated in these reported traits, but the regulatory roles of other members should be further explored in unknown properties in future.

## Fruit ripening

Fruit is one of important tissues in fruited plants, which harbors seed formation, development, and maturity. According to respiratory intensity during ripening, fruit is divided into climacteric and non-climacteric phenotypes. The climacteric fruit must release massive ethylene at ripening, also called ethylene-dependent fruit. On the contrary, the non-climacteric fruit is ethylene-independent. To date, ethylene-dependent fleshy-fruits are the primary materials for studying fruit ripening, such as tomato, apple, and banana. In ethylene-dependent fruits, *ERF*, as the final response gene in ethylene signaling pathway, directly regulate fruit ripening by binding to the promoters of their downstream genes, including *ACO*, *ACS*, *PG*, *EXP*, and *PSY* (Han et al. 2016; Lee et al. 2012; Liu et al. 2014). At present, *LeERF1*, *MaERF9*, *MdERF1*, and *MdERF3* has been reported as the positive activator (Li et al. 2007, 2016; Wang et al. 2007; Xiao et al. 2013), whereas *SlERF6*, *MaERF11*, and *MdERF2* are the negative repressors for fruit ripening (Han et al. 2016; Lee et al. 2012). Of these *ERFs*, *MaERF9* and *MaERF11* could not only regulate the transcription levels of *ACO1* and *ACS1* by binding to their promoter, but also physically interacted with ACO1 (Xiao et al. 2013). Interestingly, MaERF11 also interact with MaHDA1, the complex repress expression levels of downstream genes targeted by MaERF11 via histone deacetylation (Han et al. 2016). Moreover, the regulatory route of *ERF* genes is intricate during fruit ripening. In apple, *MdERF2* presents at least three roads in regulating *MdACS* expression. *MdERF2* repressor and *MdERF3* activator could regulate the transcription level of *MdACS* by binding to their promoter, respectively. Meanwhile, *MdERF2* inhibit *MdERF3* activity by combining to the DRE element in the promoter, indirectly suppressed the expression level of *MdACS*. Thirdly, a directly interaction between *MdERF2* and *MdERF3* restrain the binding of *MdERF3* to the *MdACS* promoter, and then suppress the *MdACS* expressed in fruit flesh (Li et al. 2016). In tomato, *SlERF. B3* has the ability to activate the regulatory network for fruit ripening. A dominant repressor version of *SlERF. B3* (*SlERF.B3-SRDX*) down-regulates ethylene receptor levels, but enhances triple response and up-regulated the expression levels of *EIN3-like* gene, contributing to an acceleration of fruit ripening (Liu et al. 2013). Further study found that *SlERF.B3-SRDX* could alter the expression pattern of other ERF family members. Most notably, *SlERF.B3-SRDX* also stimulate the transcription levels of ripening regulators, including *RIPENING INHIBITOR* (*RIN*), *NON-RIPENING* (*NOR*), *COLORLESS NON-RIPENING* (*CNR*), and *Homeodomain-leucine zipper HOMEOBOX* (*HB-1)* (Liu et al. 2014). Therefore, the regulatory role of AP2/ERF TF is multiple, and their regulated mechanism is very complex during fruit ripening.

## Integration of phytohormonal signals

Phytohormones are a group of naturally occurring, organic substances which affected plant growth, development, and senescence at low concentrations. Of these phytohormones, auxin, cytokinin, and gibberellin are reported to be involved in regulation of seed germination and plant growth (Pacifici et al. 2015; Urbanova and Leubner-Metzger 2016; Werner et al. 2001). Ethylene plays extremely important roles in climacteric fleshy fruit ripening and senescence (Hayama et al. 2006; Xiao et al. 2013; Yin et al. 2008), and together with jasmonate and abscisic acid, participate in defense response to biotic and abiotic stresses (Li et al. 2011; Lorenzo et al. 2003; Pré et al. 2008). Obviously, cross-talk among these phytohormones must be carried out in plant tissues. This cross-talk is always surveyed by many scientists, and increasing evidences are emerged to elucidate the talk mechanism. Ethylene signal transduction is a general pathway during the life cycle of plant. As the final response gene in ethylene signaling pathway, AP2/ERF are also documented to be involved in response to other hormones. In rice, an AP2/ERF TF OsCRL5 is induced by treating with exogenous auxin, and inhibits cytokinin signal transduction by enhancing the activities of two repressors (Kitomi et al. 2011). Interestingly, several AP2/ERF TFs in subgroup B-5 are responsible for exogenous cytokinin, thereby designated as cytokinin response factor (Rashotte et al. 2006). Also in rice, *OsEATB*, which is restrained by ethylene and abscisic acid in expression level, negative regulate gibberellin biosynthesis by down-regulating a pathway gene (Qi et al. 2011). In *Arabidopsis*, tobacco and tomato, however, few AP2/ERF TFs are shown to modulate abscisic acid responses, such as *AtERF11* and *TSRF1* (Li et al. 2011; Zhang et al. 2008). The ethylene-, jasmonate-, and abscisic acid-responsive *JERF1* regulates abscisic acid biosynthesis-related gene in expression level (Zhang et al. 2004; Wu et al. 2007). Moreover, *NIC2* participate in mediating jasmonate-elicited nicotine biosynthesis (De Boer et al. 2011). *ORA59*, which was induced by jasmonate and ethylene in expression level, is the key regulator of jasmonate- and ethylene-responsive PLANT DEFENSIN 1.2 expression by binding to GCC-box element in the promoter (Pré et al. 2008; Zarei et al. 2011). *AtERF2* is a positive regulator of jasmonate-responsive defense genes, while *AtERF4* negative adjust jasmonate-responsive defense gene expression (McGrath et al. 2005). Overall, AP2/ERF TF is the key regulator to integrate all kinds of phytohormonal signals.

## Conclusions

The AP2/ERF superfamily has hundreds of members in various plants, which contains at least one AP2 domain in all designated families. Generally, AP2/ERF TF mediates downstream responsible genes by binding to the GCC-box and/or DREB element in the promoter. Unless responses to ethylene signal, a large number of AP2/ERF members are stimulated by auxin, cytokinin, abscisic acid, and jasmonate signals. Meanwhile, several members also modulate gibberellin, cytokinin, and abscisic acid contents by directly regulating biosynthesis pathway genes of these phytohormones. Moreover, the stimulated genes would further regulate downstream effectors, resulting in changes of agronomic traits, including plant growth, defense responses, and fruit ripening (Fig. 1). In summary, AP2/ERF TF presents multiple regulatory roles in angiosperm.

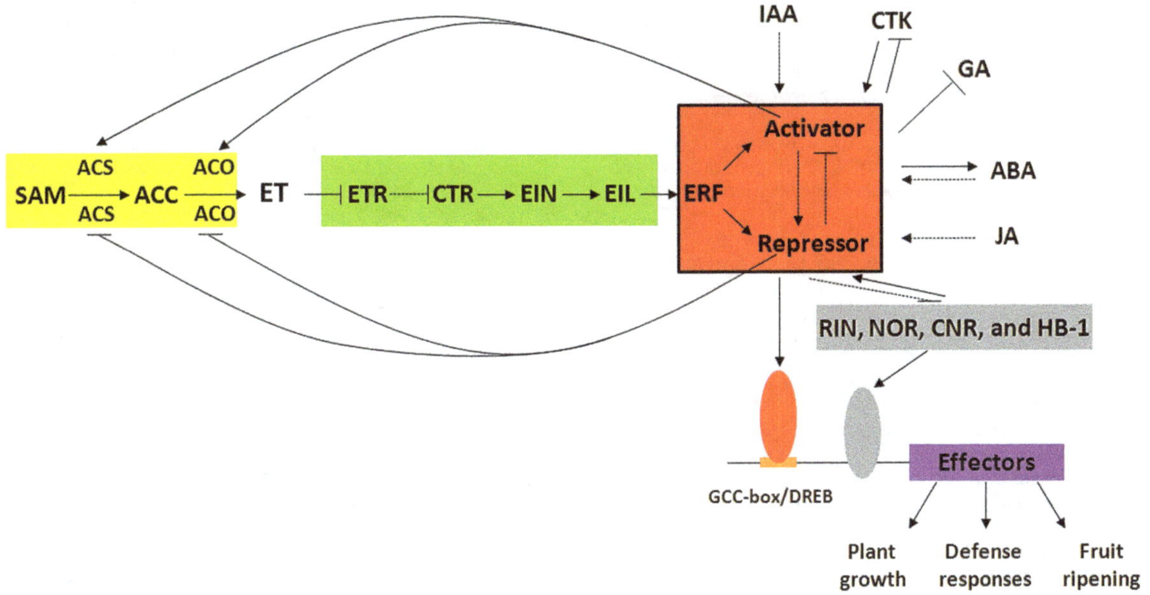


**Fig. 1** A network model for AP2/ERF genes response to phytohormones and regulating downstream effectors in angiosperm. Ethylene biosynthesis pathway, ethylene signaling pathway, and ripening regulators are indicated by *yellow*, *green*, and *gray* colors, respectively. AP2/ERF family members is *boxed* and *filled* with *red* color. In ethylene biosynthesis pathway, S-adenosine methionine (SAM) is converted to ethylene (ET) via an intermediate metabolites 1-aminocyclopropane-1-carboxyla (ACC), underlying the catalysis of the two enzymes 1-aminocyclopropane-1-carboxyla synthase (ACS) and oxidase (ACO). In ethylene signaling pathway, ET is firstly combined with ethylene receptor (ETR) to activate constitutive triple response (CTR), leading to expression of ethylene insensitive (EIN) and EIN-induced ethylene insensitive-like (EIL). EIL promote expression of ethylene response factor (ERF), including activator and repressor. The ERF activities are induced by auxin (IAA), cytokinin (CTK), abscisic acid (ABA), and jasmonate (JA), as well as ripening-related genes, such as *RIPENING INHIBITOR* (*RIN*), *NON-RIPENING* (*NOR*), *COLORLESS NON-RIPENING* (*CNR*), and Homeodomain-leucine zipper *HOMEOBOX* (*HB-1*). Meanwhile, ERF can reduce CTK and gibberellin (GA) levels but increase ABA biosynthesis. In ethylene responses, ERF also regulate ethylene level by enhancing and decreasing ACS/ACO activity mediated by the activators and repressors, respectively. As for effectors of plant growth, defense responses and fruit ripening, ERF can directly medium the expression by binding to GCC-box/ DREB element in the promoter, and have the ability to indirectly regulate it, due to few ERFs inhibit expression of *RIN*, *NOR*, *CNR*, and *HB-1* that can directly bind to the promoter of effectors

### Authors' contributions

CG and SLZ conceived and wrote the manuscript. ZHG provided the advance of classification and DNA-binding elements of AP2/ERF TF, ethylene signal transduction pathway, and ethylene-induced fruit ripening. PPH contributed the integration of phytohormone signals related to AP2/ERF. GMW and ZMJ conducted the advance of AP2/ERF TFs involved in regulating stress, plant growth and development. All authors read and approved the final manuscript.

### Acknowledgements

This work was supported by the National Natural Science Foundations of China (31471856 and 31672118).

### Competing interests

The authors declare that they have no competing interests.

### References

Abe H, Urao T, Ito T, Seki M, Shinozaki K, Yamaguchi-Shinozaki K (2003) Arabidopsis AtMYC2 (bHLH) and AtMYB2 (MYB) function as transcriptional activators in abscisic acid signaling. Plant Cell 15:63–78

Aharoni A, Dixit S, Jetter R, Thoenes E, van Arkel G, Pereira A (2004) The SHINE clade of AP2 domain transcription factors activates wax biosynthesis, alters cuticle properties, and confers drought tolerance when overexpressed in Arabidopsis. Plant Cell 16:2463–2480

Alexander L, Grierson D (2002) Ethylene biosynthesis and action in tomato: a model for climacteric fruit ripening. J Exp Bot 53:2039–2055

Anderson JP, Lichtenzveig J, Gleason C, Oliver RP, Singh KB (2010) The B-3 ethylene response factor MtERF1-1 mediates resistance to a subset of root pathogens in *Medicago truncatula* without adversely affecting symbiosis with rhizobia. Plant Physiol 154:861–873

Berrocal-Lobo M, Molina A, Solano R (2002) Constitutive expression of ETHYLENE-RESPONSE-FACTOR1 in *Arabidopsis* confers resistance to several necrotrophic fungi. Plant J 29:23–32

Broun P, Poindexter P, Osborne E, Jiang CZ, Riechmann JL (2004) WIN1, a transcriptional activator of epidermal wax accumulation in Arabidopsis. Proc Natl Acad Sci USA 101:4706–4711

Chakravarthy S, Tuori RP, D'Ascenzo MD, Fobert PR, Despres C, Martin GB (2003) The tomato transcription factor Pti4 regulates defense-related gene expression via GCC box and non-GCC box cis elements. Plant Cell 15:3033–3050

Chen TT, Yang QC, Gruber M, Kang JM, Sun Y, Ding W, Zhang TJ, Zhang XQ (2012) Expression of an alfalfa (*Medicago sativa* L.) ethylene response factor gene MsERF8 in tobacco plants enhances resistance to salinity. Mol Biol Rep 39:6067–6075

De Boer K, Tilleman S, Pauwels L, Vanden Bossche R, De Sutter V, Vanderhaeghen R, Hilson P, Hamill JD, Goossens A (2011) APETALA2/ETHYLENE RESPONSE FACTOR and basic helix-loop-helix tobacco transcription factors cooperatively mediate jasmonate-elicited nicotine biosynthesis. Plant J 66:1053–1065

Du DL, Hao RJ, Cheng TR, Pan HT, Yang WR, Wang J, Zhang QX (2013) Genome-wide analysis of the AP2/ERF gene family in *Prunus mume*. Plant Mol Biol Rep 31:741–750

Du HW, Huang M, Zhang ZX, Cheng SY (2014) Genome-wide analysis of the AP2/ERF gene family in maize waterlogging stress response. Euphytica 198:115–126

Elliott RC, Betzner AS, Huttner E, Oakes MP, Tucker WQJ, Gerentes D, Perez P, Smyth DR (1996) AINTEGUMENTA, an APETALA2-like gene of arabidopsis with pleiotropic roles in ovule development and floral organ growth. Plant Cell 8:155–168

Fischer U, Droge-Laser W (2004) Overexpression of NtERF5, a new member of the tobacco ethylene response transcription factor family enhances resistance to Tobacco mosaic virus. Mol Plant Microbe Interact 17:1162–1171

Fujimoto SY, Ohta M, Usui A, Shinshi H, Ohme-Takagi M (2000) Arabidopsis ethylene-responsive element binding factors act as transcriptional activators or repressors of GCC box-mediated gene expression. Plant Cell 12:393–404

Fukao T, Xu K, Ronald PC, Bailey-Serres J (2006) A variable cluster of ethylene response factor-like genes regulates metabolic and developmental acclimation responses to submergence in rice. Plant Cell 18:2021–2034

Fukao T, Yeung E, Bailey-Serres J (2011) The submergence tolerance regulator SUB1A mediates crosstalk between submergence and drought tolerance in rice. Plant Cell 23:412–427

Guo H, Ecker JR (2003) Plant responses to ethylene gas are mediated by SCF(EBF1/EBF2)-dependent proteolysis of EIN3 transcription factor. Cell 115:667–677

Guo HW, Ecker JR (2004) The ethylene signaling pathway: new insights. Curr Opin Plant Biol 7:40–49

Han YC, Kuang JF, Chen JY, Liu XC, Xiao YY, Fu CC, Wang JN, Wu KQ, Lu WJ (2016) Banana transcription factor MaERF11 recruits histone deacetylase MaHDA1 and represses the expression of MaACO1 and expansins during fruit ripening. Plant Physiol 171:1070–1084

Hao DY, Ohme-Takagi M, Sarai A (1998) Unique mode of GCC box recognition by the DNA-binding domain of ethylene-responsive element-binding factor (ERF domain) in plant. J Biol Chem 273:26857–26861

Hao DY, Yamasaki K, Sarai A, Ohme-Takagi M (2002) Determinants in the sequence specific binding of two plant transcription factors, CBF1 and NtERF2, to the DRE and GCC motifs. Biochem 41:4202–4208

Hattori Y, Nagai K, Furukawa S, Song XJ, Kawano R, Sakakibara H, Wu JZ, Matsumoto T, Yoshimura A, Kitano H, Matsuoka M, Mori H, Ashikari M (2009) The ethylene response factors SNORKEL1 and SNORKEL2 allow rice to adapt to deep water. Nature 460:1026–1030

Hayama H, Shimada T, Fujii H, Ito A, Kashimura Y (2006) Ethylene-regulation of fruit softening and softening-related genes in peach. J Exp Bot 57:4071–4077

Jofuku KD, Denboer BGW, Vanmontagu M, Okamuro JK (1994) Control of arabidopsis flower and seed development by the homeotic gene apetala2. Plant Cell 6:1211–1225

Kagaya Y, Ohmiya K, Hattori T (1999) RAV1, a novel DNA-binding protein, binds to bipartite recognition sequence through two distinct DNA-binding domains uniquely found in higher plants. Nucleic Acids Res 27:470–478

Kitomi Y, Ito H, Hobo T, Aya K, Kitano H, Inukai Y (2011) The auxin responsive AP2/ERF transcription factor CROWN ROOTLESS5 is involved in crown root initiation in rice through the induction of OsRR1, a type-A response regulator of cytokinin signaling. Plant J 67:472–484

Klucher KM, Chow H, Reiser L, Fischer RL (1996) The AINTEGUMENTA gene of Arabidopsis required for ovule and female gametophyte development is related to the floral homeotic gene APETALA2. Plant Cell 8:137–153

Koyama T, Nii H, Mitsuda N, Ohta M, Kitajima S, Ohme-Takagi M, Sato F (2013) A regulatory cascade involving class II ETHYLENE RESPONSE FACTOR transcriptional repressors operates in the progression of leaf senescence. Plant Physiol 162:991–1005

Krizek B (2009) AINTEGUMENTA and AINTEGUMENTA-LIKE6 act redundantly to regulate Arabidopsis floral growth and patterning. Plant Physiol 150:1916–1929

Lee JH, Hong JP, Oh SK, Lee S, Choi D, Kim WT (2004) The ethylene-responsive factor like protein 1 (CaERFLP1) of hot pepper (*Capsicum annuum* L.) interacts in vitro with both GCC and DRE/CRT sequences with different binding affinities: possible biological roles of CaERFLP1 in response to pathogen infection and high salinity conditions in transgenic tobacco plants. Plant Mol Biol 55:61–81

Lee JM, Joung JG, McQuinn R, Chung MY, Fei ZJ, Tieman D, Klee H, Giovannoni J (2012) Combined transcriptome, genetic diversity and metabolite profiling in tomato fruit reveals that the ethylene response factor SIERF6 plays an important role in ripening and carotenoid accumulation. Plant J 70:191–204

Li YC, Zhu BZ, Xu WT, Zhu HL, Chen AJ, Xie YH, Shao Y, Luo YB (2007) LeERF1 positively modulated ethylene triple response on etiolated seedling, plant development and fruit ripening and softening in tomato. Plant Cell Rep 26:1999–2008

Li ZF, Zhang LX, Yu YW, Quan RD, Zhang ZJ, Zhang HW, Huang RF (2011) The ethylene response factor AtERF11 that is transcriptionally modulated by the bZIP transcription factor HY5 is a crucial repressor for ethylene biosynthesis in Arabidopsis. Plant J 68:88–99

Li XP, Zhu XY, Mao J, Zou Y, Fu DW, Chen WX, Lu WJ (2013) Isolation and characterization of ethylene response factor family genes during development, ethylene regulation and stress treatments in papaya fruit. Plant Physiol Bioch 70:81–92

Li T, Jiang Z, Zhang L, Tan D, Wei Y, Yuan H, Li T, Wang A (2016) Apple (Malus domestica) MdERF2 negatively affects ethylene biosynthesis during fruit ripening by suppressing MdACS1 transcription. Plant J 88:735–748

Licausi F, Giorgi FM, Zenoni S, Osti F, Pezzotti M, Perata P (2010a) Genomic and transcriptomic analysis of the AP2/ERF superfamily in *Vitis vinifera*. BMC Genom 11:719

Licausi F, van Dongen JT, Giuntoli B, Novi G, Santaniello A, Geigenberger P, Perata P (2010b) HRE1 and HRE2, two hypoxia-inducible ethylene response factors, affect anaerobic responses in *Arabidopsis thaliana*. Plant J 62:302–315

Licausi F, Ohme-Takagi M, Perata P (2013) APETALA2/Ethylene Responsive Factor (AP2/ERF) transcription factors: mediators of stress responses and developmental programs. New Phytol 199:639–649

Liu Q, Kasuga M, Sakuma Y, Abe H, Miura S, Yamaguchi-Shinozaki K, Shinozaki K (1998) Two transcription factors, DREB1 and DREB2, with an EREBP/AP2 DNA binding domain separate two cellular signal transduction pathways in drought- and low-temperature-responsive gene expression, respectively, in Arabidopsis. Plant Cell 10:1391–1406

Liu M, Pirrello J, Kesari R, Mila I, Roustan JP, Li Z, Latche A, Pech JC, Bouzayen M, Regad F (2013) A dominant repressor version of the tomato Sl-ERF.B3 gene confers ethylene hypersensitivity via feedback regulation of ethylene signaling and response components. Plant J 76:406–419

Liu MC, Diretto G, Pirrello J, Roustan JP, Li ZG, Giuliano G, Regad F, Bouzayen M (2014) The chimeric repressor version of an ethylene response factor (ERF) family member, Sl-ERF.B3, shows contrasting effects on tomato fruit ripening. New Phytol 203:206–218

Lorenzo O, Piqueras R, Sanchez-Serrano JJ, Solano R (2003) ETHYLENE RESPONSE FACTOR1 integrates signals from ethylene and jasmonate pathways in plant defense. Plant Cell 15:165–178

McGrath KC, Dombrecht B, Manners JM, Schenk PM, Edgar CI, Maclean DJ, Scheible WR, Udvardi MK, Kazan K (2005) Repressor- and activator-type ethylene response factors functioning in jasmonate signaling and disease resistance identified via a genome-wide screen of *Arabidopsis* transcription factor gene expression. Plant Physiol 139:949–959

Mizoi J, Shinozaki K, Yamaguchi-Shinozaki K (2012) AP2/ERF family transcription factors in plant abiotic stress responses. Bba-Gene Regul Mech 1819:86–96

Mizukami Y, Fischer RL (2000) Plant organ size control: AINTEGUMENTA regulates growth and cell numbers during organogenesis. Proc Natl Acad Sci USA 97:942–947

Nakano T, Suzuki K, Fujimura T, Shinshi H (2006) Genome-wide analysis of the ERF gene family in Arabidopsis and rice. Plant Physiol 140:411–432

Novillo F, Medina J, Salinas J (2007) Arabidopsis CBF1 and CBF3 have a different function than CBF2 in cold acclimation and define different gene classes in the CBF regulon. Proc Natl Acad Sci USA 104:21002–21007

Oñate-Sánchez L, Anderson JP, Young J, Singh KB (2007) AtERF14, a member of the ERF family of transcription factors, plays a nonredundant role in plant defense. Plant Physiol 143:400–409

Pacifici E, Polverari L, Sabatini S (2015) Plant hormone cross-talk: the pivot of root growth. J Exp Bot 66:1113–1121

Pan Y, Seymour GB, Lu C, Hu Z, Chen X, Chen G (2012) An ethylene response factor (ERF5) promoting adaptation to drought and salt tolerance in tomato. Plant Cell Rep 31:349–360

Pirrello J, Jaimes-Miranda F, Sanchez-Ballesta MT, Tournier B, Khalil-Ahmad Q, Regad F, Latche A, Pech JC, Bouzayen M (2006) Sl-ERF2, a tomato ethylene response factor involved in ethylene response and seed germination. Plant Cell Physiol 47:1195–1205

Pirrello J, Prasad BC, Zhang W, Chen K, Mila I, Zouine M, Latche A, Pech JC, Ohme-Takagi M, Regad F, Bouzayen M (2012) Functional analysis and binding affinity of tomato ethylene response factors provide insight on the molecular bases of plant differential responses to ethylene. BMC Plant Biol 12:190

Pré M, Atallah M, Champion A, De Vos M, Pieterse CM, Memelink J (2008) The AP2/ERF domain transcription factor ORA59 integrates jasmonic acid and ethylene signals in plant defense. Plant Physiol 147:1347–1357

Puranik S, Sahu PP, Srivastava PS, Prasad M (2012) NAC proteins: regulation and role in stress tolerance. Trends Plant Sci 17:369–381

Qi WW, Sun F, Wang QJ, Chen ML, Huang YQ, Feng YQ, Luo XJ, Yang JS (2011) Rice ethylene-response AP2/ERF factor OsEATB restricts internode elongation by down-regulating a Gibberellin biosynthetic gene. Plant Physiol 157:216–228

Rao GD, Sui JK, Zeng YF, He CY, Zhang JG (2015) Genome-wide analysis of the AP2/ERF gene family in Salix arbutifolia. Febs Open Bio 5:132–137

Rashotte AM, Mason MG, Hutchison CE, Ferreira FJ, Schaller GE, Kieber JJ (2006) A subset of Arabidopsis AP2 transcription factors mediates cytokinin responses in concert with a two-component pathway. Proc Natl Acad Sci USA 103:11081–11085

Rushton PJ, Somssich IE, Ringler P, Shen QJ (2010) WRKY transcription factors. Trends Plant Sci 15:247–258

Sakuma Y, Liu Q, Dubouzet JG, Abe H, Shinozaki K, Yamaguchi-Shinozaki K (2002) DNA-binding specificity of the ERF/AP2 domain of Arabidopsis DREBs, transcription factors involved in dehydration- and cold-inducible gene expression. Biochem Bioph Res Co 290:998–1009

Sakuma Y, Maruyama K, Osakabe Y, Qin F, Seki M, Shinozaki K, Yamaguchi-Shinozaki K (2006a) Functional analysis of an Arabidopsis transcription factor, DREB2A, involved in drought-responsive gene expression. Plant Cell 18:1292–1309

Sakuma Y, Maruyama K, Qin F, Osakabe Y, Shinozaki K, Yamaguchi-Shinozaki K (2006b) Dual function of an Arabidopsis transcriptionfactor DREB2A in water-stress-responsive and heat-stress-responsive gene expression. Proc Natl Acad Sci USA 103:18822–18827

Sasaki K, Mitsuhara I, Seo S, Ito H, Matsui H, Ohashi Y (2007) Two novel AP2/ERF domain proteins interact with cis-element VWRE for wound-induced expression of the Tobacco tpoxN1 gene. Plant J 50:1079–1092

Shaikhali J, Heiber I, Seidel T, Stroher E, Hiltscher H, Birkmann S, Dietz K-J, Baier M (2008) The redox-sensitive transcription factor Rap2.4a controls nuclear expression of 2-Cys peroxiredoxin A and other chloroplast antioxidant enzymes. BMC Plant Biol 8:48

Singh KB, Foley RC, Oñate-Sánchez L (2002) Transcription factors in plant defense and stress responses. Curr Opin Plant Biol 5:430–436

Solano R, Stepanova A, Chao QM, Ecker JR (1998) Nuclear events in ethylene signaling: a transcriptional cascade mediated by ETHYLENE-INSENSITIVE3 and ETHYLENE-RESPONSE-FACTOR1. Gene Dev 12:3703–3714

Swaminathan K, Peterson K, Jack T (2008) The plant B3 superfamily. Trends Plant Sci 13:647–655

Taketa S, Amano S, Tsujino Y, Sato T, Saisho D, Kakeda K, Nomura M, Suzuki T, Matsumoto T, Sato K, Kanamori H, Kawasaki S, Takeda K (2008) Barley grain with adhering hulls is controlled by an ERF family transcription factor gene regulating a lipid biosynthesis pathway. Proc Natl Acad Sci USA 105:4062–4067

Tang YH, Qin SS, Guo YL, Chen YB, Wu PZ, Chen YP, Li MR, Jiang HW, Wu GJ (2016) Genome-wide analysis of the AP2/ERF gene family in physic nut and overexpression of the JcERF011 gene in rice increased its sensitivity to salinity stress. PLoS ONE 11:e0150879

Tian Y, Zhang HW, Pan XW, Chen XL, Zhang ZJ, Lu XY, Huang RF (2011) Overexpression of ethylene response factor TERF2 confers cold tolerance in rice seedlings. Transgenic Res 20:857–866

Trujillo LE, Sotolongo M, Menendez C, Ochogavia ME, Coll Y, Hernandez I, Borras-Hidalgo O, Thomma BPHJ, Vera P, Hernandez L (2008) SodERF3, a novel sugarcane ethylene responsive factor (ERF), enhances salt and drought tolerance when overexpressed in tobacco plants. Plant Cell Physiol 49:512–525

Urbanova T, Leubner-Metzger G (2016) Gibberellins and seed germination. Annu Plant Rev 49:253–284

van der Fits L, Memelink J (2000) ORCA3, a jasmonate-responsive transcriptional regulator of plant primary and secondary metabolism. Science 289:295–297

Wang A, Tan D, Takahashi A, Li TZ, Harada T (2007) MdERFs, two ethylene-response factors involved in apple fruit ripening. J Exp Bot 58:3743–3748

Wang YH, Wan LY, Zhang LX, Zhang ZJ, Zhang HW, Quan RD, Zhou SR, Huang RF (2012) An ethylene response factor OsWR1 responsive to drought stress transcriptionally activates wax synthesis related genes and increases wax production in rice. Plant Mol Biol 78:275–288

Welsch R, Maass D, Voegel T, DellaPenna D, Beyer P (2007) Transcription factor RAP2.2 and its interacting partner SINAT2: stable elements in the carotenogenesis of Arabidopsis leaves. Plant Physiol 145:1073–1085

Werner T, Motyka V, Strnad M, Schmulling T (2001) Regulation of plant growth by cytokinin. Proc Natl Acad Sci USA 98:10487–10492

Wilson K, Long D, Swinburne J, Coupland G (1996) A dissociation insertion causes a semidominant mutation that increases expression of TINY, an arabidopsis gene related to APETALA2. Plant Cell 8:659–671

Wu LJ, Chen XL, Ren HY, Zhang ZJ, Zhang HW, Wang JY, Wang XC, Huang RF (2007) ERF protein JERF1 that transcriptionally modulates the expression of abscisic acid biosynthesis-related gene enhances the tolerance under salinity and cold in tobacco. Planta 226:815–825

Wu LJ, Zhang ZJ, Zhang HW, Wang XC, Huang RF (2008) Transcriptional modulation of ethylene response factor protein JERF3 in the oxidative stress response enhances tolerance of tobacco seedlings to salt, drought, and freezing. Plant Physiol 148:1953–1963

Xiao YY, Chen JY, Kuang JF, Shan W, Xie H, Jiang YM, Lu WJ (2013) Banana ethylene response factors are involved in fruit ripening through their interactions with ethylene biosynthesis genes. J Exp Bot 64:2499–2510

Xu K, Xu X, Fukao T, Canlas P, Maghirang-Rodriguez R, Heuer S, Ismail AM, Bailey-Serres J, Ronald PC, Mackill DJ (2006) Sub1A is an ethylene-response-factor-like gene that confers submergence tolerance to rice. Nature 442:705–708

Yang Z, Tian L, Latoszek-Green M, Brown D, Wu K (2005) Arabidopsis ERF4 is a transcriptional repressor capable of modulating ethylene and abscisic acid responses. Plant Mol Biol 58: 585–596

Yang CY, Hsu FC, Li JP, Wang NN, Shih MC (2011) The AP2/ERF transcription factor AtERF73/HRE1 modulates ethylene responses during hypoxia in Arabidopsis. Plant Physiol 156:202–212

Yin XR, Chen KS, Allan AC, Wu RM, Zhang B, Lallu N, Ferguson IB (2008) Ethylene-induced modulation of genes associated with the ethylene signalling pathway in ripening kiwifruit. J Exp Bot 59:2097–2108

Yin XR, Xie XL, Xia XJ, Yu JQ, Ferguson IB, Giovannoni JJ, Chen KS (2016) Involvement of an ethylene response factor in chlorophyll degradation during citrus fruit degreening. Plant J 86:403–412

Zarei A, Korbes AP, Younessi P, Montiel G, Champion A, Memelink J (2011) Two GCC boxes and AP2/ERF-domain transcription factor ORA59 in jasmonate/ethylene-mediated activation of the PDF1.2 promoter in Arabidopsis. Plant Mol Biol 75:321–331

Zhang Z, Huang R (2010) Enhanced tolerance to freezing in tobacco and tomato overexpressing transcription factor TERF2/LeERF2 is modulated by ethylene biosynthesis. Plant Mol Biol 73:241–249

Zhang HW, Huang ZJ, Xie BY, Chen Q, Tian X, Zhang XL, Zhang HB, Lu XY, Huang DF, Huang RF (2004) The ethylene-, jasmonate-, abscisic acid- and NaCl-responsive tomato transcription factor JERF1 modulates expression of GCC box-containing genes and salt tolerance in tobacco. Planta 220:262–270

Zhang X, Zhang Z, Chen J, Chen Q, Wang XC, Huang R (2005) Expressing TERF1 in tobacco enhances drought tolerance and abscisic acid sensitivity during seedling development. Planta 222:494–501

Zhang HB, Yang YH, Zhang ZJ, Chen J, Wang XC, Huang RF (2008) Expression of the ethylene response factor gene TSRF1 enhances abscisic acid responses during seedling development in tobacco. Planta 228:777–787

Zhang ZJ, Zhang HW, Quan RD, Wang XC, Huang RF (2009) Transcriptional regulation of the ethylene response factor LeERF2 in the expression of ethylene biosynthesis genes controls ethylene production in tomato and tobacco. Plant Physiol 150:365–377

Zhang Z, Li F, Li D, Zhang H, Huang R (2010a) Expression of ethylene response factor JERF1 in rice improves tolerance to drought. Planta 232:765–774

Zhang H, Liu W, Wan L, Li F, Dai L, Li D, Zhang Z, Huang R (2010b) Functional analyses of ethylene response factor JERF3 with the aim of improving tolerance to drought and osmotic stress in transgenic rice. Transgenic Res 19:809–818

Zhang Z, Liu X, Wang X, Zhou M, Zhou X, Ye X, Wei X (2012a) An R2R3 MYB transcription factor in wheat, TaPIMP1, mediates host resistance to

Bipolaris sorokiniana and drought stresses through regulation of defense- and stress-related genes. New Phytol 196:1155–1170

Zhang ZJ, Wang J, Zhang RX, Huang RF (2012b) The ethylene response factor AtERF98 enhances tolerance to salt through the transcriptional activation of ascorbic acid synthesis in *Arabidopsis*. Plant J 71:273–287

Zhao Y, Wei T, Yin KQ, Chen Z, Gu H, Qu LJ, Qin G (2012) *Arabidopsis* RAP2.2 plays an important role in plant resistance to *Botrytis cinerea* and ethylene responses. New Phytol 195:450–460

Zhu XL, Qi L, Liu X, Cai SB, Xu HJ, Huang RF, Li JR, Wei XN, Zhang ZY (2014) The wheat ethylene response factor transcription factor PATHOGEN-INDUCED ERF1 mediates host responses to both the necrotrophic pathogen Rhizoctonia cerealis and freezing stresses. Plant Physiol 164:1499–1514

Zhuang J, Cai B, Peng RH, Zhu B, Jin XF, Xue Y, Gao F, Fu XY, Tian YS, Zhao W, Qiao YS, Zhang Z, Xiong AS, Yao QH (2008) Genome-wide analysis of the AP2/ERF gene family in *Populus trichocarpa*. Biochem Bioph Res Co 371:468–474

# 2

# Transcriptomic analyses reveal clathrin-mediated endocytosis involved in symbiotic seed germination of *Gastrodia elata*

Xu Zeng†, Yuanyuan Li†, Hong Ling, Sisi Liu, Mengmeng Liu, Juan Chen* and Shunxing Guo*

**Abstract**

**Background:** *Gastrodia elata* is a well-known medicinal orchid. In nature, the germination rate of *G. elata* is extremely poor, because there is no endosperm within the mature seed. It is crucial for *G. elata* to obtain nutrition from mycorrhizal fungi (*Mycena*) at the early-stage of germination. After germination, the seed gives rise to a protocorm. However, there are no "omic" studies on understanding the interaction between *Gastrodia* and *Mycena*. Here, we used transcriptomic approaches to explore changes in seed germination of *G. elata*.

**Results:** Based on RNA-Seq, a total of ~221 million clean reads were assembled *denovo* into 139,756 unigenes, including 42,140 unigenes that were annotated in public databases. Meanwhile, 1750 unigenes were identified as differentially expressed genes. Most of these differentially expressed genes were putatively involved in energy metabolism, plant defense, molecular signaling, and secondary metabolism. Additionally, numerous genes involved in clathrin-mediated endocytosis were identified from our data. Most of these genes (e.g., clathrin, adaptor protein, dynamin, HSC70) were basally expressed in seeds and highly expressed in protocorms.

**Conclusions:** Our data suggested that clathrin-mediated endocytosis could play important roles in symbiotic seed germination of *G. elata* with *Mycena* infections.

**Keywords:** *Gastrodia elata*, *Mycena*, Seed germination, Endocytosis

## Background

In nature, the establishment of orchid seedlings requires compatible fungi to colonize the seed, thus providing nutrients for the formation of a protocorm. The protocorm is a post-embryonic structure from which both shoot and root systems subsequently differentiate. After the differentiation of green leaves, most orchid seedlings acquire autotrophy, while some orchids (known as fully mycoheterotrophic plants) are achlorophyllous and obtain their entire carbon source from their mycorrhizal fungi (Leake 2004; Dearnaley 2007).

*Gastrodia elata* is a fully mycoheterotrophic orchid that associates with two groups of fungal partners, i.e., *Mycena* and *Armillaria*, to complete its ontogenesis, including seed germination, tuber formation, flowering and fruiting. *Mycena* species (e.g., *M. dendrobii*) act as a symbiont during the stages of seed germination and protocorm development (Kim et al. 2006). For further development and enlargement of tubers, flowering and fruit setting, *Armillaria* becomes essential for the nutrient supply. It is also noteworthy that *G. elata* is a well-known Chinese medicinal orchid (Tsai et al. 2016). Recently, a number of pharmacological experiments indicated that it had strong potential to combat Alzheimer's disease, Parkinson's disease and other neurodegenerative diseases (Manavalan et al. 2012).

Plants are constantly exposed to a wide variety of microorganisms with either friendly or unfriendly intentions. Indeed, some plants are engaged in mutualistic or parasitic symbioses with them, or a plant-microorganism relationship, via a molecular dialogue. The plasma

*Correspondence: kibchenjuan@126.com; sxguo2015@163.com
†Xu Zeng and Yuanyuan Li contributed equally to this work
Institute of Medicinal Plant Development, Chinese Academy of Medical Sciences & Peking Union Medical College, Beijing 100193, People's Republic of China

membrane (PM) of plants is a critical barrier that senses fungi and eventually allows their entry or the uptake of microbial molecules. Endocytosis is a process allowing extracellular particles or cargoes to enter the cell, whereby the PM invaginates and pinches off. In plants, Dhonukshe et al. (2007) have showed that clathrin-dependent endocytosis constitutes the predominant pathway for constitutive internalization of PM proteins.

In legume-*Rhizobium* symbiosis, the symbiotic relationship and nodule formation depend on mutual recognition based on a signal exchange between the two partners. Some studies revealed that *Medicago truncatula* root hair curling is associated with a strong stimulation of endocytosis at an early stage of symbiotic interaction with *Sinorhizobium meliloti*. Many plant genes associated with the initiation of endocytosis are induced, including Rab-like, Arf-like, and dynamin-like GTPases, and phosphatidylinositol 3-kinase. Most of these genes prevent root hair curling and infection thread formation (Leborgnecastel et al. 2010). However, the contribution of endocytosis to the establishment of orchid mycorrhizae has been hampered by a lack of data.

Although mycorrhizal fungi play a significant role in the symbiotic germination of orchids, the research of plant-fungi interactions in orchid mycorrhizae is very limited. In this study, we investigated changes in transcriptomic profiles during the symbiotic seed germination of *G. elata* inoculated with *M. dendrobii*. Meanwhile, endocytosis has been suggested to play a part in symbiosis between host-plant and endophyte (Wang et al. 2015). Therefore, we provided a detailed analysis of the expression patterns of genes involved in endocytosis.

Previous studies used green orchids as the experimental materials and presented some basic knowledge of the early plant-fungus interactions of orchid mycorrhizae. However, there is still no information about seed germination of achlorophyllous orchids. Since *G. elata* relies entirely on a nutrient supply from fungal partners, this model provides an ideal system to investigate the symbiotic seed germination of orchids with fungal infections. Our results have useful value for elucidating more information about the symbiotic seed germination of orchids with fungal infections.

## Methods

### Sample preparation

The flowers of *G. elata* were pollinated by hand in the production base at Shaanxi, China. Capsules were collected just prior to dehiscence. Meanwhile, the mycorrhizal fungal isolate (*M. dendrobii*) was incubated on fresh potato dextrose agar (PDA) in darkness at 25 °C. The procedure for symbiotic germination of *G. elata* was performed according to the description in a previous report (Kim et al. 2006). Germination performances were evaluated weekly under a stereo-microscope for 2 months. The protocorms were carefully collected under a stereo-microscope. For further transcriptomic studies, samples of equal fresh weight of *M. dendrobii* hyphae, *G. elata* seeds and protocorms were prepared and immediately stored at −80 °C.

### Histological study

The mycorrhizal protocorms were fixed in a solution of 2.5% glutaraldehyde and 1.6% paraformaldehyde in 0.1 M phosphate buffer (pH 6.8) overnight at 4 °C. The samples were washed with 0.1 M phosphate buffer, then dehydrated using an ethanol series, and embedded in Technovit 7100 (Kulzer & Co., Germany) as described by Yeung and Chan (2015). Then, 3 μm-thick sections were cut with glass knives using a Reichert-Jung 2040 Autocut rotary microtome. These sections were stained with periodic acid-Schiff reaction for total insoluble carbohydrates, and counterstained with either 0.05% (w/v) toluidine blue O for general histology or 1% (w/v) amido black 10B for protein. The sections were observed and the images were captured digitally using a CCD camera attached to a light microscope (AxioImager A1, Carl Zeiss AG).

### Total RNA extraction, library preparation, and transcriptome sequencing

Total RNA from each sample was extracted by using the RNeasy® Plant Mini Kit (QIAGEN, GER) according to the manufacturer's instructions. RNA degradation and contamination were checked on 1% agarose gels. In addition, the $OD_{260/230}$ ratio and the RNA integrity number (RIN) were used for assessing RNA quality and purity using a Qubit® 2.0 Fulorometer (Invitrogen, USA) and an Agilent 2100 Bioanalyzer system (Agilent Technologies, USA).

Library preparation and transcriptome sequencing were performed by the method previously described (Liu et al. 2015a). The cDNA library of each sample was prepared by using NEB Next® Ultra™ RNA Library Prep Kit for Illumina® (NEB, USA). Library quality was assessed on the Agilent 2100 Bioanalyzer system. After preparation of the library, an Illumina HiSeq 2000 platform was used for sequencing according to the manufacturer's instructions and paired-end reads were obtained.

### Sequence read mapping, assembly, annotation and classification

All the bioinformatic analyses were performed according to a previous paper (Liu et al. 2015b). The data were cleaned by removing reads containing adapter, reads containing poly-N, and reads of low quality. Meanwhile,

GC-content, Q20 and Q30 scores, and sequence duplication level were calculated. The clean reads were used in the following analyses.

As already known, it was technically impossible for us to remove the intracellular *M. dendrobii* hyphae from the *G. elata* protocorms. Unfortunately, *Mycena* and *Gastrodia* reference genomes have not yet been published now. Therefore, the hyphae group (*M. dendrobii*) was *denovo* assembled individually using Trinity with default parameters (kmer = 25) (Grabherr et al. 2011). The *Mycena* reference transcriptome was obtained. Then, the *M. dendrobii* reads derived from the protocorm group were removed by mapping all the reads against the *Mycena* reference transcriptome using Bowtie program v.0.12.8 (Langmead et al. 2009). Finally, the remaining unmapped non-BLAST and non-*Mycena* reads were prepared to detect the expression profiles of *G. elata*. All the clean reads from the seed and protocorm groups were assembled using Trinity (kmer = 25).

Distinct unigenes were indentified using the BLAST search program. All the unigenes were compared against the NCBI Nr/Nt, KOG, Swiss-Prot, KEGG and GO databases. The structure of mRNA sequences was much different between *Gastrodia* (a higher plant) and *Mycena* (fungus). Therefore, we have removed the unigenes that were not annotated with plant genes or annotated with fungus genes as reference. Finally, the remaining unigenes were used for functional analysis.

Prior to analyses of the differentially expressed genes (DEGs), the edgeR program package was used to adjust the read counts through one scaling normalized factor for each library. Then, the DEGseq (R package) was used for differential expression analysis of all libraries. The threshold for significantly differential expression was $|\log_2(\text{foldchange})| > 1$ and P value <0.05 (Li and Dewey 2011).

Gene Ontology (GO) enrichment analyses were performed by the GOseq (R packages) based Wallenius non-central hyper-geometric distribution. The results were adjusted for gene length bias in DEGs (Young et al. 2010). In addition, WEGO (Web Gene Ontology Annotation Plot) were used for plotting GO annotation results (Ye et al. 2006). KEGG is a database for recoding the collection of high-level functions and the utility of the biological system. Here, KOBAS software was used for the statistical enrichment of DEGs in KEGG pathways (Kanehisa et al. 2008).

### Quantitative PCR

RNA from the same libraries was used for quantitative PCR analyses (Livak and Schmittgen 2001; Liu et al. 2015a). Primers designed with Primer Premier 5.0 are shown in Additional file 1: Table S1. PrimeScript™ RT reagent Kit (TaKaRa, JAP) was used for reverse transcription. In total, 1 μl of RT product diluted with 20 μl of $ddH_2O$ was used as template. Then, qPCR was performed in 15 μl of reaction mixture containing 7.5 μl of 2× SYBR® Premix Ex Taq™ II (TaKaRa, JAP), 1.5 μl of cDNA template, and 0.3 μl of each gene-specific primer. In total, we performed two biological replicates and three technical replicates using the LightCycler® 480 II RT-PCR System (Roche, SWIT) and its relative quantification software. The parameters of reactions were: 95 °C for 30 s, 40 cycles of 95 °C for 5 s, and 60 °C for 30 s. The cDNA libraries were standardized to reference gene 18S. The $2^{-\Delta\Delta Ct}$ method was used for evaluating gene expression.

## Results and discussion

### Development of mycorrhizal protocorms

The mature seed of *G. elata* contains a globular embryo covered by a thin layer of seed coat (Fig. 1a). By 4 weeks after inoculation, the seed was infected by fungal hyphae and had germinated. The embryo enlarged further and mycorrhizal protocorms formed. From histological observation, the cells in the basal region of the protocorm harbored a number of coiled intracellular fungal hyphae (the pelotons), while the shoot meristem was discernible at the apical region of the protocorm (Fig. 1b).

### Transcriptome profile

We performed RNA-Seq to comprehensively explore the crucial cluster of genes associated with symbiotic seed germination of *G. elata*. To increase the accuracies of the project, two biological replicates were included in each transcriptome sample of *G. elata* (seed and early-stage protocorm). Unfortunately, it was technically impossible for us to remove the intracellular fungal hyphae (*Mycena*) from the protocorms. This meant that symbiotic cells contained transcripts produced by both partners (*G. elata* and *M. dendrobii*). Therefore, RNA from a symbiotic fungi library was used for establishing a *Mycena* reference transcriptome. All of the *M. dendrobii* reads derived from the protocorm group were removed by mapping the reads against the *Mycena* reference transcriptome. Previous studies demonstrated that a bioinformatics pipeline could be effectively applied to assign genes identified in the transcriptome of plant-fungi association (Perotto et al. 2014; Liu et al. 2015a).

In our results, the sequencing error rate did not exceed 1% (Additional file 1: Table S2). In total, 235,192,510 raw reads and 221,982,110 clean reads were acquired from the seed and protocorm libraries. Since there were no reference genomes of *G. elata*, the clean reads from all libraries were pooled together and *denovo* assembled into transcripts. Finally, we obtained 179,781 transcripts

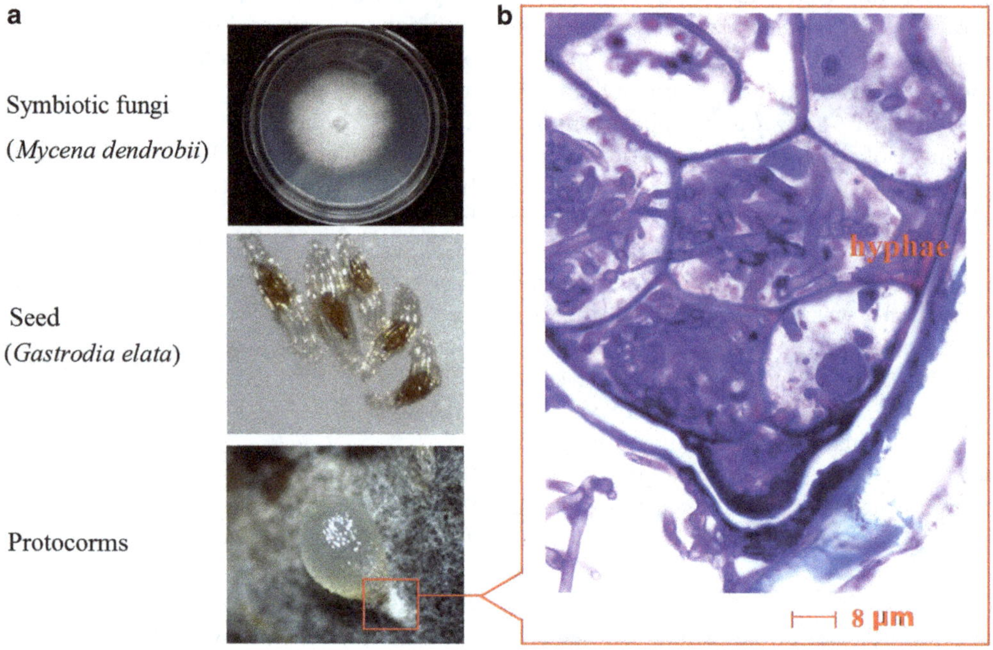


**Fig. 1** Biomorph of seed germination of *Gastrodia elata*. **a** *M. dendrobii* incubated on the fresh PDA; the mature seeds of *G. elata*; the typical protocorm of *G. elata* infected with *M. dendrobii*; **b** histological observation of the protocorms

with an N50 length of 1735 bp. We identified 139,756 unigenes with an N50 length of 1025 bp (length ranging from 201–17,061 bp). The datasets are available in the BioProject (Accession Number: PRJNA273548, mature seed and early-stage protocorm) repository of the NCBI.

## Gene annotation and function classification

Gene annotation was applied by BLAST against the NCBI Nr/Nt, Pfam, Swiss-Prot, GO, KOG and KEGG databases. In total, there were 42,140 unigenes (30.15% of all unigenes) annotated in at least one database (Table 1). Among them, a total of 30,090 unigenes (21.53%) were annotated in the Nr database. Based on Nr annotation, approximately 47.2% of annotated unigenes were assigned with a best score to a sequence from the top five species: *Elaeis guineensis*, *Musa acuminata*, *Phoenix dactylifera*, *Theobroma cacao* and *Vitis vinifera* (Additional file 1: Figure S1).

According to our results shown in Additional file 1: Figure S2, 25,105 unigenes were annotated to 56 terms of GO classification. Among these terms, "cellular process" (13,851) and "metabolic process" (13,043) were dominant within "Biological Process" (BP). For "Cellular Component" (CC), dominant subcategories were "cell" (7548) and "cell part" (7542). In the category of "Molecular Function" (MF), "binding" (12,537) and "catalytic activity" (10,439) were highly represented. Moreover, the results showed that there were 211 and 3105 unigenes annotated to "immune system process" and "response to stimulus", respectively. Similar results were reported by Zhao et al. (2013) and Perotto et al. (2014).

**Table 1 Annotation result statistics between non-redundant unigenes and public databases**

| Database | Number of unigenes | Percentage (%) |
|---|---|---|
| Annotated in NR | 30,090 | 21.53 |
| Annotated in NT | 13,503 | 9.66 |
| Annotated in KO | 9375 | 6.7 |
| Annotated in SwissProt | 22,893 | 16.38 |
| Annotated in PFAM | 24,874 | 17.79 |
| Annotated in GO | 25,105 | 17.96 |
| Annotated in KOG | 12,034 | 8.61 |
| Annotated in all databases | 3915 | 2.8 |
| Annotated in at least one database | 42,140 | 30.15 |
| Total unigenes | 139,756 | 100 |

In Additional file 1: Figure S3, there were 12,034 unigenes annotated to 26 groups based on the KOG database. Among these groups, 2132 unigenes were annotated to "General Function Prediction Only", 484 unigenes were annotated to "Carbohydrate metabolism and transport", and 1389 unigenes were annotated to "Signal transduction mechanisms". We also found that 84 unigenes were annotated to "Defence mechanisms".

As shown in Additional file 1: Figure S4, all of the unigenes associated with "human diseases" were removed, and the remaining 9375 unigenes were mapped to 19 KEGG pathways. All the pathways were divided into five categories: "cellular processes", "environmental information processing", "genetic information processing", "metabolism" and "organismal systems". The pathways with the most representation were "translation" (1387 unigenes), "carbohydrate metabolism" (817 unigenes), and "Folding, sorting and degradation" (719 unigenes). In addition, there were 227 unigenes involved in the "signal transduction" pathways.

### Differential expression gene (DEG)

The normalized-FPKM (expected number of fragments per kilobase of transcript sequence per million base pairs sequenced) was used to quantify the transcript level in reads. The normalized-FPKM is a parameter for the comparison of mRNA levels among libraries. To increase the accuracies of the measured expression levels for further analyses, data from two biological replicates were merged, and FPKM values were calculated based on the merged dataset. The resulting Pearson's correlation coefficients (R) were high between the replicates for seed ($R = 0.659$) and protocorm EP ($R = 0.697$) samples (Additional file 1: Figure S5).

As shown in Fig. 2, differential expression genes (DEGs) were analyzed between the seeds and protocorms of *G. elata*. For seed vs. protocorm, 1750 unigenes were differentially expressed, including 1439 significantly up-regulated genes and 331 obviously down-regulated unigenes in the protocorm library. Based on GO enrichment (Fig. 3), DEGs were functionally classified according to the BP, CC, and MF categories and their subcategories. The largest subcategories for each functional group were as follows: "cellular process", "metabolic process", and "response to stimulus" for BP; "cell", "cell part" and "organelle" for CC; "binding", "catalytic", and "structural molecule" for MF. According to KEGG (Fig. 4), most of the DEGs were assigned to "signal transduction", "translation", and "carbohydrate metabolism" pathways.

### Mutualistic symbiosis and endocytosis

In nature, orchid seeds possess no endosperm; therefore, they are devoid of nutrient supply. Mycorrhizal fungi provide the orchid seeds with signals and nutrients for germination, a mechanism called symbiotic germination. The process of orchid seed germination is unique among angiosperms and involves various processes such as fungi colonization, plant growth stimulation, plant response to fungi, etc. (Rasmussen 2002; Cameron et al. 2008; Rasmussen and Rasmussen 2009). In our subject, the transcriptomes of *G. elata* seeds and protocorms were sequenced. We selected 26 genes involved in clathrin-mediated endocytosis, including clathrin, adaptor protein complex (AP), AP180, adaptin-like protein, clathrin interactor epsin, dynamin-related protein (DRP), and heat shock cognate 70 kDa (HSC70). The expression of

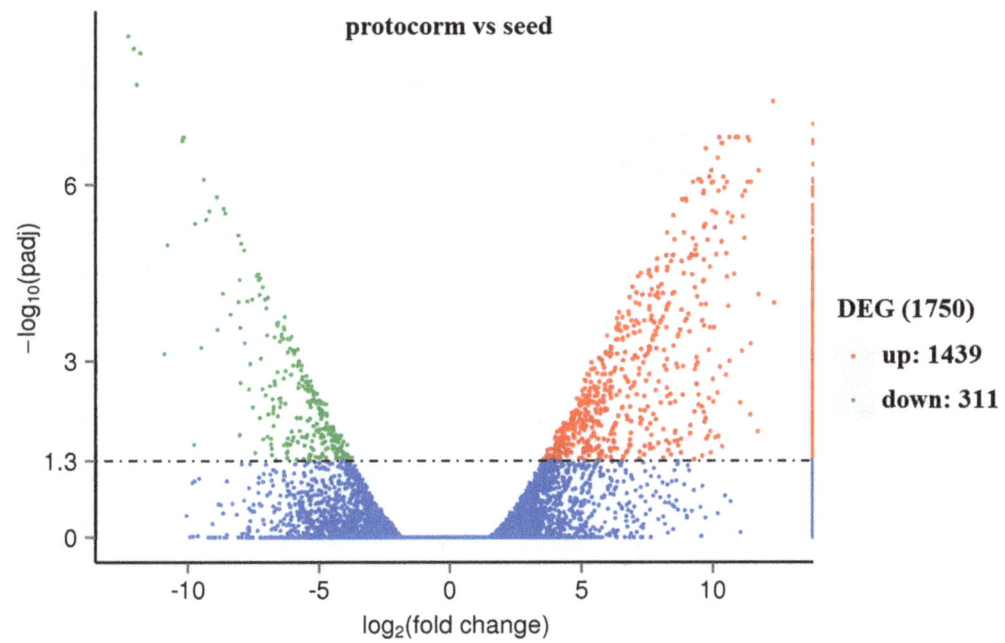


**Fig. 2** Volcano plots of differently expressed genes

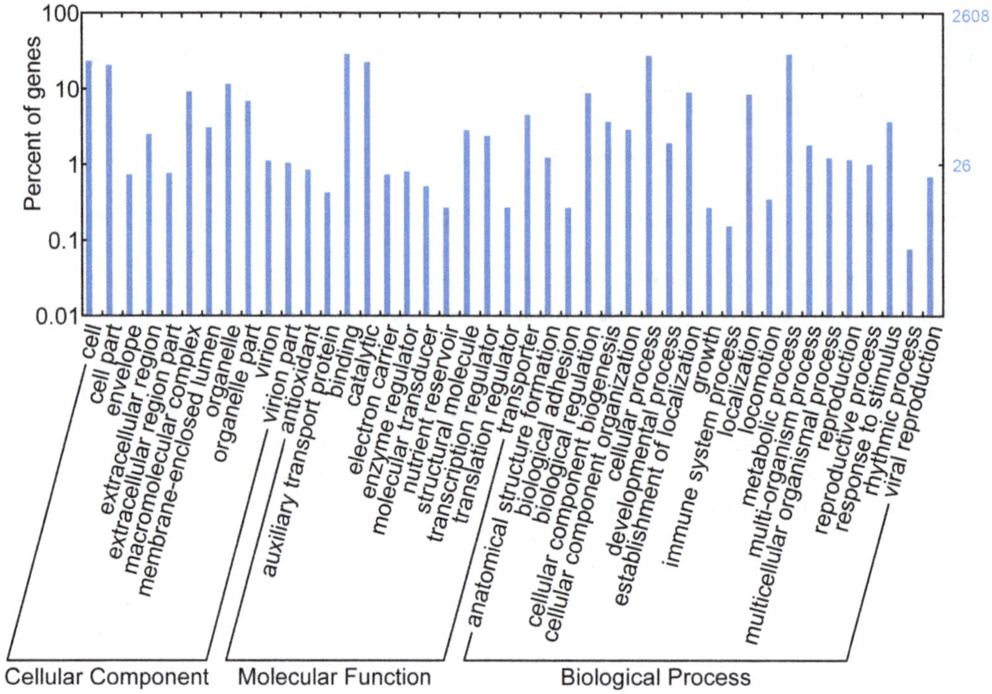


**Fig. 3** GO classification of differential expression genes (DEGs) in seeds vs. protocorms

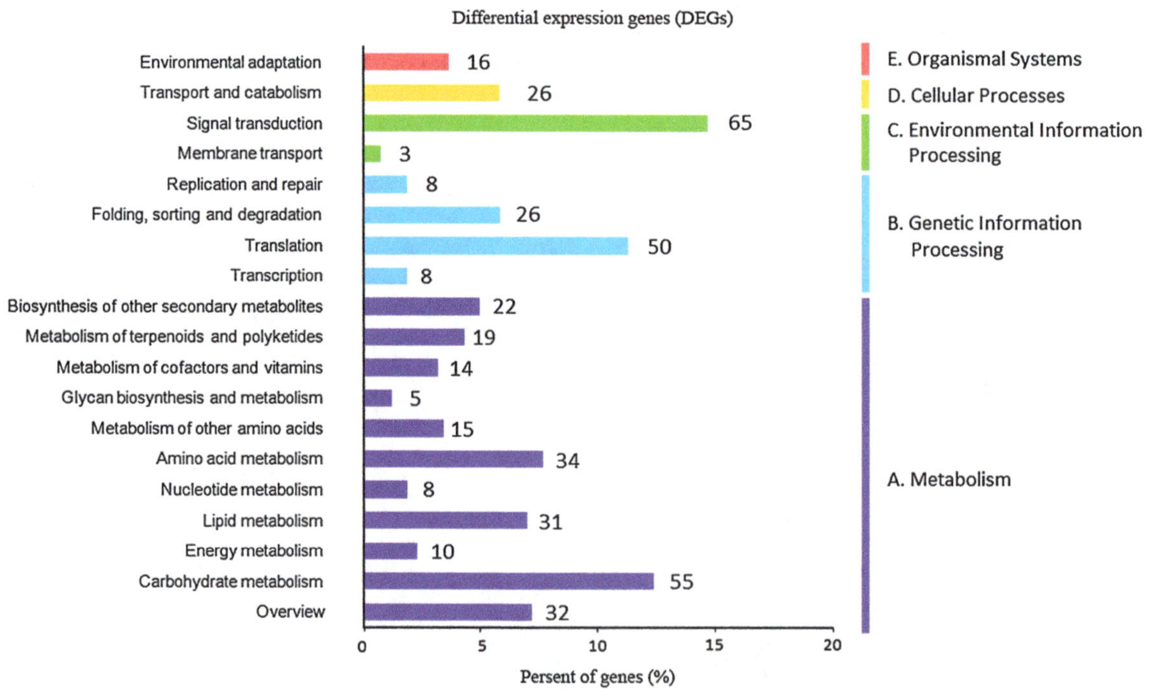


**Fig. 4** KEGG pathway enrichment scatter diagram of differential expression genes (DEGs) in seeds vs. protocorms

16 genes was significantly up-regulated in protocorms compared with seeds, according to RNA-Seq results. Our results suggested that clathrin-mediated endocytosis could play an important role in the symbiotic seed germination of orchids with fungal infections.

Endocytic mechanisms are essential for regulation of lipid and protein compositions of the plasma membrane. A previous study showed that the establishment of endocytosis was a crucial process involved in many diverse aspects of plant life, such as basic cellular functions, growth development, and hormonal signaling (Fan et al. 2015). Another important role of plant endocytosis is in communication with the environment, including nutrient uptake, toxin avoidance, and pathogen defense. Endocytosis in plants is essential for the acquisition of materials from the extracellular space. At the moment, endocytosis is generally divided into clathrin-mediated endocytosis and clathrin-independent endocytosis. The majority of endocytosis in plants depends on the coat protein clathrin (Chen et al. 2011).

A previous study showed that endocytosis played a role in symbiosis between legumes and rhizobia (Wang et al. 2015). A complex molecular mechanism was required for the establishment of symbiotic relationships between legumes and rhizobia. Legume roots released flavonoids, which triggered synthesis and secretion of rhizobial nodulation factors. These nodulation factors were recognized by legume roots and activated a symbiosis signaling pathway. The legumes allowed the rhizobia to enter root hairs and pass through plant-derived infection threads. Finally, rhizobia were released from infection threads (IT) to symbiosomes of infected cells within the root nodule. A previous study suggested that the nodulation factor signaling pathway might proceed through an endocytosis process (Timmers et al. 1998). Another study found that the *Medicago* flotillin-like proteins, lipid raft-associated integral membrane proteins involved in endocytosis, were crucial for infection thread initiation and nodule development (Haney and Long 2010).

In legume-*Rhizobium* symbiosis, a nod factor signaling pathway may proceed via ligand-induced receptor endocytosis, similar to that characterized in plant defense against pathogens. Nod factors are recognized by receptors of the LysM domain-containing receptor-like kinase family at the PM and have been detected inside root hairs, along the IT, and in the cytoplasm of infected root cells (Leborgnecastel et al. 2010).

Surface polysaccharides (SPS) from rhizobia could also be likely substrates for uptake by endocytosis. SPS are crucial signaling molecules for the development of rhizobial infection and are thought to attenuate or suppress defense reactions. Internalization of SPS has been observed in plant cells (Gross et al. 2005), thus substantiating the existence of specialized receptors, as previously proposed for legume-*Rhizobium* symbiosis.

#### Clathrin-mediated endocytosis

As shown in Fig. 5, prior observations have defined the five steps of clathrin-coated vesicle formation (Doherty and McMahon 2009; McMahon and Boucrot 2011). Clathrin-mediated endocytosis initiates at the plasma membrane with the recruitment of cargo. Cargoes that need to be internalized from the plasma membrane are packaged by co-operative assembly of coat proteins, including clathrin, adaptor protein complex and associated accessory adaptor proteins (adaptin, intersectin, AP180, clathrin interactor epsin, TPLATE-like protein, etc.). Clathrin-coated pits (CCPs) then mature, and dynamin and DRPs assemble at the neck of the bud for scission of the CCPs from the plasma membrane. Once detached from the membrane and formed into clathrin-coated vesicles (CCVs), the clathrin coat disassembles in seconds to form uncoated vesicles that fuse with the early endosome (EE), where the cargo is further sorted. ATPase HSC70 and its cofactors (auxilin, etc.) are necessary for uncoating of CCVs and clathrin component recycling.

In our RNA-Seq results (Table 2), 16 genes involved in clathrin-mediated endocytosis were significantly up-regulated in protocorms compared with those in seeds, including clathrin, adaptor protein, adaptin-like protein, A180, epsin, DRP and HSC70. These genes involved in clathrin-mediated endocytosis may therefore be associated with the symbiotic seed germination of orchids with fungal infections. In agreement with our results based on the comparative transcriptome analysis of symbiotic and asymbiotic seed germination of *Anoectochilus roxburghii* (NCBI Accession Number: PRJNA299493), we also found that many genes involved in clathrin-mediated endocytosis (e.g., clathrin, epsin, dynamin, HSC70, AP180) were significantly up-regulated in symbiotic protocorms compared with those in asymbiotic protocorms.

#### Clathrin

Clathrin plays an important role in many cellular processes, such as endosomal sorting, protein secretion, mitosis, and endocytosis. Since the subunits are recruited from the cytoplasm, clathrins form three-legged molecules, termed hexameric triskelions. In general, clathrin is composed of three heavy chains (CHCs) and three light chains (CLCs). Two CHC genes and three CLC genes exist in the *Arabidopsis* genome (Fan et al. 2015). Our results (Table 2), show that one CHC gene and one CLC gene were expressed during seed germination of *G. elata* infected with *M. dendrobii.* Of these genes, the CHC gene (c113523_g1) was up-regulated in protocorms compared to that in seeds.

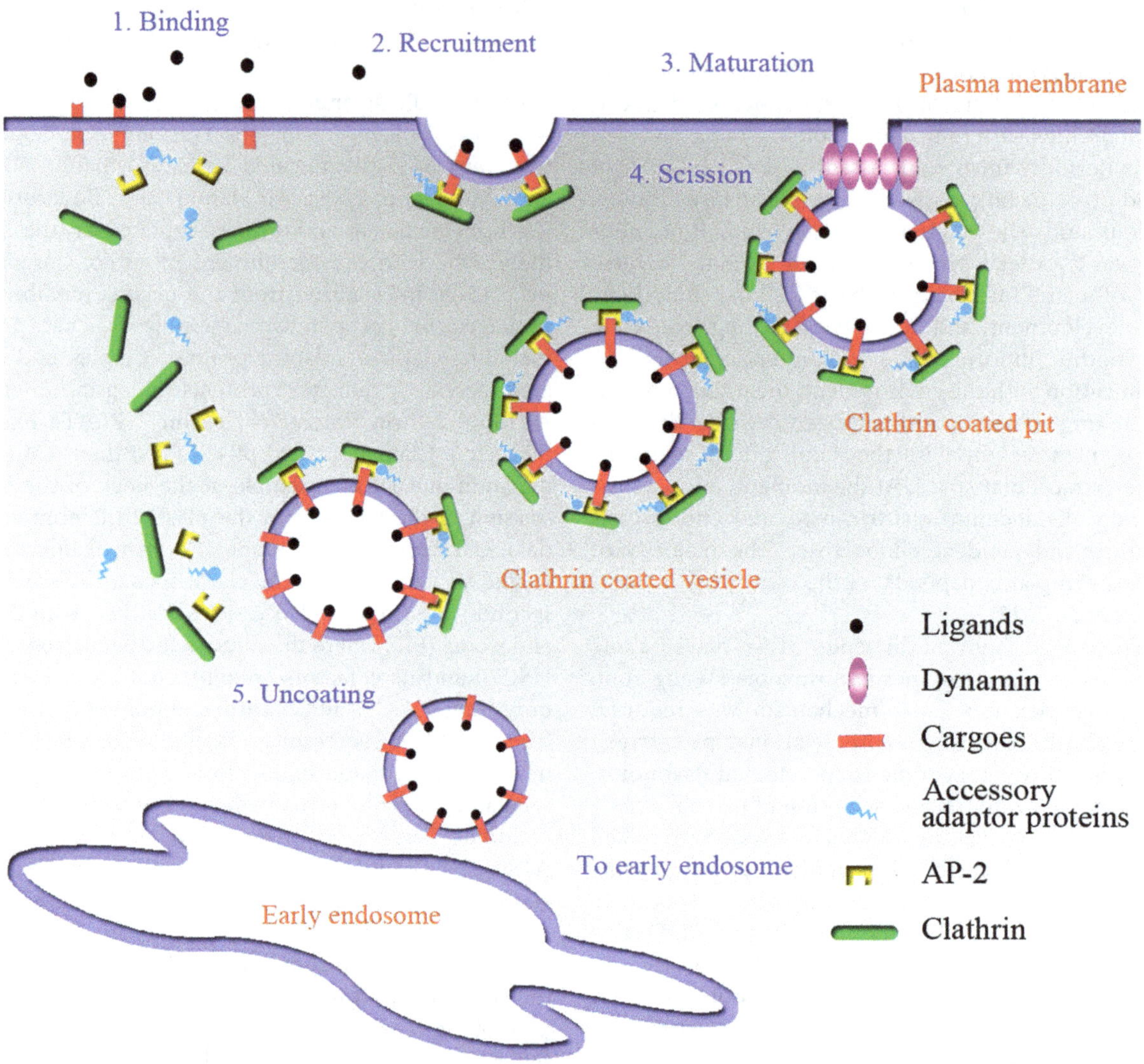


**Fig. 5** Quantitative PCR analysis of the putative genes

#### Adaptor complexes

Clathrin is not able to directly bind membranes or cargoes. At the budding site, adaptor protein complexes are recruited for linking lipids, cargo, and adaptor proteins to the clathrin triskelion. Of the four AP complexes (AP1-4), the highly conserved AP2, which comprises two large (α, β) subunits, a medium (μ) and a small (σ) subunit, serves as a central player in the initiation of CCP formation. A previous study showed that orthologs of different adaptins were present in plant genomes, including AP2 components (Chen, Irani et al. 2011). Here, we found that 12 genes encoding AP1 (β/μ/σ/γ), AP2 (α/μ), AP3 (δ/σ/β/μ) and AP4 (ε/σ) were expressed during symbiotic seed germination. Among these genes, AP1 σ and γ (c49102_g1 and c98691_g1), AP2 α (c33125_g1), and AP4 ε (c43068_g1) genes were differentially up-regulated on transcriptional levels in protocorms compared with those in seeds (Table 2).

In addition to the classical adaptins, a large number of accessory proteins have a similar function in linking cargo and lipids to the AP2-clathrin assembly, including epsin, AP180, eps15, and TPLATE. These accessory proteins were also able to influence actin dynamics and induce membrane curvature for invagination of the plasma membrane into vesicles (Fan et al. 2015). In our study, three epsin (1/2/3) genes, one AP180 gene, and one adaptin-like gene were observed in our data. Furthermore, three accessory proteins, epsin 1, AP180 and adaptin-like protein (c36930_g2, c50842_g1 and c28875_g1) showed increased accumulation in protocorms according to transcriptomic data.

**Table 2 The up- and down-regulated genes in protocorms of *G. elata* compared with seeds**

| Unigene ID | Putative gene | Refseq | Ratio change |
|---|---|---|---|
| c113523_g1 | CHC | XP_010907410.1 | 2.56* |
| c113909_g1 | CLC | XP_008802347.1 | 1.89 |
| c29883_g1 | AP-1 β | XP_009385856.1 | 1.92 |
| c36857_g1 | AP-1 μ | XP_010923038.1 | 1.28 |
| c49102_g1 | AP-1 σ | AFS50008.1 | 2.26* |
| c98691_g1 | AP-1 γ | XP_010913560.1 | 2.31* |
| c29046_g1 | AP-2 μ | XP_007039526.1 | 1.68 |
| c33125_g1 | AP-2 α | XP_010933729.1 | 2.59* |
| c126705_g1 | AP-3 δ | XP_009396369.1 | 0.9 |
| c29490_g1 | AP-3 σ | XP_010939294.1 | 1.54 |
| c46833_g1 | AP-3 β | XP_010937611.1 | 0.95 |
| c50113_g1 | AP-3 μ | XP_010921397.1 | 1.17 |
| c43068_g1 | AP-4 ε | XP_010909540.1 | 2.09* |
| c50920_g1 | AP-4 σ | XP_007152949.1 | 2.39* |
| c28875_g1 | Adaptin | XP_008792344.1 | 2.40* |
| c50842_g1 | AP180 | XP_009396332.1 | 2.18* |
| c36930_g2 | Epsin 1 | XP_008799757.1 | 2.04* |
| c50396_g1 | Epsin 2 | XP_009392171.1 | 1.62 |
| c45022_g1 | Epsin 3 | XP_010922341.1 | 0.96 |
| c29413_g1 | DRP3 | XP_008794571.1 | 3.07* |
| c32547_g1 | DRP5 | XP_010918171.1 | 3.84* |
| c43852_g2 | DRP1 | XP_010933528.1 | 5.75* |
| c56183_g1 | DRP2 | XP_008779526.1 | 2.48* |
| c29657_g1 | HSC70 | XP_011094842.1 | 2.08* |
| c33306_g1 | HSC70 | XM_009382388.1 | 2.45* |
| c7920_g1 | HSC70 | XP_011027182.1 | 2.23* |

Only differential expression genes with change ratios ≥2 (up) or ≤0.5 (down) were marked with asterisk (*). Transcriptomic analyses were performed on the seeds and protocorms of *G. elata*. Of the 26 genes involved in clathrin-mediated endocytosis, 16 genes were found to be significantly up-regulated in the protocorms based on RNA-seq analyses. *CHC* clathrin heavy chain, *CLC* clathrin light chain, *AP-1 β/μ/σ/γ* adaptor protein-1 complex subunit beta/mu/sigma/gamma, *AP-2 α/μ* adaptor protein-2 complex subunit alpha/mu, *AP-3 δ/σ/β/μ* adaptor protein-3 complex subunit delta/sigma/beta/mu, *AP-4 ε/σ* adaptor protein-4 complex subunit epsilon/sigma, *AP180* clathrin assembly protein, *espin* clathrin interactor epsin, *DRP* dynamin-related protein, *HSC70* heat shock cognate 70 kDa

### Dynamin

Following maturation of CCPs, dynamins are recruited by BAR domain-containing proteins at the CCP neck. As membrane scission proteins, dynamins are members of a large family of GTPase proteins that form a helical polymer around the constricted CCP neck and mediate fission of the vesicle from the plasma membrane, thus releasing the CCV into the cell. Of the six dynamin-related protein families in plants, DRP1 and DRP2 were demonstrated to be involved in clathrin-mediated membrane trafficking (Chen et al. 2011; McMahon and Boucrot 2011). In total, four dynamin-related genes were identified in our results (Table 2). Our study also revealed that four dynamin-related proteins, DRP1, 2, 3 and 5, were differentially up-regulated on transcriptional levels in protocorms compared with those in seeds.

### HSC70

Once detached from the plasma membrane, the clathrin coat is subsequently released from the CCVs by ATPase HSC70 and its cofactor, auxilin (or cyclin G-associated kinase in non-neuronal tissues). The uncoated CCVs then undergo further trafficking within the cell and fuse with early endosomes. According to our data (Table 2), three HSC70 genes were observed during symbiotic seed germination. All of these genes (c29657_g1, c33306_g1 and c7920_g1) were differentially up-regulated on transcriptional levels in protocorms compared with those in seeds.

### qPCR analysis of putative genes

To confirm the reliability of the RNA-Seq data, we performed quantitative PCR analysis on 7 selected genes putatively involved in the endocytosis mechanism, such as clathrin (c113523_g1), adaptor protein (c33125_g1), epsin (c36930_g2), adaptin-like protein (c28875_g1), AP180 (c50842_g1), dynamin-related protein (c43852_g2) and heat shock cognate 70 kDa (c29657_g1). The results of qPCR related to expression changes of these genes are shown in Fig. 6. In our results, most of the genes showed low expression levels in seeds and high expression levels in protocorms during the developmental process of seed germination in *G. elata*. Taken together, all the unigenes showed expression patterns that were consistent with the transcriptomic data, indicating that our experimental results were valid.

## Conclusions

In conclusion, transcriptomic analyses provide a powerful method for investigating putative genes involved in the symbiotic seed germination of orchids. In this study, we performed transcriptome sequencing of mature seeds and early-stage protocorms from *G. elata* to identify these genes and quantify their expression in seed germination. Subsequently, we analyzed putative differently expressed genes of clathrin, adaptor protein, AP180, adaptin-like protein, clathrin interactor epsin, dynamin-related protein, and HSC70 that were produced in the process of plant endocytosis. Our results indicated that the up-regulation of expression of genes related to clathrin-mediated endocytosis could play an important role in symbiotic germination in *G. elata* infected by *M. dendrobii*, especially at the early stage of protocorm development. The RNA-Seq data from our study provide an important resource for studying interactions between plant seeds and symbiotic fungi. Further

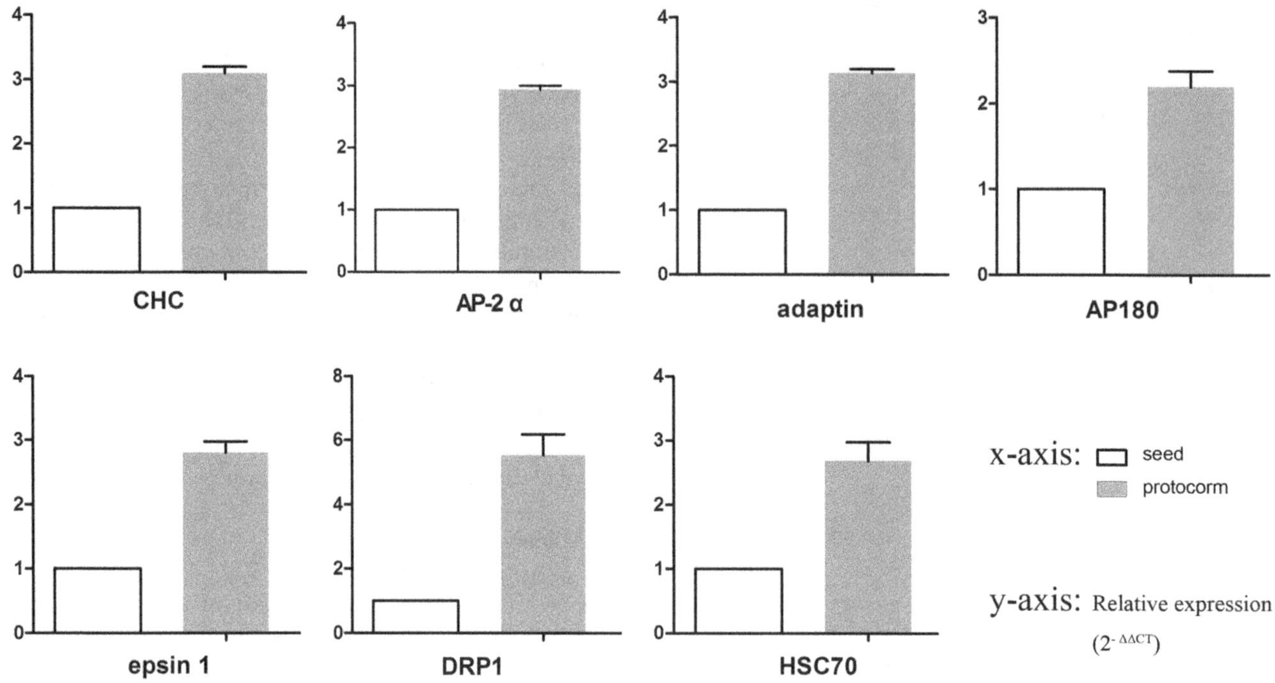


**Fig. 6** Clathrin-mediated endocytosis

genomic research of orchids and mycorrhizal fungi will provide new insights into the interactions between fungi and plants.

#### Authors' contributions

Conceived and designed the experiments: XZ, YYL and SXG. Performed the experiments: XZ, YYL and JC. Analyzed the data: XZ and HL. Contributed reagents/materials/analysis tools: SSL and MML. Wrote the paper: XZ and YYL. All authors read and approved the final manuscript.

#### Acknowledgements

This study was financially supported by the Program for Innovative Research Team in IMPLAD (PIRTI-IT 1302). The funders had no role in study design, data collection and analysis, decision to publish, or preparation of the manuscript.

#### Competing interests

The authors declare that they have no competing interests.

#### Funding

This study was financially supported by the Program for Innovative Research Team in IMPLAD (PIRTI-IT 1302).

#### References

Cameron DD, Johnson I, Read DJ, Leake JR (2008) Giving and receiving: measuring the carbon cost of mycorrhizas in the green orchid, *Goodyera repens*. New Phytol 180:176–184

Chen X, Irani NG, Friml J (2011) Clathrin-mediated endocytosis: the gateway into plant cells. Curr Opin Plant Biol 14:674–682

Dearnaley JD (2007) Further advances in orchid mycorrhizal research. Mycorrhiza 17:475–486

Dhonukshe P, Aniento F, Hwang I, Robinson DG, Mravec J, Stierhof Y-D, Friml J (2007) Clathrin-mediated constitutive endocytosis of PIN auxin efflux carriers in *Arabidopsis*. Curr Biol 17(6):520–527

Doherty GJ, McMahon HT (2009) Mechanisms of endocytosis. Annu Rev Biochem 78:857–902

Fan L, Li R, Pan J, Ding Z, Lin J (2015) Endocytosis and its regulation in plants. Trends Plant Sci 20:388–397

Grabherr MG, Haas BJ, Yassour M, Levin JZ, Thompson DA, Amit I, Adiconis X, Fan L, Raychowdhury R, Zeng Q (2011) Full-length transcriptome assembly from RNA-Seq data without a reference genome. Nat Biotechnol 29:644–652

Gross A, Kapp D, Nielsen T, Niehaus K (2005) Endocytosis of *Xanthomonas campestris* pathovar campestris lipopolysaccharides in non-host plant cells of *Nicotiana tabacum*. New Phytol 165:215–226

Haney CH, Long SR (2010) Plant flotillins are required for infection by nitrogen-fixing bacteria. Proc Natl Acad Sci 107:478–483

Kanehisa M, Araki M, Goto S, Hattori M, Hirakawa M, Itoh M, Katayama T, Kawashima S, Okuda S, Tokimatsu T (2008) KEGG for linking genomes to life and the environment. Nucleic Acids Res 36(Suppl 1):D480–D484

Kim YI, Chang KJ, Ka KH, Hur H, Hong IP, Shim JO, Lee TS, Lee JY, Lee MW (2006) Seed germination of *Gastrodia elata* using symbiotic fungi, *Mycena osmundicola*. Mycobiology 34:79–82

Langmead B, Trapnell C, Pop M, Salzberg SL (2009) Ultrafast and memory-efficient alignment of short DNA sequences to the human genome. Genome Biol 10:R25

Leake JR (2004) Myco-heterotroph/epiparasitic plant interactions with ectomycorrhizal and arbuscular mycorrhizal fungi. Curr Opin Plant Biol 7:422–428

Leborgnecastel N, Adam T, Bouhidel K (2010) Endocytosis in plant-microbe interactions. Protoplasm 247(3):177–193

Li B, Dewey CN (2011) RSEM: accurate transcript quantification from RNA-Seq data with or without a reference genome. BMC Bioinform 12:1–3

Liu MM, Xing YM, Zhang DW, Guo SX (2015a) Transcriptome analysis of genes involved in defence response in Polyporus umbellatus with Armillaria mellea infection. Sci Rep 5(3):16075

Liu SS, Chen J, Li SC, Zeng X, Meng ZX, Guo SX (2015b) Comparative transcriptome analysis of genes involved in GA-GID1-DELLA regulatory module in symbiotic and asymbiotic seed germination of *Anoectochilus roxburghii* (Wall.) Lindl. (Orchidaceae). Int J Mol Sci 16:30190–30203

Livak KJ, Schmittgen TD (2001) Analysis of relative gene expression data using real-time quantitative PCR and the $2^{-\Delta\Delta CT}$ method. Methods 25:402–408

Manavalan A, Feng L, Sze SK, Hu JM, Heese K (2012) New insights into the brain protein metabolism of *Gastrodia elata*-treated rats by quantitative proteomics. Journal of Proteomics 75:2468–2479

McMahon HT, Boucrot E (2011) Molecular mechanism and physiological functions of clathrin-mediated endocytosis. Nat Rev Mol Cell Biol 12:517–533

Perotto S, Rodda M, Benetti A, Sillo F, Ercole E, Rodda M, Girlanda M, Murat C, Balestrini R (2014) Gene expression in mycorrhizal orchid protocorms suggests a friendly plant-fungus relationship. Planta 239:1337–1349

Rasmussen HN (2002) Recent developments in the study of orchid mycorrhiza. Plant Soil 244:149–163

Rasmussen HN, Rasmussen FN (2009) Orchid mycorrhiza: implications of a mycophagous life style. Oikos 118:334–345

Timmers AC, Auriac MC, Billy FD, Truchet G (1998) Nod factor internalization and microtubular cytoskeleton changes occur concomitantly during nodule differentiation in alfalfa. Development 125:339–349

Tsai CC, Wu KM, Chiang TY, Huang CY, Chou CH, Li SJ, Chiang YC (2016) Comparative transcriptome analysis of *Gastrodia elata* (Orchidaceae) in response to fungus symbiosis to identify gastrodin biosynthesis-related genes. BMC Genomics 17:1–16

Wang C, Zhu M, Duan L, Yu H, Chang X, Li L, Kang H, Feng Y, Zhu H, Hong Z (2015) *Lotus japonicus* clathrin heavy chain 1 is associated with Rho-Like GTPase ROP6 and involved in nodule formation. Plant Physiol 167:1497–1510

Ye J, Fang L, Zheng H, Zhang Y, Chen J, Zhang Z, Wang J, Li S, Li R, Bolund L (2006) WEGO: a web tool for plotting GO annotations. Nucleic Acids Res 34(suppl 2):W293–W297

Yeung EC, Chan CK (2015) The glycol methacrylate embedding resins-technovit 7100 and 8100. Plant microtechniques and protocols. Springer, New York

Young MD, Wakefield MJ, Smyth GK, Oshlack A (2010) Method Gene ontology analysis for RNA-seq: accounting for selection bias. Genome Biol 11:R14

Zhao MM, Zhang G, Zhang DW, Hsiao YY, Guo SX (2013) ESTs analysis reveals putative genes involved in symbiotic seed germination in *Dendrobium officinale*. PLoS ONE 8:e72705

# 3

# Enzymatic and non-enzymatic comparison of two different industrial tomato (*Solanum lycopersicum*) varieties against drought stress

Özge Çelik*, Alp Ayan and Çimen Atak

**Abstract**

**Background:** The aim of this study is to compare the tolerance mechanisms of two industrial tomato varieties (X5671R and 5MX12956) under drought stress. 14 days-old tomato seedlings were subjected to 7 days-long drought stress by withholding irrigation. The effects of stress were determined by enzymatic and non-enzymatic parameters. The physiological damages were evaluated via lipid peroxidation ratio, total protein content, relative water content, chlorophyll content and proline accumulation. Enzymatic responses were determined by biochemical analysis and electrophoresis of SOD, APX, POX and CAT enzymes.

**Results:** Relative water contents of X5671R and 5MX12956 varieties at 7th day of drought were decreased to 8.4 and 12.2%, respectively. Applied drought decreased all photosynthetic pigments of X5671R and 5MX12956 varieties during the treatment period significantly comparing to the Day 0 as the control. Total protein content, lipid peroxidation and proline accumulation presented increased values in both varieties in accordance with the increasing stress intensity. According to lipid peroxidation analysis, 5MX12956 tomato variety was found more drought sensitive than X5671R variety. Antioxidative enzyme activities showed increases in both varieties as a response to drought stress, although CAT and APX activities presented decrease on the 7th day of applied stress. 7 days long drought stress differentially altered POX, APX and SOD isozyme patterns. Same POX bands were observed in both varieties with different band intensities.

**Conclusions:** However, main isozyme pattern differences were obtained for SOD and APX. APX1, Fe-SOD and Cu/Zn-SOD2 isozyme bands should be evaluated to define their main role in the tolerance mechanism of both tomato varieties.

**Keywords:** *Solanum lycopersicum*, Drought stress, Antioxidant enzyme activities, Isozyme pattern

## Background

Drought stress is described as one of the most harmful natural hazards that limits the growth, productivity and crop yield of plants. The aggressive usage of natural resources and increase in the global climate temperatures are two main factors causing drought. The general effects of drought stress have been studied in different plant species. The main inhibitory features of drought stress were observed on cell proliferation and expansion, leaf size, stem elongation, root proliferation alongside with crop growth and biomass accumulation (Harrison et al. 2014; Sekmen et al. 2014; Shanker et al. 2014; Zdravkovic et al. 2013). On the other hand, plants involve specific protective adaptation mechanisms against short term and long term drought. Plants can either enhance response mechanisms to avoid the effects of drought or tolerate the adverse effects of the stress. Plants can reduce leaf size and number, increase the area and length of roots

*Correspondence: ocelik@iku.edu.tr
Department of Molecular Biology and Genetics, Faculty of Science and Letters, Istanbul Kultur Univesity, Ataköy, 34156 Istanbul, Turkey

to absorb more water, produce thicker cuticula and wax layers, control stomatal closure rates, induce cell turgor maintenance mechanisms such as accumulation of compatible solutes, reduce light absorbance by reflecting the exposed sun light to reduce the water loss by transpiration and the alternative mechanisms to reduce toxic oxidative radicals which were produced by the stress (Harb et al. 2010; Shanker et al. 2014). Therefore, they can avoid the short term drought until the life cycle is completed to produce next the generation. Short term drought stress depends on random climate changes and is mostly predictable. Long term drought stress depends on more complex weather features and needs advanced molecular mechanisms to enhance increased transcription rates of functional and regulatory genes involved in drought tolerance (Harb et al. 2010; Jeanneau et al. 2002; Shanker et al. 2014). In next decades, the drought effected area is expected to enlarge around the world following the reducing sustainable water resources and increasing long term water deficiency situations. Comprehensive physiological and molecular studies are necessary for both in vitro and field conditions to evaluate existing crop varieties and improving them.

Tomato (*Solanum lycopersicum*) which is a member of Solanaceae family, is one of the most important agricultural plant worldwide, because of its industrial products and nutritional content (George et al. 2013; Zdravkovic et al. 2013). According to the 2012 world tomato production report of FAO, Turkey was in fourth place by 11,350,000 million tons of production following China (50,000,000), India (17,500,000) and United States (13,206,950) (FAO 2012). Tomato is a sensitive plant against various environmental abiotic stress factors. It is known fact that heat and drought stresses have the most limiting effects on tomato varieties among all other abiotic stress factors. Especially, drought stress during vegetative and early reproductive periods of tomato life cycle reduces yield dramatically (George et al. 2013; Wahb-Allah et al. 2011; Zdravkovic et al. 2013).

As the other abiotic stress factors, drought causes imbalance on equilibrium between reactive oxygen species (ROS) such as hydroxyl radicals ($OH^{\cdot}$), hydrogen peroxide ($H_2O_2$), singlet oxygen ($^1O_2$), superoxide radical ($^-O_2$) and scavenging activities of antioxidant system elements. Reactive oxygen species are products of metabolic processes taking place in different compartments of cell such as chloroplasts, mitochondria and peroxisomes. ROS, which are the group of free radicals derived from $O_2$, are consisted of reactive molecules and ions (Gill and Tuteja 2010). These free radicals which have high reaction potential are harmless at ground levels and have lifetime in unit of micro seconds until they are scavenged in cell. This short lifetime makes ROS effective only in several 100 nm radius. However, when cellular components undergo to oxidative stress, biomolecules as proteins, unsaturated fatty acids, enzymes and nucleic acids are exposed to damaging effects of ROS. Loss or malfunction of key biomolecules may even trigger death of cells depending on the duration and intensity of stress. The relation between the antioxidant system elements to suppress toxic levels of ROS within the cell, is physiologically balanced. Therefore, plants initiate enzymatic (POX, EC1.11.1.7; APX, EC1.11.1.1; SOD, EC1.15.1.1; CAT, EC1.11.1.6) and non-enzymatic (carotenoids, ascorbate, glutathione, tocopherols) defense mechanisms to obtain oxidative homeostasis (Bartosz 1997; Mittler 2002; Sekmen et al. 2014). As known from previous reports, different plant varieties may differ on metabolic pathways. Biochemical strategies are getting important to enhance tolerance capacities of plants under drought conditions. Induction of antioxidant enzyme activities is known to have the major role in survival against drought. As a result, it is getting important to detect changes on enzymatic and non-enzymatic antioxidant systems to reveal metabolic pathways of drought stress tolerance of different plant varieties. Therefore, target oriented breeding and selection studies can be conducted to improve economically important crop plants such as tomato against several environmental abiotic stress factors as drought.

In the present study, antioxidant systems of two economically important industrial tomato varieties (X5671R and 5MX12956) were investigated against drought stress. Physiological and biochemical characteristics of these two varieties were evaluated under intensive drought conditions according to the activities of antioxidants as POX, APX, CAT, SOD and their isozyme profiles. Both varieties were evaluated for their isozyme band diversities to identify their tolerance responses against drought.

## Methods

### Plant material

X5671R and 5MX12956 industrial tomato seeds were provided from Agromar Brand Seeds Company in Bursa/Turkey. Seeds were germinated in 252-well styrofoam filled with perlite and watered by Scotts fertilizer (1.2 µS EC) for 14 days. Drought stress was applied by withholding watering starting from 14th day for all experimental groups, except the control group (Day 0). Samples of both varieties were collected daily during 7 days period and stored at −20 °C for further analysis. All treatments were designed as three independent replicates and each treatment contained 36 plantlets.

### Relative water content

Three random leaves from the same part of each treatment group were collected to determine relative water contents. After measuring the fresh weights (FW), leaves were placed into the distilled water for 8 h in order to obtain turgid weight. Following the turgid weight (TW) measurement, leaves were heated in dry heat incubator for 24 h to obtain dry weights. Relative water contents (RWC) were calculated according to the formula as reported by Smart (1974):

$$\text{RWC}\ (\%) = \frac{\text{FW} - \text{DW}}{\text{TW} - \text{DW}} \times 100$$

### Chlorophyll and carotenoid content

Chlorophyll a, chlorophyll b, total chlorophyll and carotenoid contents of X5671R and 5MX12956 tomato varieties were calculated according to the spectrophotometric method described by Arnon (1949). Leaf tissues were grinded by cold %80 acetone and homogenate filtered through Whatman No:1 filter paper. Absorbance values were measured at 663 and 645 nm wavelengths by using acetone as blank.

### Lipid peroxidation

Drought stress is known to induce lipid peroxidation which indicates the level of oxidative damage on cellular membranes. MDA which is a decomposition product of lipid peroxidation was measured according to method of Stewart and Bewley (1980) for all experimental groups. MDA concentrations were calculated by using an extinction coefficient of 155 $mM^{-1}$ $cm^{-1}$. The results were expressed as μmol MDA $g^{-1}$ FW.

### Proline levels

Proline levels were determined according to the method of Bates et al. (1973). Extracts were prepared in 3% sulphosalicylic acid. Filtered extracts were boiled at 100 °C for 1 h following the addition of acid ninhydrin reagent (ninhydrin and glacial acetic acid). Reaction was stopped in ice bath and fractions were separated by using toluene. Absorbance values were measured at 520 nm wavelength. Proline levels were calculated by using standard calibration curve in unit of μg $\mu L^{-1}$ proline $g^{-1}$ FW.

### Total protein extraction and enzyme activity assays

Total protein extractions for all enzymatic analysis (activity and isozyme assays) were performed by using the same buffer except ascorbate peroxidase (APX) assays. 0.5 g of tomato leaves were grounded by liquid nitrogen and homogenized in 1.25 mL of 50 mM Tris–HCl (pH 7.8) containing 0.1 mM EDTA, 0.2% Triton X-100, 1 mM PMSF and 2 mM DTT. For APX enzyme extraction, 5 mM sodium ascorbate was added to the buffer and DTT was replaced by 2% PVP. All extractions were performed at 4 °C. Samples were centrifugated at 14,000 rpm for 30 min after the extraction. Supernatants were stored at 4 °C until the analysis. Total protein contents were calculated by using bovine serum albumin as protein standard according to the method described by Bradford (1976) and described as μg $g^{-1}$ FW.

The soluble protein profiles were visualized by SDS-PAGE. 75 μg of each sample were loaded to the 4% stacking gel and 12% separating gels according to the method of Laemmli (1970). Electrophoresis was continued at 30 mA constant current until the tracking dye reached to the bottom of the gel. Separate native-PAGE gels were used to identify the isozymes for each enzyme.

POX activity (EC1.11.1.7) was calculated by the increase in absorbance at 470 nm followed for 90 s according to the method of Seevers et al. (1971). The reaction mixture of 0.1 M phosphate buffer (pH 7.0) contained 20 mM guaiacol and 20 mM $H_2O_2$. A unit of POX activity was defined as mmol $H_2O_2$ decomposed $ml^{-1}$ $min^{-1}$.

APX (EC1.11.1.1) was measured due to decrease in absorbance at 290 nm as a result of the oxidization of ascorbate according to the method of Nakano and Asada (1981). The reaction mixture contained 50 mM Na-phosphate buffer (pH 7.0) including 50 mM ascorbate, 0.1 mM $Na_2$-EDTA, 1.2 mM $H_2O_2$. APX was calculated by using the coefficient of 2.8 $mM^{-1}$ $cm^{-1}$. One unit of APX was defined as 1 mmol $ml^{-1}$ ascorbate oxidized $min^{-1}$.

SOD (EC1.15.1.1) activity was assayed according to the method of Beauchamp and Fridovich (1971) based on the photochemical reduction ability of NBT by SOD. Assays were carried out at 25 °C by using the reaction buffer of 50 mM Na-phosphate buffer (pH 7.8) including 0.033 mM NBT, 10 mM L-methionine, 0.66 mM $Na_2$EDTA and 0.0033 mM riboflavin. Reactions were initiated by addition of riboflavine to the mixture at last. After 10 min of incubation under intense light, the absorbance values were determined at 560 nm. SOD activity was defined as % inhibition of NBT by photoreduction activity of the enzyme.

CAT (EC1.11.1.6) activity was determined by the method of Aebi (1984) due to the disappearance of $H_2O_2$ at 240 nm. Reaction mixture contained 50 mM Na-phosphate buffer (pH 7.0) including 30 mM $H_2O_2$. The decrease in absorbance was followed for 90 s. 1 mmol $H_2O_2$ $ml^{-1}$ $min^{-1}$ was defined as 1 U of CAT activity.

### Identification of isozymes

Electrophoretic separations were performed by using 4% stacking and 12% separating native polyacrylamide gels under constant current of 120 mA at 4 °C. To determine the SOD isozymes, equal amount of protein samples (75 μg/well) were loaded as described by Laemmli (1970). After electrophoretic separation, SOD isozymes were identified according to the photochemical staining method described by Beauchamp and Fridovich (1973). SOD isozymes were differentiated by using SOD inhibitors. The selective inhibitions were performed by 2 mM KCN for Cu/Zn-SOD activity and 3 mM $H_2O_2$ for Cu/Zn-SOD and Fe-SOD. Mn-SOD is resistant to both inhibitors.

POX isozymes were detected by the method of Seevers et al. (1971). 50 μg/well protein extracts were loaded. Gels were stained by 0.2 M Na-phosphate buffer (pH 7.0) containing 20 mM $H_2O_2$ and 20 mM guaiacol after separation. Reaction was stopped by incubating the gels in 7% acetic acid.

APX gel staining was performed according to the method described by Mittler and Zilinskas (1991). Method is based on inhibition of NBT by ascorbate. Gels were pre-run without loading samples to gel in running buffer (pH 7.0) containing 2 mM sodium ascorbate for an hour to equilibrate gel before the electrophoretic separation. After the separation, gels were incubated in 50 mM potassium phosphate buffer (pH 7.0) containing 2 mM sodium ascorbate for 30 min, then transferred to potassium phosphate buffer (pH 7.8) containing 4 mM sodium ascorbate and 2 mM $H_2O_2$ for 20 min. Finally, gels were stained by 50 mM potassium phosphate buffer (pH 7.8) containing 28 mM TEMED and 2.5 mM NBT for 20 min.

The isoenzyme bands were named in the order to their migration distances. Gels were visualized by Bio-RAD Doc XR Image system. Band intensity profiles were quantified by GelQuant.NET version 1.8.2 software.

### Statistical analysis

The experiments were repeated three times, and each data point was the mean of three replicates. The results were expressed as mean and error bars were used for showing standard error of mean values (±SEM). Statistical evaluation of the means was performed by using one-way ANOVA analysis and statistically meaningful data were compared using Student–Newman Keuls test performed by using GraphPad Prism version 4.00 for Windows statistics software. Different letters in graphs indicates significant differences between treatments ($P < 0.05$).

## Results and discussion

Drought stress effects various metabolic processes adversely via altering key biomolecules by causing oxidative stress. Different plant species and varieties may involve alternative response strategies against drought stress. Studying on defense mechanisms against drought stress for several plant species might have importance to reveal new breeding strategies (Farooq et al. 2009; Griffiths and Parry 2002). The main aspect of this study is to evaluate enzymatic and non-enzymatic responses of two industrial tomato varieties against 7 days long drought conditions.

### Relative water contents

The inadequate water uptake caused by drought stress leads plants to lose water as a result of metabolic processes depending on subjected stress level. Therefore, plants involve various strategies to conserve cellular water in response to increasing drought intensity. The availability of the controlled mechanism is an important factor to define tolerance of different varieties against drought. In this context, many studies reported alterations on relative water contents of various crop plants depending on the duration of applied drought (Alexieva et al. 2001; Bai et al. 2006; Keyvan 2010; Saeidi and Zabihi-e-Mahmoodabad 2009).

After 7 days of drought treatment, both varieties lost relative water contents dramatically. RWC of X5671R variety was decreased to the level of 8.4% at 7 day of stress, which was 68.5% before the treatment. 5MX12956 variety presented relatively lower decrease on RWC compared to the X5671R variety. RWC of 5MX71R was decreased to the level of 12.2% at 7th day from the level of 66.4% measured at Day 0 (Fig. 1A).

In the present study, relative water contents (RWC) of the leaves of X5671R and 5MX12956 varieties decreased significantly. Decrease was gradual in response to the increasing drought stress through 7 days period. Saruhan et al. (2010) reported decreases in relative water content in the leaf and petiole during the drought period.

### Lipid peroxidation

Abiotic stress factors as drought cause oxidative stress by disturbing the balance between antioxidants and reactive oxygen species. MDA which is a product of peroxidation of lipid membranes in cells can be used as an indicator of oxidative stress in plants. In our study, lipid peroxidation levels started to increase from 1st day for both varieties. The lipid peroxidation rates were given in Fig. 1B. The lipid peroxidation rates were found 52.34 and 57.28% increased at 7th day of stress as an indicator of oxidative stress in X5671R and 5MX12956 tomato varieties, respectively (Fig. 1B). The regression analysis between the MDA levels of both varieties were found statistically significant ($P < 0.05$). According to the experimental results of lipid peroxidation, 5MX12956 tomato variety was determined as more drought sensitive than X5671R variety.

MDA concentration was gradually increased and statistical significance was determined from the beginning of the drought treatment with respect to Day 0 for both varieties. According to the results of MDA content, X5671R tomato variety was found as more drought tolerant than

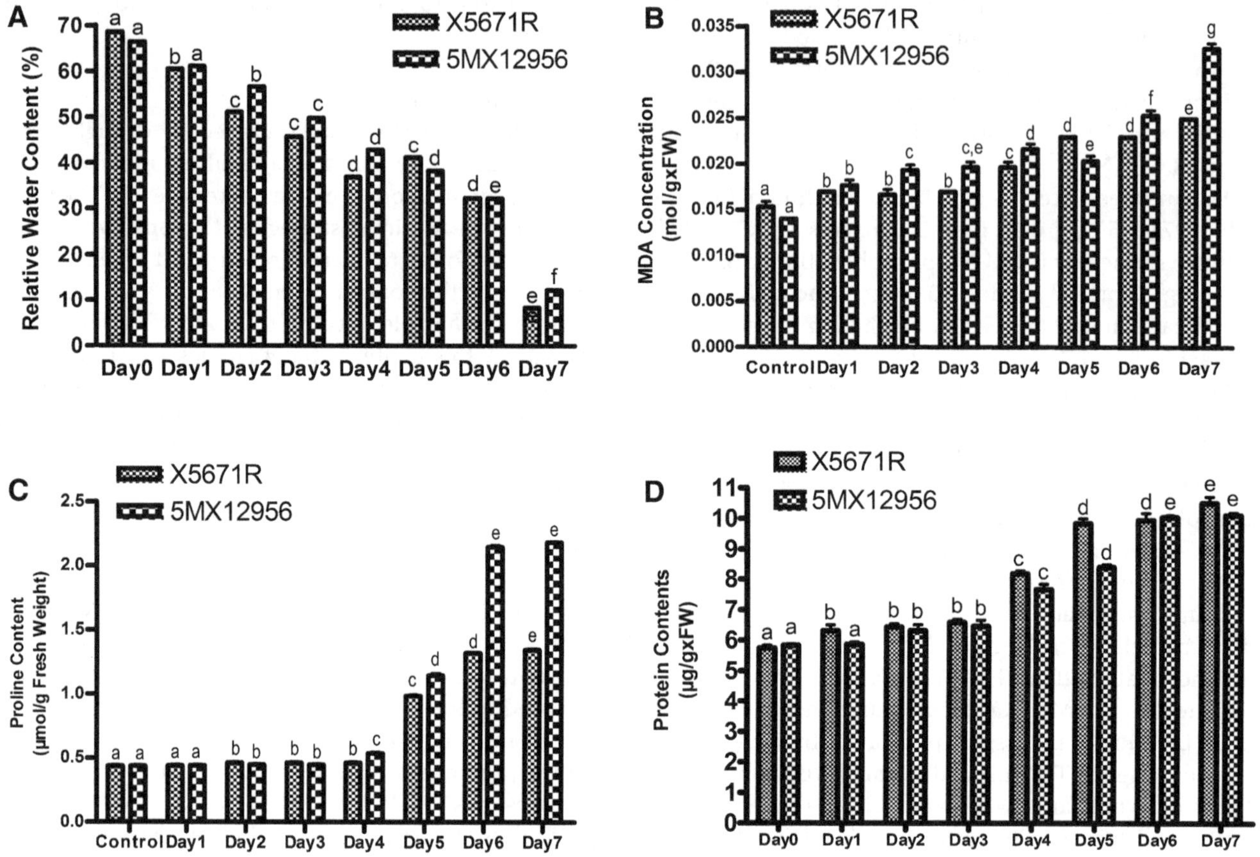


**Fig. 1** Effects of 7 day drought stress on relative water content (**A**), lipid peroxidation rate (**B**), proline content (**C**) and total protein concentration (**D**) of X5671R and 5MX12956 tomato seedlings. Data represent the average of three independent experiments with three replicates. *Vertical bars* indicate ± SEM and the *differentially given letters* represent significance at the 0.05 level

5MX12956 variety. Furthermore, the increase in MDA levels was proportional to the linear decrease in the relative water contents which offers related pattern between increasing oxidative stress and lipid peroxidation.

Sánchez-Rodríguez et al. (2012) applied drought by non-irrigation to 48 day-old tomato plants sown in perlite under greenhouse conditions. They reported MDA increase for unirrigated tomato varieties comparing to the irrigated control group. Also applied drought stress to two different tomato varieties by withholding irrigation on 22 days old plants. They reported significant MDA increase after drought treatment.

### Proline content

Plants gain tolerance by osmolyte accumulation under stress conditions. Proline is one of the compatible solutes that plays roles in plants as osmotic regulator during water deficiency and also takes place in defense systems against other environmental stress factors. Because of its osmotic regulation function during drought, many studies reported drought induced accumulation of proline (Hong-Bo et al. 2006; Mohammadkhani and Heidari 2008a).

In our study, proline levels were increased in both tomato varieties as response to drought stress during 7 days and the values were presented in Fig. 1C. The increase in proline levels was found statistically significant at 2nd day of the stress for 5MX12956 and X5671R varieties, respectively. A sharp increase was observed at the 5th day of stress for both varieties. At the 5th day of stress, the proline level of X5671R was 2.25-fold increased while 2.63-fold increase was observed for 5MX12956 variety in comparison to Day 0 ($P < 0.05$). Proline levels were 3.08- and 5.01-fold increased after 7 days of drought treatment for X5671R and 5MX12956 tomato varieties, respectively. Mohammadkhani and Heidari (2008a) were also reported a statistically significant proline accumulation after 2nd day of drought application suggesting increase on proline synthesis enzymes and decrease on proline degradation ($P < 0.05$).

### Soluble protein content

Various alterations occur on the physiological and biochemical pathways of the plants during environmental stresses as drought. Drought stress may effect protein content of the plants differently depending on the duration and the severity of the stress. Mohammadkhani and Heidari (2008b) reported that accumulation of some proteins and enzymes especially related to glycolysis, Krebs cycle and lignin synthesis increased during drought stress. However, total soluble protein content of plants can also be reduced when the level of oxidative stress was higher than the tolerance capacity of defense system of the plants (Celik and Atak 2012).

Changes in total protein content caused by drought stress treatment were given in Fig. 1D. Statistically significant increase in protein contents was measured from the first day of stress for X5671R variety, while the significance was observed from the 2nd day for 5MX12956 variety with respect to Day 0 ($P < 0.05$). In our present study, linear increase on total protein contents of both varieties in response to increasing drought severity was observed. This result suggest involvement of increased enzymatic and non-enzymatic proteins as a response to drought in both varieties. Several studies showed relationship between accumulation of drought-induced proteins such as dehydrins and dehydrin-like proteins and physiological adaptations in water limitations (Mohammadkhani and Heidari 2008b).

### Chlorophyll and carotenoid contents

Photosynthetic pigments play important roles in harvesting photons for energy metabolism of plants and also have role in stress response against oxidative stress conditions. In many studies chlorophyll and carotenoid contents of different crop plants reported to stay unchanged or decreased depending on the drought application and duration of stress (Farooq et al. 2009). Talebi et al. (2013) reported statistically significant decrease on photosynthetic pigment contents in 35 different Kabuli chickpea varieties which were subjected to drought stress by non-irrigation for 3 weeks. We observed decrease in chlorophyll a, chlorophyll b, total chlorophyll and carotenoid levels of both varieties in a linear manner by increasing drought intensity. The decrease of photosynthetic pigment concentration was one of the signs of reduction in growth and metabolic processes after water loss (Farooq

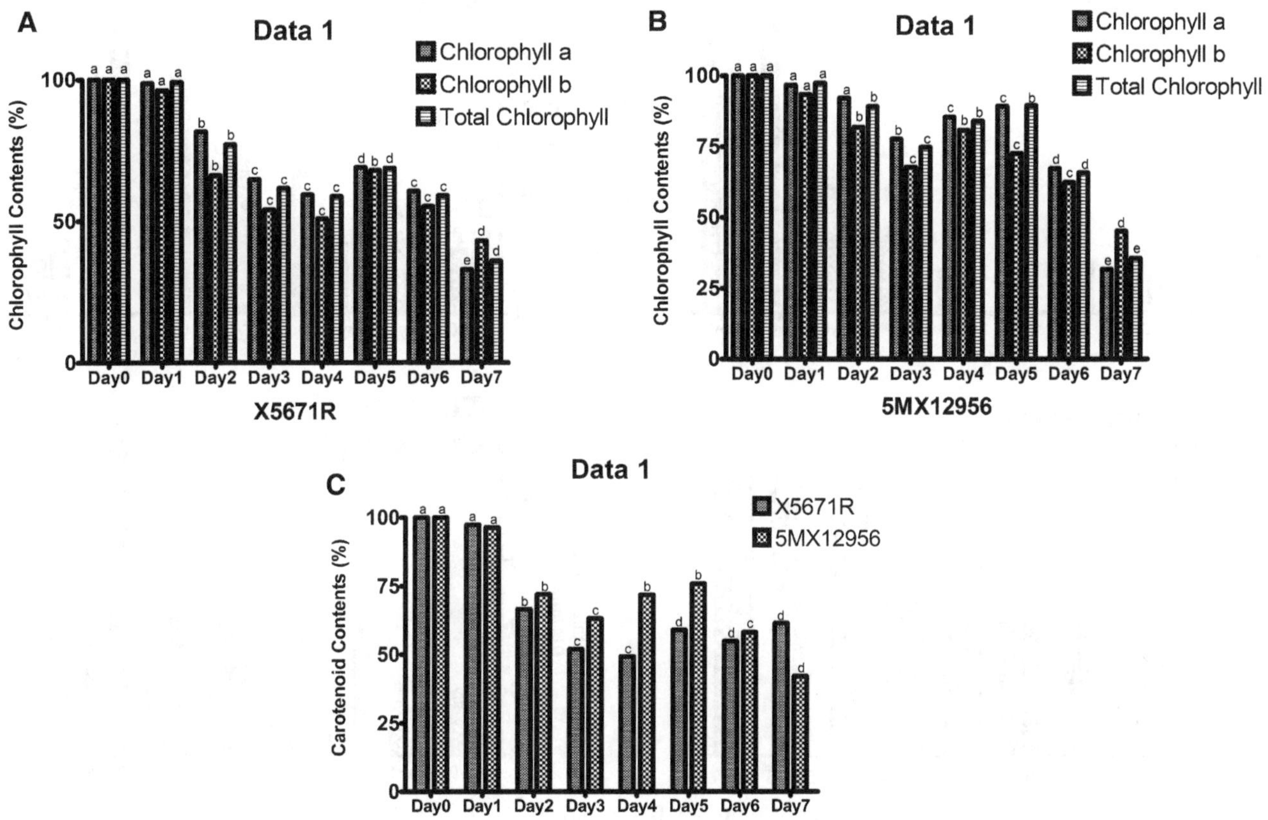


**Fig. 2** Effects of 7 days-long drought stress on photosynthetic pigment concentrations of X5671R (**A**) and 5MX12956 (**B**) and carotenoid (**C**). In all results, Day 0 is estimated as 100% and calculations presented as percentage changes comparing to the Day 0

et al. 2009). Drought treatment decreased all photosynthetic pigments of X5671R and 5MX12956 varieties significantly, comparing to the Day 0. Chlorophyll a content decreased to the levels of 33.0 and 31.7% at the 7th day of the drought stress, respectively (Fig. 2A, B). Similar decrease pattern was observed for chlorophyll b and total chlorophyll contents (Fig. 2A, B). Chlorophyll b decreased to the levels of 43.1 and 45.1% while total chlorophyll levels were 36.0 and 35.6%, respectively at the 7th day of the applied drought ($P < 0.05$). Effects of drought were also observed on carotenoid content. Carotenoid levels decreased in all treatment days and decreased 38.6 and 57.9% at 7th day of the stress respectively, compared to the Day 0 ($P < 0.05$) (Fig. 2C).

### Enzymatic parameters

Another important stress scavenging mechanism is related to antioxidant enzyme activity profiles during stress. Increased stress tolerance is mostly seen as a result of increased antioxidant enzyme activities. To evaluate the enzymatic responses of X5671R and 5MX12956 tomato varieties, we measured enzymatic activities of SOD, POX, APX and CAT.

#### *Antioxidant enzyme activities*

Peroxidase is an antioxidant enzyme which catalyse $H_2O_2$ dependent oxidation reactions (Acar et al. 2001; Demiral and Turkan 2005). Peroxidase enzyme activities of two varieties subjected to the 7 days of drought stress are presented in Fig. 3A. Peroxidase activities of both varieties presented increasing profile during the 7 days of stress. However, in both varieties, statistically significant POX activity increases were determined on 5th day of the treatment ($P < 0.05$). The highest activity increase was observed on the 6th day of stress for 5MX12956 variety by 7.4-fold. While this activity decreased to 6.4-fold, X5671R variety presented gradual increase and POX activity reached to 4.76-fold on 7th day of treatment, with respect to the Day 0.

Bai et al. (2006) applied drought stress to maize plants in soil conditions by reducing relative water of soil. They reported increasing peroxidase enzyme activity related

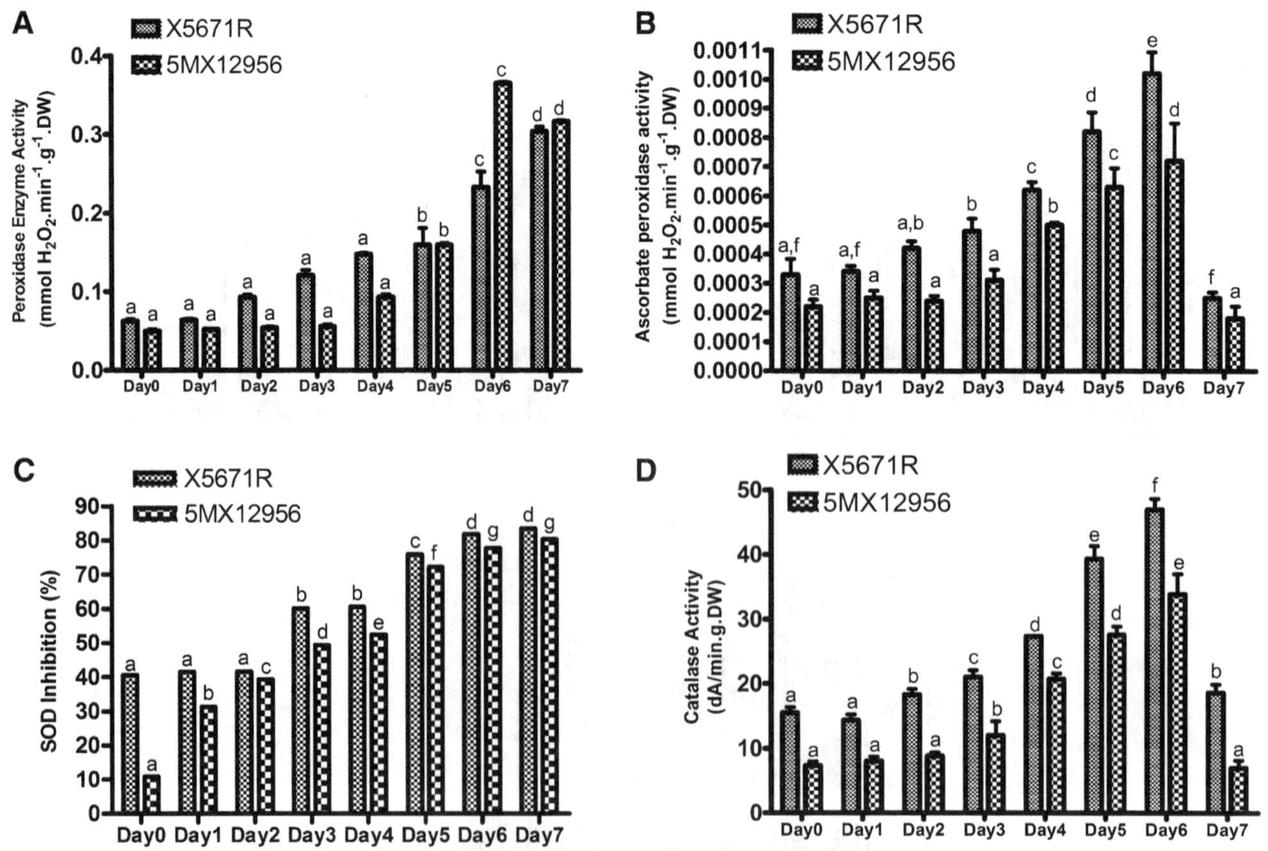


**Fig. 3** Effects of 7 day drought stress on the activities of POX (**A**), APX (**B**), SOD (**C**) and CAT (**D**) of X5671R and 5MX12956 tomato seedlings. Data represent the average of three independent experiments with three replicates. *Vertical bars* indicate ± SEM and the *differentially given letters* represent significance at the 0.05 level

to the drought severity. Chakraborty and Pradhan (2012) experimented enzymatic performance of five different wheat varieties in response to 3, 6 and 9 days of drought stress applied by withholding watering in pots. They reported statistically significant increases on peroxidase activities of different wheat varieties depending on the duration of stress. According to our results, while the first statistically significant activity increases presented in 5th day, peroxidase activity was extended until the end of applied stress period for X5671R variety, although 5MX12956 variety presented the highest activity in 6th day and the activity was decreased in 7th day of treatment.

Ascorbate peroxidase enzyme uses ascorbic acid to reduce $H_2O_2$ in plants which is generated in increased levels by many environmental stress factors as drought, salt, high and low temperatures (Caverzan et al. 2012; Shigeoka et al. 2002). As shown in the Fig. 3B, APX enzyme activities were increased due to the increasing drought stress through 6 days. After 6th day, APX enzyme activities of both varieties were decreased below the untreated control group by the increasing oxidative stress. APX activity of 5MX12956 variety was not statistically significant for Day 0 and Day 7 of treatment. At 6th day of stress treatment APX activities of X5671R and 5MX12956 varieties were reached to the maximum levels of 3.1- and 3.3-fold, respectively.

Zlatev et al. (2006) applied drought stress to three different 14-days old bean varieties by withholding irrigation until RWC of leaves reach to 65%. They reported increased APX enzyme activities for all three varieties suggesting APX is involving $H_2O_2$ scavenging in bean. Chugh et al. (2011) reported increased APX activities for drought tolerant maize varieties while sensitive varieties presented decreased APX. These results suggested that increased APX activity might be adaptive response against the drought. Zhang and Kirkham (1996) applied drought stress to sunflower and sorghum seeds in pot conditions by withholding watering. They reported sunflower APX enzyme activity was not effected from applied drought stress, while sorghum plants decreased APX activity greatly. They suggested this may indicate greater accumulation of $H_2O_2$ in cells that can result decreased synthesis of APX and increased degradation of enzyme. In the present study, drought conditions were more intense and RWC of both tomato varieties decreased to the levels below 20%. Both tomato varieties presented increased APX enzyme activity ($P < 0.07$), however the oxidative stress level was beyond the scavenging capacity of the enzyme at 7th day.

Figure 3C shows the results of SOD enzyme activities of X5671R and 5MX12956 varieties which were subjected to drought stress by non-watering for 7 days. Both varieties increased SOD activities following the increasing oxidative stress induced by the drought conditions. X5671R variety showed no statistically significant changes in enzyme activity until the 3rd day of the treatment, while 5MX12956 variety was effected from the 1st day and started to increase SOD enzyme activity ($P < 0.05$). At the third day of the stress, SOD activities were changed 19.6 and 38.5% for X5671R and 5MX12956, respectively. At the 7th day, when the enzyme levels reached to the maximum level, these increase levels were determined as 42.9 and 69.4% respectively. As a result, SOD activities of both tomato varieties presented linear increase during the applied drought period. Similar results were found for sunflower (Zhang and Kirkham 1996), rice (Sharma and Dubey 2005), sesame (Fazeli et al. 2007), cucumber (Liu et al. 2009), wheat (Baisak et al. 1994) and many other crop plants.

Catalase enzyme catalysis increasing $H_2O_2$ levels to water and oxygen in plant cells produced as a side product of photoreactions (Luna et al. 2005). Catalase (CAT) enzyme activities are presented in Fig. 3D. Catalase enzyme activities of both varieties showed similar trendline as ascorbate peroxidase activities. Significant increase on X5671R enzyme activity observed from 2nd day, while 5MX12956 variety presented from the 3rd day of stress. At the 6th day of the stress when the highest CAT activity levels were observed for both varieties, 3.02- and 4.62-fold increase measured compared to Day 0 values of X5671R and 5MX12956 varieties, respectively. At the 7th day, catalase enzyme activities showed decrease in both varieties due to increasing oxidative damage. The measured CAT activity of X5671R was statistically significant at Day 7 and found 1.19-fold higher than Day 0, whereas no statistically significant activity change was determined between Day 0 and Day 7 for 5MX12956 variety ($P < 0.05$). Wang et al. (2009) applied drought stress by 35% PEG to six alfalfa tissue cultures for 7 days. They reported increased CAT enzyme activity on leaf samples of all six alfalfa varieties. Chugh et al. (2011) applied drought to 7 days old plantlets of six different maize varieties by withholding watering for 3 days. They reported increased CAT activity for two drought tolerant varieties and decreased CAT activities for 4 sensitive maize varieties. In our research, CAT activities of both varieties presented statistically significant increase for 6 days of drought ($P < 0.05$). However, in 7th day, plants could not tolerate the adverse effects of drought and CAT activities regressed to the Day 0 values. Similarly, Chakraborty and Pradhan (2012) applied drought stress to 5 different 30 days-old wheat varieties by withholding watering for 3, 6 and 9 days. They reported increased CAT activity at 3rd day for three varieties. After 3rd day, CAT activities of these same varieties decreased below

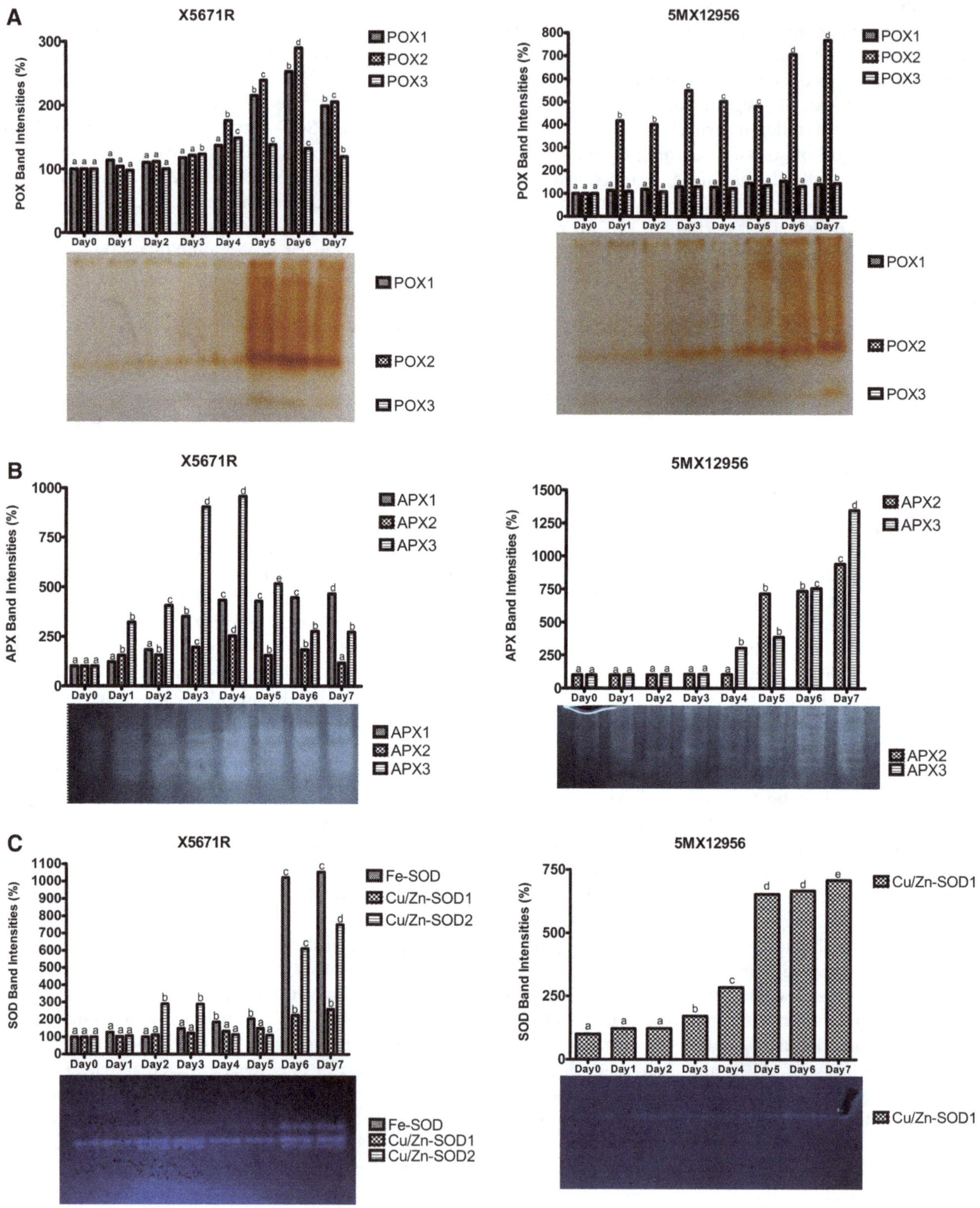


**Fig. 4** Effects of 7 day drought stress on the activity staining and % changes of POX (**A**), APX (**B**) and SOD (**C**) isozymes of X5671R and 5MX12956 tomato seedlings. Samples applied to the gels contained 50 µg protein for POX and APX and 75 µg for SOD. The *differentially given letters* represent significance at the 0.05 level

control values suggesting that severity of stress is above the tolerance capacity of these varieties. Luna et al. (2005) applied drought stress to wheat by decreasing soil water content to 36–55% (mild drought) and 11–18% (severe drought). They reported significant CAT activity increase only in severe drought conditions.

***Enzyme isozyme profiles***

Isozymes are known to play a vital role in plant defense mechanisms (Johnson 2012). In this study, isozyme diversity of two industrial tomato varieties were investigated to identify band profiles as biochemical marker for the drought tolerance. Various crop plants also present different isozyme patterns during environmental stress conditions.

Peroxidase activity staining and band intensities belonging to the isozyme profiles of both varieties were presented on Fig. 4A, respectively. POX1, POX2 and POX3 isozymes were detected for both varieties. In this present study, POX isozyme activities of both tomato varieties increased due to the increasing drought especially after 4th day of treatment. All detected POX isozymes of X5671R variety increased for 6 days of drought treatment. After 6th day of treatment, expression levels of the isozymes were started to decrease in both varieties. POX1 and POX2 isozyme expressions reached to the maximum level at 6th day. However, at 7th day all isozyme band intensities showed decrease. Despite of this decrease all band intensities of Day 7 were higher than Day 0. The band intensities of POX1, POX2 and POX3 were 1.99-, 2.06- and 1.20-fold increased in comparison to the Day 0 values at 7th day of stress, respectively. However, POX isozyme band intensities were found higher than the values determined at Day 0 during the stress for X5671R variety. POX2 band intensity showed increasing pattern during the drought stress treatment in 5MX12956 variety. POX2 band intensity reached to the maximum level at 7th day of stress and was recorded as 7.68-fold higher than Day 0. According to the native PAGE results, increases in POX1 and POX2 isozymes are detected as the main reasons behind the increase of POX activity for X5671R variety, while only POX2 isozyme seems to effect the activity of POX enzyme for 5MX12956 variety.

Sen and Alikamanoglu (2014) applied drought to sugar beet tissue cultures by different concentrations of PEG6000. They detected eight POX isozymes after native PAGE separation. Among these eight isozymes, POX1 and POX5 were not detected in bands belonging to the control plants. Abedi and Pakniyat (2010) applied drought stress to 3 weeks old Brassica plantlets by reducing soil water capacity to 60 and 30%. They detected five POX isozyme bands after drought treatment. They detected increase in various POX isozymes suggesting the role of different POX isozymes to superoxide scavenging in response to drought conditions. Srivalli et al. (2003) applied drought stress to 50 days old rice plants by withholding water in 10, 13, 16 days periods. They detected four POX isozymes in response to drought treatment. They reported significant increase in POX3 isozyme which was higher than the control even after rewatering. POX2 and POX4 were not effected from application among the treatment but POX1 presented low abundance. They also indicated the importance of POX and SOD as key components against drought stress.

As another peroxide detoxifying enzyme APX activities of X5671R and 5MX12956 varieties presented via increase in isozyme pattern following the treatment days. Ascorbate peroxidase enzyme isozyme profiles of 7 days drought treated X5671R and 5MX12956 varieties were presented on Fig. 4B, respectively. APX1, APX2, APX3 isozymes were detected for X5671R variety, while 5MX12956 variety presented only APX2 and APX3 isozymes. For X5671R variety, band densities belonging to the APX1 isozyme increased at 7th day of the stress by 4.66-fold rate. While APX2 and APX3 band intensities reached the maximum levels at 4th day and then showed a decreasing profile until 7th day. Band intensity of APX3 isozyme showed a sharp increase rate by 3.22-fold on Day 1 and the intensity of this isozyme reached to 9.58-fold on 4th day of stress. At the 7th day of treatment APX2 and APX3 isozymes were determined higher than Day 0 by the 1.16- and 2.74-fold, respectively. Only APX2 and APX3 isozymes were detected for 5MX12956 variety after the activity staining of native polyacrylamide gels. Results are given in Fig. 4B. The band densities were increased in accordance with increasing drought intensity. We observed a sharp increase on Day 5 in APX2 isozyme by 7.16-fold, whereas APX3 isozyme showed 3.05-fold increase on Day 4. Although the band intensities showed similar increase rates by 7.36 and 7.56-fold at the 6th day of stress, responses of isozyme profiles presented differences on Day 7. APX2 band intensity reached to the maximum level by 9.4-fold while the density of APX3 isozyme was 13.48-fold at the 7th day of stress compared to Day 0. The variations in APX isozyme densities of both tomato varieties were found statistically significant ($P < 0.05$). Among detected APX1, APX2 and APX3 isozymes, APX1 and APX3 were more significant for X5671R variety. However, only APX2 and APX3 isozymes detected in 5MX12956 variety. X5671R isozymes were more intense comparing to the 5MX12956 variety. Their response profiles against drought stress were also different. In 5MX12956 variety, APX3 isozyme showed increase at 4th day initially; at 5th day, the increase rate of APX2 isozyme was higher than APX3 isozyme and on the 6th day, their activities were detected nearly the same. On the 7th day of the stress, APX3 isozyme density was observed 1.43-fold higher than APX2 isozyme. This isozyme difference between varieties also presented in accordance with APX activity results. Increase in the activities of APX in drought stressed plants are indicative of activation of ascorbate–glutathione cycle in tomato plants.

Şen and Alikamanoglu (2014) detected APX1 and APX2 in all samples while APX3 observed in most of the samples and APX4 was observed in all samples except the control. Srivalli et al. (2003) applied drought stress to 50 days old rice plants by withholding water in 10, 13, 16 days period. Six different APX isozymes were detected in native PAGE separation. APX2 isozyme was detected in all groups while APX3 and APX4 were only increased after drought treatment and decreased after re-watering. APX5 was effected negatively from drought and decreased after treatment. APX6 was found most dominant isozyme and kept uniform profile during the drought period. They suggested that differential appearance of APX isozymes may be result of differential functions of various APX isozymes in plants.

Another important element of antioxidative system is superoxide dismutase. SOD isozymes play their protective roles in different plant cell compartments. SOD isozyme band profiles showed differences between the tomato varieties as response to 7 day long drought stress. We detected Fe-SOD, Cu–Zn/SOD1 and Cu–Zn/SOD2 isozymes for X5671R variety, while only Cu–Zn/SOD1 isozyme was detected for 5MX12956 variety. For X5671R tomato variety, all SOD isozyme densities were increased during the stress treatment, but their increment rates were showed alterations among treatment days. Fe-SOD isozyme presented a sharp increase on Day 6 by 10.22-fold and continued to increase on Day 7. Cu–Zn/SOD1 showed gradual increase at 7th day of the stress and band density increased 2.60-fold compared to Day 0. The band density profiles of Cu–Zn SOD2 showed sharp increase at the 6th day. At the 7th day of the stress, activity of this enzyme was found as 7.51-fold increased with respect to Day 0. In 5MX12956 tomato variety, the only detected SOD isozyme was Cu–Zn/SOD1 and it showed increased pattern during the stress. After 5th day of the stress rapid increase was detected by 6.53-fold in comparison to Day 0. Last 3 days, the band intensities were increased severely by 7.09-fold, compared to Day 0 ($P < 0.05$). The changes in SOD isozyme activities were given in Fig. 4C.

Cu/Zn-SOD2 and Fe-SOD were determined as the most responsive isozymes against drought stress for X5671R variety. This result suggests the involvement of chloroplast to the early protection against oxidative stress caused by drought for this variety. However, in contrast with the X5671R variety, 5MX12956 variety presented only Cu/Zn-SOD1 isozyme increase while Cu/Zn-SOD2 and Fe-SOD isozymes were not detected. Fe-SOD is seem as the most responsive isozyme against drought in tolerant variety. Boaretto et al. (2014) applied drought stress by reducing relative water of soil in pot conditions to two different sugarcane cultivars. In water deficiency, they reported increased SOD enzyme activities for both varieties. They also reported two Mn-SOD and six Cu/Zn-SOD isozymes after drought treatment suggesting involvement of an extra SOD VI (Cu/Zn-SOD) to be involved in drought tolerance. Sekmen et al. (2014) applied combined drought and heat stress to two different cotton varieties. They reported one Mn-SOD, one Fe-SOD and two Cu/Zn-SOD isozymes for both varieties. They also suggested Cu–Zn/SOD2 to be the most responsive isozyme against drought in cotton. Abedi and Pakniyat (2010) applied drought stress to 3 weeks old Brassica plantlets by reducing soil water capacity to 60 and 30%. They detected two Mn-SOD isozymes and two Cu/Zn-SOD isozymes after stress application and suggested that the main reason in SOD activity increase may be related to Mn-SOD isozyme increase. Fe-SOD isozyme was not detected in this application. They discussed that Mn-SOD mainly function in mitochondrial electron transport chain and this cell compartment can be crucial for superoxide protection under stress.

## Conclusions

In conclusion, all the enzyme systems reveals the difference between the drought tolerant/sensitive varieties. The analysis results showed that the antioxidative system responses are different between two tomato varieties. The present study demonstrated specific changes in antioxidative system in two varieties of *S. lycopersicum*. The changes in isozyme patterns and enzyme activities were supported the sensitivity of 5MX12956 and tolerance of X5671R tomato varieties against 7 days long drought stress. The differences between antioxidative mechanism activities of two industrial tomato varieties that were initiated as a series of responses against drought stress has been proven via biochemical and electrophoretic analyses.

According to our results, increased POX1 and POX3 isozyme bands seem to have role in drought tolerance. Also, the main difference between the varieties is the absence of APX1 isozyme band in 5MX12956 tomato variety. Although there are differences in band intensities of APX2 and APX3 isozymes between the varieties during the stress treatment, the relation between the tolerance and APX1 isozyme should be evaluated. Another important isozymic difference was also observed between SOD banding patterns. Fe-SOD and Cu/Zn-SOD2 isozymes seem to be more effective in drought tolerance of X5671R tomato variety.

In further studies, the exact relations between tolerance isozyme differences need to be defined and their availability should be evaluated as potential biochemical markers for drought tolerance studies in breeding programs of these tomato varieties.

**Authors' contributions**

ÖÇ and ÇA conceived and designed the experiments; ÖÇ and AA performed the experiments and analyzed the data; ÖÇ and ÇA wrote the manuscript. All authors read and approved the final manuscript.

**Acknowledgements**

Not applicable.

**Funding**

Not applicable.

**References**

Abedi T, Pakniyat H (2010) Antioxidant enzyme changes in response to drought stress in ten cultivars of oilseed rape (*Brassica napus* L.). Czech J Genet Plant Breed 46:27–34

Acar O, Turkan I, Ozdemir F (2001) Superoxide dismutase and peroxidase activities in drought sensitive and resistant barley (*Hordeum vulgare* L.) varieties. Acta Physiol Plant 23:351–356

Aebi H (1984) Catalase in vitro. Methods Enzymol 105:121–126

Alexieva V, Sergiev I, Mapelli S, Karanov E (2001) The effect of drought and ultraviolet radiation on growth and stress markers in pea and wheat. Plant Cell Environ 24:1337–1344

Arnon DI (1949) Copper enzymes in isolated chloroplasts. Polyphenoloxidase in *Beta vulgaris*. Plant Physiol 24:1–15

Bai LP, Sui FG, Ge TD, Sun ZH, Lu YY, Zhou GS (2006) Effect of soil drought stress on leaf water status, membrane permeability and enzymatic antioxidant system of maize. Pedosphere 16:326–332

Baisak R, Rana D, Acharya PBB, Kar M (1994) Alterations in the activities of active oxygen scavenging enzymes of wheat leaves subjected to water stress. Plant Cell Physiol 35:489–495

Bartosz G (1997) Oxidative stress in plants. Acta Physiol Plant 19:47–64

Bates LS, Waldren RP, Teare ID (1973) Rapid determination of free proline for water-stress studies. Plant Soil 39:205–207

Beauchamp C, Fridovich I (1971) Superoxide dismutase: improved assays and an assay applicable to acrylamide gels. Anal Biochem 44:276–287

Beauchamp CO, Fridovich I (1973) Isozymes of superoxide dismutase from wheat germ. Biochim Biophys Acta 317:50–64

Boaretto LF, Carvalho G, Borgo L, Creste S, Landell MG, Mazzafera P, Azevedo RA (2014) Water stress reveals differential antioxidant responses of tolerant and non-tolerant sugarcane genotypes. Plant Physiol Biochem 74:165–175

Bradford MM (1976) A rapid and sensitive method for the quantitation of microgram quantities of protein utilizing the principle of protein-dye binding. Anal Biochem 72:248–254

Caverzan A, Passaia G, Rosa SB, Ribeiro CW, Lazzarotto F, Margis-Pinheiro M (2012) Plant responses to stresses: role of ascorbate peroxidase in the antioxidant protection. Genet Mol Biol 35:1011–1019

Celik O, Atak C (2012) The effect of salt stress on antioxidative enzymes and proline content of two Turkish tobacco varieties. Turk J Biol 36:339–356

Chakraborty U, Pradhan B (2012) Oxidative stress in five wheat varieties (*Triticum aestivum* L.) exposed to water stress and study of their antioxidant enzyme defense system, water stress responsive metabolites and $H_2O_2$ accumulation. Braz J Plant Physiol 24:117–130

Chugh V, Kaur N, Gupta AK (2011) Evaluation of oxidative stress tolerance in maize (*Zea mays* L.) seedlings in response to drought. Indian J Biochem Biophys 48:47–53

Demiral T, Turkan I (2005) Comparative lipid peroxidation, antioxidant defense systems and proline content in roots of two rice cultivars differing in salt tolerance. Environ Exp Bot 53:247–257

FAO. 2012. FAOSTAT

Farooq M, Wahid A, Kobayashi N, Fujita D, Basra SMA (2009) Plant drought stress: effects, mechanisms and management. Agron Sustain Dev 29:185–212

Fazeli F, Ghorbanli M, Niknam V (2007) Effect of drought on biomass, protein content, lipid peroxidation and antioxidant enzymes in two sesame cultivars. Biol Plant 51:98–103

George S, Jatoi SA, Siddiqui SU (2013) Genotypic differences against peg simulated drought stress in tomato. Pak J Bot 45:1551–1556

Gill SS, Tuteja N (2010) Reactive oxygen species and antioxidant machinery in abiotic stress tolerance in crop plants. Plant Physiol Biochem 48:909–930

Griffiths H, Parry M (2002) Plant responses to water stress. Ann Bot 89:801–802

Harb A, Krishnan A, Ambavaram MMR, Pereira A (2010) Molecular and physiological analysis of drought stress in Arabidopsis reveals early responses leading to acclimation in plant growth. Plant Physiol 154:1254–1271

Harrison MT, Tardieu F, Dong ZS, Messina CD, Hammer GL (2014) Characterizing drought stress and trait influence on maize yield under current and future conditions. Global Change Biol 20:867–878

Hong-Bo S, Xiao-Yan C, Li-Ye C, Xi-Ning Z, Gang W, Yong-Bing Y, Chang-Xing Z, Zan-Min H (2006) Investigation on the relationship of proline with wheat anti-drought under soil water deficits. Colloids Surf B Biointerfaces 53:113–119

Jeanneau M, Gerentes D, Foueillassar X, Zivy M, Vidal J, Toppan A, Perez P (2002) Improvement of drought tolerance in maize: towards the functional validation of the Zm-Asr1 gene and increase of water use efficiency by over-expressing C4-PEPC. Biochimie 84:1127–1135

Johnson M (2012) Studies on intra-specific variation in a multipotent medicinal plant *Ocimum sanctum* Linn. using isozymes. Asian Pac J Trop Biomed 2:S21–S26

Keyvan S (2010) The effects of drought stress on yield, relative water content, proline, soluble carbohydrates and chlorophyll of bread wheat cultivars. J Anim Plant Sci 8:1051–1060

Laemmli UK (1970) Cleavage of structural proteins during the assembly of the head of bacteriophage T4. Nature 227:680–685

Liu ZJ, Zhang XL, Bai JG, Suo BX, Xu PL, Wang L (2009) Exogenous paraquat changes antioxidant enzyme activities and lipid peroxidation in drought-stressed cucumber leaves. Sci Hortic 121:138–143

Luna CM, Pastori GM, Driscoll S, Groten K, Bernard S, Foyer CH (2005) Drought controls on H2O2 accumulation, catalase (CAT) activity and CAT gene expression in wheat. J Exp Bot 56:417–423

Mittler R (2002) Oxidative stress, antioxidants and stress tolerance. Trends Plant Sci 7:405–410

Mittler R, Zilinskas BA (1991) Purification and characterization of pea cytosolic ascorbate peroxidase. Plant Physiol 97:962–968

Mohammadkhani N, Heidari R (2008a) Drought-induced accumulation of soluble sugar and proline in two maize varieties. World Appl Sci J 3:448–453

Mohammadkhani N, Heidari R (2008b) Effects of drought stress on soluble proteins in two maize varieties. Turk J Biol 32:23–30

Nakano Y, Asada K (1981) Hydrogen-peroxide is scavenged by ascorbate-specific peroxidase in spinach-chloroplasts. Plant Cell Physiol 22:867–880

Saeidi M, Zabihi-e-Mahmoodabad R (2009) Evaluation of drought stress on relative water content and chlorophyll content of sesame (*Sesamum indicum* L.) genotypes at early flowering stage. Res J Environ Sci 3:345–350

Sánchez-Rodríguez E, Rubio-Wilhelmi MDM, Blasco B, Leyva R, Romero L, Ruiz JM (2012) Antioxidant response resides in the shoot in reciprocal grafts of drought-tolerant and drought-sensitive cultivars in tomato under water stress. Plant Sci 188–189:89–96

Saruhan N, Terzi R, Saglam A, Kadioglu A (2010) Scavenging of reactive oxygen species in apoplastic and symplastic areas of rolled leaves in *Ctenanthe setosa* under drought stress. Acta Biol Hung 61:282–298

Seevers PM, Daly JM, Catedral FF (1971) The role of peroxidase isozymes in resistance to wheat stem rust disease. Plant Physiol 48:353–360

Sekmen AH, Ozgur R, Uzilday B, Turkan I (2014) Reactive oxygen species scavenging capacities of cotton (*Gossypium hirsutum*) cultivars under combined drought and heat induced oxidative stress. Environ Exp Bot 99:141–149

Sen A, Alikamanoglu S (2014) Characterization of drought-tolerant sugar beet mutants induced with gamma radiation using biomedical analysis and isozyme variations. J Sci Food Agric 94:367–372

Shanker AK, Maheswari M, Yadav SK, Desai S, Bhanu D, Attal NB, Venkateswarlu B (2014) Drought stress responses in crops. Funct Integr Genom 14:11–22

Sharma P, Dubey RS (2005) Modulation of nitrat reductase activity in rice seedlings under aluminium toxicity and water stress: role oof osmolytes as enzyme protectant. J Plant Physiol 162:854–864

Shigeoka S, Ishikawa T, Tamoi M, Miyagawa Y, Takeda T, Yabuta Y, Yoshimura K (2002) Regulation and function of ascorbate peroxidase isoenzymes. J Exp Bot 53:1305–1319

Smart RE (1974) Rapid estimates of relative water content. Plant Physiol 53:258–260

Srivalli B, Sharma G, Khanna-Chopra R (2003) Antioxidative defense system in an upland rice cultivar subjected to increasing intensity of water stress followed by recovery. Physiol Plant 119:503–512

Stewart RRC, Bewley JD (1980) Lipid-peroxidation associated with accelerated aging of soybean axes. Plant Physiol 65:245–248

Talebi R, Ensafi MH, Baghebani N, Karami E (2013) Physiological responses of chickpea (*Cicer arietinum*) genotypes to drought stress. Environ Exp Biol 11:9–15

Wahb-Allah MA, Alsadon AA, Ibrahim AA (2011) Drought tolerance of several tomato genotypes under greenhouse conditions. World Appl Sci J 15:933–940

Wang WB, Kim YH, Lee HS, Kim KY, Deng XP, Kwak SS (2009) Analysis of antioxidant enzyme activity during germination of alfalfa under salt and drought stresses. Plant Physiol Biochem 47:570–577

Zdravkovic J, Jovanovic Z, Djordjevic M, Girek Z, Zdravkovic M, Stikic R (2013) Application of stress susceptibility index for drought tolerance screening of tomato populations. Genetika-Belgrade 45:679–689

Zhang J, Kirkham MB (1996) Antioxidant responses to drought in sunflower and sorghum seedlings. New Phytol 132:361–373

Zlatev Z, Lidon F, Ramalho J, Yordanov I (2006) Comparison of resistance to drought of three bean cultivars. Biol Plant 50:389–394

# 4

# Collecting near mature and immature orchid seeds for ex situ conservation: 'in vitro collecting' as a case study

Jonathan P. Kendon[1], Landy Rajaovelona[2], Helen Sandford[1], Rui Fang[1], Jake Bell[1] and Viswambharan Sarasan[1*]

**Abstract**

**Background:** Lack of phenological information and efficient collecting methods are considered impediments for orchid seed collecting. This leads to opportunistic collecting as part of general seed collecting schedules that may last few weeks especially in remote areas. The study explored the feasibility of collecting near mature and immature seeds to support conservation action plans. Mature, near mature and immature seeds of orchids were collected from the wild in the Central Highlands of Madagascar (CHM). Seed capsules were collected in sterile culture medium in the wild, to prevent deterioration of seeds inside the capsule after collecting, later to be cultured under laboratory conditions.

**Results:** Seed capsules collected by the in vitro collecting (IVC) method were kept in very good condition for up to 4 weeks before germination under in vitro conditions. Significantly faster and higher germination rate ($p < 0.001$) than mature seeds was observed in a number of taxa collected during a 3 year-long study. In some taxa even immature seeds, with no apparent sign of testa covering the embryo, germinated following IVC where mature seeds failed to germinate.

**Conclusions:** We propose that IVC method has potential to complement conventional seed collecting by increasing the germplasm that can be used in integrated conservation action plans. Improvements can be made in developing collections for taxa from biodiversity hotspots and remote areas where collecting requires considerable resources. This method can further be used on a wider selection of plants from different geographic areas and on embryo rescue programmes for economically important plants.

## Background

Orchids are the most complex and enigmatic group of flowering plants, with a global distribution, yet are considered to be the most threatened family. A large proportion of orchids are endemic in areas where habitat loss, illegal collecting and loss of pollinators are prevalent (Merritt et al. 2014). Due to the threats to orchids worldwide and the adverse effect on recruitment in the wild they have been included in Appendix II of the Conservation on International Trade in Endangered Species of Wild Fauna and Flora (CITES). Orchid seeds appear to be short-lived compared to crop plant seeds, even if considered desiccation tolerant (Pritchard et al. 1999; Seaton and Pritchard 2003). Seed dormancy in many plant families seriously impacts outcomes at the time seeds are sown (Merritt et al. 2005). Lack of seed information and handling practices continuously affect outcomes in restoration programmes. This includes data deficiencies pertaining to phenology of seed development and maturation for the majority of wild species (Broadhurst et al. 2008; Mortlock 2000) and this can lead to asynchronous timing of seed collection (Merritt and Dixon 2011). In orchids, irregular germination in nature as well as under in vitro conditions is a major bottleneck for conservation of rare and threatened taxa (Barsberg et al. 2013). In particular, terrestrial orchid seeds are even more difficult to work with due to quick seed maturation,

*Correspondence: v.sarasan@kew.org
[1] Natural Capital and Plant Health, Royal Botanic Gardens Kew, Richmond, Surrey TW9 3AB, UK
Full list of author information is available at the end of the article

dormancy-related germination issues, and desiccation sensitivity. Therefore, terrestrial orchids present significant conservation challenges.

The number of ex situ conservation facilities worldwide has grown dramatically over the years (Wyse-Jackson 2001), but have become specifically integrated to achieve objectives under national and regional conservation priorities (Maunder et al. 2001). Nevertheless, biodiversity conservation in resource poor areas of the world still faces an uncertain future. One of the key priorities of ex situ conservation is long term storage in seed banks and cryopreservation facilities which require substantial resources for capacity building. Long term storage of orchids requires cryopreservation at ultra-low temperature as a general rule because longevity under conventional seed banking conditions is not completely reliable (Merritt et al. 2014).

There are nearly 1000 orchid taxa in Madagascar, of which 85% are endemic—an exceptional area of biodiversity for orchids (Moat and Smith 2007; Tyson 2000). They are found in several phytogeographical regions with unique climatic conditions. This global biodiversity hotspot is a perfect study area to test orchids for collecting methods and assessing viability. We selected the Central Highlands of Madagascar (CHM) as our study area with over 50 species of orchids. The habitats of the study comprised inselbergs, savannah, grasslands, montane rocky scrubland and gallery forests. This area has granitic rocks, marble and quartzite with the majority of orchids being lithophytes and the rest true epiphytes and terrestrials.

In the present paper we describe methods for collecting seeds at different maturity levels from the CHM habitats using IVC methods. IVC is the method of initiating in vitro cultures from plants in the wild. The method has been used in the past for a variety of functions ranging from horticulture to conservation (Warren 1983; Alvarenga et al. 2002; Henao et al. 2002; Saldana et al. 2002; Brenes et al. 2002; Engelmann 2002; Sandoval 2002). Propagation of orchids from seeds has long been achieved through the use of capsules which are either fully matured or green, un-dehisced and near-mature. Collected dry seeds are cleaned and dried before either storage or culturing for germination. Seed viability upon storage under standard seed bank conditions is still not completely reliable for orchids (Merritt et al. 2014). In many cases, especially in temperate terrestrial orchids, higher germination rates have been achieved using green capsules than mature seeds. This method is dependent on the immediate culturing of seed after harvest from green capsules. Seeds collected from plants in the wild, spread over large swathes of land mass, from remote locations usually perish due to moisture loss and microbial contamination (unpublished results). The critical aspect of working towards developing a successful IVC protocol is the control of contamination (Pence 2005) along with maintaining the viability of the differentially matured seeds. In many cases the techniques were specifically developed for only a limited number of taxa as described in the present study. We discuss the importance of this new method to improve collecting and germination efficiency of differentially matured orchid seeds. The implications of this approach to cryopreservation and living collection development for reintroduction and assisted colonisation are discussed in detail.

## Methods

### Study area and seed collecting

Studies in the field were conducted in the CHM habitats where a large number of endemic orchids are found in discontinuous populations. The orchid seed capsules were collected from plants in the wild in inselberg, savannah, montane rocky grassland, and gallery forests (Fig. 1).

Only a small number of capsules were allowed to be collected due to the fragmented nature of populations and few available capsules in individual plants. Orchids from these diverse habitats exist as epiphytes, lithophytes, and terrestrials. Different habitats and life forms contribute to the diversity of seed capsules which were available for this study. Seed capsules were categorised into five groups based on their size.

Seed capsules were considered mature when they were yielding to the touch and had yellow, red or brown colouration, and contained mature seeds. Seed capsules were collected in vitro if they contained near-mature or immature seeds. These capsules were green, firm and of similar dimensions to mature capsules. Mature capsules were cut from the plant, wrapped in paper in a sealed seed envelope and placed in a field desiccator. This consisted of a sealable plastic box containing orange silica gel. Several collections were made per species for both IVC and mature capsules, and collection was performed in an opportunistic manner in accordance with guidelines laid down by the seed conservation department, RBG Kew (Way 2003). On arrival at RBG Kew the seeds were equilibrated over a saturated solution of $CaCl_2$ inside a glass desiccator, before being sealed in glass vials and stored in a fridge at 3 °C.

The following three-step procedure was employed for the IVC of seed capsules from all species collected from the wild in Itremo, CHM.

### Surface sterilisation in the field

Flower parts were removed by scalpel blade and the capsules were cleaned with moist sterile towels immediately after collection. These seed capsules were then sterilised

**Fig. 1** Orchids and their habitats in Itremo, Central Highlands, Madagascar. **a** *Aerangis ellisii* and **b** *Angraecum rutenbergianum* with near-mature seed capsules; **c** rocky montane grassland; **d** savannah; **e** inselberg; **f** gallery forest. Note smoke visible in tile (**c**) as a result of extensive grass fires

in 0.5% (w/v) sodium dichloroisocyanurate (NaDCC) solution (Sigma Chemicals, UK) for a period depending on size, nature of seed capsule and where it was collected from (Table 1). Using results of the preliminary IVC studies, on two terrestrial orchid species at RBG Kew, the concentrations of NaDCC and PPM to be used in this study were optimised. For sizes 1–3 plastic disposable containers (150 ml volume) with screw cap lids (Sterilin, UK) were used to culture the capsules following surface sterilisation; for size 4 and 5 Dilu vials (27 ml volume, Alpha Laboratories, UK) with flip tops were used. Sterilised seed capsules were washed in one change of 0.01% NaDCC before being transferred to IVC medium containing NADCC and plant preservative mixture (PPM, Apollo Scientific, UK). The culture medium selected for Madagascar IVC was ¼ MS (Murashige and Skoog 1962) supplemented with 1% sucrose, 0.1% activated charcoal and 1% agar to which was added 0.03% (w/v) NaDCC, once the medium had cooled to around 40 °C to avoid evolution of chlorine gas, and 2 ml/l of PPM (v/v).

**Table 1 List of taxa from Itremo, Central Highlands, Madagascar: seed capsule size and seed capsule colour 2–3 weeks after in vitro collecting and results of seed capsule sterilisation that yielded clean seeds 3 weeks after sowing**

| Taxon | Life form | Colour of seed pod on arrival at RBG Kew | Size group | Init ial pod steri-lisat ion t ime (mins) | Response at step 1, 2 and 3 [Sterile (S)/ Contaminated (C)] | | | Response after seed sowing |
|---|---|---|---|---|---|---|---|---|
| | | | | | 1 | 2 | 1 | |
| *Aerangis citrata* | *EP* | Green | 4 | 45 | S | – | – | S |
| *Aerangis ellisii* | *LP* | Green + brown | 2 | 80 | C | S | – | S |
| *Aerangis ellisii* | *LP* | Green, majority brown | 2 | 80 | C | S | – | C |
| *Aerangis punctata* | *EP* | Reddish tinge | 4 | 60 | S | – | – | S |
| *Aerangis punctata* | *EP* | Brown | 4 | 50 | C | S | – | S |
| *Aerangis sp.* | *LP* | Green | 4 | 40 | S | – | S | |
| *Aerangis sp.* | *LP* | Green | 3 | 40 | S | – | S | |
| *Aerangis sp.* | *LP* | Green | 3 | 40 | S | – | S | |
| *Aerangis sp.* | *LP* | Yellowish brown | 3 | 60 | C | C | S | S |
| *Angraecum calceolus* | *LP* | Green | 3 | 50 | S | – | – | S |
| *Angraecum magdalenae* | *LP* | Green/brown | 1 | 90 | C | S | – | S |
| *Aerangis protensum* | *LP, TR* | Greenish/yellow brownish | 2 | 60 | C | S | – | S |
| *Aerangis rutenbergianum* | *LP* | Brown | 3 | 60 | C | S | – | S |
| *Aerangis rutenbergianum* | *LP* | Green | 4 | 50 | S | – | – | S |
| *Aerangis sororium* | *LP* | Green | 1 | 90 | S | – | – | S |
| *Aerangis sororium* | *LP* | Brown | 1 | 90 | C | S | – | S |
| *Aerangis sp.* | *LP* | Greenish brown | 5 | 40 | S | – | – | S |
| *Bulbophyllum peyrotii* | *EP* | Brownish black | 5 | 60 | C | S | – | S |
| *Habenaria ambositrana* | *TR* | Green | 5 | 60 | S | – | – | S |
| *Oeceoclades calcarata* | *LP* | Brown | 1 | 80 | C | S | – | S |
| *Polystachya fusiformis* | *LP, EP* | Green | 4 | 45 | C | S | – | S |
| *Polystachya fusiformis* | *LP, TR* | Green/brown | 5 | 60 | S | – | – | S |

Capsule size group: above 1500 mm$^2$ (1), above 750 mm$^2$ (2), above 300 mm$^2$ (3), above 150 mm$^2$ (4), below 150 mm$^2$ (5)

*TR* terrestrial, *EP* epiphyte, *LP* lithophyte as found in the field. Response: *C* contaminated, *S* sterile

### Rescue of contaminated seed capsules at the CHM base

Seed capsules were observed under a stereo field microscope for contamination within 48 h and rescued if necessary using the following procedure. Capsules were carefully removed from the vial, wiped with a sterile antiseptic wipe and gently agitated in 0.5% (w/v) NaDCC plus a drop of hand soap for at least 1 h (small, little contamination) and up to 2 h (large, moderate contamination) depending on size of the capsule and level of contamination. Re-sterilised capsules were placed in vials containing IVC medium as described above.

### Final surface sterilisation before culturing at RBG Kew

Once the seed capsules arrived at RBG Kew, seeds were cultured directly into the culture medium containing MS nutrients, Phytamax (Sigma Chemicals, UK) or other appropriate media under sterile conditions depending on the species. All culture media were supplemented with 2% sucrose with pH set at 5.8 and solidified with 0.8% agar. Media were autoclaved at 121 °C for 15 min. Before pouring the media into 5 cm Petri dishes, 5 ml/l (v/v) PPM and 0.015% (w/v) NaDCC were added once the medium had cooled to around 40 °C. Seed capsules were cut open and seeds transferred to culture medium and incubated at 22 °C.

### Sonication, viability and carapace staining

This study was conducted to see the effect of seed coats on germination. Approximately 350–450 *Angraecum protensum* seeds each were suspended in water in 10 ml screw cap centrifuge tubes and sonicated for 0, 1 and 3 min in a Decon F5300b ultrasonic bath operating at a frequency of 40 kHz (Decon laboratories, Sussex UK). Sonicated seeds were assessed for viability by 2, 3, 5 triphenyl-2H-tetrazolium chloride (TTC) (VWR Leicestershire, UK) staining test (Hosomi et al. 2011). Staining to observe lignin on the carapace (covering of the embryo) was done by embedding seeds into JB-4 resin and sectioning using an ultra–microtome (Sullivan-Brown et al. 2011). Sectioned seed samples were left to dry at 60 °C and stained in Nile red (Sigma-Aldrich Co. LLC., UK) at 0.01 mg/ml and slides were left in the stain for 10 min (Yeung et al. 1996). The stain was washed off with deionised water and the sample mounted with 0.1% N-propyl-gallate (Sigma Chemicals, UK) before cover slips 24 mm × 50 mm were added. Photographs were taken with a fluorescence microscope (LEICA CTR 6000, LEICA microsystems, Germany).

### Data collection

Number of seed capsules contaminated at stage 1, stage 2 and stage 3 were recorded. For both IVC and mature seed, numbers of total full seeds within randomly selected fields under microscope (approximately 100 seeds in each field) were used to calculate the percentage of full seeds. Germination was defined as emergence of the protocorm from the testa. Germination was recorded based on the number of full seeds, discounting seeds with poorly formed or no embryos and four replicates were used. Percent of clean cultures and percent of seed germination data were analysed against the seed capsule size and geographical location of the plant. Kruskal–Wallis test was employed as a non-parametric method to look for the difference of germination rate between species because it does not assume a normal distribution of the residuals, and is applicable to comparing groups with different sample sizes (Theodorsson-Norheim 1986). The p-values of pairwise comparisons in the post hoc Dunn's test were adjusted using Bonferroni-type adjustment to avoid type I error inflation that leads to a false positive discovery rate. Kruskal–Wallis test with post hoc Dunn's test was performed using PMCMR package for R (Pohlert 2014) while correlation analyses used the stats package for R (R 2015).

## Results

Small numbers of capsules were collected from fragmented populations, single plants in some cases, from the CHM following the guidelines stipulated in the collecting permit from Madagascar Conservation authorities. The majority of seed capsules collected by the IVC method yielded seeds in a good state of freshness that stayed sterile inside the capsules. There was some discoloration in a small percentage of small seed capsules. Based on the size, colour and texture of the seed capsules the sterilisation regime was followed as in Table 1, which shows that almost all the seed capsules collected in Madagascar by IVC gave rise to sterile seed cultures. In total 87.5% of the capsules were sterile and the seeds were also clean after 3 months of culture. Both NaDCC and PPM were essential to suppress microbial growth at various steps of the process from wild collecting to culture of seeds under laboratory conditions.

### Major taxa studied

*Angraecum* **spp**

There are 133 identified species of *Angraecum* from Madagascar of which 125 are endemic and 15 (12%) species are found in the study area of Itremo (CHM) alone (Hermans et al. 2007). We have selected this genus to study in length about the seed behaviour of both mature and near mature seeds. This was primarily due its abundance in the granite/marble outcrops where anthropogenic pressures exist, such as fire and illegal mining.

IVC seed appeared to germinate much more quickly than mature seed. In the case of *Angraecum magdalenae* 20% of seeds germinated after 2 weeks. More than 60% average germination was achieved with IVC seeds when mature seeds showed only 20% germination after four weeks (Fig. 2).

Mature seeds of many *Angraecum* species had a good proportion of full seeds but when compared against IVC seeds there are significant differences in germination (Fig. 3). Up to 2–3 times increase in germination was recorded when *Angraecum magdalenae* IVC seeds were germinated compared to mature seeds under in vitro culture conditions (Fig. 3). The results of the germination data of *Angraecum rutenbergianum* show the importance of collecting seed capsules at the right maturity. None of the seeds from IVC seed capsule 1 (Fig. 3) germinated because the seeds were too immature. Seeds from seed capsules 2 and 3 germinated at a reasonable percentage which is rather better than the corresponding values for mature seeds (Fig. 3).

Further collections of seed capsules of *Angraecum protensum* were made and we have found wide

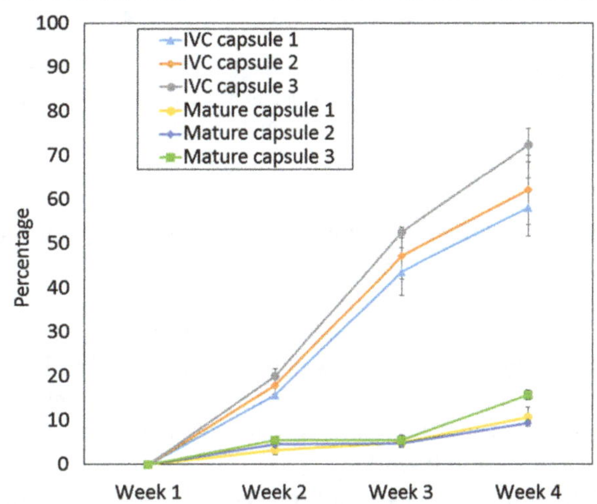


**Fig. 2** Germination of mature seeds from mature capsules versus seeds of in vitro-collected near-mature capsules of *Angraecum magdalenae*, recorded over a 4-week period. Means of four replicates ±standard error

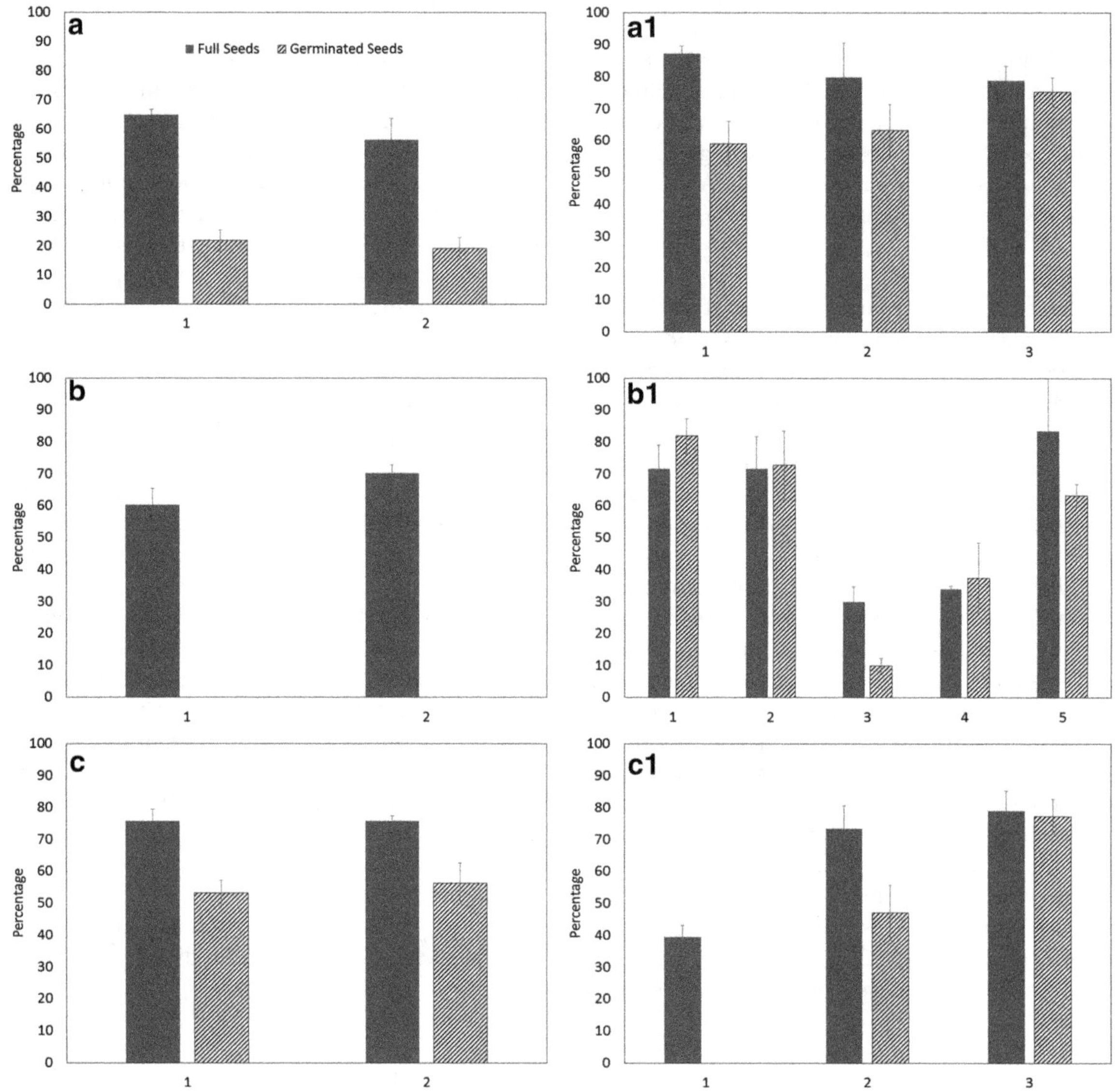


**Fig. 3** Percent full seed and seed germination after 6 months for seeds from mature capsules (*letter*) and seeds from in vitro collected near-mature capsules (*letter* + *1*) seeds. **a**, **a1** *Angraecum magdalenae*; **b**, **b1** *Angraecum calceolus*; **c**, **c1** *Angraecum rutenbergianum*; *Bar numbers* refer to mature and in vitro collected seed capsule numbers. Means of four replicates ±standard error

differences in germination between seed capsules of different maturity (all seeds have coloured testa and opaque embryo but capsule colour varied from green to brown) (Fig. 4). We have observed very low germination from this species as described earlier. The histogram shows a clear difference in germination between seed capsules although most capsules had very similar percentage of full seed.

Immature and near-mature seeds of *Angraecum* spp. had clear, near colourless testae in contrast to mature seeds, which have a dark brown, barely translucent testa (Fig. 5).

### Sonication, viability and carapace staining

There was an increased viability and germination, compared to the control, when seeds were sonicated for 3 min. Different sonication times (0.5–5 min) damaged the seed coat, from making perforations on the seed coat to completely opening up the seed releasing the embryo at 5 min. However, the carapace was damaged by the process of physical scarification through sonication treatments.

There were some variations in full seed percentage in both *Angraecum calceolus* and *Angraecum rutenbergianum* (Fig. 3) but not in *Angraeum protensum* (Fig. 4).

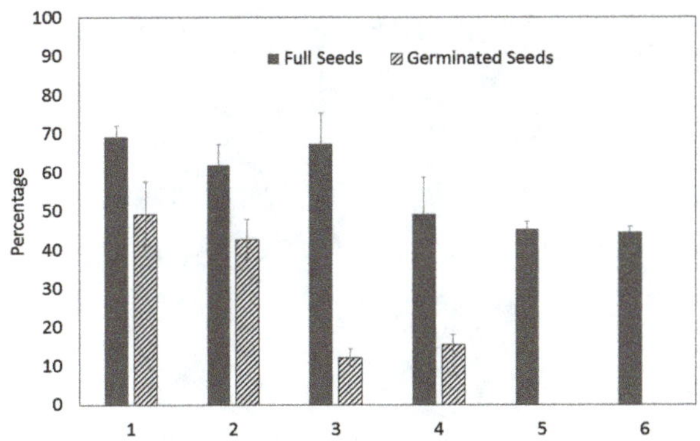


**Fig. 4** Germination percentage of mature seeds of *Angraecum protensum* originating from different seed capsules collected from Itremo, Central Highlands, Madagascar. *Each numbered bar* represents an individual capsule. Means of four replicates ±standard error

Germination in mature seeds was nil (Fig. 3) in *Angraecum calceolus* but varied in both *Angraecum rutenbergianum* and *Angraecum protensum*. Seed germination was erratic in *Angraecum protensum* and for that reason we have conducted the above pilot study to understand the nature of the covering of the embryo and its effect on seed germination. The carapace was found to be a significant layer of lignin surrounding the embryo of this species. This carapace can be removed (partially or fully) by prolonged pre-treatment with different bleach concentrations and exposure times (overnight incubation in 0.1% sodium hypochlorite or sodium dichloroisocyanurate was optimum, data not included). However, we did not find a statistically significant improvement in germination when bleach was used. This could be due to harmful effects of bleach on the embryo proper.

### *Aerangis* spp

The genus *Aerangis* has 20 species in Madagascar with 14 endemic taxa and 6 species (42%) are found in the Itremo region of CHM alone (Cribb and Hermans 2009). *Aerangis ellisii* displayed significant variation in percentage of full seeds across both IVC and conventionally collected capsules derived from different plants. This may be explained by incomplete pollination, post-pollination barriers or embryo development problems. Even freshly collected full seeds failed to germinate when the seed capsules had above 70% full seeds. With a high percentage germination through IVC this species could be a major beneficiary of this process, especially given its low abundance and sensitivity to anthropogenic fire.

The maturity of *Aerangis ellisii* seeds that were collected by IVC fell into three categories i.e. (a) immature with transparent endosperm and indistinct testa, (b) immature with marginally opaque endosperm and clear testa, (c) near mature with opaque endosperm and testa. The embryos of these seeds were opaque and covered more than 2/3rd of the seed volume compared to just half in the case of the seeds with transparent embryos. Depending on the maturity of seeds germination percentage was varied. Full germination was not observed from the third category of seeds (most mature) but signs of seed germination/initiation (Fig. 6) were noticed in four out of six seed capsules ranging from 1.8 to 5.1% from thousands of seeds. Interestingly this was noticed in seed capsules with very low percentage of full seeds. On the other hand in two mature seed capsules, with more than 70% full seeds, no seeds showed any sign of germination. Most immature seeds with a testa (group 1) germinated

**Fig. 5** Light microscope images of in vitro collected seeds on in vitro germination medium **a** *Angraecum magdalenae*; **b** *Angraecum calceolus*; **c** *Angraecum rutenbergianum*; **d** *Angraecum rhynchoglossum* (*scale bar* 500 μm). The *lower row* (**a1**–**d1**, *scale bar* 1 mm) shows protocorms of in vitro collected seeds

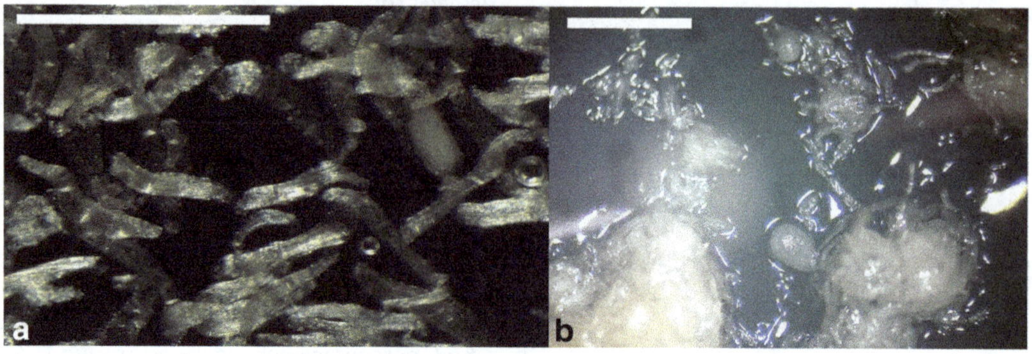


**Fig. 6** Light microscope images of **a** in vitro collected seed (*scale bar* 500 μm) in germination medium; **b** germinated protocorms (*scale bar* 1 mm) of *Aerangis ellisii*, 2 months after sowing

while more mature embryos with clear testa (group 2) failed to germinate.

Many capsules of *Aerangis ellisii*, both collected by IVC and as mature, had a very low percentage of full seeds (Fig. 7). *Aerangis ellisii* displayed remarkable variation between seed capsules of different maturity with mature seeds showing less than 5% germination in the majority while more than 70% germination was observed from immature seeds (IVC seeds). There were capsules in both mature and IVC which had more than 70% full seeds but germination was nil (Fig. 7).

The two mature seed capsules had 80% or more full seeds but failed to germinate except in one seed capsule where 0.3% seeds germinated (Fig. 7). However, all three seed capsules of *Aerangis* sp (yet to be identified) show exceptionally high seed germination (70% or more) from IVC seeds where nearly 90% of the seeds were full.

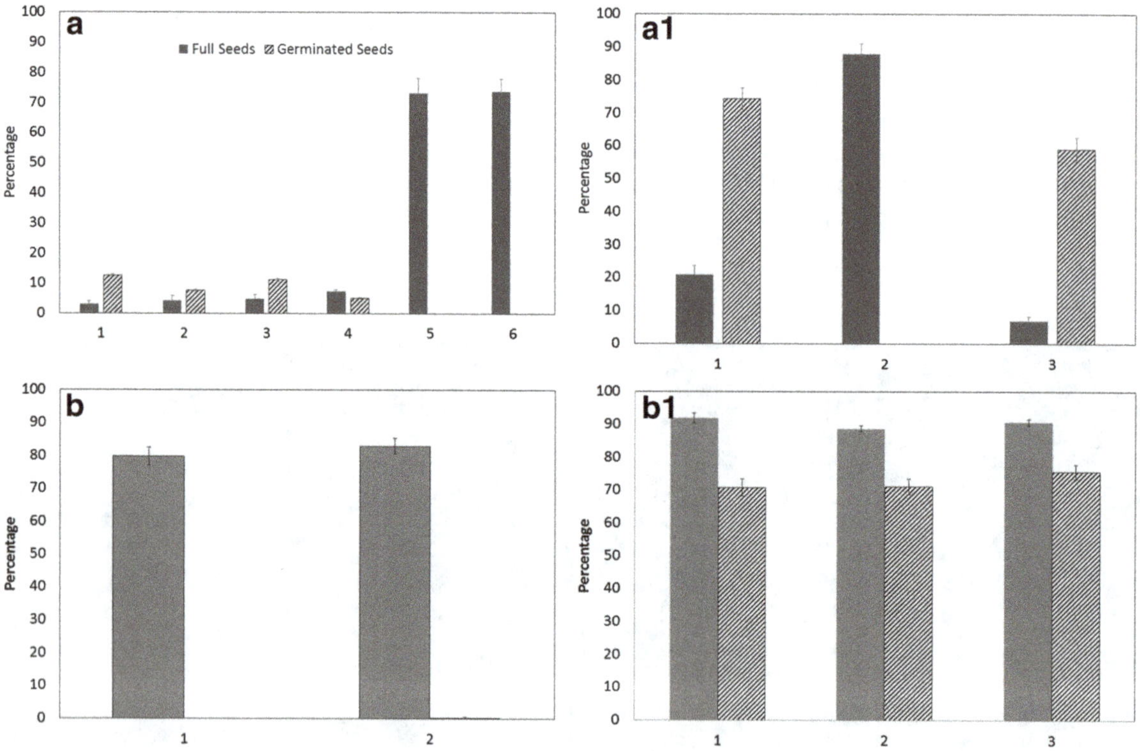


**Fig. 7** Percent full seed and germination of *Aerangis* species after 6 months. **a** *Aerangis ellisii* mature capsule; **a1** *Aerangis ellisii* in vitro collected capsule; **b** *Aerangis* sp. mature capsule; **b1** *Aerangis* sp. in vitro collected capsule. *Each numbered bar* represents a single seed capsule. Mean of four replicates ±standard error

**Prioritisation of taxa for conservation action plans based on threat, viability and germination bottlenecks**

Orchids in the study area of CHM are subjected to threats associated with anthropogenic fires, over-grazing, mining and illegal collecting. The impact of these threats on particular habitats associated with each species, together with the poor quality of mature seeds have helped us develop our assessment in Table 2. Tables 1 and 2 represent samples of capsules measured and recorded. For clarity, not all capsules collected for every species have been included. Where low numbers of individuals occur in areas of high threat levels and low mature seed germination percentage was recorded, species have been given a 'high priority' conservation category.

## Discussion

To our knowledge this is the first documented use of NaDCC (Fig. 8) as a medium additive to prevent microbial contamination during storage. Use of a lower concentration of NaDCC as a successful sterilant is due to the greater proportion of HOCl which is pH dependant (White 1986). Unlike hypochlorites NaDCC can stay stable in liquid form (Parkinson et al. 1996) due to its chemical structure and release chlorine slowly as has been reported before (Sarasan et al. 2006). This quality is useful for in vitro (Niedz and Bausher 2002) and post-harvest studies (Nicholl and Prendergast 1998) in plants. With low phytotoxicity, and damage restricted to old living tissue, NaDCC makes a suitable antimicrobial substance to be used for longer treatment as discussed in this study. Use of PPM in STEP 2 and STEP 3 was crucial to keep the capsule and seeds under sterile conditions (Fig. 8). PPM has been widely used for a range of species as an effective antimicrobial substance to reduce contamination of plan tissue cultures (Sarasan et al. 2006).

Collecting mature seeds is the generally accepted approach as seeds remain viable for a reasonable period of time for use in culture compared to less mature seeds as reported before (Stoutamire 1992). As seeds mature the seed coat becomes hydrophobic which limits water and nutrient absorption (Zhang et al. 2013) and, as reported before, in terrestrial taxa accumulation of abscisic acid (ABA) occurs as seeds mature (Kinderen 1987; Lee 2003; Lee et al. 2007). This must be the case with some of the species to prepare them to face adverse environmental conditions before they can be germinated with the help of mycorrhizal fungi in the wild. The testa of near mature seeds have living cells and are yet to acquire their hydrophobic nature. Accordingly, there were not many barriers to absorb water/nutrients from the media to germinate in vitro. In addition to the physical barrier the high endogenous ABA level in mature seeds may

**Table 2 Threat assessment of selected species from Itremo, Central Highlands, Madagascar, based on mature seed traits**

| Taxon | Life form | Geographic distribution | Percent average full seeds | Percent germination (mature) | Percent TTC viability | Conservation priority |
|---|---|---|---|---|---|---|
| *Aerangis ellisii* | *LP* | Core | 27.7 | 2.3 | 1.9 | High priority |
| *Aerangis* sp | *EP* | Peripheral | 81.3 | 0.16 | N/A | High priority |
| *Angraecum calceolus* | *LP* | Core | 58.8 | 0 | N/A | Medium priority |
| *Angraecum rhynchoglossum* | *EP* | Core | 85.3 | 0 | N/A | Medium priority |
| *Angraecum rutenbergianum* | *LP* | Core | 78.5 | 39.4 | 65.2 | Medium priority |
| *Benthamia cinnabarina* | *TR* | Core | 85.6 | 4.7 | 85.7 | High priority |
| *Benthamia rostrata* | *TR* | Core | 86.4 | 8.2 | 10.5 | Medium priority |
| *Bulbophyllum bicoloratum* | *EP* | Peripheral | 88 | 0 | 0 | High priority |
| *Bulbophyllum peyrotii* | *EP* | Core | 98.4 | 92 | 87.3 | Not under threat |
| *Disa incarnata* | *TR* | Core | 60.5 | 8 | 0.3 | Medium priority |
| *Habenaria ambositrana* | *TR* | Core | 92 | 15.3 | 44.4 | Medium priority |
| *Jumellea amplifolia* | *LP* | Peripheral | 88.3 | 0 | 0 | High Priority |
| *Liparis* sp | *TR* | Core | 68.8 | 6.1 | N/A | Medium priority |
| *Oeceoclades calcarata* | *LP* | Peripheral | 12.6 | 0 | 65.9 | High priority |
| *Polystachya fusiformis* | *LP* | Core | 70.1 | 65.13 | 43.8 | Not under threat |
| *Polystachya.* sp | *LP* | Core | 84.8 | 0 | 0 | High priority |
| *Polystachya.* sp | *EP* | Core | 17.7 | 0 | N/A | High priority |
| *Satyrum trinerve* | *TR* | Core | 77.7 | 14.9 | 0 | Medium priority |
| *Tylostigma nigrescens* | *TR* | Core | 76.9 | 28.6 | 36.9 | Not under threat |

Categorisation: High Priority, Medium Priority and Not Under Threat currently. Geographic distribution refers to the range within core protected area or peripheral protected area, Itremo, Madagascar

*TR* terrestrial, *EP* Epiphyte, *LP* Lithophyte as found in the field, *N/A* data not available

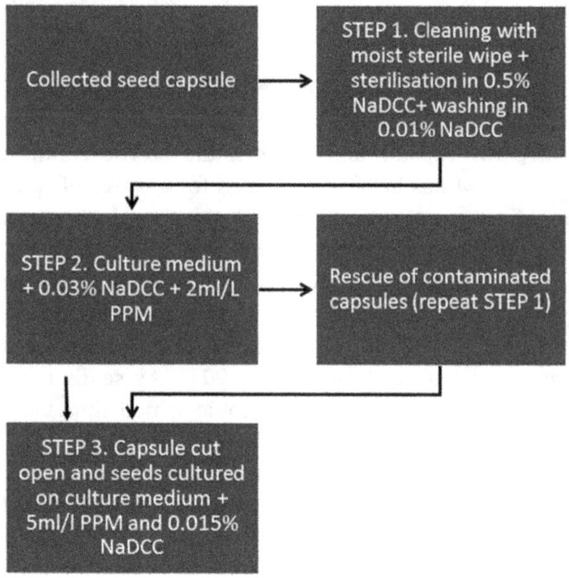


**Fig. 8** Flow chart explaining the different steps of sterilisation and culture in vitro of different types of wild collected orchid seed capsules, and final culture under laboratory conditions. Sodium dichloroisocyanurate (NaDCC) was used as sterilant and antimicrobial at different concentrations in culture while plant preservative mixture (PPM) was used at different concentrations as an antimicrobial substance

cause poor germination in orchids, especially terrestrial orchids (Kinderen 1987). In the cases of *Calanthe tricarinata* (Lee et al. 2007) and *Cypripedium formosanum*, the endogenous ABA level is low in immature seeds, and increases rapidly as the seeds approach maturity, coinciding with a rapid decrease in seed germination. Collecting immature seeds at 6 weeks after pollination (WAP) could give high seed permeability and low endogenous ABA, which results in improved seed germination in vitro (Lee 2003).

Mature orchid seeds showing varying germination frequencies may be demonstrating dormancy induced by inhibitory substances as reported before (Nagashima 1982a). However, immature seeds are only available during a short of window of time and need to be sown immediately because of the lack of an effective preservation method (Steele 1996; Hirano et al. 2005). It has been predicted that populations occurring in marginal habitats may be more vulnerable to repeated population crashes which can also result in a significant reduction in genetic diversity (Cozzolino et al. 2003), which may be the case in many orchids of Madagascar where only a few individuals are left in the wild.

Expanding IVC along with conventional collecting for seed banking will significantly improve the success of ex situ conservation of Madagascan orchids. Ultimately there is a great need to understand the importance of mycorrhizal fungi in epiphytes and, especially, lithophytes to improve reintroduction/restoration to support integrated conservation for meaningful outcomes in the longer term. At the moment very little is known about the mycorrhizae of lithophytes and epiphytes of Madagascan orchids except the preliminary studies conducted in our laboratory (Yokoya et al. 2015).

During embryo development in orchids the inner integument shrinks and forms a tight layer, which encloses the embryo, termed 'carapace' (Lee et al. 2005), mainly described in terrestrial taxa. The cuticular stain Nile Red detects this layer which may play a role in the hydrophobic nature of orchid seed and may help the seeds survive in harsh conditions. The carapace is visible in the seeds of *Angraecum protensum* as reported in terrestrial orchids (Zhang et al. 2013; Lee et al. 2005), when seeds were sectioned and stained with Nile Red. The carapace is quite prominent in lithophytic orchids, as in terrestrial taxa, and can act as a robust protective cover for the embryo. When seeds were scarified mechanically, germination percentage achieved was close to the viability percentage recorded by TTC staining. This could be the reason why IVC collected seeds performed better, with carapace under-developed, compared to fully mature seeds with well-developed carapace. In lithophytic species this may be a distinctive feature as they endure long periods of harsh conditions in a dry lithophytic habitat but detailed screening is essential to find out the trend within this group of plants. According Lee et al. (2005) the timing of seed collection outweighs the composition of the culture medium and the seed pre-treatments in a terrestrial orchid, where they have used bleach as chemical scarification. When sonication (0.5–5 min) was used, however, 3 min exposure improved seed germination by more than 30%. This treatment made perforations in the carapace which improved germination, either by reducing the hydrophobic nature of the seed, aiding moisture absorption and/or breaking the dormancy. Although this is a small study in a lithophytic orchid, the results indicate that the carapace is a barrier which protects the seeds in order to overcome adverse environmental conditions and keep seeds viable until germination can take place under favourable conditions with the help of orchid mycorrhizal fungus/fungi.

*Angraecum rutenbergianum* appeared to be a good coloniser of its habitat in the study area of Itremo. Although the species has both epiphytic and lithophytic habit, plants were found, in most cases, growing in exposed cracks and bare faces of crags and boulders. Seedling recruitment appeared to be high as large groups of seedlings could be found growing in lichen within a few metres of groups of adult plants. This would tally with a reasonable germination percentage of mature seed (Fig. 3). In contrast, *Aerangis ellisii*, which is also found as either epiphyte or lithophyte but found predominantly growing as a lithophyte in the CHM, existed in extremely small populations of large adult plants with few seedlings evident in their vicinity. Erratic pollination has been recorded in two separate studies in *Aerangis ellisii* (Nilsson and Rabakonandrianina 1988; Nilsson et al. 1992) where the pollinator activity is geographically limited. It is striking that, in both IVC and mature capsules with high percentage of full seeds, germination percentage was nil. This discovery raises further concerns for this CITES Appendix I species.

The low germination in mature seeds of *Aerangis ellisii* may be showing the consequences of inbreeding as only isolated plants were encountered in each location. This has been reported before in this species (Nilsson et al. 1988). The same kind of response was noticed in the case of mature seeds of an unidentified species of another *Aerangis*. Full seed percentage was consistent in all the capsules collected, both mature and IVC (Fig. 8), although mature seeds failed to germinate while IVC seeds germinated very well (more than 60% from all seed capsules). We have conducted further studies with small seed samples in another *Aerangis* species and noticed a similar trend. This means there may be an onset of dormancy/development of carapace layer/s as soon as the seeds start maturing. IVC can be used as a complementary tool as part of an integrated conservation action plan.

In both *Aerangis* and *Angraecum* a trend exists where IVC seeds performed far better than mature seeds and this cannot be ignored. As Madagascar is home to about 133 *Angraecum* species and more than 20 species of *Aerangis* their ex situ conservation action plans can be built around collecting seeds at immature/near mature stages to support asymbiotic models for living collection development and cryopreservation. The number of full seeds and viability of mature seeds all point to the fact that *Aerangis* is a genus which has serious conservation issues. The reported pollination behaviour of *Aerangis ellisii* (Nilsson and Rabakonandrianina 1988; Nilsson et al. 1992) explains the low full seed percentage, and the fragmentation we have found in the wild shows very little recruitment, all of which could drive the species to serious population decline in the coming decades. Detailed studies are in progress in our laboratory to understand the symbiotic relationship of this species using fungi isolated from seedlings collected from the wild.

Harding et al. (2013) reviewed the ex situ conservation of endemic flora from biodiversity hotspots, specifically using Brazil as an example and are of the opinion that ex situ conservation of endemic taxa in gene banks is often mired by information deficiency in areas of molecular genetics, seed storage and related aspects. Changing climate, population fragmentation and a lack of phenology data means reliable methods are required to collect seeds from as many species as possible to conduct trials for germination, cryopreservation and reintroduction programmes. The recent recommendation to cryopreserve orchid seeds rather than attempting to store them in conventional seed banks at −20 °C (Merritt et al. 2014) requires urgent attention. Mature good quality seed collections are a pre-requisite to achieve this system of storage. However, there are some orchid taxa which can only be collected opportunistically and in small numbers which need an alternative approach. According to Nagashima (1982b) some mature seeds have intrinsic germination problems due to accumulation of inhibitory substances and ensuing dormancy or potentially by the impermeability of the seed to a level that reduces germination following full seed maturation (Miyoshi and Mii 1988). Several methods such as sonication (Miyoshi and Mii 1988), scarification by chemicals (Mweetwa et al. 2008), and chilling (Kauth et al. 2011) have been reported to improve germination by degrading the physical barrier. According to Rasmussen (1995) the alternative to bypass this bottleneck is by sowing immature seeds without any pre-treatment. For example, seeds of *Cymbidium goeringii* showed higher germination when they were harvested close to the completion of embryogenesis but not at seed maturation Nagashima (1982b). As seeds are available only during a specific window of time they should be collected at the right time and sown immediately. When an in vitro laboratory is remotely located and access to facilities is limited it is difficult to collect plant material at the right maturity, therefore, IVC has applications to circumvent this bottleneck.

Statistical analysis showed that the tribe Vandeae, comprising the genera *Aerangis, Angraecum, Jumellea* and *Polystachya* had significantly higher germination of seeds collected through IVC than conventionally collected mature seeds. This suggests a trend in seed behaviour in this taxon. In other tribes (here including Podochileae, Cymbidieae and Orchideae), there was no significant difference between the germination of mature versus IVC seeds. The majority of epiphytic species from Madagascar can be collected by the standard collecting method as we have noticed in several species of *Bulbophyllum* and *Polystachya*. This shows that in these taxa, for seed that is viable, IVC does not lead to deterioration of seed and can complement conventional collecting as part of collecting for conventional seed banking.

Due to restricted availability of orchid capsules from the wild the numbers of capsules obtained for each species varied, as seen in Figs. 3, 4 and 7. Only mature seeds were available to collect in the case of *Angraecum protensum*. In order to justify the improved orchid seed germination by IVC method a data set consisting of full seed ratio and germination rate for collections of 6 species was subject to Kruskal–Wallis one-way analysis of variance. Comparing with mature seed the collections by IVC method promoted the germination rate of overall collection of six species in the present study with high significance ($p < 0.001$), No difference of germination rate observed between species ($p = 0.29$). The unverified species *Aerangis* sp. showed significantly higher full seed ratio than *Aerangis ellisii*. ($p < 0.01$) (Kruskal–Wallis test results available in Additional file 1). Pearson's product-moment correlation test on the total 37 observations in the overall data set revealed a weak positive linear relationship ($r = 0.35$) between full seed ratio and germination rate of those collected seed ($p < 0.05$; Additional file 1).

One of the principal threats in the CHM habitats is fire (Hermans et al. 2007; Whitman et al. 2011) and a number of lithophytic orchids may be more prone to threats of extinction from the studied habitats than terrestrial orchids. This group of vulnerable taxa from the CHM should be prioritised and targeted for ex situ collections with IVC as a potential tool to complement conventional seed collecting.

## Conclusions

There are several impediments to work on species of high conservation value especially in small islands and biodiversity hotspot countries. The main problems are remoteness of the area for international teams who collaborate with local partners to access plants in the wild, few seed samples to collect, and in particular only limited information available on the phenology of seed capsule development and maturity. Seed storage of recalcitrant crop wild relatives, which underpins food security in the coming decades, and threatened plants (Li and Pritchard 2009) requires high quality materials. Collecting of recalcitrant seeds for large scale storage programmes require tangible sterilisation models to keep the embryonic axis/embryos in good condition until they can be processed in the lab. Our method using NaDCC offers great potential to work on storage, especially cryopreservation. Berjak et al. (2014) demonstrated the efficiency of NaDCC to keep materials in good condition before cryostorage.

For orchids our work done in the Central Highlands of Madagascar points to the fact that large genera such as *Angraecum* and *Aerangis* are facing serious threats to their existence due to low seed set, poor quality of seeds, and intrinsic poor recruitment in addition to the well-documented problems of habitat loss and climate change. As terrestrial taxa have specific maturity periods for their peak viability IVC offers great potential for collecting seeds at the right maturity for species where phenological data is available.

The method described here has the potential to aid species with recalcitrant seeds, and vegetative materials from critically endangered species. We believe our findings also have the potential to harvest seeds straight from the plant to growth medium without creating a major shock to the physiology of the seed which is essential for both applied research in groups of high conservation and crop value, and for experimental studies.

#### Authors' contributions

JK collected the seed capsules in Madagascar, conducted germination trials, collected data, and wrote the article. LR identified study areas and orchids, and collected seed capsules. HS conducted sonication and electron microscopic studies. RF analysed data and wrote the article. JB analysed data and wrote the article. VS conceived the idea, designed the project, and wrote the article. All authors read and approved the final manuscript.

#### Author details

[1] Natural Capital and Plant Health, Royal Botanic Gardens Kew, Richmond, Surrey TW9 3AB, UK. [2] Kew Madagascar Conservation Centre, Lot II J 131 B Ambodivoanjo-Ivandry, 101 Antananarivo, Madagascar.

#### Acknowledgements

We kindly acknowledge the support received throughout the study period from Stuart Cable (Kew's Madagascar Team), Kew Madagascar Conservation Centre staff (Gaetan Ratovonirina) and PBZT staff (Solo Rapanarivo, Jacky Andriantiana). We thank Margaret Ramsay, Edward Jones, Claudius Grehl and Nicola Bates for assisting in culturing, staining and photography of IVC seed at RBG Kew.

#### Competing interests

The authors declare that they have no competing interests.

#### An explanation of why your manuscript should be published in Botanical Studies

This manuscript reports the use of a novel method of seed collecting of orchids for conservation. The special issue on Orchid Conservation by Botanical Studies is a suitable place for it to be published as it reach a wide range of readers who are interested in orchids.

#### An explanation of any issues relating to journal policies

Not applicable.

#### Confirmation that all authors have approved the manuscript for submission

All authors have approved the manuscript for submission to Botanical Studies.

#### Confirmation that the content of the manuscript has not been published, or submitted for publication elsewhere

The whole or part of the manuscript has not been published or under consideration for publication anywhere else.

**Funding**
The authors acknowledge financial support from the Sainsbury Orchid Endowment and Bentham-Moxon Trust.

**References**

Alvarenga SV, Bianchetti BL, Gonzalez LPE, Sandoval OE, Martınez ZMB (2002) Cacao. In: Pence VC, Sandoval JA, Villalobos VM, Engelmann F (eds) *In vitro* collecting techniques for germplasm conservation. IPGRI technical bulletin, 7, International Plant Genetic Resources Institute, Rome, Italy, pp 47–51

Barsberg S, Rasmussen HN, Kodahl N (2013) Composition of *Cypripedium calceolus* (Orchidaceae) seeds analyzed by attenuated total reflectance IR spectroscopy: in search of understanding longevity in the ground. Am J Bot 100:2066–2073. doi:10.3732/ajb.1200646

Berjak P, Cherian J, Makhathini AP, Sershen PNW (2014) Embryonic axes of a tropical recalcitrant-seeded species: successful elimination of micro-organisms and potential for zygotic synthetic seed (synseed) production. Seed Sci Tech 42:150–160

Brenes HA, Garcia TV, Velasco UE (2002) Citrus. In: Pence VC, Sandoval JA, Villalobos VM, Engelmann F (eds) *In vitro* collecting techniques for germplasm conservation. IPGRI technical bulletin, 7, International Plant Genetic Resources Institute, Rome, Italy, pp 56–60

Broadhurst LM, Lowe A, Coates DJ, Cunningham SA, McDonald M, Vesk PA, Yates C (2008) Seed supply for broadscale restoration: maximizing evolutionary potential. Evol Appl 1:587–597

Cozzolino S, Noce ME, Musacchio A, Widmer A (2003) Variation at a chloroplast minisatellite locus reveals the signature of habitat fragmentation and genetic bottlenecks in the rare orchid *Anacamptis palustris* (Orchidaceae). Am J Bot 90:1681–1687. doi:10.3732/ajb.90.12.1681

Cribb P, Hermans J (2009) Field guide to the orchids of Madagascar. Kew Publishing, Royal Botanic Gardens

Engelmann I (2002) Coconut. In: Pence VC, Sandoval JA, Villalobos VM, Engelmann F (eds) *In vitro* collecting techniques for germplasm conservation. IPGRI technical bulletin, 7, International Plant Genetic Resources Institute, Rome, Italy, p 68–71

Harding K, Benson EE, Nunes EC, Pilatti FK, Lemos J, Viana AM (2013) Can biospecimen science expedite the ex situ conservation of plants in megadiverse countries? A focus on the flora of Brazil. Crit Rev Plant Sci 32:411–444. doi:10.1080/07352689.2013.800421

Henao MLM, Tapia BC, Gil EFL, Sandoval JA (2002) Musa. In: Pence VC, Sandoval JA, Villalobos VM, Engelmann F (eds) *In vitro* collecting techniques for germplasm conservation. IPGRI technical bulletin, 7, International Plant Genetic Resources Institute, Rome, Italy, pp 52–55

Hermans J, Hermans C, Du Puy D, Cribb PJ, Bosser J (2007) Orchids of Madagascar. Royal Botanic Gardens, Kew

Hirano T, Godo T, Mii M, Ishikawa K (2005) Cryopreservation of immature seeds of *Bletilla striata* by vitrification. Plant Cell Rep 23:534–539. doi:10.1007/s00299-004-0893-9

Hosomi ST, Santos RB, Custodio CC, Seaton PT, Marks TR, Machado-Neto NB (2011) Preconditioning *Cattleya* seeds to improve the efficacy of the tetrazolium test for viability. Seed Sci Tech 39:178–189. doi:10.15258/sst.2011.39.1.15

Kauth PJ, Kane ME, Vendrame WA (2011) Comparative in vitro germination ecology of *Calopogon tuberosus* var. *tuberosus* (Orchidaceae) across its geographic range. In Vitro Cell Dev Biol Plant 47:148–156. doi:10.1007/s11627-010-9316-5

Kinderen VD (1987) Abscisic acid in terrestrial orchid seeds: a possible impact on their germination. Lindleyana 2:84–87

Lee Y-I (2003) Growth periodicity, changes of endogenous abscisic acid during embryogenesis, and in vitro propagation of *Cypripedium formosanum Hay*. Dissertation, National Taiwan University, Taipei

Lee Y-I, Lee N, Yeung EC, Chung MC (2005) Embryo development of *Cypripedium formosanum* in relation to seed germination *in vitro*. J Am Soc Hort Sci 130:747–753

Lee Y-I, Lu CF, Chung MC, Yeung EC, Lee N (2007) Developmental changes in endogenous abscisic acid concentrations and asymbiotic seed germination of a terrestrial orchid, *Calanthe tricarinata* Lindl. J Am Soc Hort Sci 132:246–252

Li D-Z, Pritchard HW (2009) The science and economics of ex situ plant conservation. Trends Plant Sci 14:614–621

Maunder M, Lyte B, Dransfield J, Baker W (2001) The conservation value of botanic garden palm collections. Biol Conserv 98:259–271. doi:10.1016/S0006-3207(00)00160-9

Merritt DJ, Dixon KW (2011) Restoration Seed banks—a matter of scale. Science 332:424–425. doi:10.1126/science.1203083

Merritt DJ, Turner SR, Commander LE, Dixon KW (eds) (2005) Proceedings of the fifth Australian workshop on native seed biology Brisbane, Australia

Merritt DJ, Hay FR, Swarts ND, Sommerville KD, Dixon KW (2014) *Ex situ* conservation and cryopreservation of orchid germplasm. Int J Plant Sci 175:46–58. doi:10.1086/673370

Miyoshi K, Mii M (1988) Ultrasonic treatment for enhancing seed germination of terrestrial orchid, *Calanthe discolor*, in asymbiotic culture. Sci Hort 35:127–130. doi:10.1016/0304-4238(88)90044-1

Moat J, Smith P (2007) Atlas of the vegetation of Madagascar. Royal Botanic Gardens, Kew

Mortlock W (2000) Local seed for revegetation. Ecol Manag Restor 1:93–101. doi:10.1046/j.1442-8903.2000.00029.x

Murashige T, Skoog F (1962) A revised medium for rapid growth and bio assays with tobacco tissue cultures. Physiol Plant 15:473–497. doi:10.1111/j.1399-3054.1962.tb08052.x

Mweetwa AM, Welbaum GE, Tay D (2008) Effects of development, temperature, and calcium hypochlorite treatment on in vitro germinability of *Phalaenopsis* seeds. Sci Hort 117:257–262

Nagashima T (1982a) Studies on the seed germination and embryogenesis in the *Bletilla striata* Rchb-F and *Calanthe discolor* Lindl. J JPN Hortic Sci 51:82–93

Nagashima T (1982b) Studies on the seed-germination and embryogenesis in the *Cymbidium goeringii* Rchb-F and *Paphiopedilum insigne* var *Sanderae* Rchb-F. J JPN Hortic Sci 51:94–105

Nicholl P, Prendergast M (1998) Disinfection of shredded salad ingredients with sodium dichloroisocyanurate. J Food Process Pres 22:67–79. doi:10.1111/j.1745-4549.1998.tb00805.x

Niedz RP, Bausher MG (2002) Control of in vitro contamination of explants from greenhouse- and field-grown trees. In Vitro Cell Dev. Biol Plant 38:468–471. doi:10.1079/ivp2002316

Nilsson L, Rabakonandrianina E (1988) Hawk-moth scale analysis and pollination specialization in the epilithic Malagasy endemic *Aerangis ellisii* (Reichenb Fil) Schltr (Orchidaceae). Bot J Lin Soc 97:49–61. doi:10.1111/j.1095-8339.1988.tb01686.x

Nilsson LA, Rabakonandrianina E, Pettersson B (1992) Exact tracking of pollen transfer and mating in plants. Nature 360:666–668. doi:10.1038/360666a0

Parkinson M, Prendergast M, Sayegh AJ (1996) Sterilisation of explants and cultures with sodium dichloroisocyanurate. Plant Growth Regul 20:61–66. doi:10.1007/bf00024060

Pence VC (2005) In vitro collecting (IVC) I. The effect of collecting method and antimicrobial agents on contamination in temperate and tropical collections. In Vitro Cell Dev Biol Plant 41:324–332. doi:10.1079/Ivp2004629

Pohlert T (2014) The pairwise multiple comparison of mean ranks package (PMCMR). In: R package. http://CRAN.R-project.org/package=PMCMR. Accessed 12 Feb 2017

Pritchard HW, Poynter ALC, Seaton PT (1999) Inter-specific variation in orchid seed longevity in relation to ultra-dry storage and cryopreservation. Lindleyana 14:92–101

Rasmussen HN (1995) Terrestrial orchids: from seed to mycotrophic plant. Cambridge University Press, Cambridge

Saldaña LH, Oicata ML, Ponce BMM, Díaz CJH (2002) Coffee. In: Pence VC, Sandoval J.A.; Villalobos, V.M.; Engelmann, F (Eds) *In vitro* collecting techniques for germplasm conservation. IPGRI technical bulletin, 7, International Plant Genetic Resources Institute, Rome, Italy, p 42–46

Sandoval JA, Villalobos VM (2002) Avocado. In: Pence VC, Sandoval JA, Villalobos VM, Engelmann F (eds) *In vitro* collecting techniques for germplasm conservation. IPGRI technical bulletin 7, International Plant Genetic Resources Institute, Rome, Italy, pp 61–34

Sarasan V, Cripps R, Ramsay MM, Atherton C, McMichen M, Prendergast G, Rowntree JK (2006) Conservation in vitro of threatened plants—progress

in the past decade. In Vitro Cell Dev Biol Plant 42:206–214. doi:10.1079/ivp2006769
Seaton PT, Pritchard HW (2003) Orchid germplasm collection, storage and exchange. In: Dixon KW, Kell SP, Barrett RL, Cribb PJ (eds) Orchid conservation. Natural History Publications, Kota Kinabalu, pp 227–258
Steele WK (1996) Large scale seedling production of North American *Cypripedium* species. In: Allen, C (ed) North American native terrestrial orchids propagation and production. North American Native Terrestrial Orchid Conference. Germantown, Maryland, p 11–26
Stoutamire WP (1992) Orchid seeds versus the mails. AOSB 61:578–581
Sullivan-Brown J, Bisher ME, Burdine RD (2011) Embedding, serial sectioning and staining of zebrafish embryos using JB-4 resin. Nat Protoc 6:46–55. doi:10.1038/nprot.2010.165
R core team, (2015) R, A language and environment for statistical computing, version 3.4.0
Theodorsson-Norheim E (1986) Kruskal-Wallis test: BASIC computer program to perform nonparametric one-way analysis of variance and multiple comparisons on ranks of several independent samples. Comput Meth Prog Biol 23:57–62
Tyson P (2000) The eighth continent: life, death and discovery in the lost world of Madagascar. William Morrow (Harper Collins) Publishers, New York
Warren R (1983) Tissue culture. Orchid Rev 91:306–308
Way MJ (2003) Collecting seed from non-domesticated plants for long-term conservation. In: Smith RD, Dickie JD, Linington SH, Pritchard HW, Probert RJ (eds) Seed Conservation: turning science into practice. Royal Botanic Gardens, Kew, pp 163–201
White G (1986) Handbook of chlorination. Van Reinhold Co., New York
Whitman M, Medler M, Randriamanindry JJ, Rabakonandrianina E (2011) Conservation of Madagascar's granite outcrop orchids: the influence of fire and moisture. Lankesteriana 11:55–67
Wyse-Jackson P (2001) An international review of the ex situ plant collections of the botanic gardens of the world. Bot Gard Conserv News 3:22–33
Yeung EC, Zee SY, Ye X (1996) Embryology of *Cymbidium sinense*: embryo development. Ann Bot 78:105–110. doi:10.1006/anbo.1996.0101
Yokoya K, Zettler LW, Kendon JP, Bidartondo MI, Stice AL, Skarha S, Corey LL, Knight AC, Sarasan V (2015) Preliminary findings on identification of mycorrhizal fungi from diverse orchids in the Central Highlands of Madagascar. Mycorrhiza 25:611–625. doi:10.1007/s00572-015-0635-6
Zhang Y, Lee Y-I, Deng L, Zhao SW (2013) Asymbiotic germination of immature seeds and the seedling development of *Cypripedium macranthos* Sw., an endangered lady's slipper orchid. Sci Hort 164:130–136. doi:10.1016/j.scienta.2013.08.006

# 5

# A perspective on orchid seed and protocorm development

Edward C. Yeung*

**Abstract**

This perspective draws attention to the functional organization of orchid seed and protocorm during the course of development. The orchid embryos have a well-organized developmental plan generating a blue-print of a protocorm as they mature. The different phases of embryo development in orchids, i.e. histodifferentiation, storage product synthesis and accumulation, and maturation are essentially similar to other flowering plants. The protocorm is considered as a unique structure designed to establish symbiotic association with mycorrhizal fungi and with the primary goal to form a shoot apical meristem. This perspective brings forth arguments that the processes of embryo and protocorm development are highly programmed events, enhancing survival of orchid seeds and plantlets in their natural habitats. Furthermore, the ability of protocorm cells to divide, makes them ideal explants for micropropagation and transformation studies. Through seed germination and micropropagation using protocorms as explants, orchid conservation efforts are greatly enhanced.

**Keywords:** Embryo, Suspensor, Endosperm, Seed coat, Protocorm, Phytohormones, Shoot apical meristem, Regeneration, Protocorm-like bodies, Orchid mycorrhizae, Micropropagation

## Background

An understanding of plant reproductive biology is essential to plant conservation efforts. In orchids, after a successful pollination event, numerous seeds are produced within a single capsule. Utilizing seeds through asymbiotic and symbiotic seed germination for plant maintenance and propagation are key approaches in orchid conservation efforts. In order to further optimize current methodologies on seed germination and plantlet establishment, we need to have a proper understanding of seed biology, as well as the structure and function of protocorm.

Key features of orchid embryos, seeds, and protocorms are summarized in this article. The globular-shape and small sizes of orchid embryos imply that the embryos are simple and not as well developed as in other flowering plants. On the contrary, the author sees that orchid embryos have developed a plan, generating a blue-print of a protocorm as they mature. The different phases of embryo development in orchids, i.e. histodifferentiation, storage product accumulation, and maturation are essentially similar to other flowering plants. The protocorm formed at the time of seed germination is considered as a unique structure designed to establish symbiotic association with mycorrhizal fungi and with the primary goal to form a shoot apical meristem. This perspective brings forth arguments that the process of embryo and protocorm formation involves highly programmed events, enhancing survival of orchid seeds and plantlets in their natural habitats. The intent of this article is to generate additional interests, discussions, and further studies on orchid seed biology in the near future.

## Embryo development

### The zygote and the proembryo

In flowering plants, a successful fertilization event results in seed formation in which an ovule becomes a seed, the egg cell becomes a zygote, the central cell forms the endosperm, and the integuments develop into the seed coat. Detailed accounts on changes of the egg cell before and after fertilization about orchids are few. In *Epidendrum scutella*, after fertilization, the size of the zygote is

*Correspondence: yeung@ucalgary.ca
Department of Biological Sciences, University of Calgary, Calgary, AB T2N 1N4, Canada

similar to the egg cell and there is little change to the vacuolar system within its cytoplasm (Cocucci and Jensen 1969). Rough endoplasmic reticulum and polysomes become more abundant, indicating an increase in metabolic activities within the cell. It is interesting to note that dictyosomes could not be detected in the zygote after fertilization (Cocucci and Jensen 1969). At the light microscopic level, the zygote maintains a uniform dense cytoplasm, e.g. *Cymbidium sinense* (Yeung et al. 1996) and *Calypso bulbosa* (Yeung and Law 1992), or it can have a polarized appearance with a distinct vacuole(s) located at the micropylar end such as in *Cypripedium passerinum* (Law and Yeung 1993) and *Phaius tankervilliae* (Ye et al. 1997). The phenomenon of 'zygote shrinkage' (see Natesh and Rau 1984) prior to the first zygote mitosis appears to be absent. Changes in microtubule cytoskeleton have been noted, reflecting dynamic changes within the zygote prior to its first mitotic division. Soon after zygote formation, the microtubules are randomly arranged. As the zygote increases in size, the microtubules organize into a meshwork in the cytoplasm and a dense array of microtubules begins to appear in the cortical region near the chalazal pole (Huang et al. 1998). Moreover, additional studies are needed to decipher cellular events leading to the establishment of cell polarity within the zygote at this critical stage of development.

In the zygote, varying degree of vacuolation appears within its cytoplasm. In *C. bulbosa* (Yeung and Law 1992) and *C. sinense* (Yeung et al. 1996) the cytoplasm of the zygote remains dense through the first mitotic division. In these examples, the first zygote division is usually symmetrical or slightly asymmetrical in appearance. For orchids having a polarized zygote with large vacuole(s) located at the micropylar end of the cell, such as those found in *C. passerinum* (Law and Yeung 1993) and *Phaius tankervilliae* (Ye et al. 1997), the first division is asymmetrical giving rise to two cells of unequal sizes. Changes to the orientation and distribution of the microtubule cytoskeleton are observed at this time (Huang et al. 1998). Irrespective of the cell division symmetry, cell division results in the formation of a terminal and a basal cell, each having a distinct fate. The basal cell can differentiate into a short-lived embryonic organ, known as the suspensor, and the terminal cell will give rise to the embryo proper.

One of the key events occurring after zygote division is the establishment of a polarized embryo axis leading to cell fate determination and the formation of a body plan (Ueda and Laux 2012; Wabnik et al. 2013). Such information similar to the study of *Arabidopsis* is lacking in orchids. The oblique divisions observed in the formation of the proembryo in *C. sinense* (Yeung et al. 1996), the changes in microtubule organization (Huang et al. 1998), the varied patterns of suspensor formation, and limited number of cells in the embryo proper, suggest new regulatory controls in proembryo establishment in orchid.

Based on the cell division pattern in the zygote and proembryo, as well as the varied forms of suspensor morphologies, different embryo types were proposed (Swamy 1949; Johansen 1950). Moreover, the suggested embryological system of classification sees limited use by embryologists. More recently, Clements (1999) provided a more extensive description of embryo developmental types, incorporating the classification system proposed by Dressler (1993).

### The suspensor

The different morphologies of the suspensor within Orchidaceae generate a lot of interest in the study of embryo development (see Yam et al. 2002). The suspensor can appear as a single cell, highly elongated haustorial structures, or multicellular swollen forms that fill the embryo cavity (Swamy 1949). What directs the growth and morphogenesis of the varied suspensor forms? In the study of the nun orchid suspensor development, cytoskeletal elements, i.e. microtubules and microfilaments, play an important role in suspensor morphogenesis (Ye et al. 1997), especially in the elongation of suspensor cells. Dynamic changes in the microtubule cytoskeleton are also observed in *C. sinense* during the course of suspensor formation and elongation (Huang et al. 1998). However, other extrinsic and intrinsic signals regulating suspensor development are not known at present.

With the small size of the embryo proper and the lack of an endosperm, it is logical to speculate that the varied forms of suspensors can play important roles in nutrient uptake. Histochemical staining of cell walls indirectly supports the notion that orchid suspensor can serve as a nutrient uptake site. The polychromatic stain toluidine blue O (TBO) can distinguish lignin, cellulose, and pectic substances on the basis of color differences (O'Brien et al. 1964; Feder and O'Brien 1968; O'Brien and McCully 1981). The fact that the suspensor cell wall stains pinkish-purple with no detectable lignin and cutin deposits, suggests that water and water soluble substances should be able to move apoplastically from the seed coat to the developing embryo proper via suspensor cells (Lee and Yeung 2010). The pinkish-purple color of the suspensor wall after TBO stain also suggests the presence of negatively charged groups which can increase cation-binding capacity of a wall (Smith 1972), and this can aid in solute uptake. A simple plasmolysis experiment indicates that the suspensor cell in the nun orchid has a more negative osmotic potential than the surrounding seed coat cells (Lee and Yeung 2010). The negative osmotic potential provides a driving force for water uptake by the suspensor. In order to visualize the uptake pathway, a fluorescent tracer 6-carboxyfluorescein diacetate (CFD) was applied to developing seeds. Once taken up by the cell, CFD would be hydrolyzed to carboxyfluorescein, giving a fluorescence signal within the cytoplasm. In this experiment, the fluorescence signal was first detected in the suspensor cell before appearing in the embryo proper (Lee and Yeung 2010). This work clearly demonstrates the pathway of solute movement and the importance of suspensor in embryo nutrition.

In flowering plants, the suspensor can take on structural and biochemical specializations (Yeung and Meinke 1993; Kawashima and Goldberg 2010). Transfer cell morphology is one of the common structural specializations observed in suspensor cells. In *Paphiopedilum delenatii*, wall ingrowths are found in suspensor cells (Lee et al. 2006). Wall ingrowths increase the surface area of a cell which can aid in short distance transport. Hence, the formation of wall ingrowths further strengthens the notion that orchid suspensor is specialized in nutrient uptake. The membrane associated with wall ingrowths is unique, e.g. with a higher density of proton pumps (Bouché-Pillon et al. 1994). Future studies demonstrating the presence of biochemical specialization(s) will serve as proof of orchid suspensor function.

In *Phaseolus* species, having large suspensors, high quantities of different phytohormones have been detected (see Yeung and Meinke 1993; Kawashima and Goldberg 2010). Although orchid suspensor is small, it may have a similar biosynthetic ability. If this is true, the orchid suspensor will have additional roles to play during embryogenesis.

Not all orchids have a well-developed suspensor (see Swamy 1949; Clements 1999). In genera such as *Spiranthes*, many species are considered to be suspensorless. At present, the nutrient uptake pathway by these embryos is not known. In the absence of a suspensor, it is logical to speculate that the entire surface of the embryo proper must be able to absorb nutrients directly from surrounding maternal tissues, i.e. the seed coat. In *Cyrtosia javanica*, the seed coat is multilayered and the cells maintain their cytoplasm until seed maturation (Yang and Lee 2014). Furthermore, dense cytoplasmic accessory and antipodal cells are present, locating at the chalazal end of the seed coat, near the developing embryo (Yang and Lee 2014). Can these cells function like the haustoria-type suspensor cells by 'extracting' nutrients from the seed coat and further supply them to the developing embryo? In flowering plant, a cuticle is present on the surface of the embryo proper (Rodkiewicz et al. 1994; Lackie and Yeung 1996) and this can impede diffusion of nutrients into the embryo. Moreover, in *C. javanica*, the cuticle has a patchy appearance and it is initially absent from the walls at the chalazal end (Yang and Lee 2014). These structural features indicate that the *C. javanica* embryo is capable of nutrient uptake, especially at the chalazal end near the accessory and antipodal cells. In order to determine how nutrients enter the embryo in the absence of a suspensor, it is necessary to study the microenvironment surrounding the developing embryo and to perform tracer experiments similar to the study for the nun orchid.

**The embryo proper**

The pattern of embryo development in flowering plants can be divided into several phases (Goldberg et al. 1994). Once the polar axis is established with the formation of the proembryo, histodifferentiation begins with the formation of a shoot and a root apical meristems, the primary meristems (protoderm, ground meristem, and procambium), and the cotyledon(s). Upon the completion of the histodifferentiation phase, storage product synthesis and accumulation begins within embryo cells. Concomitant with storage product accumulation, the maturation phase sets in and the embryo will acquire desiccation tolerance and prepares for developmental arrest and germination events. This pattern of development contributes to the success and survival of flowering plants in the natural habitats. In lower vascular plants such as bryophytes and ferns, once the histodifferentiation phase is completed, the embryo will germinate immediately without a resting phase, as they do not have the ability to undergo storage product biosynthesis and developmental arrest. Moreover, in flowering plants, given the right conditions, developing embryo can also precociously germinate (Raghavan 1997) after the completion of the histodifferentiation phase, as the embryo has already established a functional body plan ready for germination.

Structural descriptions of mature orchid embryos are detailed by Yam et al. (2002). The number of cells and embryo sizes vary among species of Orchidaceae. Only 8 cells were reported for *Epipogium aphyllum* and approximately 700 cells for *Bletilla striata* (see Yam et al. 2002). The mature *B. striata* embryo appears to be one of the largest in the orchid family. Does histodifferentiation occur within a tiny orchid embryo proper? Although there is no obvious tissue pattern formed, histodifferentiation can still be identified (Vinogradova and Andronova 2002). The following discussion argues that histodifferentiation is an integral and a key step during early embryogeny. In orchid embryos, a gradient of cell size is often recognized with smaller cells located at the apical (chalazal) end and larger cells located at the basal (micropylar) end of the embryo. This represents the existence of a structural polarity. The cells at the apical pole of the embryo are destined to form a meristematic zone and the basal cells are designed to house the symbiont upon seed germination. Since the small and large cells have distinct cell fate upon seed germination, a physiological polarity must also be present within the embryo prior to seed germination. It is interesting to note that for easy-to-germinate species such as *Phalaenopsis amabilis*, a marked gradient of cell size exists within the embryo (Lee et al. 2008). This is in contrast to the difficult-to-germinate species such as *C. bulbosa* in which there is no marked difference in cell size (Yeung and Law 1992). Another indication of histodifferentiation is the formation a protoderm. Judging from the uniform cell profile of the surface layer, the embryo proper has acquired protoderm characteristics. A cuticle is found to envelop the embryo proper such as in *C. sinense* (Yeung et al. 1996) and *P. delenatii* (Lee et al. 2006). The ability to form a cuticle is an indication that the surface cells have acquired epidermal cell characteristics. The best argument that histodifferentiation has indeed occurred in an orchid embryo is that 'immature' orchid seeds can germinate asymbiotically in vitro, in a simple medium as long as sucrose is present (Knudson 1922). The orchid embryo has to establish a protocorm body plan during early embryogeny, in order to germinate precociously.

In orchids, the embryo is small with a reduced number of cells. Mitotic activities are arrested early. One can encounter mitotic figures readily within the embryo proper mainly soon after fertilization. Since the embryo proper has few cells, especially when compared to other flowering plants, this suggests that the rate of cell division is slow and/or the cells have a long cell cycle time. A recent study on the expression patterns of the cell-cycle genes in *Phalaenopsis aphrodite* indicates that genes are coordinately regulated from ovule development to embryogenesis (Lin et al. 2014). It would be interesting to determine the intrinsic factors within the embryo which regulate the cell cycle program and mitotic activity as embryo develops.

The absence of a cotyledon is a common characteristic of orchid species. Not more than 10 species have been reported to have a small protrusion that appears to be a rudimentary cotyledon (see Nishimura 1991). In a majority of cases, the protrusions fail to develop further in a protocorm and appear to have no contribution to protocorm development (Nishimura 1991). *Bletilla striata* has the most distinct protrusion at the apical end of the embryo; can this protrusion be considered a cotyledon? It is important to note that *B. striata* has one of the largest embryos in Orchidaceae (Yam et al. 2002). Judging from scanning electron micrographs, the expanding rudimentary cotyledon of *B. striata* closely encircles the developing shoot apical meristem (SAM) (see Figs. 4–12 in Vinogradova and Andronova 2002). Unlike other orchid protocorms, an expanded meristematic zone is not present prior to SAM formation. The close physical proximity between the protrusion and the SAM suggests that it might be a cotyledon. As shown in *Arabidopsis* embryo development, the SAM and cotyledons are formed at the time during histodifferentiation (Laux et al. 2004). If SAM cells or cells having SAM characteristics can be demonstrated in *B. striata* embryo during embryogeny, the protrusion should be regarded as a cotyledon. Future molecular genetics studies on embryo development will provide new insight into the identity of *B. striata* cotyledon-like protrusion and why cotyledons are absent in Orchidaceae.

In flowering plants, storage product deposition begins soon after the completion of the histodifferentiation phase. In dicots such as canola and *Arabidopsis*, the embryo proper begins to expand and fills the endosperm cavity and mitotic activities soon subside. This is followed by the appearance of starch deposits. Starch granules are subsequently replaced by the formation of storage protein and lipid bodies. In orchids, a similar pattern is observed. In *C. sinense* (Yeung et al. 1996), starch granules first appear within the cytoplasm. The large vacuoles are gradually replaced by small ones and storage protein and lipid bodies begin to appear. In a mature orchid embryo, there are abundant storage protein and lipid deposits; starch granules are rare. Besides histological studies, little is known about the physiology of storage product deposition and how storage products influence germination behavior and long term seed storability (Schwallier et al. 2011; Colville et al. 2016). In canola microspore-derived embryos, abscisic acid (ABA) has been shown to induce lipid and oleosin biosynthesis (Zou et al. 1995). ABA levels begin to increase in *Cypripedium formosanum* (Lee et al. 2015) approximately at the same time of storage product accumulation, indicating that ABA can play a role in storage product deposition in orchids.

Are there sufficient storage reserves to sustain seed germination? Orchid embryos though small, have cells packed with storage products. In their natural habitat, it has been reported that the storage products, especially lipid reserves are not utilized unless mycorrhizal establishment has initiated. Harrison (1977) suggested that "this slow expenditure of reserve metabolites could permit the orchid seedling to survive for a longer period until an appropriate endophytic infection is established". In *Arabidopsis*, ABA levels can influence availability of energy and nutrients in seeds during germination (Garciarrubio et al. 1997). A high level of ABA can prevent the breakdown of storage proteins (Garciarrubio et al. 1997) and lipids (Graham 2008).

The final stage of embryo development is considered as the maturation phase. Seeds need to be desiccation resistant in order to survive the natural environment. Recent evidence indicates that a desiccation period is necessary, preparing seeds for subsequent germination (Holdsworth et al. 2008; Angelovici et al. 2011). In *P. amabilis* hybrids, the decline in relative water content between 150 and 165 day after pollination coincides with the acquisition of desiccation tolerance (Schwallier et al. 2011). The presence of high levels of ABA in mature orchid seeds may be essential for the acquisition of desiccation tolerance, as demonstrated in the study of alfalfa seeds (Xu and Bewley 1995). In order to respond to potential abiotic stresses, a common biochemical marker, the late embryogenesis abundant (LEA) proteins, are shown to accumulate at the time of seed maturation in *Dendrobium officinale* (Ling et al. 2016). The presence of long lived mRNAs in mature seeds is a well-documented phenomenon in many flowering plants (Dure 1979, 1985; Sano et al. 2012). Recent proteomic analyses in rice indicated the upregulation of 20 proteins even in the presence of a transcription inhibitor, actinomycin D, indicating that long lived mRNAs must be present in mature seeds (Sano et al. 2012). In *Spathoglotis plicata*, Raghavan and Goh (1994) used the $^3$H-poly-(U) in situ hybridization method to demonstrate that poly(A)-RNAs are present and uniformly distributed throughout the mature embryo. This study indicates the potential presence of long-lived mRNAs in mature embryo cells. Future proteomic analysis, similar to that performed in rice, may enable the identification of proteins essential to early stages of germination. From the available information, preparations during seed maturation in orchids appear to be similar to other flowering plants.

The tiny orchid embryo is often described as "globular", implying that orchid embryo is poorly developed and organized when compare to embryos of other flowering plants. Judging from the cell arrangement and their ability to form protocorms upon germination, histodifferentiation has occurred and the 'protocorm body plan' is laid down ready for germination. As seeds mature, additional changes have taken place in preparation for developmental arrest and subsequent germination events. Therefore, one should consider that the pattern of embryo development in orchids is similar to that of other flowering plants; not simple, nor primitive.

### The endosperm

Yam et al. (2002) provided a detailed account summarizing the presence, absence and/or development of an endosperm in orchid embryos. Although some degrees of endosperm development, i.e. with a multinucleate stage in some species (Yam et al. 2002), continual development similar to other flowering plants is absent. The failure to form a functional endosperm is one of the distinctive features in orchid seed development. Why does endosperm formation fail? Anomalous development in the ovules and/or the male gametes can contribute to endosperm failure in orchids. The formation of the central cell in an embryo sac is complex (Lopez-Villalobos et al. 2016). The competition for nutrients among numerous ovules, may have a negative effect on central cell development. Is the reduction in the number of nuclei present within an embryo sac, commonly referred as the 'striking phenomenon' (Yeung and Law 1997; Batygina et al. 2003), an early indication of central cell abnormality? In *E. scutella*, the polar nuclei fuse before fertilization; however, fusion is not complete (Cocucci and Jensen 1969). In *C. sinense* a microtubular network can be found in the cortex of the endosperm cell up to the four-cell stage of embryo development; however, microtubules are not found to be associated with the nuclei present (Huang et al. 1998). The lack of a functional cytoskeletal network may be partially responsible for the absence of mitotic activities within the endosperm cell.

Can the male gametes contribute to endosperm failure? In *E. scutella*, once the sperm nucleus enters the central cell, the chromatin appears less dense in the embryo sac than in the pollen tube (Cocucci and Jensen 1969). In the nun orchid, the gamete nucleus that fuses with the polar nuclei of the central cell also has a lower staining intensity, prior to fertilization (Ye et al. 2002). Gamete dimorphism has been reported in sperm cells of a number of species such as tobacco late in sperm development (Tian et al. 2001). Preferential fertilization of one of the sperm cells with the egg has been noted. In *Plumbago zeylanica*, the plastid-rich, mitochondrion-poor sperm cell tends to fuse with the egg cell (Russell 1985) and a preferential transmission of supernumerary B chromosomes to the egg cells during sexual reproduction in maize has been reported (see Weterings and Russell 2004). Although gamete dimorphism is not a universal phenomenon in flowering plants, if there is defect in the male gamete destined for the central cell, endosperm development could fail. In orchids, since a massive number of pollen tubes is present within a capsule prior to fertilization, this is an excellent experimental system to study sperm cell structure and function and the results can provide clues as to whether the male gametes can contribute to endosperm failure in orchids.

A varied number of nuclei is present within the primary endosperm cell after fertilization. The origin of these nuclei has generated a lot of discussion. The nuclei within the endosperm cell can represent unfused nuclei from the central cell together with one of the gamete nuclei, or as mitotic products of the primary endosperm nucleus after fertilization. It is generally agreed that there is no fusion of nuclei within the central cell and the nuclei soon degenerate within the primary endosperm cell as the embryo begins to develop. In *Vanilla planifolia*, Swamy (1947) reported the formation of a primary endosperm nucleus and having several rounds of division prior to degeneration; however, a recent study by Kodahl et al. (2015) indicated that such a process did not occur. It is also possible that "remnants of the primary chalazal nuclei persist in the organized embryo sac" as in *Epipogium roseum* (Arekal and Karanth 1981), contributing to the confusion of nuclear number within the primary endosperm cell. Even with improved optical instrumentation, the origin of various nuclei is difficult to determine. To confirm that the nuclei observed are indeed originated from the primary endosperm nucleus, one needs to clearly demonstrate the presence of mitotic figures within the primary endosperm cell.

The lack of a functional endosperm may be the cause of "tiny" embryo formation. In the majority of flowering plants, after fertilization, endosperm develops rapidly aiding in seed enlargement. The endosperm also serves as a nutrient depot and is also a rich source of phytohormones such as gibberellins and cytokinins. All these factors can promote and regulate embryo development (Lopes and Larkins 1993). In the absence of an endosperm, the orchid embryo cannot expand and hence remains small.

In general, when the endosperm fails to develop, the embryo also aborts (see Vijayaraghavan and Prabhakar 1984). The fact that the orchid embryo continues to develop and survive without the presence of an endosperm indicates modifications to the embryo developmental program. The ability of the suspensor to acquire nutrients and the presence of a cuticle to prevent rapid desiccation, are possible strategies that enable continual embryo development in the absence of an endosperm. The ability of orchid embryos to develop without an endosperm, removes one of the potential barriers for hybrid failure (van der Pilj and Dodson 1966).

One can envisage that orchids have devised a new strategy in seed development during their evolution. Since the success of plantlet formation rests on the success of protocorms in establishing proper mycorrhizal interactions, endosperm formation is deemed unnecessary.

## Seed development-regulation by phytohormones

Phytohormones play important roles in all aspects of plant growth and development. Similar to other flowering plants, phytohormones play important roles in orchids. In *E. ibaguense*, following pollination, changes in phytohormones are noted during fruit and seed development with a continual increase in cytokinin activity and a detectable gibberellin activity, indicating the involvements of phytohormones in orchid fruit and seed development (Taylor et al. 1982). At present, studies on phytohormones focusing on orchid reproductive development are few. More efforts are needed in the future.

### Auxin

In orchids, auxin has been shown to play essential roles during pollination and early stages in fruit development (see Novak et al. 2014). Changes in ethylene and auxin are noted at the time of pollination in *Phalaenopsis*. Ethylene in the presence of auxin is required for initiating ovule development and pollen germination and growth (Zhang and O'Neill 1993). During protocorm development, polar auxin transport inhibitors can influence leaf formation and protocorm morphogenesis (Novak et al. 2014). Expression analysis of fertilization and early embryogenesis-associated genes in *Phalaenopsis* indicate that auxin and ethylene can play a role at the critical stage of seed formation (Chen et al. 2016). Although the precise action of auxin is not clear during seed development and germination, the information at hand indicates that auxin must has an important role to play.

### Cytokinins

Cytokinins are essential to plant growth and development, especially in regulating mitotic activities (Emery and Atkins 2006). In *E. ibaguense*, an increase of cytokinin activity coincides with fertilization and commencement of embryo development (Taylor et al. 1982). Since by that time cell division in the fruit wall has more or less completed, the increase in cytokinin activity is most likely related to embryo development (Taylor et al. 1982). However, unlike the majority of flowering plants, cytokinin activity remains high throughout maturation. Is this a unique phenomenon common to orchids? Do the high levels of endogenous cytokinins play a role in seed germination? Many questions remain to be answered.

### Gibberellins

In seeds such as cereals and legumes, major peaks of biologically active gibberellins are found during early seed development, and little activity can be detected as seeds mature (see Taylor et al. 1982). In *E. ibaguense*, although there is a gradual reduction in GA activities, a significant level can still be detected in mature seeds, similar to the pattern observed for cytokinin activity. Again, this begs the question as to whether gibberellins have a role to play during germination and protocorm development.

### Abscisic acid

It is well documented that abscisic acid (ABA) plays important regulatory roles in different aspects of seed development such as storage product biosynthesis and seed dormancy (Nambara et al. 2010). Endogenous ABA has been reported by Van der Kinderen (1987) using chromatographic methods and by Lee et al. (2007b; 2015) using the technique of ELISA and immunostaining. In *Cypripedium formosanum* (Lee et al. 2015), ABA level remains low during early seed development and rises rapidly mid-way during seed maturation. The ABA level follows changes in water content suggesting changes in water content can serve as an endogenous signal for ABA biosynthesis. The importance of ABA is clearly demonstrated experimentally through the use of the ABA biosynthesis inhibitor, fluoridone. Enhanced germination of mature seeds was noted after injecting fluoridine into developing seed capsules (Lee et al. 2015). For difficult to germinate species, the low percentage of seed germination is attributed to their endogenous ABA content (Van der Kinderen 1987). Lowering of ABA level through seed stratification and pre-treatment procedures prior to seed germination can enhance seed germination (Linden 1992). Hence, in vitro culture of immature seeds, prior to the increase in ABA level, improves successes to asymbiotic germination of orchid seeds.

## The physical environment surrounding the embryo

Orchid seeds are small and dust-like, and can have a variety of surface features, i.e. smooth or sculptured. The seed is air-filled and houses an even smaller embryo. The physical properties of orchid seeds and the biological implications of various characteristics have been detailed in an extensive review by Arditti and Ghani (2000) and a summary of seed structures can be found in the monograph by Dressler (1993). The developmental features of seed coats are discussed by Molvray and Chase (1999) and Lee et al. (2007a).

Although orchid seeds are small, they have distinct seed coat features. In general, the seed coat is composed of a few layers of cells with no vascular tissue present, and it develops mainly from the outer integument of the ovule after fertilization. The inner integument tends to degenerate. Exceptions are noted as in the subfamily Vanilloideae; a number of species, e.g. *Vanilla planifolia* (Swamy 1947) and *Cyrtosia javanica* (Yang and Lee 2014) have a multi- layered seed coat and the outermost layer is heavily lignified. For the majority of orchid seeds, even though the seed coat is thin, the addition of secondary walls, phenolic substances, and cuticular materials offer additional protection to the embryo within.

The inner integument of the ovule usually fails to develop much further after fertilization, the cells often degenerate or collapse around the embryo proper forming a 'shell', termed as 'carapace' (Veyret 1969; also see Yam et al. 2002). In *Dactyloriza majalis*, a carapace is readily detected and it wraps tightly around the embryo (Rasmussen 1995). The carapace can offer further protection to the embryo. Additional deposition of phenolic compounds and cuticular materials further strengthens the carapace. The low percentage of mature seed germination observed in *Cephalanthera falcata* is attributed to the accumulation of substances such as lignin in the inner integument, a possible cause of germination inhibition (Yamazaki and Miyoshi 2006). In easy-to-germinate species such as *Orchis morio* and *Serapias lingua*, a carapace is incompletely formed around the embryo (Veyret 1969). In the natural environment, the presence of added protection from carapace may prolong seed survival, enhancing their chance of establishing mycorrhizae association with an appropriate symbiont. However, it can be one of the major causes in inhibiting germination of mature seeds in vitro.

## Knowledge of seed development aids in asymbiotic seed germination efforts

The features discussed above enable orchid seeds to germinate symbiotically in their natural habitats. Moreover, features such as high ABA levels and the formation of a thick carapace with phenolic deposits will have negative impacts on asymbiotic seed germination. It is well established that selecting 'green' capsules at a certain time of fruit development increases the rate of asymbiotic seed germination. This phenomenon can now be explained by the fact that the embryo has completed the histodifferentiation program with the formation of a blue-print for protocorm formation. Also at this time, ABA level is low and only a thin cuticle is present around the embryo proper. Additional physical barriers such as a thick seed coat and carapace have not yet fully developed. All these features facilitate asymbiotic seed germination of immature seeds.

The accumulation of high levels of ABA and secondary metabolites are recognized as key negative factors in asymbiotic germination of mature seeds. Methods are now available to reduce the levels of inhibitory substances using solvents in conjunction with seed sterilization protocols (Kauth et al. 2008; Lee 2011). Additional methods, i.e. low temperature, including freezing (Rasmussen 1995; Mweetwa et al. 2008), and sonication (Miyoshi and Mii 1988) are some of the methods that can improve asymbiotic mature seed germination (see Lee and Yeung 2017).

Although successes are regularly reported in the literature, it is important to note that when comparing asymbiotic and symbiotic germination of the same species, the development of the protocorms and plantlets can differ (Vinogradova and Andronova 2002) and growth is usually better in the latter (Johnson et al. 2007). This indicates that current in vitro protocols and/or culture media still need to be improved and the conditions optimized.

## Protocorm

Upon seed germination, the orchid embryo first develops into a protocorm before forming a plantlet. The term protocorm was introduced by Treub in 1890 to describe the early stages of in the germination of lycopods (Arditti 1992). Veyret (1974) and Vinogradova and Andronova (2002) gave detailed accounts on the varied forms of protocorms and plantlets in their extensive reviews.

From a developmental and functional point of view, orchid protocorm is a unique structure designed to establish symbiotic association with a compatible fungus, and with the primary goal to form a shoot apical meristem (SAM) for plantlet growth. The body plan of a protocorm is established during embryogeny, with cells having distinct cell fates. Hence the protocorm stage should not be considered as an extension of embryonic development. A recent molecular biology study also indicates that protocorms are molecularly distinct from zygotic embryonic tissues in *Phaleaenopsis aphrodite* (Fang et al. 2016). In order to avoid confusion in description, one should reframe from using the term 'embryo' when referring to a protocorm.

Upon placement on appropriate medium, the embryo within the seed coat begins to swell in size and becomes a protocorm. Cells at the apical end divide rapidly, while those at the basal end enlarge and increase their ploidy level during asymbiotic (see Chen et al. 2009) and symbiotic (Rasmussen 1990) germination. The early events in protocorm development are likely regulated by the preprogrammed long-lived mRNAs formed during seed maturation as indirectly demonstrated by Raghavan and Goh (1994). The unique properties of protocorms are further illustrated during symbiotic seed germination. Recent EST analysis in *Dendrodium officinale* reveals the involvement of putative genes (Zhao et al. 2013). Genes and proteins related to defence and stress response, metabolism, transcriptional regulation, transport process and signal transduction pathway are upregulated (Zhao et al. 2013; Valadares et al. 2014; López-Chávez et al. 2016). These studies indicate that the protocorm has the ability to response and prepare to interact with mycorrhizal fungi.

### Protocorm is prepared to receive its symbiont

Fungal hyphae enter the imbibed embryos or developing protocorms at their basal end. For newly imbibed seeds, the degenerated suspensor appears to be the preferred site of hyphae entry (see Wright et al. 2005). Although the suspensor has degenerated, because its wall is primary in nature, penetration through the suspensor or nearby cells is ensured. The presence of the carapace and the cuticular covering of the embryo proper ensure that the apical protocorm cells are protected and not physically damage by the hyphae. For protocorms with more advanced development, i.e. with rhizoid formation, epidermal cells and rhizoids appear to be the preferred site of hyphae entry (see Wright et al. 2005). Stoutamire (1983) suggests that the trichomes may have a function to play in establishing mycorrhizal association.

In *Caladenia tentaculata*, the compatible fungal isolate always infects the seed through the suspensor, resulting in a successful mycorrhizal establishment in protocorms (Wright et al. 2005). In this orchid, phenolic material produced by the embryo diffuses towards its suspensor end; this substance may guide the hyphae towards the embryo. Pelotons appear within the enlarged basal cells of the protocorm and they are digested by protocorm cells at regular intervals for nutrients. For incompatible strains, fungal hyphae do not infect protocorms. They enter through rhizoids or other epidermal cells rather than through the degenerated suspensor. In this case, peloton fails to form and the protocorms are often destroyed by the fungal hyphae (Wright et al. 2005). Moreover, in *Dactyloriza majalis*, fungal hyphae have to enter the protocorm through rhizoids in order to be successful (Rasmussen 1990); entry of hyphae through the suspensor fails to establish mycorrhizal association.

The site of hyphae entry may have important consequences in the success or failure in establishing mycorrhizal association. Since some orchids are 'suspensorless' while others may have taken a longer time to germinate forming rhizoids, the subtle structural and physiological differences in protocorm development may determine which symbiont is best suited for a particular orchid. Do different mycorrhizal fungi have their own preferred site of entry into a protocorm in order to establish a successful interaction? When comparing in vitro vs. in situ studies, selectivity toward fungal symbionts differs for the same species (Masuhara and Katsuya 1994; see also Smith and Read 2008). Is that due to minor differences such as the presence or absence of rhizoids in protocorm development?

The ability to digest pelotons to obtain nutrient supplies is an important strategy in orchids in order to achieve successful plantlet formation. For mycotrophic orchids, nutrients have to come from the mycorrhizal fungi. A recent study by Fochi et al. (2017) demonstrated that a bidirectional nutrient transfer can occur between orchids and its symbionts. Future studies using combined techniques of molecular genetics, cell biology, biochemistry, and physiology will provide additional insights and possible mechanisms into nutrient transfer between them. Besides contributing nutrient supplies to the orchid host, mycorrhizal fungi are known to produce phytohormones (Liu et al. 2010). Can they contribute to morphogenesis and subsequent plantlet formation of protocorms through phytohormone supplies? Although there are still many unanswered questions, the above examples highlight the unique properties of protocorms and the process of mycorrhizal establishment and their interactions.

### The ultimate goal of a protocorm is to form a shoot apical meristem (SAM)

Meristems make the plants (Sussex 1989). Without a functional SAM, there is no further development into a plantlet. Figure 1a–i detail SAM formation in *E. ibaguense*. The embryo at the time of germination has a dense cytoplasm, especially at the future shoot pole. Protein bodies are abundant at this time (Fig. 1a). Within 96 h of germination, storage products are mobilized and judging from the cell profiles, mitotic activities have initiated (Fig. 1b). In *Spathoglottis plicata*, active mitotic activities can be found in the apical pole of the protocorm as early as 4 days after sowing (Raghavan and Goh 1994). With the disappearance of storage products, plastids become abundant within protocorm cells and starch granules are readily detected within plastids (Fig. 1c). The cells at the shoot pole remain small and have similar cytological features (Fig. 1c). Soon after, for cells that are destined to form the future SAM initials, they are characterized by having a large nucleus to cytoplasm ratio (Fig. 1d). The shoot apical meristem initials soon become more distinct as cells begin to take on a square to angular shape and starch granules are not as abundant as surrounding cells (Fig. 1e). These cytological features clearly mark the initiation of a SAM.

The next event is the formation of the first leaf primordium. The apical meristem zone first broadens and in the peripheral region, cells forming the first leaf primordium begin to enlarge (Fig. 1f) and as a result, a leaf primordium begins to protrude above the surface (Fig. 1g–i). This results in generating a small depression with the SAM cells located at the center of the depression (Fig. 1g, h). The 'dimple'-shaped apex has been reported in the literature, e.g. see figures in Leroux et al. (1997) and Vinogradova and Andronova (2002). The SAM continues to mature and increase in size and becomes a dome-shaped structure, initiating leaf primordia at regular intervals (Fig. 1i). It is only after the formation of a well-defined SAM, adventitious roots begin to form near the SAM endogenously.

The nature and origin of the protrusion from a protocorm is a subject of discussion (Vinogradova and Andronova 2002; Batygina et al. 2003; Yam et al. 2002). Should it be considered as a cotyledon or a leaf? Since a protocorm is a post-embryonic structure, the initial leaf-like protrusion formed at the protocorm stage cannot be regarded as cotyledons (Batygina and Andronova 2000). As shown in Fig. 1, the initiation of the first leaf primordium takes place after the formation of a SAM in its peripheral region, even though the SAM has not yet taken on a mature domed-shaped structure.

Although one can pin-point the location of a SAM in the protocorm, the molecular genetics involve in SAM initiation are not known. In Arabidopsis, SAM genes such as *WUSCHEL* and *CLAVATA3* are expressed early at the histodifferentiation stage (Laux et al. 2004). The importance of *WUS* in meristem formation and function is well established. In monocot species, such as corn and rice, the orthologues of *WUS* appear to be regulated differently from *Arabidopsis* (Nardmann and Werr 2006). What are the genes involved in SAM formation in orchid protocorm? Do SAM genes regulated differently from the rest of the flowering plants? To truly understand the origin of a SAM and the nature of the first leaf primordium, additional detailed cellular and molecular genetic studies are needed. Many important questions still need to be answered.

### When does the protocorm stage end and plantlet stage begin?

Although the formation of a protocorm leading to plantlet formation is a continuous developmental process, for the ease of description and discussion between studies, the author would like to suggest that protocorm stage ends once a SAM is clearly established and the plantlet stage begins. After SAM formation, the protocorm can degenerate or enlarge forming a 'corm-like' structure, taking on varied forms (Veyret 1974; Batygina et al. 2003). The timing of root emergence also differs widely between species. Furthermore, the entire plantlet can remain subterranean for quite some time or elongate rapidly, emerges from the substratum and turns green (see Veyret 1974; Batygina and Andronova 2000; Batygina et al. 2003; Vinogradova and Andronova 2002). With such diverse morphogenetic changes, referring these varied forms as protocorms can be confusing.

For the ease of description and comparative evaluations from different published results, morphological stages of a protocorm or a plantlet in one's study should be clearly defined. The proposed stages of germination and protocorm development by Zettler and Hofer (1998) are useful schemes detailing changes in seeds during the course of germination and protocorm formation. Vinogradova and Andronova (2002) also recognized five shoot types to describe the complexity of shoots formed by the apical buds of protocorms.

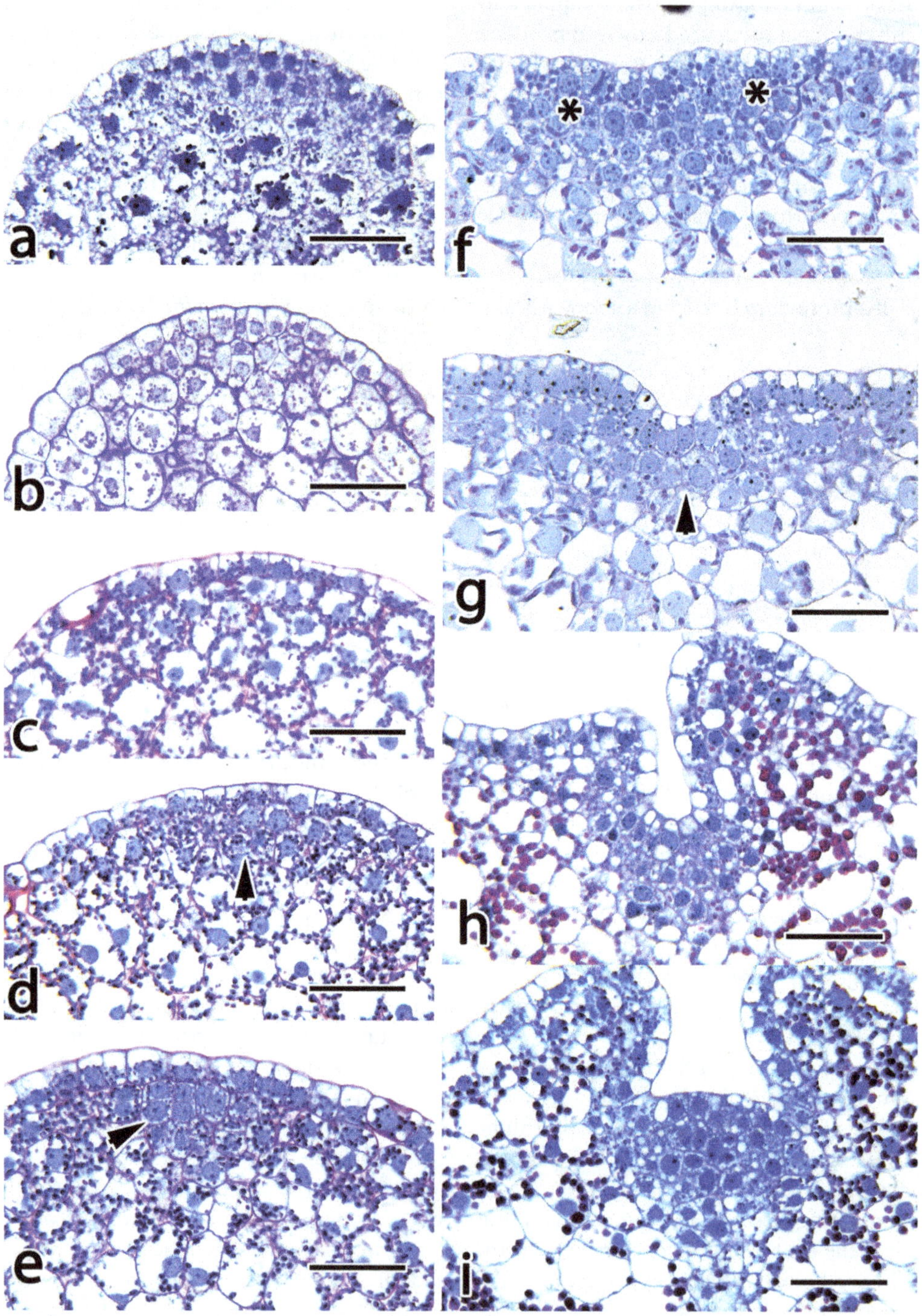


**Fig. 1** Shows the structural ontogeny of a shoot apical meristem in *Epidendrum ibaguense*. **a** At the time of seed germination, the embryo cells have abundant storage product deposits. **b** Within 96 h, storage products have been mobilized and cells are becoming vacuolated; judging from the cell profile of the apical layer, mitotic divisions have begun. **c** Approximately 7 days after germination, protocorms start to turn green, plastids become abundant with starch within protocorm cells. Protocorm cells at the future shoot pole, albeit smaller, have similar cytological features. **d** Soon after, future shoot meristem cells (*arrowhead*) can be identified as they are smaller in size and with a large nucleus to cytoplasm ratio. **e** The SAM initials (*arrowhead*) take on a *square-shape* and starch granules become less abundant. These features clearly mark the structural identity of SAM cells. **f** The SAM increases in size and starch granules remain less abundant when compare to the rest of the protocorm cells. In the peripheral region of the SAM, the cells (*Asterisk*) start to expand. **g** Continual expansion of peripheral meristem cells result in the formation of a protrusion; this results in the formation of a 'dimple' with the SAM cells (*arrowhead*) located at the depression. **h** The SAM continues to grow in size, surrounded by a developing leaf primordium. **i** The SAM in a mature plantlet takes on a slightly domed-shaped structure, forming leaf primordia at a regular interval. All *scale bar* = 50 µm

## Protocorms as explants for micropropagation and transformation studies

In a micropropagation system, initiating mitotic activities of an explant and generating cells with 'organogenic' or 'embryogenic' properties can be a challenge. The ability of apical protocorm cells to divide make them ideal explants for micropropagation and transformation studies. Rapid divisions occur 'naturally' at the meristematic zone allowing for better responses to exogenously applied plant growth regulators. Protocorm cells still retain 'embryogenic' properties, under appropriate in vitro culture conditions. Cells at the apical end can give rise to protocorm-like bodies (PLBs) directly from the surface of protocorms or indirectly through the formation of organogenic calli. The structural organization and developmental properties of PLBs are similar to protocorms in general. PLBs are considered as 'somatic embryos' of orchids. Lee et al. (2013) demonstrated that during PLB formation, there is a transient formation of storage proteins within PLB cells. This clearly indicates that PLBs retain features of zygotic embryo development, albeit transient in nature. In an in vitro environment, PLBs quickly "germinate" and function as protocorms, giving rise to plantlets.

Protocols utilizing protocorms for further micropropagation purposes are readily available in the literature (see reviews by Chugh et al. 2009; Teixeira da Silva 2013). Continual optimization of protocols such as thin section techniques (Teixeira da Silva 2013) and new approaches such as mass culture via bioreactor (Park et al. 2000) will be key to both commercialization and conservation of orchids. Descriptive information concerning PLB initiation from different explants is available in the literature, e.g. Paek et al. (2011); however, mechanistic processes regulating PLB formation are still lacking. Further studies will provide needed information on how the process of PLB regeneration is regulated.

Protocorms and PLBs are excellent explants for transformation studies (Hossain et al. 2013). Plant transformation technologies augment conventional breeding program. Due to the long generation time of orchids, transformation may reduce the time needed to produce a desirable trait. The key event in any transformation protocol is the ability of explants to regenerate after transformation. The ability for protocorms or protocorm sections to generate PLBs, make them ideal candidates for transformation studies. Successes are now reported regularly in the literature using protocorms and PLBs as explants (Phlaetita et al. 2015). One can expect rapid progress in this area of research in the near future.

## Protocorm explants can generate polyploidy plants

Successful polyploid induction can improve plant characteristics. For orchids, polypoid production can improve floral or growth characteristics (Chen et al. 2011; Miguel and Leonhardt 2011). In *Vanda*, embryo cells are diploid maintaining a 2C complement of DNA; however, upon in vitro culture, up to 32C DNA content can be observed in some protocorm cells (Lim and Loh 2003). Chen et al. (2009; 2011) took advantage of the endopolyploidy properties of *Phalaenopsis* protocorm in which basal protocorm cells have a DNA content ranging from 2 to 8C. Through regeneration of PLBs from protocorm thin sections and subsequent selection, polyploid plants can be identified and propagated (Chen et al. 2009; 2011).

Due to the small size and the exposed meristematic surface of protocorms and PLBs, polyploid plants can also be generated using anti-microtubule agents such as colchicine and oryzalin (Miguel and Leonhardt 2011; Chung et al. 2014). After treating with these anti-microtubule agents, followed by selection protocols, desirable polyploid plants can be generated, multiplied and maintained (Miguel and Leonhardt 2011; Chung et al. 2014).

## Conclusions and future perspectives

Although it is minute in size, orchid embryo has carefully planned developmental programs ensuring successful symbiotic seed germination in their natural habitats. The outcome of embryo development is to generate a protocorm which has a unique structural designed for mycorrhizal establishment and a goal to form a shoot apical meristem for plantlet establishment.

At present, successes have been recorded for asymbiotic and symbiotic germination of seeds and micropropagation and transformation studies using protocorms (Chugh et al. 2009; Teixeira da Silva 2013; Lee and Yeung 2017). Many protocols are available which will aid to maintain and propagate important orchid species for further conservation purposes. In recent years, genomic information is becoming available for key commercial species, e.g. Cai et al. (2015), Niu et al. (2016), and Zhang et al. (2016). These studies will provide important data bases to decipher further molecular genetic information. In addition to taking advantage of this information, it is important to combine other approaches such as physiological, biochemical and cell biological methods in studying the seed and protocorm biology in order to gain a comprehensive picture of the entire process.

It is clear from this overview that much more information is needed. Many questions remain. From a developmental biology point of view, orchid embryo and protocorm are excellent and exciting experimental systems for plant biology research. Chances are, we will learn new regulatory mechanisms which will provide new perspectives to plant biology as a whole.

**Acknowledgements**

The author thanks Professor C.C. Chinnappa for the critical reading of the manuscript. This research was supported by Discovery grants from the Natural Sciences and Engineering Research Council of Canada to E. C. Yeung. Dedicated to the memory of Professor Ian M. Sussex.

**Competing interests**

The author declares that he has no competing interests.

**Funding**

Not applicable.

**References**

Angelovici R, Fait A, Fernie AR, Galili G (2011) A seed high-lysine trait is negatively associated with the TCA cycle and slows down *Arabidopsis* seed germination. New Phytol 189:148–159

Arditti J (1992) Fundamentals of orchid biology. Wiley, New York

Arditti J, Ghani AKA (2000) Numerical and physical properties of orchid seeds and their biological implications. New Phytol 145:367–421

Arekal GD, Karanth KA (1981) The embryology of *Epipogium roseum* (Orchidaceae). Plant Syst Evol 138:1–7

Batygina TB, Andronova EV (2000) The orchid protocorm: an opinion. In: Mukerji K (ed) Glimpses in botany. APH Publishing Corporation, New Delhi

Batygina TB, Bragina EA, Vasilyeva VE (2003) The reproductive system and germination in orchids. Acta Biol Craco Ser Bot 45:21–34

Bouché-Pillon S, Fleurat-Lessard P, Serrano R, Bonnemain J-L (1994) Asymmetric distribution of the plasma-membrane H+-ATPase in embryos of *Vicia faba* L. with special reference to transfer cells. Planta 193:392–397

Cai J, Liu X, Vanneste K, Proost S, Tsai WC, Liu KW, Chen LJ, He Y, Xu Q, Bian C et al (2015) The genome sequence of the orchid *Phalaenopsis equestris*. Nat Genet 47:65–72

Chen WH, Tang CY, Kao YL (2009) Ploidy doubling by in vitro culture of excised protocorms or protocorm-like bodies in *Phalaenopsis* species. Plant Cell Tissue Organ Cult 98:229–238

Chen W-H, Kao Y-L, Tang C-Y, Lean G-T (2011) Endopolyploidy in *Phalaenopsis orchids* and its application in polyploid breeding. In: Chen W-H, Chen H-H (eds) Orchid biotechnology II. World Scientific Publishing Co Ltd, Singapore

Chen JC, Wei MJ, Fang SC (2016) Expression analysis of fertilization/early embryogenesis-associated genes in *Phalaenopsis orchids*. Plant Signal Behav 11:e1237331

Chugh S, Guha S, Rao IU (2009) Micropropagation of orchids: a review on the potential of different explants. Sci Hortic 122:507–520

Chung MY, Kim CY, Min JS, Lee DJ, Naing AH, Chung JD, Kim CK (2014) In vitro induction of tetraploids in an interspecific hybrid of *Calanthe* (*Calanthe discolor* X *Calanthe sieboldii*) through colchicine and oryzalin treatments. Plant Biotechnol Rep 8:251–257

Clements MA (1999) Embryology. In: Pridgeon AM, Cribb PJ, Chase MW, Rasmussen FN (eds) Genera Orchidacearum, Vol. 1, General introduction, Apostasioideae, Cyripedioideae. Oxford University Press, Oxford

Cocucci AE, Jensen WA (1969) Orchid embryology: the mature megagametophyte of *Epidendrum scutella* following fertilization. Am J Bot 56:629–640

Colville L, Marks TR, Pritchard HW, Custodio CC, Machado-Neto NB (2016) Development of a reliable GC-MS method for fatty acid profiling using direct transesterification of minimal quantities of microscopic orchid seeds. Seed Sci Res 26:84–91

Dressler RL (1993) Phylogeny and classification of the orchid family. Cambridge University Press, Cambridge

Dure LS III (1979) Role of stored messenger RNA in late embryo development and germination. In: Rubenstein I, Phillips RL, Green CE, Gengenbach BG (eds) The plant seed. Academic Press, New York

Dure L (1985) Embryogenesis and gene expression during seed formation. Oxf Survey Plant Mol Cell Biol 2:179–197

Emery RJN, Atkins CA (2006) Cytokinins and seed development. In: Basra AS (ed) Seed science and technology: trends and advances. Haworth Press Inc., Binghamton

Fang SC, Chen JC, Wei MJ (2016) Protocorms and protocorm-like bodies are molecularly distinct from zygotic embryonic tissues in *Phalaenopsis aphrodite*. Plant Physiol 171:2682–2700

Feder N, O'Brien TP (1968) Plant microtechnique: some principles and new methods. Am J Bot 55:123–142

Fochi V, Chitarra W, Kohler A, Voyron S, Singan VR, Lindquist EA, Barry KW, Girlanda M, Grigoriev IV, Martin F et al (2017) Fungal and plant gene expression in the *Tulasnella calospora–Serapias vomeracea* symbiosis provides clues about N pathways in orchid mycorrhizas. New Phytol 213:365–379

Garciarrubio A, Legaria JP, Covarrubias AA (1997) Abscisic acid inhibits germination of mature *Arabidopsis* seeds by limiting the availability of energy and nutrients. Planta 203:182–187

Goldberg RB, De Paiva G, Yadegari R (1994) Plant embryogenesis: zygote to seed. Science 266:605–614

Graham IA (2008) Seed storage oil mobilization. Annu Rev Plant Biol 59:115–142

Harrison CR (1977) Ultrastructural and histochemical changes during the germination of *Cattleya aurantiaca* (Orchidaceae). Bot Gaz 138:41–45

Holdsworth MJ, Bentsink L, Soppe WJJ (2008) Molecular networks regulating *Arabidopsis* seed maturation, after-ripening, dormancy and germination. New Phytol 179:33–54

Hossain MM, Kant R, Van PT, Winarto B, Zeng S-J, Teixeira da Silva JA (2013) The application of biotechnology to orchids. Crit Rev Plant Sci 32:69–139

Huang BQ, Ye XL, Yeung EC, Zee SY (1998) Embryology of *Cymbidium sinense*: the microtubule organization of early embryos. Ann Bot 81:741–750

Johansen DA (1950) Plant embryology. Chronica Botanica, Waltham

Johnson TR, Stewart SL, Dutra D, Kane ME, Richardson L (2007) Asymbiotic and symbiotic seed germination of *Eulophia alta* (Orchidaceae)—preliminary evidence for the symbiotic culture advantage. Plant Cell Tissue Organ Cult 90:313–323

Kauth PJ, Dutra D, Johnson TR, Stewart SL, Kane ME, Vendrame W (2008) Techniques and applications of in vitro orchid seed germination. In: Teixeira da Silva JA (ed) Floriculture, ornamental and plant biotechnology: advances and topical Issues (Vol. V). Global Science Books, Ltd., Isleworth

Kawashima T, Goldberg RB (2010) The suspensor: not just suspending the embryo. Trends Plant Sci 15:23–30

Knudson L (1922) Non-symbiotic germination of orchid seeds. Bot Gaz 73:1–25

Kodahl N, Johansen BB, Rasmussen FN (2015) The embryo sac of *Vanilla imperialis* (Orchidaceae) is six-nucleate, and double fertilization and formation of endosperm are not observed. Linn Soc Bot J 177:202–213

Lackie S, Yeung EC (1996) Zygotic embryo development in *Daucus carota*. Can J Bot 74:990–998

Laux T, Würschum T, Breuninger H (2004) Genetic regulation of embryonic pattern formation. Plant Cell 16(Suppl):S190–S202

Law SK, Yeung EC (1993) Embryology of *Cypripedium passerimum* (Orchidaceae): ovule development. Lindleyana 8:139–147

Lee YI (2011) In vitro culture and germination of terrestrial Asian orchid seeds. In: Thorpe TA, Yeung EC (eds) Plant embryo culture: methods and protocols. Humana Press, New York

Lee Y-I, Yeung EC (2010) The osmotic property and fluorescent tracer movement of developing orchid embryos of *Phaius tankervilliae* (Aiton) Bl. Sex Plant Reprod 23:337–341

Lee Y-I, Yeung EC (eds) (2017) Orchid propagation: from laboratories to greenhouses: methods and protocols. Springer, New York [in press]

Lee Y-I, Yeung EC, Lee N, Chung M-C (2006) Embryo development in the lady's slipper orchid, *Paphiopedilum delenatii* with emphases on the ultrastructure of the suspensor. Ann Bot 98:1311–1319

Lee Y-I, Yeung EC, Chung M-C (2007a) Embryo development of orchids. In: Chen W-H, Chen H-H (eds) Orchid biotechnology. World Scientific Publishing Co. Pte. Ltd., Singapore

Lee YI, Lu CF, Chung MC, Yeung EC, Lee N (2007b) Developmental changes in endogenous abscisic acid concentrations and asymbiotic seed germination of a terrestrial orchid, *Calanthe tricarinata* Lindl. J Am Soc Hortic Sci 132:246–252

Lee Y-I, Yeung EC, Lee N, Chung MC (2008) Embryology of *Phalaenopsis amabilis* var. formosa: embryo development. Bot Stud 49:139–146

Lee Y-I, Hsu S-T, Yeung EC (2013) Orchid protocorm-like bodies are somatic embryos. Am J Bot 100:2121–2131

Lee Y-I, Chung M-C, Yeung EC, Lee N (2015) Dynamic distribution and the role of abscisic acid during seed development of a lady's slipper orchid, *Cypripedium formosanum*. Ann Bot 116:403–411

Leroux GL, Barabe D, Vieh J (1997) Morphogenesis of the protocorm of *Cypripedium acaule* (Orchidaceae). Plant Syst Evol 205:53–72

Lim WL, Loh CS (2003) Endopolyploidy in *Vanda* miss Joaquim (Orchidaceae). New Phytol 159:279–287

Lin HY, Chen JC, Wei MJ, Lien YC, Li HH, Ko SS, Liu ZH, Fang SC (2014) Genome-wide annotation, expression profiling, and protein interaction studies of the core cell-cycle genes in *Phalaenopsis aphrodite*. Plant Mol Biol 84:203–226

Linden B (1992) Two new methods for pretreatment of seeds of Northern orchids to improve germination in axenic culture. Ann Bot Fenn 29:305–313

Ling H, Zeng X, Guo S (2016) Functional insights into the late embryogenesis abundant (LEA) protein family from *Dendrobium officinale* (Orchidaceae) using an *Escherichia coli* system. Sci Rep 6:39693

Liu H, Luo Y, Liu H (2010) Studies of mycorrhizal fungi of Chinese orchids and their role in orchid conservation in China—a review. Bot Rev 76:241–262

Lopes MA, Larkins BA (1993) Endosperm origin, development, and function. Plant Cell 5:1383–1399

López-Chávez MY, Guillén-Navarro K, Bertolini V, Encarnación S, Hernández-Ortiz M, Sánchez-Moreno I, Damon A (2016) Proteomic and morphometric study of the in vitro interaction between *Oncidium sphacelatum* Lindl. (Orchidaceae) and *Thanatephorus* sp. RG26 (Ceratobasidiaceae). Mycorrhiza 26:353–365

Lopez-Villalobos A, Lopez-Quiroz AA, Yeung EC (2016) Asymmetric cell division in the zygote of flowering plants: the continuing polarized event of embryo sac development. In: Rose RJ (ed) Molecular cell biology of the growth and differentiation of plant cells. CRC Press, Boca Raton

Masuhara G, Katsuya K (1994) In situ and in vitro specificity between *Rhizoctonia* spp. and *Spiranthes sinensis* (Persoon) Ames. var. amoena (M. Bieberstein) Hara (Orchidaceae). New Phytol 127:711–718

Miguel TP, Leonhardt KW (2011) In vitro polyploid induction of orchids using oryzalin. Sci Hortic 130:314–319

Miyoshi K, Mii M (1988) Ultrasonic treatment for enhancing seed germination of terrestrial orchid, *Calanthe discolor*, in asymbiotic culture. Sci Hortic 35:127–130

Molvray M, Chase MW (1999) Seed morphology. In: Pridgeon AM, Cribb PJ, Chase MW, Rasmussen FN (eds) Genera Orchidacearum Volume 1 General introduction, Apostasioideae, Cypripedioideae. Oxford University Press Inc., New York

Mweetwa AM, Welbaum GE, Tay D (2008) Effects of development, temperature, and calcium hypochlorite treatment on in vitro germinability of *Phalaenopsis* seeds. Sci Hortic 117:257–262

Nambara E, Okamoto M, Tatematsu K, Yano R, Seo M, Kamiya Y (2010) Abscisic acid and the control of seed dormancy and germination. Seed Sci Res 20:55–67

Nardmann J, Werr W (2006) The shoot stem cell niche in angiosperms: expression patterns of WUS orthologues in rice and maize imply major modifications in the course of mono- and dicot evolution. Mol Biol Evol 23:2492–2504

Natesh S, Rau MA (1984) The embryo. In: Johri BM (ed) Embryology of angiosperms. Springer-Verlag, Berlin

Nishimura G (1991) Comparative morphology of cotyledonous orchid seedlings. Lindleyana 6:140–146

Niu SC, Xu Q, Zhang GQ, Zhang YQ, Tsai WC, Hsu JL, Liang CK, Luo YB, Liu ZJ (2016) De novo transcriptome assembly databases for the butterfly orchid *Phalaenopsis equestris*. Sci Data 3:160083

Novak SD, Luna LJ, Gamage RN (2014) Role of auxin in orchid development. Plant Signal Behav 9:2972277-1

O'Brien TP, McCully ME (1981) The study of plant structure: principles and selected methods. Termarcarphi Pty, Ltd, Melbourne

O'Brien TP, Feder N, McCully ME (1964) Poly-chromatic staining of plant cell walls by toluidine blue O. Protoplasma 59:368–373

Paek KY, Hahn EJ, Park SY (2011) Micropropagation of *Phalaenopsis* orchids via protocorms and protocorm-like bodies. Methods Mol Biol 710:293–306

Park SY, Murthy HN, Yoeup PK (2000) Mass multiplication of protocorm-like bodies using bioreactor system and subsequent plant regeneration in *Phalaenopsis*. Plant Cell Tissue Organ Cult 63:67–72

Phlaetita W, Chin DP, Otanga NV, Nakamura I, Mii M (2015) High efficiency *Agrobacterium*-mediated transformation of *Dendrobium orchid* using protocorms as a target material. Plant Biotech 32:323–327

Raghavan V (1997) Molecular embryology of flowering plants. Cambridge University Press, Cambridge

Raghavan V, Goh CJ (1994) DNA synthesis and mRNA accumulation during germination of embryos of the orchid *Spathoglottis plicata*. Protoplasma 183:137–147

Rasmussen HN (1990) Cell differentiation and mycorrhizal infection in *Dactylorhiza majalis* (Rchb. F.) Hunt & Summerh. (Orchidaceae) during germination in vitro. New Phytol 116:137–143

Rasmussen HN (1995) Terrestrial orchids: from seed to mycotrophic plant. Cambridge University Press, Cambridge

Rodkiewicz B, Fyk B, Szczuka E (1994) Chlorophyll and cutin in early embryogenesis in *Capsella*, *Arabidopsis*, and *Stellaria* investigated by fluorescence microscopy. Sex Plant Reprod 7:287–289

Russell SD (1985) Preferential fertilization in *Plumbago*: ultrastructural evidence for gamete-level recognition in an angiosperm. Proc Nat Acad Sci USA 82:6129–6132

Sano N, Permana H, Kumada R, Shinozaki Y, Tanabata T, Yamada T, Hirasawa T, Kanekatsu M (2012) Proteomic analysis of embryonic proteins synthesized from long-lived mRNAs during germination of rice seeds. Plant Cell Physiol 53:687–698

Schwallier R, Bhoopalan V, Blackman S (2011) The influence of seed maturation on desiccation tolerance in *Phalaenopsis amabilis* hybrids. Sci Hortic 128:136–140

Smith DL (1972) Staining and osmotic properties of young gametophytes of *Polypodium vulgare* L. and their bearing on rhizoid function. Protoplasma 74:465–479

Smith SE, Read DJ (2008) Mycorrhizal symbiosis. Academic Press, New York

Stoutamire WP (1983) Early growth in North American terrestrial orchid seedlings. In: Plaxton EH (ed), North American terrestrial orchids symposium II, proceedings and lectures, 14–24. Michigan Orchid Society, Livonia, MI

Sussex IM (1989) Developmental programming of the shoot meristem. Cell 56:225–229

Swamy BGL (1947) On the life-history of *Vanilla planifolia*. Bot Gaz 108:449–459

Swamy BGL (1949) Embryological studies in the Orchidaceae. II. Embryology. Am Midl Nat 41:202–232

Taylor JS, Blackman SJ, Yeung EC (1982) Hormonal and structural aspects of fruit development in the orchid, *Epidendrum*. J Exp Bot 33:495–505

Teixeira da Silva JA (2013) Orchids: advances in tissue culture, genetics, phytochemistry and transgenic biotechnology. Floric Ornam Biotechnol 7:1–52

Tian HQ, Zhang ZJ, Russell SD (2001) Sperm dimorphism in *Nicotiana tabacum* L. Sex Plant Reprod 14:123–125

Ueda M, Laux T (2012) The origin of the plant body axis. Curr Opin Plant Biol 15:578–584

Valadares RB, Perotto S, Santos EC, Lambais MR (2014) Proteome changes in *Oncidium sphacelatum* (Orchidaceae) at different trophic stages of symbiotic germination. Mycorrhiza 24:349–360

Van der Kinderen G (1987) Abscisic acid in terrestrial orchid seeds: a possible impact on their germination. Lindleyana 2:84–87 van der Pilj L, Dodson CH (1966) Orchid flowers: their pollination and evolution. University of Miami Press, Coral Gables

Veyret Y (1969) La structure des semences des *Orchidaceae* et leur aptitude à la germination in vitro en cultures pures. In: Travaux du Laboratorie de la Jaysinia, vol 3. Laboratorie de Biologie Végétale du Muséum, Paris, pp 89–98

Veyret Y (1974) Development of the embryo and the young seedling stages of orchids. In: Withner CL (ed) The orchids: scientific studies. Wiley, New York, pp 223–265

Vijayaraghavan MR, Prabhakar K (1984) The endosperm. In: Johri BM (ed) Embryology of angiosperms. Springer, Berlin

Vinogradova TN, Andronova EV (2002) Development of orchid seeds and seedlings. In: Kull T, Arditti J (eds) Orchid biology: reviews and perspectives, VIII. Kluwer Academic Publishers, Dordrecht

Wabnik K, Robert HS, Smith RS, Friml J (2013) Modeling framework for the establishment of the apical-basal embryonic axis in plants. Curr Biol 23:2513–2518

Weterings K, Russell SD (2004) Experimental analysis of the fertilization process. Plant Cell 16:S107–S118

Wright M, Guest D, Cross R (2005) Development of mycorrhiza association in *Caladenia tentaculata*. Selbyana 26:114–124

Xu N, Bewley JD (1995) The role of abscisic acid in germination, storage protein synthesis and desiccation tolerance in alfalfa (*Medicago sativa* L.) seeds, as shown by inhibition of its synthesis by fluridone during development. J Exp Bot 46:687–694

Yam TW, Yeung EC, Ye X-L, Zee S-Y, Arditti J (2002) Embryology-seeds. In: Arditti J, Kull T (eds) Orchid biology: reviews and perspectives VIII. Kluwer Academic Publishers, Dordrecht

Yamazaki J, Miyoshi K (2006) In vitro asymbiotic germination of immature seed and formation of protocorm by *Cephalanthera falcate* (Orchidaceae). Ann Bot 98:1197–1206

Yang CK, Lee YI (2014) The seed development of a mycoheterotrophic orchid, *Cyrtosia javanica* Blume. Bot Stud 55:44

Ye XL, Zee SY, Yeung EC (1997) Suspensor development in the nun orchid, *Phaius tankervilliae*. Int J Plant Sci 158:704–712

Ye X-L, Yeung EC, Zee S-Y (2002) Sperm movement during double fertilization of a flowering plant, *Phaius tankervilliae*. Planta 215:60–66

Yeung EC, Law SK (1992) Embryology of *Calypso bulbosa*. II. Embryo development. Can J Bot 70:461–468

Yeung EC, Law SK (1997) Ovule development. In: Arditti J, Pridgeon AM (eds) Orchid biology: reviews and perspectives VII. Kluwer, Dordrecht

Yeung EC, Meinke DW (1993) Embryogenesis in angiosperms: development of the suspensor. Plant Cell 5:1371–1381

Yeung EC, Zee SY, Ye XL (1996) Embryology of *Cymbidium sinense*: embryo development. Ann Bot 78:105–110

Zettler LW, Hofer CJ (1998) Propagation of the little club-spur orchid (*Platanthera clavellate*) by symbiotic seed germination and its ecological implications. Environ Exp Bot 39:189–195

Zhang XS, O'Neill SD (1993) Ovary and gametophyte are coordinately regulated by auxin and ethylene following pollination. Plant Cell 5:403–418

Zhao M-M, Zhang G, Zhang D-W, Hsiao Y-Y, Guo S-X (2013) ESTs analysis reveals putative genes involved in symbiotic seed germination in Dendrobium officinale. PloS One 8:e72705

Zou J, Abrams GD, Barton DL, Taylor DC, Pomeroy MK, Abrams SR (1995) Induction of lipid and oleosin biosynthesis by (+)-Abscisic acid and its metabolites in microspore-derived embryos of *Brassica napus* L.cv Reston (Biological responses in the presence of 8[prime]-Hydroxyabscisic acid). Plant Physiol 108(2):563–571

Zhang GQ, Xu Q, Bian C, Tsai WC, Yeh CM, Liu KW, Yoshida K, Zhang LS, Chang SB, Chen F, Shi Y, Su YY, Zhang YQ, Chen LJ, Yin Y, Lin M, Huang H, Deng H, Wang ZW, Zhu SL, Zhao X, Deng C, Niu SC, Huang J, Wang M, Liu GH, Yang HJ, Xiao XJ, Hsiao YY, Wu WL, Chen YY, Mitsuda N, Ohme-Takagi M, Luo YB, Van de Peer Y, Liu ZJ (2016) The *Dendrobium catenatum* Lindl. genome sequence provides insights into polysaccharide synthase, floral development and adaptive evolution. Sci Rep 6:19029. doi:10.1038/srep19029

# 6

# Photosynthetic acclimation of *Grammatophyllum speciosum* to growth irradiance under natural conditions in Singapore

Jie He[1*], Regina M. P. Lim[1], Sabrina H. J. Dass[1] and Tim W. Yam[2]

**Abstract**

*Grammatophyllum speciosum*, a native species to Singapore, have become extinct mainly due to habitat loss. Recently, Singapore has reintroduced *G. speciosum* into the natural environment under the orchid conservation programme. In this study, leaves of *G. speciosum* grown under low light (LL) under natural conditions had faster expansion rate and higher specific leaf area than leaves grown under intermediate light (IL) and high light (HL). All leaves had more than 95% midday relative water content. Although midday $F_v/F_m$ ratios were lower in HL leaves than in IL and LL leaves, none of them exhibited chronic photoinhibition. HL leaves had upregulated their light utilization through higher photochemical quantum yield ($\Delta F/F_m'$) and greater electron transport rate. HL leaves also had higher non-photochemical quenching, indicating that they had higher capability to dissipate excess light as heat, which was supported by their lower chlorophyll but higher carotenoids content. Although there was a linear correction between leaf temperature and photosynthetic photon flux density (PPFD), no correlations were found between stomatal conductance ($g_s$) and PPFD, $g_s$ and leaf temperature. Light-saturated photosynthetic $CO_2$ assimilation rate ($A_{sat}$) was significantly higher in HL leaves than those of IL and LL leaves. However, all leaves had similar light-saturated stomatal conductance. Although LL leaves had higher leaf total reduced nitrogen that those of IL and HL leaves, none of them seemed to suffer from nitrogen deficiency during the experimental period. To conclude, *G. speciosum* is able to survive under different growth irradiances without watering and adding fertilizers.

**Keywords:** $CO_2$ assimilation, Photosynthetic pigments, $F_v/F_m$ ratio, Leaf temperature, Relative water content, Specific leaf area, Stomatal conductance, Total reduced nitrogen

## Background

Singapore houses 226 species of native tropical orchids. However, many of our natural habitats have been destroyed due to urbanisation. Besides the five common species, 178 species are extinct while the rest are either critically endangered or vulnerable (Ng 1994). One of the extinct species is *Grammatophyllum speciosum*, also known as the tiger orchid, which is the largest epiphytic $C_3$ orchid plant in the world. Since 1995, The National Parks Board (NParks) in Singapore has been efficacious in propagating and reintroducing *G. speciosum* in its orchid conservation programme (Yam et al. 2010). More than 80% of native orchids planted on trees and ground in parks and nature reserve areas under different growth irradiances have survived (Yam 2013; Yam et al. 2011). Those could not have suffered from the complex interactions of high irradiances, high temperature, drought stress and nutrient deficiency.

Under tropical field conditions, orchids cope with variation of light intensities daily. Excessive light is one of the environmental stresses experienced by orchid plants (He et al. 1998, 2004, 2014; Koh et al. 1997; Tay et al.

*Correspondence: jie.he@nie.edu.sg
[1] Natural Sciences and Science Education Academic Group, National Institute of Education, Nanyang Technological University, 1 Nanyang Walk, Singapore 637 616, Singapore
Full list of author information is available at the end of the article

2015). When they absorb excess light energy, if not dissipated as heat, it can be detrimental to photosystem II (PSII), known as dynamic or chronic photoinhibition (He et al. 1996). Dynamic photoinhibition is a reversible downregulation mechanism to reduce the light utilisation efficiency by diverting the excessive energy to the xanthophyll cycle so as to protect PSII reaction centres from photodamage (Demmig-Adams and Adam III 1992; Chow 1994; Osmond 1994). On the other hand, the excess energy could cause damage to the photosynthetic reaction centres, particularly of PSII, leading to sustained photoinhibition and then reduction of plant growth and productivity (Powles 1984; Osmond 1994; Barber 1995; Adams III et al. 2006). Chronic photoinhibition is the slower reversible loss of the function of PSII which is dependent on the repair and recovery rates of D1 protein in PSII reaction centres (Osmond 1994). Long term exposure to high irradiance would lead to irreversible oxidation of the chlorophyll (Chl) and destruction of chloroplasts due to formation of reactive oxygen species (ROS) (Lichtenthaler and Wellburn 1983). Thus, orchids grown under different growth irradiances, have different amount of Chl a, b and carotenoids (Car) (Lichtenthaler and Burkart 1999) with plants under high irradiance having more Car especially xanthophyll to photoprotect PSII from damage.

When the absorbed light energy exceeds the requirement of carbon fixation for light energy under high temperature or water deficit condition, photoinhibition is exacerbated (Lawlor and Tezara 2009). With water deficit, there is a reduction in leaf turgor potential and water content (Gomes et al. 2008); formation of ROS, which destroys the cell membrane, nucleic acid and protein (Ashraf 2009) such as Rubisco, an important carboxylase in $C_3$ plants, thus decreasing carboxylation efficiency. Also, stomata close with water deficit to reduce transpiration which affects the leaf temperature (Farquhar and Sharkey 1982). Stomatal closure could also reduce photosynthesis by direct effects on the photosynthetic capacity of the mesophyll cells (Jones 1998). Closure of stomata results in reduced light-saturated stomatal conductance ($g_{s\,sat}$) and photosynthetic $CO_2$ assimilation, ($A_{sat}$) affecting leaf growth and productivity (He et al. 2001).

Photoinhibition can also be aggravated by the lack of nutrients. Nitrogen (N) is often considered as the most limiting nutrient for the growth and yield of plants worldwide (Angus et al. 1993). N is a key component for protein synthesis and specifically for the maintenance of the abundant proteins associated with the photosynthetic apparatus (Loebl et al. 2010). Since PSII reaction centers are constantly damaged by high irradiances, cells need N to synthesis new proteins required for PSII repair (Nishiyama et al. 2006). If PSII repair cannot counteract photoinactivation due to N depletion, then the cells are subjected to chronic photoinhibition (Ragni et al. 2008).

Very little is known about the growth and photosynthetic performances of *G. speciosum* grown under natural tropical conditions. Furthermore, leaf photosynthetic gas exchange of plants subjected to mild drought stress may be largely due to stomatal limitation, rather than biochemical factors such as Rubisco protein (Lawlor and Cornic 2002; Bota et al. 2004; Flexas et al. 2006). Since all *G. speciosum* plants grown under natural conditions used for this project were maintenance free without watering and fertilization, it was often questioned that if N deficiency had occurs in these plants. Thus, using the native orchid species, *G. speciosum*, this project aimed to study (1) the effects of different growth irradiances on leaf growth, water relations and photosynthetic gas exchanges, (2) photosynthetic utilization of light energy and, (3) the concentration of leaf total reduced nitrogen (N) from leaves grown under different growth irradiances.

## Methods

### Plant materials

*Grammatophyllum speciosum* plants were planted on the grounds of National Institute of Education, Singapore in October 2012 by NParks under different growth irradiances, namely high light (HL, open field), intermediate light (IL) and low light (LL) respectively. The IL and LL conditions were obtained from different levels of shading by the surrounding trees. The average maximal photosynthetic photon flux density (PPFD) during midday at the top of the canopy were *ca.* 2000, 1200 and 600 under HL, IL and LL, respectively on sunny days. About 30 plants were planted under each of the light conditions. The 5th fully expanded leaves from the top of the plants were tagged for measurements from January to November 2013. These orchids were neither watered nor fertilized.

### Measurement of leaf length and specific leaf area (SLA)

The lengths of the tagged leaves were measured thrice by inserting the stick abaxially. Six weeks later, leaves were harvested. To estimate SLA, six leaf discs (3.4 $cm^2$) per individual condition were collected with a cork borer. The leaf's midveins were avoided to reduce variations. The discs were dried to constant mass at 80 °C, and then weighed. Dry weight (DW) was used to calculate the SLA, the ratio of leaf area to DW. The measurements were carried out from February to March 2013.

### Measurement of midday leaf relative water content (RWC)

On sunny days during March and October, leaves were harvested during midday. Six small squares (1 cm by 1 cm) were cut from one leaf before weighing the fresh weight (FW) instantly with an analytical balance (Sartorius). The samples were floated on water in the dark for 24 h before measuring their saturated weight (SW). They were then dried in the oven at 80 °C for 72 h before DW was obtained. RWC was calculated as RWC = (FW − DW)/(SW − DW) × 100%.

#### Measurements of photosynthetic photon flux density (PPFD), predawn and midday $F_v/F_m$ ratios

PPFDs were measured using a photosynthetically available radiation quantum sensor and reading unit (Skye Instruments Ltd, Llandrindod, UK). They were measured from six different positions above the leaves for HL, IL and LL respectively. Chl fluorescence $F_v/F_m$ ratios measured the potential efficiency of excitation energy captured by PSII. The non-destructive measurements were carried out predawn and during midday (1200–1300 h) with the Plant Efficiency Analyser, PEA (Hansatech Instruments Ltd, England) on sunny days in March and October. Attached leaves were pre-darkened with clips for 15 min before measurements. Dark-adapted leaves were placed under the light pipe and irradiated with the pulsed lower intensity-measuring beam to measure $F_0$, basal Chl fluorescence. $F_m$, maximum Chl fluorescence was assessed by 0.8 s of saturated pulse (> 6000 mol $m^{-2}$ $s^{-1}$). The variable fluorescence yield, $F_v$, was determined by $F_m$–$F_o$. The efficiency of excitation energy captured by open PSII reaction centres in dark adapted leaves was estimated by the fluorescence $F_v/F_m$ ratio.

#### Measurements of different Chl fluorescence parameters

These measurements were carried out in October. Newly fully expanded leaves were harvested at 0800 h for Chl fluorescence analysis. The effective photochemical quantum yield ($\Delta F/F_m'$), electron transport rate (ETR) and non-photochemical quenching (NPQ) of Chl fluorescence were determined using the Imaging PAM Chl Fluorometer (Walz, Effeltrich, Germany) at 25 °C in the laboratory. Leaf discs were pre-darkened for 15 min before the measurements. Images of fluorescence emission from the PAM Chl Fluorometer, were digitized within the camera via a Firewire interface (400 Mb/s) (Firewire-1394, Austin, TX, USA) to a personal computer for storage and analysis. The details of measuring light pulses, actinic illumination and saturation pulses were described previously (He et al. 2011). Rapid light curve measurements in the presence of actinic illuminations (Schreiber et al. 1997) were obtained through the application of a series of 10-s light exposures with increasing irradiance from 1 to 1585 µmol photons $m^{-2}$ $s^{-1}$. In the presence of actinic illumination, the current fluorescence yield ($F_t$ = F), and the maximum fluorescence ($F_m'$) at the steady state, were determined, from which the effective PSII quantum yield, $\Delta F/F_m'$ $[(F_m' - F)/F_m')]$ and ETR (PPFD × $\Delta F/F_m'$ × 0.5 × 0.84) could be calculated. The use of two photons is necessary to transport one electron (factor 0.5). Correction factor 0.84 takes into account that only a fraction of incident light is really absorbed by photosynthesis. NPQ was defined as: NPQ = $(F_m - F_m')/F_m'$ (Rascher et al. 2000).

#### Determination of photosynthetic pigments

These measurements were carried out in both March and October. Newly fully expanded leaves were harvested at 0800 h. Samples of 0.05 g were weighed and cut into very small pieces before soaking in 5 ml of *N,N*-dimethylformamide in the dark for 48 h at 4 °C. They were then quantified spectrophotometrically (Du 650, Beckman, USA) using the procedure of Wellburn (1994).

#### Measurements of diurnal changes in $g_s$ and leaf temperature

The $g_s$ and leaf temperature were measured every 2 h from 0800 to 1800 h using the leaf porometer chamber (SC-1, Decagon, U.S.) with a fixed diffusion path to the leaf surface. These measurements were carried out in March on three sunny days.

#### Measurements of $A_{sat}$ and $g_{s\,sat}$

These measurements were carried out in both March and October on three sunny days. They were measured simultaneously with an open infrared gas analysis system with a 6 $cm^2$ chamber (LI-6400, Biosciences, US) between 0900 and 1030 h. Readings were measured with a LED light source, which supplied a saturated PPFD of 1000 mol $m^{-2}$ $s^{-1}$. The light source emitted light in the wavelength range of 660–675 nm. Average ambient $[CO_2]$ and relative humidity in the chamber were 400 ± 5 µmol $mol^{-1}$ and 70%, respectively.

#### Determination of leaf total reduced nitrogen (TRN)

In October, the same leaves used for the measurements of $A_{sat}$ and $g_{s\,sat}$ were harvested immediately after recording the values of $A_{sat}$ and $g_{s\,sat}$. The leaves were then used for the analysis of TRN. Leaf TRN concentration was determined by Kjeldahl digestion of dried samples in concentrated $H_2SO_4$. Dry samples of 0.15 were placed into a digestion tube with a Kjeldahl tablet and 2.5 ml of concentrated $H_2SO_4$. The mixture was then digested (about 90 min) until clear. After the digestion was completed, the mixture was allowed to cool for 30 min and the TRN was determined by with a Kjeltec auto 2300 analyser.

#### Statistical analysis

One-way ANOVA was used to test for significant differences among different growth irradiances, using Tukey's multiple comparison tests to discriminate the means (MINITAB, Inc., Release 15, 2007).

## Results

### Leaf length, SLA and midday leaf RWC

All leaves increased gradually in length over 6 weeks of measurement. Leaves of *G. speciosum* plants grown under LL seemed to grow faster than leaves grown under IL and HL, which were the longest after 6 weeks (Fig. 1A, $p < 0.05$). However, there was no significant difference in leaf length between IL and HL leaves after 6 weeks of expansion (Fig. 1A, $p > 0.05$). SLA was not significantly different between leaves grown under HL and IL but LL grown leaves had significantly higher SLA (Fig. 1B). No significant differences in midday leaf RWC were observed across the plants grown under three different growth irradiances (Fig. 2, $p > 0.05$). Results of midday leaf RWC shown in Fig. 2 were obtained on sunny days in October. Same measurements were carried out on sunny days in March and all leaves had midday RWC greater than 95%.

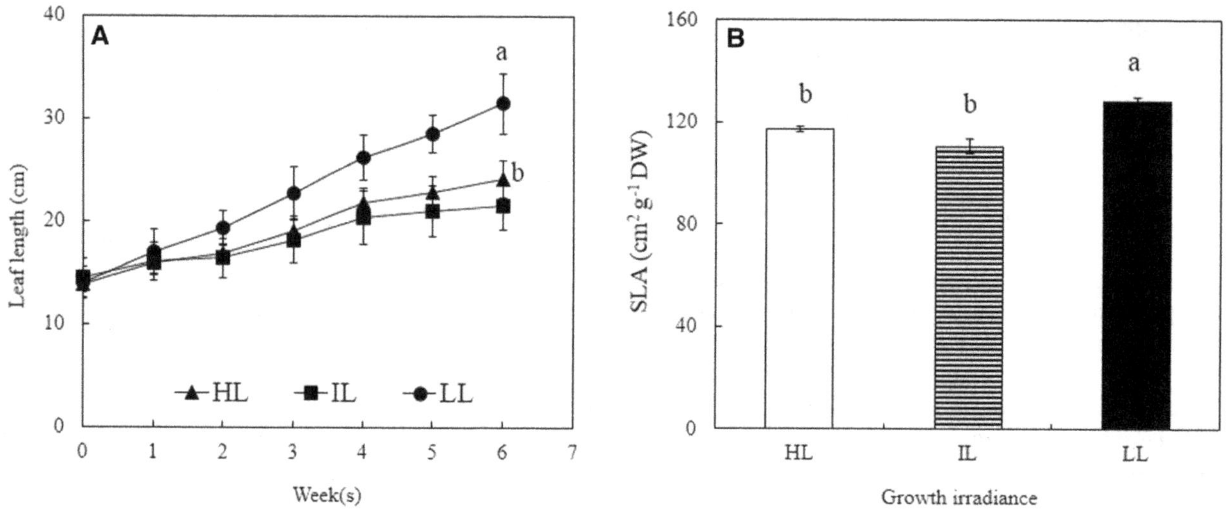


**Fig. 1** Leaf length (**A**) and SLA (**B**) of young expanding leaves of *G. speciosum* under different growth irradiances. Each value is the means from 6 different leaves of 6 different plants. Vertical bars represent SE. Means with same letter above the bars are not statistically different ($p > 0.05$) as determined by Tukey's test

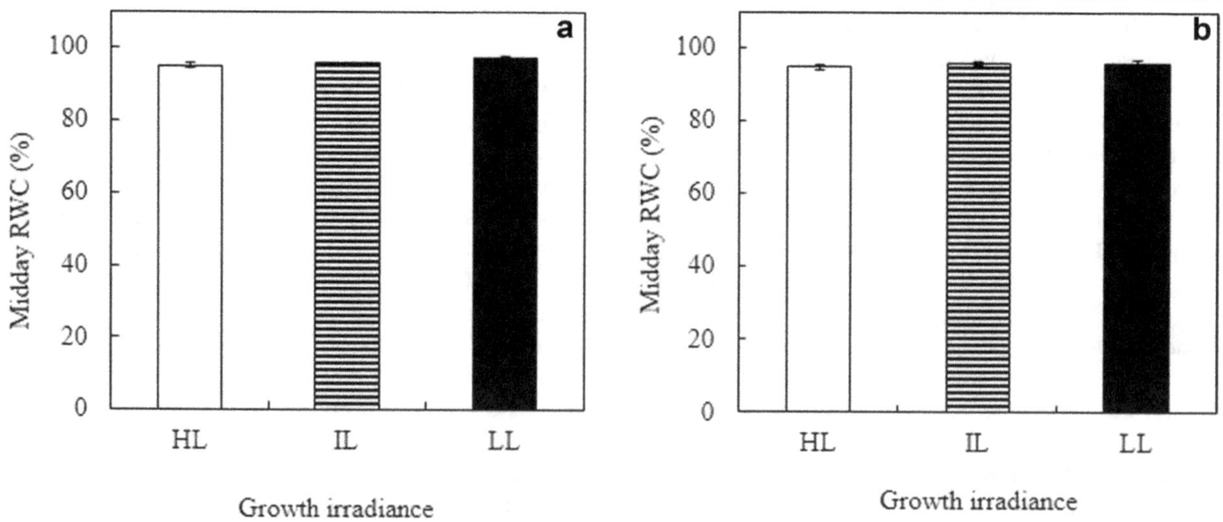


**Fig. 2** Midday RWC of young expanding (**a**) and fully expanded (**b**) leaves of *G. speciosum* under different growth irradiances. Each value is the mean from 6 different leaves of 6 different plants. Vertical bars represent SE

### Midday and predawn $F_v/F_m$ ratios measured under field conditions

These measurements were carried out in both March and October on sunny days. Similar results were obtained and thus, only data from October are shown in Fig. 3. Among all leaves, the lowest midday $F_v/F_m$ ratio was recorded from the leaves that were grown under HL followed by those grown under IL and the highest $F_v/F_m$ ratio was found in leaves exposed to LL (Fig. 3B, $p < 0.05$). Differences in midday $F_v/F_m$ ratios corresponded with the levels of light to which they were exposed (Fig. 3A), that was, the higher the PPFD illuminated the leaves, the lower the midday $F_v/F_m$ ratio. However, there were no significant differences in predawn $F_v/F_m$ ratio and all the leaves had predawn $F_v/F_m$ ratio $\geq 0.8$ (Fig. 3C).

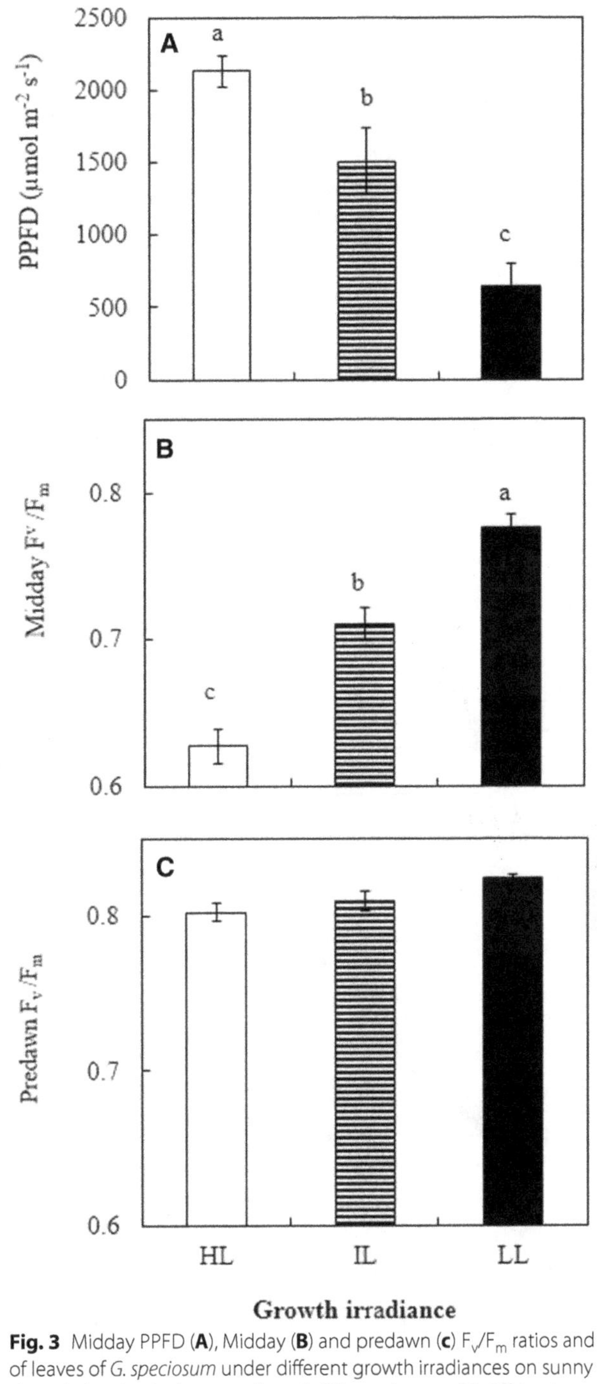


**Fig. 3** Midday PPFD (**A**), Midday (**B**) and predawn (**c**) $F_v/F_m$ ratios and of leaves of *G. speciosum* under different growth irradiances on sunny days. Each value is the mean from 6 different leaves of 6 different plants measured from three sunny days. Vertical bars represent standard errors. Means with same letter above the bars are not statistically different ($p > 0.05$) as determined by Tukey's test

### Photochemical efficiency and photosynthetic pigments measured in the laboratory

Parameters such as $F/F_m'$ (photochemical quantum yield at actinic light) ETR and NPQ were used to explore the utilization of light energy by leaves grown under different growth irradiances. All results shown in Fig. 4 were obtained in October. The light response curves of these parameters were measured under different PPFDs in the laboratory from 1 to 1585 μmol $m^{-2}$ $s^{-1}$. When measured at the lowest PPFD of 1 μmol $m^{-2}$ $s^{-1}$, $\Delta F/F_m'$ values were not significantly different among the different leaves ranging from 0.788 to 0.754. However, the values of $\Delta F/F_m'$ decreased to 0.637, 0.549 and 0.468 measured at PPFD of 415 μmol $m^{-2}$ $s^{-1}$ for leaves grown under HL, IL and LL, respectively (Fig. 4a). The differences in $\Delta F/F_m'$ among the different leaves at this PPFD were significantly different ($p < 0.05$). A further decreases of $\Delta F/F_m'$ in all leaves were observed when PPFD was higher than 415 μmol $m^{-2}$ $s^{-1}$ with their value almost zero at PPFD of 1585 μmol $m^{-2}$ $s^{-1}$. However, the rate of deceases in $\Delta F/F_m'$ was the fastest in leaves grown under LL followed by that of leaves under IL with increasing PPFD. Leaves grown under HL exhibited the slowest decrease rate of $\Delta F/F_m'$ with increasing PPF. The HL leaves had the highest values of $\Delta F/F_m'$ measured at higher PPFDs followed by the IL leaves and the LL leaves had the lowest values at each PPFD from 415 to 1255 μmol $m^{-2}$ $s^{-1}$ (Fig. 4a, $p < 0.05$).

Initially, there were steep increases of ETR values in all leaves with increasing PPFD until 715 μmol $m^{-2}$ $s^{-1}$, after which they all decreased sharply (Fig. 4b). Significant differences in ETR values were obtained among leaves grown under different grown irradiances from PPFD of 605 to 1585 μmol $m^{-2}$ $s^{-1}$ ($p < 0.05$). Although the changes of ETR to increasing PPFD were similar in all leaves, the ETR of LL leaves was the lowest at any given PPFD followed by that of IL leaves. The HL leaves had the highest ETR compared to those of IL and LL from 605 to 1585 μmol $m^{-2}$ $s^{-1}$.

There were gradual increases in NPQ of IL and HL leaves from 1 to 835 μmol photon $m^{-2}$ $s^{-1}$ after which

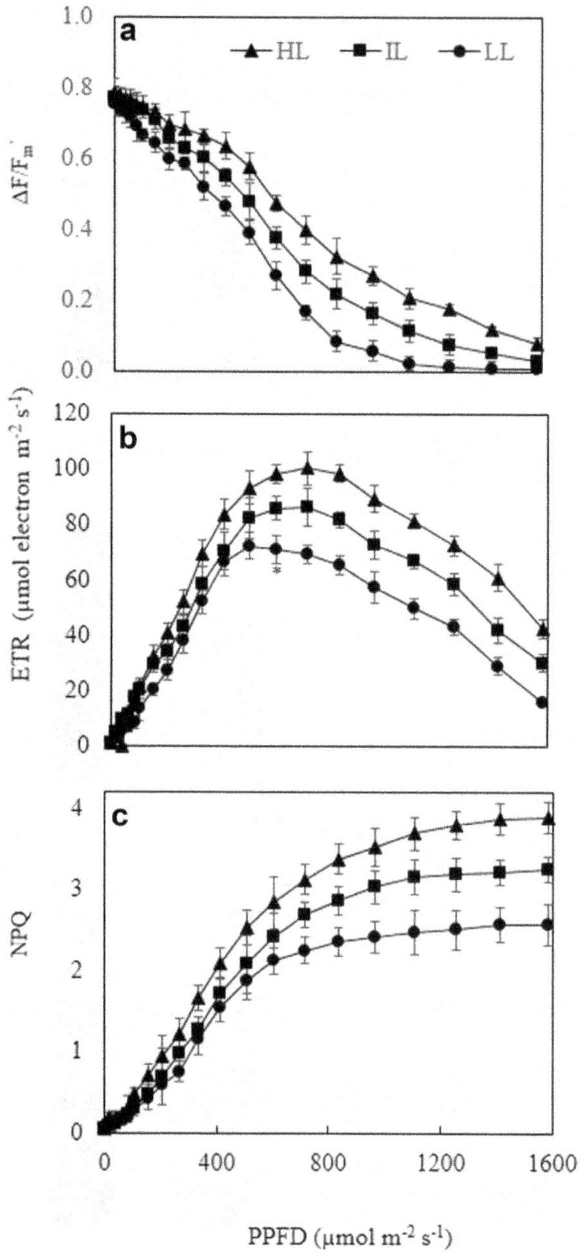


**Fig. 4** $\Delta F/F_m'$ ratio (**a**), ETR (**b**) and NPQ (**c**) of leaves of *G. speciosum* grown under different growth irradiances. Each value is the mean from 30 readings of 6 different leaves from 6 different plants. Vertical bars represent standard errors. When the standard error bars cannot be seen, they are smaller than the symbols

they plateau around 2.69 and 3.12, respectively (Fig. 3C). The increase of NPQ was faster in HL leaves than in IL at the any given PPFD from 715 to 1585 μmol photon $m^{-2}$ $s^{-1}$ ($p < 0.05$). For LL leaves, although its NPQ also increased with increasing PPFD, it plateaus at much lower PPFD of 605 μmol photon $m^{-2}$ $s^{-1}$ compared to those of IL and HL leaves. At each given PPFD, LL leaves had the significantly lowest NPQ that those of IL and HL leaves from PPFD of 715 μmol photon $m^{-2}$ $s^{-1}$ onwards ($p < 0.05$).

Photosynthetic pigments of leaves of *G. speciosum* grown under different growth irradiances were analyzed in both March and October. Similar trends were observed and thus, only results obtained from Oct are presented in Fig. 5. The total Chl content was significantly higher in LL leaves than those of IL and HL leaves (Fig. 5A, $p < 0.05$). However, there was no significant difference in Chl a/b ratio across the different growth irradiances (Fig. 5B, $p > 0.05$). Total Car content of LL and IL leaves was not significantly different ($p > 0.05$) but they were much lower than that of HL leaves (Fig. 5C, $p < 0.05$). Chl/Car ratios of LL and IL leaves were higher than that of HL (Fig. 5D, $p < 0.05$) due to their higher Chl content (Fig. 5A) or lower Car content (Fig. 5C).

#### Diurnal changes of PPFD, $g_s$ and leaf temperature and correlations among them

These measurements were carried out in March on three sunny days. Similar results were obtained among the three different sunny days. The values of PPFD (Fig. 6A) and leaf temperature (Fig. 6B) increased parallel from 0800 to 1400 h and then decreased after that. There were significant differences in PPFD (Fig. 6A), $g_s$ leaf temperature (Fig. 6B) and $g_s$ (Fig. 6C) across the different growth irradiances at 1000 h, when HL leaves had the highest values followed by those of IL leaves ($p < 0.05$). The LL leaves had the lowest leaf temperature and $g_s$ at the same given time compared to those of IL and HL leaves ($p < 0.05$). It was interesting to see that $g_s$ decreased from 1000 h onwards in all leaves. A close linear correlation between leaf temperature PPFD was established in leaves grown under different growth irradiance (Fig. 7a, $r^2 = 0.9106$). However, there were no correlations between $g_s$ and PPFD (Fig. 7b, $r^2 = 0.2757$), leaf temperature and $g_s$ (Fig. 7c, $r^2 = 0.1054$).

#### $A_{sat}$, $g_{s\,sat}$ and Leaf TRN

$A_{sat}$ of HL leaves was significantly higher than those of IL and LL leaves but not significantly different between IL and LL leaves (Fig. 8A, $p < 0.05$). However, there was no significant difference in the $g_{s\,sat}$ among all leaves (Fig. 8B, $p > 0.05$). For the leaf TRN concentration, there was no significant difference between HL and IL leaves (Fig. 8C, $p > 0.05$) but LL leaves had significant higher leaf TNR concentration compared to those of HL and IL leaves (Fig. 8C; $p < 0.05$). All results shown in Fig. 8 were obtained in October. $A_{sat}$ and $g_{s\,sat}$ were also measured in March and the trends were similar to those presented in Fig. 8A, B.

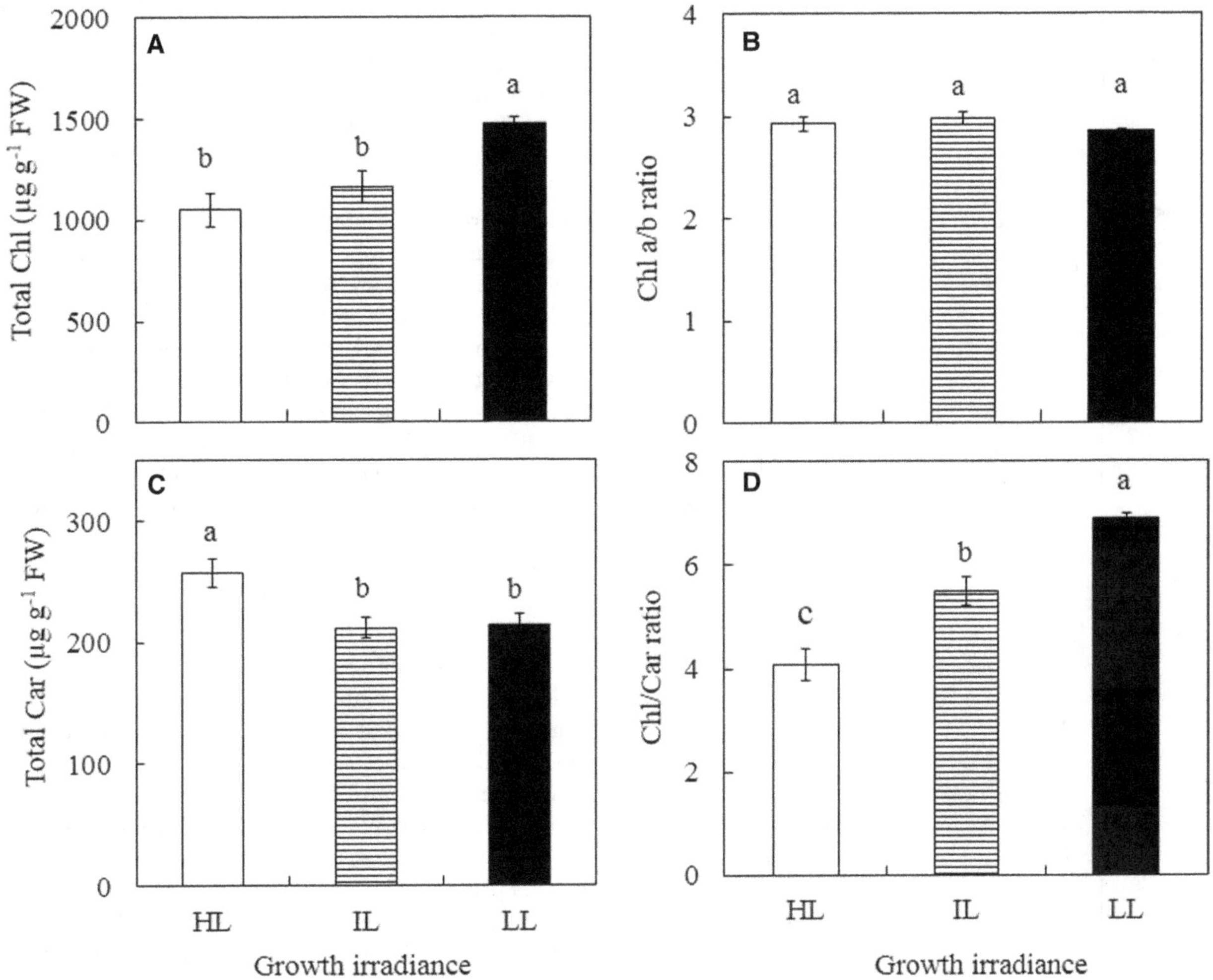


**Fig. 5** Total Chl content (**A**), Chl a/b ratio (**B**), total Car content (**C**) and Chl/Car ratio (**D**) of leaves of *G. speciosum* grown under different growth irradiances. Each value is the mean from 4 different leaves of 4 different plants. Vertical bars represent standard errors. Means with same letter above the bars are not statistically different ($p > 0.05$) as determined by Tukey's test

## Discussion

The light environments for different leaves on the same plants vary according to the position of the sun and variable cloud cover. Furthermore, every leaf on the same plant was in different light environment due to its position on the plant in relation to the other leaves. These factors explained the differences in the variation in leaf development on the same plant. The extension zone of the *G. speciosum*, being a monocot, was enclosed by older leaves sheath shielding it from light (Dale 1988). This could be an added constraint. There might also be a wide variation in growth rate for leaves on the same plant. Too much light can cause detrimental damages to the photosynthetic apparatus (Lambers et al. 2008). On the other hand, insufficient light could reduce photosynthetic rates and a subsequent reduction in overall plant growth (He et al. 1998; Koh et al. 1997; He and Teo 2007). In the present study, leaves grown under LL seemed to expand significantly faster (Fig. 1A) and thinner reflected by higher SLA (Fig. 1B) than leaves grown under IL and HL. Under shady conditions, to optimise photosynthetic efficiency, SLA of leaves increased, resulting in longer and thinner leaves (Dale 1988). For plants grown under HL and IL conditions, reducing the surface area of leaves, leading to lesser water loss by evapotranspiration could be a strategy for water conservation (Tay et al. 2015).

Water deficit is one of the greatest limitations to the growth of epiphytic orchids (Laube and Zotz 2003; Zotz et al. 2010). However, in the present study, the midday RWC of all leaves grown under three growth irradiances were more than 95% for both young expanding (Fig. 2a) and fully expanded (Fig. 2b) leaves indicating that this

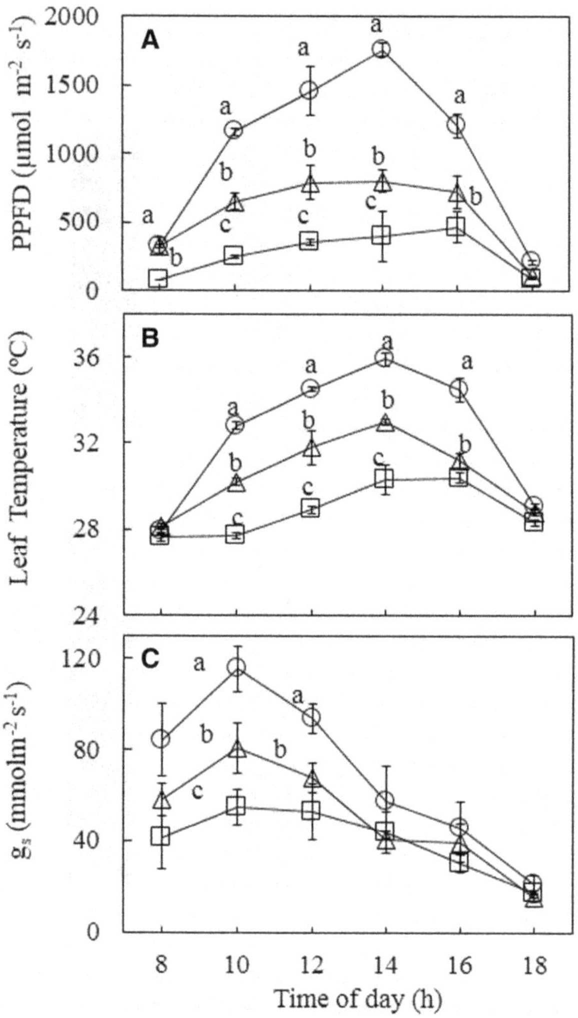


**Fig. 6** Diurnal changes in PPFD (**A**), leaf temperature (**B**) and $g_s$ (**C**) of leaves of *G. speciosum* grown under different growth irradiances. Each value is the mean from 6 different leaves of 6 different plants. Vertical bars represent standard errors. Means with same letter are not statistically different (p > 0.05) as determined by Tukey's test

orchid species did not suffer drought stress (Lizana et al. 2006; Lugojan and Ciulca 2011). Even though they were not irrigated, they could have a large supply of water from their pseudobulbs (Arditti 1992; He et al. 2011, 2014; Tay et al. 2015). Moreover, the study was carried out during the Northeast monsoon (October to November). The fat egg-shaped pseudobulbs of this tropical orchid store large amount of water during rainy season. Similar measurements were also carried out during the dry season (February to March). All leaves had more 95% midday RWC regardless of growth irradiances (data not shown). Stomatal regulation is one of the strategies preventing water loss (Yordanov et al. 2000). According to Cornic and Fresneau (2000), up to leaf RWC of about 70%, it is likely that stomatal closure plays the main role in preventing water loss (He et al. 2014; Tay et al. 2015). In the present study, decreased $g_s$ from 1000 h onwards (Fig. 6C) supported this postulation.

In this study, on sunny days, leaves grown under HL had much lower midday $F_v/F_m$ ratio than those under IL and LL (Fig. 3B). HL leaves, being in open field, were exposed to maximum PPFD of 2133 μmol $m^{-2}$ $s^{-1}$, which were much higher than those of IL and LL leaves (Fig. 3A). The lowest midday $F_v/F_m$ ratio due to higher PPFD indicated that dynamic PSII photoinhibition was much more severe in HL leaves than in IL and LL leaves. Photoinhibition could even occur at moderate or low light when other adverse conditions such as high temperature is present (He et al. 1996, 1998; He and Teo 2007). Leaf temperature (Fig. 6B) increased parallel with PPFD (Fig. 3A) from 0800 to 1400 h and the close linear correlation between leaf temperature and PPFD (Fig. 7a), suggested that dynamic PSII photoinhibition under tropical natural conditions resulted from the combination of high PPFD and high leaf temperature (He et al. 1996, 1998). It was reported that a large supply of water allowed leaves under all growth irradiances to modulate their leaf temperature, preventing them from overheating achieved through the evaporation of water from stomata known

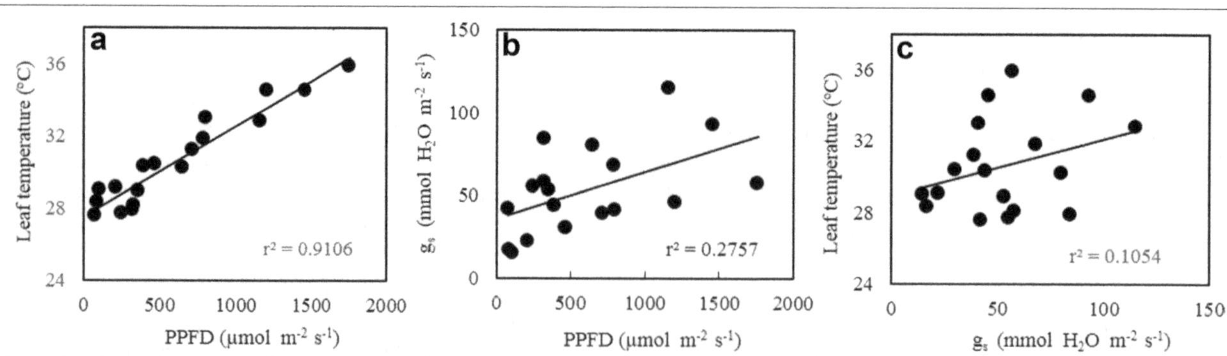


**Fig. 7** Correlations between leaf temperature and PPFD (**a**), $g_s$ and PPFD (**b**), leaf temperature and $g_s$ (**c**) of leaves of *G. speciosum* grown under different growth irradiances. All data derived from Fig. 6

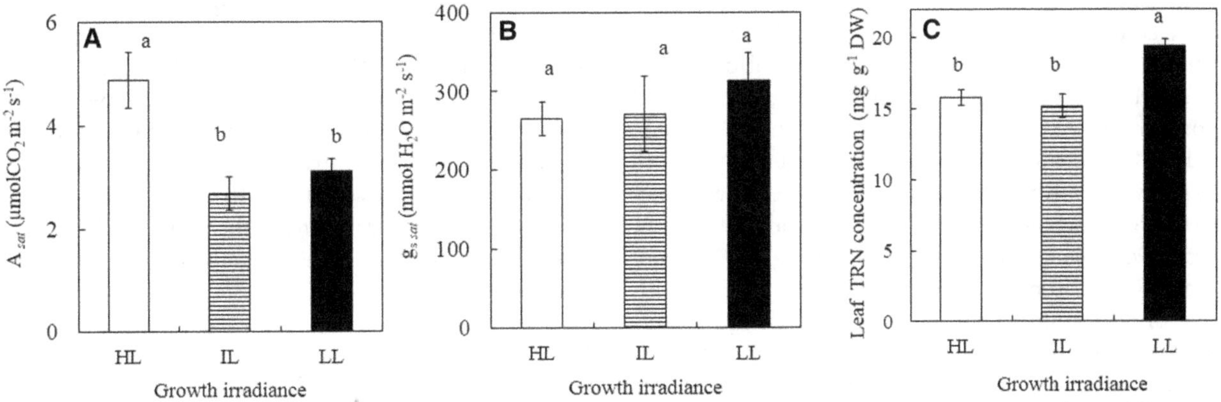


**Fig. 8** $A_{sat}$ (**A**), $g_{s\,sat}$ (**B**) and TRN (**C**) of leaves of *G. speciosum* grown under different growth irradiances. Each value is the mean from 4 different leaves of 4 different plants. Vertical bars represent standard errors. Means with same letter above the bars are not statistically different ($p > 0.05$) as determined by Tukey's test

transpiration (He et al. 2001; Mohotti and Lawlor 2002; Hetherington and Woodward 2003; Crawford et al. 2012) and thus, protecting their PSII from photoinhibitory damages (Jagtap et al. 1998; He et al. 2001). However, in the present study, $g_s$ (Fig. 6C) increased from 0800 to 1000 h with increasing PPFD (Fig. 6A) in all leaves and it did not increase further with increasing PPFD and leaf temperature from 1000 h onwards. This result implied that *G. speciosum* had developed the avoidance of drought stress by widely opening stomata only for a short period of time in the early morning to conserve water (Ort 2001). Thus, the results of the present study did not support strategy of transpiration cooling as stomata of all leaves were partially closed with decreased $g_s$ (Fig. 6C) under high temperatures during midday (Fig. 6B). Although there were no correlations between the stomatal conductance and PPFD (Fig. 7b); or leaf temperature (Fig. 7c); across the different irradiances and across the whole day, $g_s$ (Fig. 6C) increased from 0800 to 1000 h with increasing PPFD (Fig. 6A). Stomata remain closed in the afternoon and opening only in the morning may play the main role in preventing water loss for this epiphytic tropical orchid (He et al. 2014; Tay et al. 2015). Furthermore, other environmental conditions such as humidity (Grantz 1990; Peak and Mott 2011) and vapour pressure deficit (VPD) (Day 2000; Shirke and Pathre 2004) may play crucial roles in regulating stomatal conductance.

Stomatal closure during midday could result in potential depletion of internal $CO_2$. Low availability of $CO_2$ at high light intensities may lead to the decrease of photosynthetic electron consumption, causing a dynamic photoinhibition (Osmond 1994; Cornic and Fresneau 2000). In the present study, all leaves had experienced dynamic photoinhibition measured by midday $F_v/F_m$ ratios that were below 0.8 (Fig. 3B). However, all leaves completely recovered through the night from dynamic photoinhibition as they all had predawn $F_v/F_m$ ratios greater than 0.8 (Osmond 1994). Rapidly reversible decreases in maximal PS II efficiency (i.e. lowered $F_v/F_m$ ratios of pre-darkened leaves) is a photoprotective energy dissipation process, even though rates of photosynthetic electron transport remain maximal and photosynthesis is not inhibited (Adams et al. 1999). In the present study, leaves grown under HL had greater dynamic photoinhibition, these leaves, however, had higher $\Delta F/F_m'$ (Fig. 4a), ETR (Fig. 4b), and NPQ (Fig. 4c), compared to those grown under IL and LL exhibiting their higher capacities of utilizing and dissipating light energy (He and Lee 2004; He et al. 2014). When a proportionally greater amount of the absorbed light cannot be utilized in photochemical activity at higher light levels, leaves normally upgregulated their Car level to dissipate excess photons through the xanthophyll cycle (Adams and Demmig-Adams, 1992; Demmig-Adams and Adams III 1992, 2006). When leaves grown under HL, there was an accompanied increase in Car (Fig. 5C), decrease in Chl (Fig. 5A) and thus, higher ratio of Car/Chl (Fig. 5D) to offer photoprotection (Armstrong and Hearst 1996; Puthur 2005). Shading increased Chl content of LL grown leaves (Fig. 5A, Anderson et al. 1991; Newman and Follett 1998). Plants adapted to HL are well known to have high Chl a/b ratio (Anderson 1986; Anderson and Osmond 1987). However, no significant differences were observed in Chl a/b ratios among all leaves. The low Chl content in leaves of HL plants exposing to maximal PPFD above 1500 μmol $m^{-2}$ $s^{-1}$ could be a result of adaptation (Anderson 1986).

Photosynthesis of healthy leaves increases proportionally to the increases in PPFD until its rate begins to saturate. Increases of photosynthetic $CO_2$ assimilation rate with increasing PPFD for *G. speciosum* grown under all light conditions were very slow and they all began to show saturation under a PPFD about 600 μmol photon $m^{-2}$ $s^{-1}$ (data not shown). Using tropical crop species, Da Matta and colleagues (2001) studied the actual photosynthetic rate ($A$). They reported that $A$, determined under non-limiting light at ambient temperature and $CO_2$, varied from 5.0 up to 26.3 μmol $CO_2$ $m^{-2}$ $s^{-1}$. It was also reported previously, on an area basis, $A$ covers a wide range, from less than 2 up to 70 μmol $CO_2$ $m^{-2}$ $s^{-1}$ (Larcher 1995) and may even higher with a value of 80 μmol $CO_2$ $m^{-2}$ $s^{-1}$ from *Amaranthus retroflexus* (Pearcy and Ehleringer 1984). In the present study, $A_{sat}$ (Fig. 8A) was higher in leaves grown under HL than under IL and LL from *G. speciosum*, a tropical native $C_3$ orchid species. Higher leaf photosynthetic rate normally parallels its higher stomatal conductance at either high or low VPD environments (Shirke and Pathre 2004). In this study, $g_{s\,sat}$ did not seem to increase with increasing growth irradiance (Fig. 8B). VPD could affect the internal $CO_2$ concentration ($C_i$) however, all leaves had similar $C_i$ with average values around 280–300 mol $CO_2$ $mol^{-1}$ between 0900 and 1030 h (data not shown). $A_{sat}$ of 5 μmol $CO_2$ $m^{-2}$ $s^{-1}$ (Fig. 8A) from HL grown *G. speciosum* leaves was much lower than most tropical $C_3$ plants. The differences in $A$ among species depend on the proportion of the absorbed excitation energy that is used for photosynthesis (Adams and Demmig-Adams 1992; Adams et al. 2002; Demmig-Adams and Adams 1992). On other hand, N availability is a major factor limiting growth and development of plants (Kraiser et al. 2011). It is a key component for protein synthesis and for the maintenance of the abundant proteins associated with the photosynthetic apparatus (Loebl et al. 2010). Although the LL grown leaves had higher total leaf TRN compared to HL and IL plants (Fig. 8C), none of them was suffered from N deficiency as all plants that had more than 1.5% N without supplying fertilizers to plants, during the experiment period (12 months after transplanting). Lower $A_{sat}$ was not caused by degradation of Rubisco either as all leaves had similar levels of Rubisco protein (data not show). Based on the above discussion, low photosynthetic rate and slow growth of *G. speciosum* could be mainly due to the limitation of internal $CO_2$ concentration resulting from closing stomata for a long period of time during the day to conserve water (Ort 2001).

## Conclusion

Leaves grown under LL expanded faster and they were thinner than those grown under IL and HL. Water deficit was not observed in any leaves as they could avoid drought stress by opening their stomata only during early morning and closing or partially closing them from midday onwards. Although HL grown leaves experienced dynamic photoinhibition but not chronic photoinhibition as these leaves had higher capacities of utilizing and dissipating light energy. Low photosynthetic rate of all leaves of *G. speciosum* was mainly due to the limitation of internal $CO_2$ concentration instead of N deficiency. Slow but healthy growth of all *G. speciosum* under natural conditions was due to their different physiological acclimations to different growth environments.

### Abbreviations

$A_{sat}$: light-saturated photosynthetic $CO_2$ assimilation; Car: carotenoids; Chl: chlorophyll; DW: dry weigh; ETR: electron transport rate; $F_o$: minimal fluorescence yield of a "dark-adapted" sample; $F_m$ and $F_v$: maximal and variable fluorescence yields obtained from a dark-adapted sample upon application of a saturation pulse of radiation, respectively; $F_m'$: the maximum fluorescence at the steady state; $F_t$: the current fluorescence yield; $F/F_m'$: the effective photochemical quantum yield; $F_v/F_m$: ratio to estimate the maximum portion of absorbed quanta used in PSII reaction centers; FW: fresh weight; HL: high light; IL: intermediate light; LL: low light; $g_s$: stomatal conductance; $g_{s\,sat}$: light-saturated stomatal conductance; NPQ: non-photochemical quenching; PPFD: photosynthetic photon flux density; PS II: photosystem II; SLA: specific leaf area; RWC: relative water content; TRN: total reduced nitrogen.

### Authors' contributions

JH and TWY proposed the project based on TWY's project on "Orchid conservation in Singapore". JH and TWY carried out the first field study. JH supervised the laboratory analysis and repeated certain analysis. She also plotted all figures and wrote the manuscript. RMPL and SHJD, as MSc students carried out the experiments designed by HJ and TWY and analyzed the data. TWY planted all materials and also corrected the manuscript. All authors read and approved the final manuscript.

### Author details

[1] Natural Sciences and Science Education Academic Group, National Institute of Education, Nanyang Technological University, 1 Nanyang Walk, Singapore 637 616, Singapore. [2] Singapore Botanic Gardens, National Parks Board, 1 Cluny Road, Singapore 25956, Singapore.

### Acknowledgements

This project was supported by the teaching vote, Ministry of Education, Singapore. We also thank Singapore Botanic Gardens and National Parks Board for providing the plant materials.

### Competing interests

The authors declare that they have no competing interests.

### Funding

This project was supported by the teaching vote, Ministry of Education, Singapore.

### References

Adams WW III, Demmig-Adams B (1992) Operation of the xanthophyll cycle in higher plants in response to diurnal changes in incident sunlight. Planta 186:390–398

Adams WW III, Demmig-Adams B, Logan BA, Barker DH, Osmond CB (1999) Rapid changes in xanthophyll cycledependent energy dissipation and photosystem II efficiency in two vines, *Stephania japonica* and *Smilax australis*, growing in the understory of an open *Eucalyptus* forest. Plant Cell Environ 22:125–136

Adams WW III, Demmig-Adams B, Rosenstiel TN, Brightwell AK, Ebbert V (2002) Photosynthesis and photoprotection in overwintering plants. Plant Biol 4:545–557

Adams WW III, Zarter CR, Mueh KE, Amiard V, Demmig-Adams B (2006) Energy dissipation and photoinhibition: a continuum of photoprotection. In: Demmig-Adams B, Adams WW, Mattoo AK (eds) Photoprotection, photoinhibition, gene regulation, and environment. Springer, The Netherlands, pp 49–64

Anderson JM (1986) Photoregulation of the composition, function, and structure of thylakoid membranes. Annu Rev Plant Physiol 37:93–136

Anderson JM, Osmond CB (1987) Shade-sun responses: compromises between acclimation and photoinhibition. In: Kyle DJ, Osmond CB, Arntzen CJ (eds) Photoinhibition. Elsevier, Amsterdam, pp 1–38

Anderson PC, Norcini JG, Knox GW (1991) Influence of irradiance on leaf physiology and plant growth characteristics of *Rhododendron x* 'Pink Ruffles'. J Amer Soc Hort Sci 116:881–887

Angus JF, Bowden JW, Keating BA (1993) Modelling nutrient responses in the field. Plant Soil 156:57–66

Arditti J (1992) Fundamentals of orchid biology. Wiley, New York

Armstrong GA, Hearst JE (1996) Carotenoids 2: genetics and molecular biology of carotenoid pigment biosynthesis. FASEB J 10:228–237

Ashraf M (2009) Biotechnological approach of improving plant salt tolerance using antioxidants as markers. Biotech Adv 27:84–93

Barber J (1995) Molecular-basis of the vulnerability of photosystem-II to damage by light. Aust J Plant Physiol 22:201–208

Bota J, Medrano H, Flexas J (2004) Is photosynthesis limited by decreased Rubisco activity and RuBP content under progressive water stress? New Phytol 162:67–681

Chow WS (1994) Photoprotection and photoinhibitory damage. Adv Mol Cell Biol 10:15–196

Cornic G, Fresneau C (2000) Photosynthetic carbon reduction and carbon oxidation cycles are the main electron sinks for photosystem II activity during a mild drought. Ann Bot 89:887–894

Crawford AJ, Mclachlan DH, Herington AM, Franklin KA (2012) High temperature exposure increases plant cooling capacity. Curr Biol 22:R396–R397

Da Matta FM, Loos RA, Rodrigues R, Barros RS (2001) Actual and potential photosynthetic rates of tropical crop species. Rev Bras Fisiol Veg 13:24–32

Dale JE (1988) The control of leaf expansion. Ann Rev Plant Physiol Plant Mol Biol 39:267–295

Day MD (2000) Influence of temperature and leaf-to-air vapor pressure deficit on net photosynthesis and stomatal conductance in red spruce (*Picea rubens*). Tree Physiol 20:57–63

Demmig-Adams B, Adams WW III (1992) Photoprotection and other responses of plants to high light stress. Ann Rev Plant Physiol Plant Mol Biol 43:599–626

Demmig-Adams B, Adams WW III (2006) Photoprotection in an ecological context: the remarkable complexity of thermal energy dissipation. New Phytol 172:11–21

Farquhar GD, Sharkey TD (1982) Stomatal conductance and photosynthesis. Ann Rev Plant Physiol Plant Mol Biol 33:317–345

Flexas J, Ribas-Carbó M, Bota J, Galmés J, Henkle M, Martínez-Cañellas S, Medrano H (2006) Decreased Rubisco activity during water stress is induced by stomatal closure, not by decreased relative water content. New Phytol 172:73–82

Gomes FP, Oliva MA, Mielke MS, Almeida AAFD, Leite HG, Aquino LA (2008) Photosynthetic limitations in leaves of young Brazilian green dwarf coconut (*Cocos nucifera L.*'nana') palm under well-watered conditions or recovering from drought stress. Environ Exp Bot 62:195–204

Grantz DA (1990) Plant response to atmospheric humidity. Plant Cell Environ 13:667–679

He J, Lee SK (2004) Photosynthetic utilization of radiant energy by temperate lettuce grown under natural tropical condition with manipulation of root-zone temperature. Photosynthetica 42:457–463

He J, Teo LCD (2007) Susceptibility of CAM *Dendrobium* Burana Jade green leaves and green flower petals to high light under tropical natural conditions. Photosynthetica 45:214–221

He J, Chee CW, Goh CJ (1996) "Photoinhibition" of *Heliconia* under natural tropical conditions—importance of leaf orientation for light interception and leaf temperature. Plant Cell Environ 19:1238–1248

He J, Kooh GH, Hew CS (1998) Susceptibility of CAM *Dendrobium* leaves and flowers to high light and high temperature under natural tropical conditions. Environ Exp Bot 40:255–264

He J, Lee SK, Dodd IC (2001) Limitations to photosynthesis of lettuce grown under tropical conditions: alleviation by root-zone cooling. J Exp Bot 52:1323–1330

He J, Ouyang W, Chia TF (2004) Growth and photosynthesis of virus-infected and virus-eradicated orchid plants exposed to different growth irradiances under natural tropical conditions. Physiol Plant 121:612–619

He J, Tan BHG, Qin L (2011) Source-to-sink relationship between green leaves and green pseudobulbs of $C_3$ orchid in regulation of photosynthesis. Photosynthetica 49:209–218

He J, Teh ZY, Yam TW (2014) Orchid conservation in Singapore under natural conditions: responses of *Grammatophyllum speciosum* to growth irradiances. Plant Sci Interna 1:11–23

Hetherington AM, Woodward I (2003) The role of stomata in sensing and driving environmental changes. Nature 424:901–908

Jagtap V, Bhargava S, Streb P, Feierabend J (1998) Comparative effect of water, heat and light stresses on photosynthetic reactions in *Sorghum bicolor* (L.) Moench. J Exp Bot 49:1715–1721

Jones HG (1998) Stomatal control of photosynthesis and transpiration. J Exp Bot 49:387–398

Koh GH, He J, Hew CS (1997) Photosynthetic light utilization by CAM *Dendrobium* flowers. Photosyntetica 34:367–376

Kraiser T, Gras DE, Gutiérrez AG, González B, Gutiérrez RA (2011) A holistic view of nitrogen acquisition in plants. J Exp Bot 62:1455–1466

Lambers H, Chapin FS III, Pons TL (2008) Plant physiological ecology. Springer, The Netherlands, pp 75–81

Larcher W (1995) Physiological plant ecology. Springer, Berlin

Laube S, Zotz G (2003) Which abiotic factors limit vegetative growth in a vascular epiphyte? Funct Ecol 17:598–604

Lawlor DW, Cornic G (2002) Photosynthetic carbon assimilation and associated metabolism in relation to water deficits in higher plants. Plant Cell Environ 25:275–294

Lawlor DW, Tezara W (2009) Causes of decreased photosynthetic rate and metabolic capacity in water-deficient leaf cells: a critical evaluation of mechanisms and integration of processes. Ann Bot 103:561–579

Lichtenthaler HK, Burkart S (1999) Photosynthesis and high light stress. Bulg J Plant Physiol 25:3–16

Lichtenthaler HK, Wellburn AR (1983) Determinations of total carotenoids and chlorophylls *a* and *b* of leaf extracts in different solvents. Biochem Soc Trans 11:591–592

Lizana C, Wentworth M, Martinez JP, Villegas D, Meneses R, Murchis EH, Pastenes C, Lercari B, Vernieri P, Horton P, Pinto M (2006) Differential adaptation of two varieties of common bean to abiotic stress. Effects of drought on yield and photosynthesis. J Exp Bot 57:685–697

Loebl M, Cockshutt AM, Campbell DA, Finkel ZV (2010) Physiological basis for high resistance to photoinhibition under nitrogen depletion in *Emiliania huxleyi*. Limnol Oceanogr 55:2150–2160

Lugojan C, Ciulca S (2011) Evaluation of relative water content in winter wheat. J Horti For Biotech 15:173–177

Mohotti AJ, Lawlor DW (2002) Diurnal variation of photosynthesis and photoinhibition in tea: effects of irradiance and nitrogen supply during growth in the field. J Exp Bot 53:313–322

Newman SE, Follett MW (1998) Irrigation frequency and shading influences on water relations and growth of container-growth *Euonymus japonica* 'Aureo-marginat'. J Environ Horti 6:96–100

Ng PKL (1994) The Singapore red data book: Threatened plants and animals of Singapore, A Community Service Project by Asia Pacific Breweries. Nature Society, Singapore

Nishiyama Y, Allakhverdiev S, Murata N (2006) A new paradigm for the action of reactive oxygen species in the photoinhibition of photosystem II. BBA Bioenerg 1757:742–749

Ort DR (2001) When there is too much light. Plant Physiol 125:29–32

Osmond CB (1994) What is photoinhibition? Some insights from comparisons of shade and sun plants. In: Baker NR, Bowyer, JR (eds) Photoinhibition of photosynthesis in the field. Oxford: Bios Scientific Publishing Oxford, pp 1–24

Peak D, Mott KA (2011) A new, vapour-phase mechanism for stomatal responses to humidity and temperature. Plant Cell Environ 34:162–178

Pearcy RW, Ehleringer J (1984) Comparative ecophysiology of $C_3$ and $C_4$ plants. Plant Cell Environ 7:1–13

Powles SB (1984) Photoinhibition of photosynthesis induced by visible-light. Ann Rev Plant Physiol Plant Mol Biol 35:15–44

Puthur J (2005) Influence of light intensity on growth and crop productivity of *Vanilla planifolia* Andr. Gen Appl Plant Physiol 31:215–224

Ragni M, Airs R, Leonardos N, Geider R (2008) Photoinhibition of PSII in *Emiliania huxleyi* (*Haptophyta*) under highlight stress: the roles of photoacclimation, photoprotection and photorepair. J Phycol 44:670–683

Rascher U, Liebig M, Lüttge U (2000) Evaluation of instant light-response curves of chlorophyll fluorescence parameters obtained with a portable chlorophyll fluorometer on site in the field. Plant Cell Environ 23:1397–1405

Schreiber U, Gademann R, Ralph PJ, Larkum AWD (1997) Assessment of photosynthetic performance of Prochloron in *Lissoclinum patella* by in situ and in hospite chlorophyll fluorescence measurements. Plant Cell Physiol 38:945–951

Shirke PA, Pathre UV (2004) Influence of leaf-to-air vapour pressure deficit (VPD) on the biochemistry and physiology of photosynthesis in *Prosopis juliflora*. J Exp Bot 55:2111–2120

Tay S, He J, Yam TY (2015) Photosynthetic light utilization efficiency, water relations and leaf growth of $C_3$ and CAM tropical orchids under natural conditions. Am J Plant Sci 6:2949–2959

Wellburn AR (1994) The spectral determination of chlorophylls A and B, as well as total carotenoids, using various solvents with spectrophotometers of different resolution. J Plant Physiol 144:307–313

Yam TW (2013) Native orchids of Singapore. Division, identification and conservation. National Parks Board, Singapore

Yam TW, Chua J, Tay F, Ang P (2010) Conservation of the native orchids through seedling culture and reintroduction—A Singapore experience. Bot Rev 76:263–274

Yam TW, Tay F, Ang P, Soh W (2011) Conservation and reintroduction of native orchids of Singapore—the next phase. Eur J Environ Sci 1:27–38

Yordanov I, Velikova V, Tsonev T (2000) Plant responses to drought, acclimation, and stress tolerance. Photosynthetica 38:171–186

Zotz G, Bogusch W, Hietz P, Ketteler N (2010) Growth of epiphytic bromeliads in a changing world: the effects of CO2, water and nutrient supply. Acta Oecol 36:659–665

7

# Photosynthetic gas exchange responses of *Swietenia macrophylla* King and *Melia azedarach* L. plantations under drought conditions

Hong-Chyi Jhou[1], Ya-Nan Wang[2], Chung-Shien Wu[3], Jui-Chu Yu[1] and Chung-I Chen[2*]

**Abstract**

**Background:** The environmental stresses caused by climate change have become more severe in recent decades, affecting tree growth and physiology. Tropical forests have great potential for global carbon sequestration. However, they suffer from heavy rainfall and prolonged dry periods due to climate change. *Swietenia macrophylla* King and *Melia azedarach* L. are economically valuable trees that are widely planted in southern Taiwan. Plantations are exposed to either prolonged dry periods or heavy rainfall within the seasons of tropical monsoon areas. Photo-physiological comparisons may provide information that can improve management of *S. macrophylla* and *M. azedarach* plantations in tropical regions.

**Results:** Both species exhibited a midday depression in leaf photosynthesis regardless of the season. The net photosynthetic rate ($P_N$), stomatal conductance ($g_s$), and transpiration rate ($E$) in the dry season all significantly decreased in both tree species. In addition, *M. azedarach* used water more efficiently than did *S. macrophylla* during the dry season, but *S. macrophylla* had higher $P_N$ compared with that in *M. azedarach* during the wet season. Temperature and vapor pressure deficit influenced $P_N$ variation in *S. macrophylla* and *M. azedarach*, respectively. Our data suggested that the $P_N$ and $g_s$ of *M. azedarach*, but not of *S. macrophylla*, were linearly correlated during the dry season. The reduction of the leaf area was more sever in *M. azedarach* than in *S. macrophylla*, thus preventing water loss more efficiently.

**Conclusions:** *M. azedarach* adapted to drought by reducing total leaf area and maintaining higher $P_N$, $g_s$, $E$, and WUE compared with those measured in *S. macrophylla* during the dry season. *M. azedarach* is more drought adaptation and more suitable for both humid and semi-humid areas than *S. macrophylla*, whereas the latter should be limited to more humid areas.

**Keywords:** Photosynthesis, Prolonged drought, Soil water content, Stomatal conductance, Transpiration rate, Vapor pressure deficit

## Background

Global climate change has increased the incidence and severity of extreme climate events such as droughts and high temperatures (IPCC 2015), which strongly affect plant physiology and growth. Numerous reports have emphasized that water stress negatively affects gas exchange, which often results in lower leaf photosynthesis, stomata conductance, and transpiration (Ogaya and Peñuelas 2003; Leuzinger et al. 2005; Guerfel et al. 2009; Li et al. 2016). Decreased leaf photosynthesis can be caused by either stomatal or non-stomatal limitations (Taiz and Zeiger 2002). Therefore, the measurement of gas exchange in plants could reflect the physiological response to environmental conditions and the relationship between gas exchange and environmental variables

*Correspondence: rk.aee@msa.hinet.net
[2] Department of Forestry and Resource Conservation, National Taiwan University, 10617 Taipei, Taiwan
Full list of author information is available at the end of the article

could provide a more detailed understanding of how environmental variables limit photosynthesis, growth performance, and biomass (Ishida et al. 1996).

Tropical forests play an important role in the global carbon cycle because they account for a large part of terrestrial net primary productivity, while covering a lesser part of the global land surface (Slot and Winter 2017). Tropical areas experience wet seasons (in the summer) with heavy rainfall and prolonged dry seasons (in the winter), which can strongly affect plant distribution and growth (Engelbrecht et al. 2006). Taiwan's subtropical climate is characterized by an average annual rainfall of > 2500 mm, with heterogeneous distribution across the seasons and geographical areas. In particular, southern Taiwan has had increasingly frequent high temperatures and rainstorms followed by extended droughts. These conditions greatly affect tree growth in the area. To better understand how these changing environmental factors affect plant growth in this region, we estimated gas-exchange variables in two Meliaceae tree species that are widely planted in southern Taiwan: *Swietenia macrophylla* King and *Melia azedarach* L. The semi-deciduous *S. macrophylla* originated from Central and South America, and it is a major forestation species of high economic value in the Taiwanese lumber industry (Grogan et al. 2005). Although generally characterized by shade intolerance, *S. macrophylla* seedlings are able to grow in the forest understory (Grogan et al. 2005; Cordeiro et al. 2009). The deciduous *M. azedarach*, a native to Taiwan and South China, is fast-growing and shade-intolerant, with leaves and fruits that contain insecticidal compounds. The species is also harvested for its high-quality wood (Kuo et al. 2004; Liao et al. 2004), making it a popular species to cultivate. Both species have shallow root systems that are characteristic of many tropical trees (Toky and Bisht 1992; Dünisch et al. 2003).

Several researches demonstrated that different species adopt different strategies in response to seasonal environmental conditions (Sun et al. 2011; Koller et al. 2013; Arndt et al. 2015). For example, gas exchange rate, water use efficiency, chlorophyll fluorescence and seasonal defoliation fluctuate among seasons and species. Thus, preliminary observations revealed that both species exhibited stronger growth than other species in the study site in southern Taiwan (Chen et al. 2013, 2016a), but still showed a poor growth performance compared with those cultivated in other regions in Taiwan, Brazil, and China. The aim of the present study was to assess the vulnerability of both commercial species to environmental stresses. We hypothesized that the two species may adopt different strategies to adapt to water stress that affect the growth and physiology of plantations under field conditions. Rather than performing a single, instantaneous measurement, we chose to chart the daily and seasonal variation in gas exchange because time-course data provide more reliable information on photosynthetic productivity (Yin et al. 2006). Our results will contribute information that can improve management of *S. macrophylla* and *M. azedarach* plantations in tropical regions.

## Methods

### Plant material and growing conditions

The experimental plantation used for this study is located on the Wan-Long Farm, which belongs to the Pingtung operation branch, Taiwan Sugar Corporation in Sinpi Township, Pingtung, Taiwan (120°36′30″ E, 22°31′26″ N; 69 m above sea level). The total area of the plantation in the farm is approximately 291 ha, and the soil consists of a sandy loam. The study site experiences a typical tropical monsoon climate, with a high frequency of typhoons and late thundershowers in the summer and little precipitation in other seasons. The mean monthly daytime air temperatures for March, June, September, and December in 2010 were 25.1, 29.1, 28.6, and 23.1 °C, respectively; they were measured at a meteorological station located 50 m west of the experimental site. The annual precipitation from 2005 to 2010 was estimated to be 2300–3800 mm, with the majority of the rainfall concentrated in the months between May and September. For example, 92% of the total annual precipitation in 2010 happened between May and September, and the dry season lasted for over half a year (accumulated rainfall between October 2009 and April 2010 was only ~ 70 mm). Based on the climate diagrams of this site, the wet season was defined to last from May to September and the dry season in the remaining months.

At the experimental plantation, 14 broadleaved tree species were planted between 2002 and 2005, and *S. macrophylla* and *M. azedarach* were planted in 2002. Irrigation and fertilization were applied in the first few years after afforestation. Additionally, pruning and mowing were also applied annually in late summer. Generally, *M. azedarach* defoliates in February and sprouts new leaves and buds in March, whereas *S. macrophylla* defoliates partially in March. The trees were planted in the same plot, which was subdivided by species. For the experiment, we selected trees located at a subplot boundary so that the two species were side-by-side with 2.5 m distance between each other. The average diameter at breast height (DBH) and total height of *S. macrophylla* and *M. azedarach* in the experimental plot are presented in Table 1. In 2009, *S. macrophylla* and *M. azedarach* stand densities were 1111 and 1420/ha, and in 2010, they were 1070 and 1413/ha, respectively. The survival rates of *S. macrophylla* and *M. azedarach* on the plantation were 96.3 and 99.5%, respectively.

**Table 1 Average values (± standard error) of diameter at breast height (DBH), height (H), and number of samples (N) for *Swietenia macrophylla* and *Melia azedarach* in 2010**

| | *S. macrophylla* | *M. azedarach* |
|---|---|---|
| DBH (cm) | 9.90 ± 2.88b | 11.86 ± 4.14a |
| H (m) | 9.12 ± 1.96a | 9.43 ± 1.90a |
| N | 69 | 104 |

Means followed by different lowercase letters differ significantly between seasons at $P < 0.05$

Leaf area index (LAI) is a dimensionless quantity variable of the forest canopy defined as the one-sided leaf surface area per ground surface area. Monthly measurements of the LAI per species were taken simultaneously for the top and under canopy of three sample trees at dusk using a plant canopy analyzer (LAI-2200, LI-COR, Lincoln, NE, USA).

### Measurement of gas-exchange variables

Diurnal gas exchange was measured once a month in 2010. We measured a group of ambient variables [photosynthetic photon flux densities (PPFD), leaf-air vapor pressure deficit ($VPD_l$), leaf temperature ($T_l$), ambient $CO_2$ concentration ($CO_2$)] that could influence diurnal gas exchange and estimated physiological variables [net photosynthetic rate ($P_N$), stomatal conductance ($g_s$), transpiration rate ($E$), intercellular $CO_2$ concentration ($C_i$), water use efficiency (WUE)] on four dates within a typical season during 2010: late dry season (27 March), early wet season (26 June), late wet season (25 September), and early dry season (30 December). Data were recorded in the field using two portable photosynthesis systems (LI-6400-sun-sky, LI-COR, USA). For all measurements, the leaves were placed in in the chamber (LI-6400-08), using a clump-on leaf cuvette; the chamber was covered with transparent film at the top to ensure that ambient light of equal intensity reached the leaves. WUE was estimated as $P_N/E$.

Gas-exchange variables were measured in three fully expanded leaves per sampled tree under direct sunlight each hour from 08:00 to 16:00, on the same day in the three sampled trees per species. Because of the limited time available for each measurement, we sampled three nearby trees per species in the same plot. The ambient air temperature was recorded during each sampling, as was the ambient $CO_2$, which was approximately 394, 376, 389, and 395 ppm on March 27, June 26, September 25, and December 30, 2010, respectively. The average daytime PPFD above canopy was 1739 ± 667, 1762 ± 800, 1520 ± 856, and 1413 ± 528 μmol photons/$m^2$/s in March, June, September, and December, 2010, respectively. The light saturation points of *S. macrophylla* during winter, spring, summer, and autumn were 500, 750, 1000, and 500 μmol photons/$m^2$/s, respectively, and in *M. azedarach*, they were 500, 500, 1200, and 500 μmol photons/$m^2$/s during the same respective seasons. These values were estimated by measuring the seasonal light response curve in 2011 (data not shown). Soil water content (SWC) and relative humidity (RH) were continuously recorded at a meteorological station 50 m west of the sampled tree plantation using a time-domain reflectometer (CS616; Campbell Scientific Inc., Logan, UT, USA) and a relative humidity probe (HMP45C; Vaisala, Finland). The soil water content device was placed 20 cm underground and the humidity probe was set 9 m aboveground. According to the previous research conducted in this site, the soil profile was shallow (< 40 cm depth) (Cheng et al. 2016). The optimum temperature of photosynthesis is generally between 25 and 30 °C. Thus, we set 30 °C as the break point.

### Data analysis

Collected data were analyzed in *SAS 9.3* (SAS Institute Inc., Cary, NC, USA). Pearson's correlation was used to examine the relationship between photosynthetic variables and environmental factors. Duncan's multiple range tests were run a posteriori to test the difference among the seasons and species. Significance was considered at $P < 0.05$ and 0.01. Regression and curve-fitting analyses were performed using SigmaPlot 12.0 (Systat Software Inc, San Jose, CA, USA). The relation among gas exchange parameters was analyzed using the monthly measurement data. Data are presented as mean ± standard error.

## Results

### Environmental conditions

The annual rainfall was 2848.5 mm in 2010, and the average annual rainfall was 2764.5 mm during the 2008–2010 period. The recorded rainfall varied considerably in 2010 from a maximum of 1178.5 mm in September to a minimum of 2.5 mm in March. A typhoon was the primary cause of the high rainfall volume in September. The accumulated rainfall from May to September accounted for more than 92% of the annual precipitation, confirming that these months constituted the wet season (Fig. 1). In addition, the fluctuations in SWC and RH also reflected the rainfall volume. Although the precipitation was mainly concentrated between May and September, the SWC and RH were increased from May to October represented the pattern of time lag. For example, both SWC and RH were the highest in September but then decreased drastically in November and December when the precipitation was approximately zero.

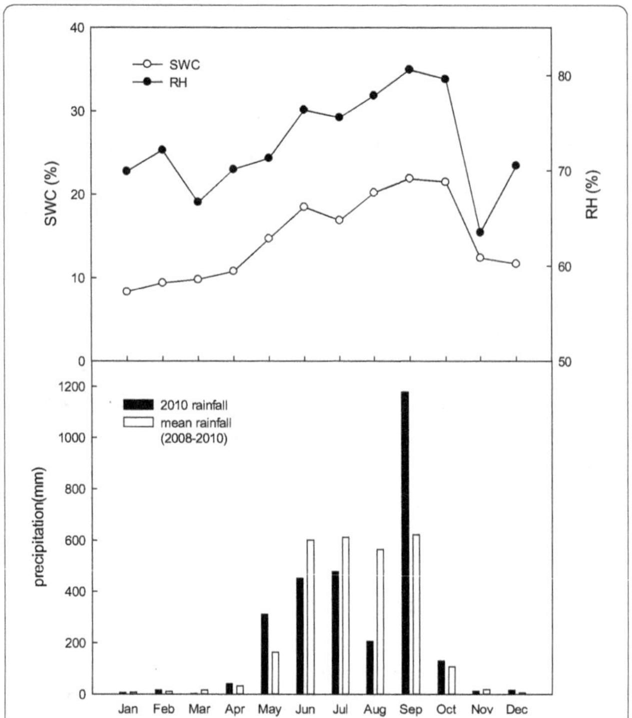


**Fig. 1** Monthly changes in soil water content (SWC), relative humidity (RH), and rainfall in 2010, and the mean rainfall between 2008 and 2010

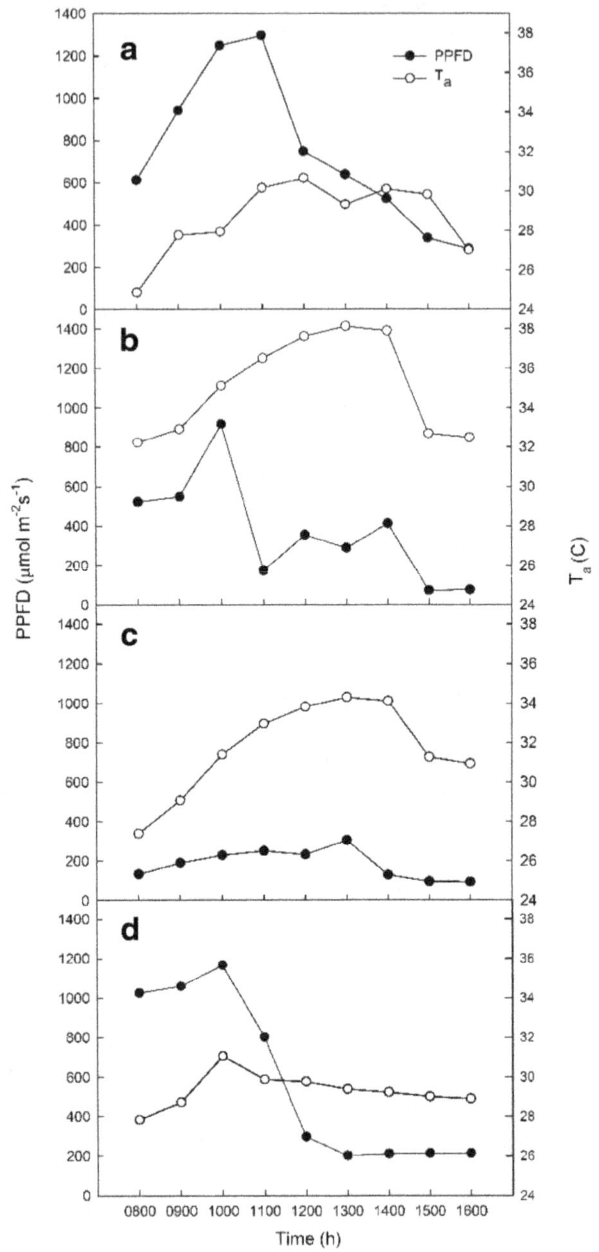


**Fig. 2** Diurnal variation in photosynthetic photon flux density (PPFD) and air temperature ($T_a$) in 2010: late dry season (27 March; **a**), early wet season (26 June; **b**), late wet season (25 September; **c**), and early dry season (30 December; **d**)

Across all seasons, PPFD gradually increased throughout the morning and then decreased after noon. March and December air temperatures ($T_a$) peaked at 30.7 °C (12:00) and 31.1 °C (10:00), respectively. The highest $T_a$ of the year (38.1 °C) occurred in June at 13:00, but the highest PPFD was measured in September at 13:00, when the maximum $T_a$ was 34.3 °C (Fig. 2). The highest $VPD_a$ for each season was 3.31 kPa at 11:00, 4.22 kPa at 11:00, 3.88 kPa at 13:00, and 3.38 kPa at 11:00 on 27 March, 26 June, 25 September, and 30 December, respectively.

## Gas exchange in leaves

Overall, the $g_s$ and $E$ values of both species were higher in the wet season than in the dry season. Interestingly, the two species had distinct $P_N$ values across all seasons. In the dry season, the *S. macrophylla* $P_N$ was low, but in the wet season, it increased by nearly fourfold from 08:00 to 13:00, and then decreased drastically (Fig. 3). In *M. azedarach*, however, the seasonal variation of $P_N$ was indistinct (Fig. 3). As a result, *M. azedarach* had higher $P_N$ than *S. macrophylla* in the dry season, but lower $P_N$ in the wet season.

The two species had similar $g_s$ values in the dry season, but in the wet season, *M. azedarach* had higher $g_s$ than *S. macrophylla* throughout the day (Fig. 3). The lowest $g_s$ values were 0.0041 and 0.0024 mmol $H_2O/m^2/s$ in the late dry season and early dry season, respectively. Variation in $C_i$ was the opposite of $P_N$, with *S. macrophylla* exhibiting higher $C_i$ values in the dry season than in the wet season (Fig. 3). A comparison of diurnal WUE revealed that *M. azedarach* had significantly higher values than *S. macrophylla* in the dry season, but the differences between the species were less in the wet season (Fig. 3). The $VPD_l$ increased from 8:00 to 11:00 in the dry

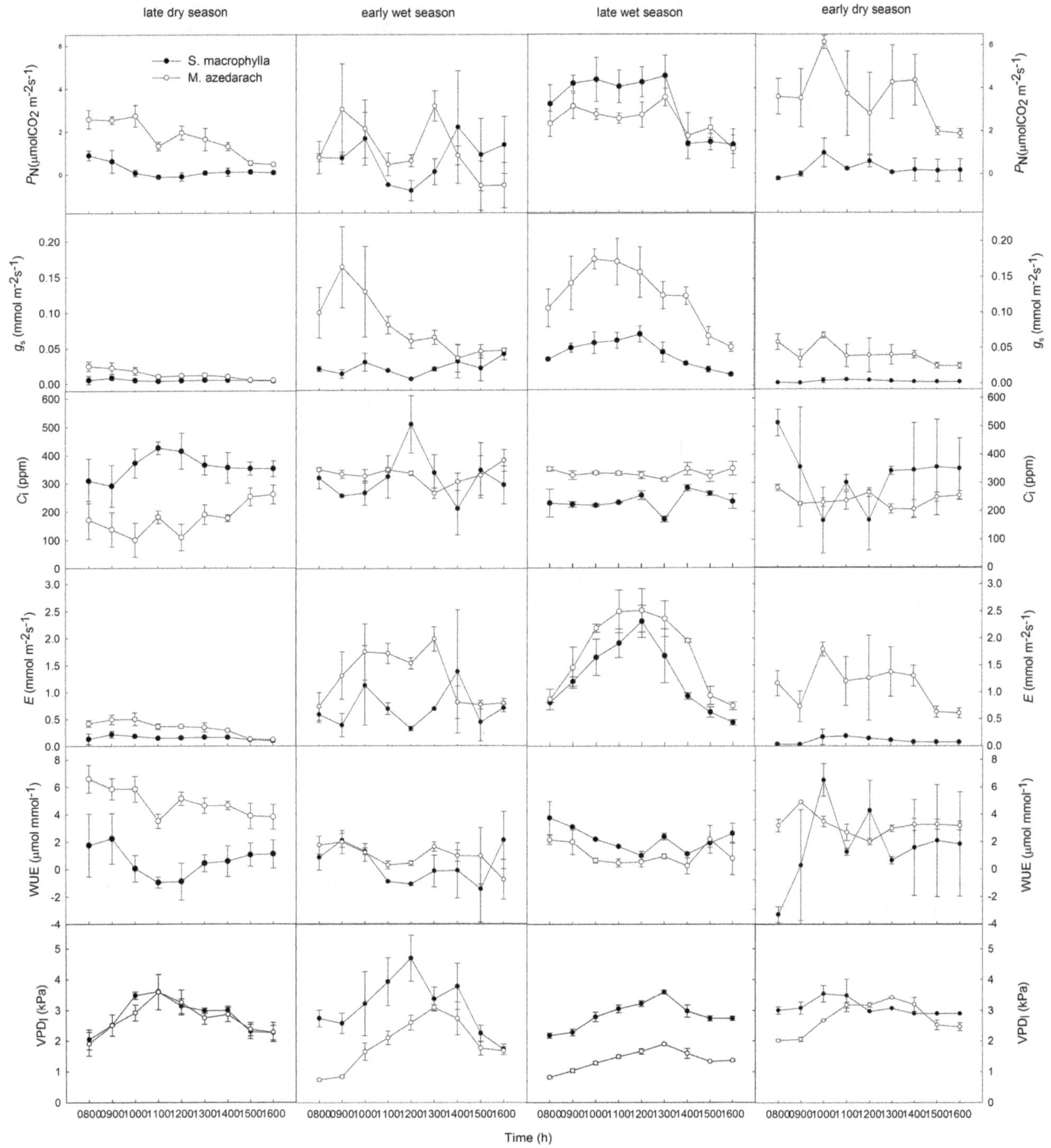


**Fig. 3** Diurnal variation in gas exchange in *Swietenia macrophylla* and *Melia azedarach* in 2010. Net photosynthetic rate ($P_N$), stomatal conductance ($g_s$), intercellular $CO_2$ concentration ($C_i$), water use efficiency (WUE), and leaf-air vapor pressure deficit ($VPD_l$), late dry season (27 March), early wet season (26 June), late wet season (25 September), and early dry season (30 December). Data are represented as the mean ± SE ($N = 3$)

season, whereas the increase continued until 13:00 in the wet season (Fig. 3). These results show that the gas exchange of both species was representative of different patterns in response to drought conditions.

During the dry season, $P_N$ had a significantly negative correlation ($P < 0.01$) with $C_i$ in both species. Furthermore, in S. *macrophylla*, the $P_N$ was negatively correlated ($P < 0.05$) with $T_l$, while in *M. azedarach*, the $P_N$ value was positively correlated with $VPD_l$ and PPFD ($P < 0.01$). During both seasons, the correlation between $g_s$ and $C_i$ was significantly positive in *M. azedarach* ($P < 0.01$) but not in *S. macrophylla*. During the wet season, *S. macrophylla* $P_N$ exhibited a significant positive correlation ($P < 0.01$) with $g_s$ and $E$, but a negative correlation ($P < 0.01$) with $T_a$ and $T_l$ (Table 2). There was no significant correlation between the gas exchange variables and $CO_2$ in *S. macrophylla* regardless of the season. However, in *M. azedarach* the $CO_2$ content significantly and positively correlated with $g_s$ ($P < 0.05$) and $C_i$ ($P < 0.01$) during the dry season, but exhibited a negative correlation with $E$ in the wet season ($P < 0.01$).

Seasonal LAI variation was different between the species (Table 3). LAI was higher in *S. macrophylla* than in *M. azedarach* throughout all seasons. *S. macrophylla* exhibited no significant difference between seasons, although this difference was slightly higher during the wet season. In contrast, *M. azedarach* exhibited a significantly higher LAI in the early wet season than in the late dry season.

Figure 4 depicts the relationship between $g_s$ and the gas-exchange variables. We observed a linear relationship between $P_N$ and $g_s$ in *S. macrophylla*, but a parabolic relationship between these factors in *M. azedarach* (Fig. 4a). The $g_s$ and $E$ values in both species exhibited a significant linear correlation ($P < 0.01$), with the slope for *S. macrophylla* being higher than that for *M. azedarach* (Fig. 4b). The relationship between $g_s$ and $C_i$ was distinctly parabolic in both species; in *S. macrophylla* for example, the relationship was a quadratic polynomial, with a decrease in $C_i$ when $g_s$ was less than 0.05 $mol/m^2/s^1$ and an increase in $C_i$ when $g_s$ was greater than 0.05 $mol/m^2/s$. In contrast, $g_s$ and $C_i$ in *M. azedarach* were positively correlated under drought conditions, whereas, in the wet season, $C_i$ remained constant as $g_s$ increased (Fig. 4c). We observed a curvilinear relationship between $g_s$ and $VPD_l$ in *M. azedarach* ($P < 0.01$), while $g_s$ decreased sharply when $VPD_l$ was over 1.5 kPa. This relationship was not significant in *S. macrophylla* (Fig. 4d).

The linear relationship between $P_N$ and $g_s$ in both species was divided into two groups by leaf temperatures above and below 30 °C in the dry and wet seasons (Fig. 5). During the dry season, in *S. macrophylla*, a significantly positive linear correlation and similar slope was observed between $P_N$ and $g_s$ regardless of the $T_l$. In the wet season, however, the linear slope at low $T_l$ was higher than that at high $T_l$. In contrast, $P_N$ and $g_s$ in *M. azedarach* had a similar slope regardless of the $T_l$ in the wet season.

**Table 3 Mean values (± standard error) of leaf area index (LAI) of *Swietenia macrophylla* and *Melia azedarach* in different seasons**

| Seasons | *S. macrophylla* | *M. azedarach* |
|---|---|---|
| Late dry | 2.53 ± 0.18a | 1.30 ± 0.55b |
| Early wet | 3.21 ± 0.42a | 3.04 ± 0.24a |
| Late wet | 3.26 ± 0.96a | 2.77 ± 0.77ab |
| Early dry | 2.75 ± 0.19a | 1.79 ± 0.18ab |

Means followed by different lowercase letters differ significantly between seasons at $P < 0.05$

**Table 2 Correlation between photosynthetic variables and environmental factors in *Swietenia macrophylla* and *Melia azedarach* during the dry and wet seasons**

| | Environmental factors[a] | | | | | | | | $CO_2$ |
|---|---|---|---|---|---|---|---|---|---|
| | $g_s$ | $C_i$ | $E$ | $VPD_a$ | $VPD_l$ | $T_a$ | $T_l$ | PPFD | |
| Dry season | | | | | | | | | |
| $P_N$ | ns/ns | −**/−** | ns/ns | ns/ns | ns/+* | ns/ns | −*/ns | ns/+** | ns/ns |
| $g_s$ | | ns/+** | +**/+** | ns/ns | ns/−** | ns/ns | ns/ns | ns/ns | ns/+* |
| $C_i$ | | | ns/ns | ns/−* | ns/−** | ns/ns | ns/ns | ns/−** | ns/+** |
| $E$ | | | | ns/ns | ns/ns | ns/ns | ns/+* | ns/ns | ns/ns |
| Wet season | | | | | | | | | |
| $P_N$ | +**/ns | ns/−* | +**/ns | ns/ns | ns/ns | −**/ns | −**/ns | ns/ns | ns/ns |
| $g_s$ | | ns/+** | +**/+** | ns/−** | ns/−** | ns/ns | −*/+* | ns/ns | ns/ns |
| $C_i$ | | | ns/+* | ns/−** | ns/−** | ns/ns | ns/+* | ns/−** | ns/ns |
| $E$ | | | | ns/ns | ns/ns | ns/+** | ns/+** | ns/ns | ns/−** |

+ and − symbols represent positive and negative correlation, respectively

ns indicates no significant correlation

*,** Indicate significance at $P < 0.05$ and $P < 0.01$, respectively. N = 18

[a] $g_s$ (stomatal conductance), $C_i$ (intercellular $CO_2$ concentration), $E$ (transpiration rate), $VPD_a$ (air vapor pressure deficit), $VPD_l$ (leaf-air vapor pressure deficit), $T_a$ (air temperature), $T_l$ (leaf temperature), PPFD (photosynthetic photon flux density), $CO_2$ (ambient $CO_2$ concentration), *S. macrophylla*/*M. azedarach*

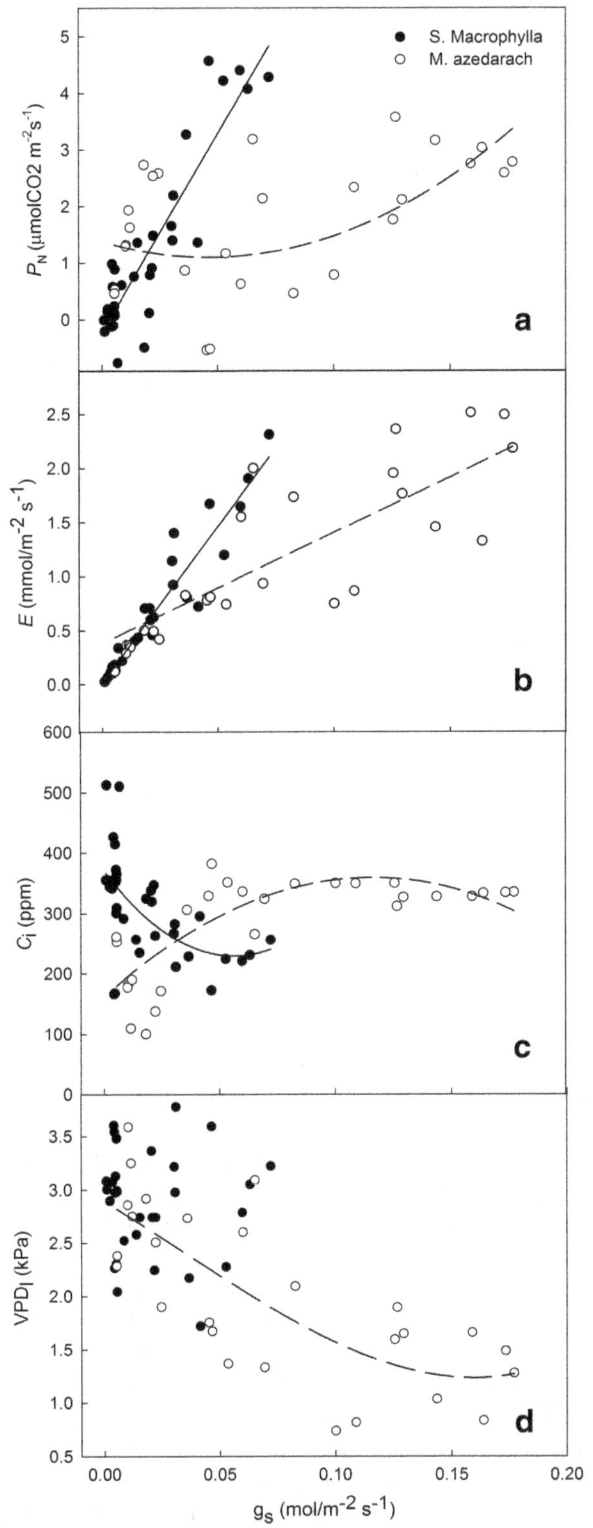


**Fig. 4** Relationship between stomatal conductance ($g_s$) and other gas exchange variables in *Swietenia macrophylla* and *Melia azedarach* ($N = 36$). Net photosynthetic rate ($P_N$; **a**), transpiration ($E$; **b**), intercellular $CO_2$ concentration ($C_i$; **c**), and vapor pressure deficit ($VPD_l$; **d**)

## Discussion

The influence of many environmental factors on photosynthesis and plant growth have been well researched in recent decades (Colom and Vazzana 2003; Ogaya and Peñuelas 2003; Li and Chen 2009). Zhang et al. (2009) suggested that a cool dry season of 4 months may have long-lasting negative effects on the physiology of Dipterocarpaceae trees in southern China. Similarly, the prolonged drought in 2010 may have influenced the $P_N$ and $g_s$ of the trees in our experimental plantation. Gas-exchange dynamics may reflect the ability of a plant to maintain its photosynthetic apparatus under various environmental conditions (Ngugi et al. 2004; Ding et al. 2006). Therefore, we compared diurnal gas-exchange variables across dry and wet seasons in the two tree species.

Midday depression in leaf photosynthesis is common in the canopy leaves of tropical forests which is caused by high temperature, irradiance, or VPD (Ding et al. 2006; Li and Chen 2009; Zhang et al. 2009; Gao et al. 2015). Both species examined in this study exhibited a midday depression in leaf photosynthesis and recovered after 14:00, regardless of the season. These patterns indicate that environmental factors may limit photosynthesis in tropical forests during the midday period, which is consistent with the observations in other evergreen communities during seasonal drought (Koch et al. 1994). Under typical conditions, photosynthetic rates usually increase with PPFD, but in both seasons in our study, the $P_N$ in *S. macrophylla* was negatively correlated ($P < 0.01$) with $T_l$ and not correlated with PPFD. However, temperature directly influenced VPD which is an important environment factor affected the leaf gas exchange. Previous estimations based on gas-exchange in peach-palm trees under subtropical conditions indicated that temperature and VPD have stronger effects on $P_N$ compared with PPFD (Tucci et al. 2010).

During the late wet season in our study, the variation in $P_N$ was similar in both species—the $P_N$ decreased as PPFD decreased after 13:00. Our results concur with a research on *S. macrophylla* seedlings in northern Brazil, where under well-watered conditions gas exchange decreased as PPFD decreased after 13:00, reaching a minimum at 17:00, whereas $g_s$ remained low under drought conditions (Cordeiro et al. 2009). Although comparisons of plant physiology between greenhouse and field conditions are limited due to differences in plant age, root depth, and the degree of control of environmental factors (Cordeiro et al. 2009), the similar decreases in $P_N$ and $g_s$ measured in the present field study and previous greenhouse studies suggest that stomatal conductance dominates transpiration and strongly affects photosynthesis in both our experimental species.

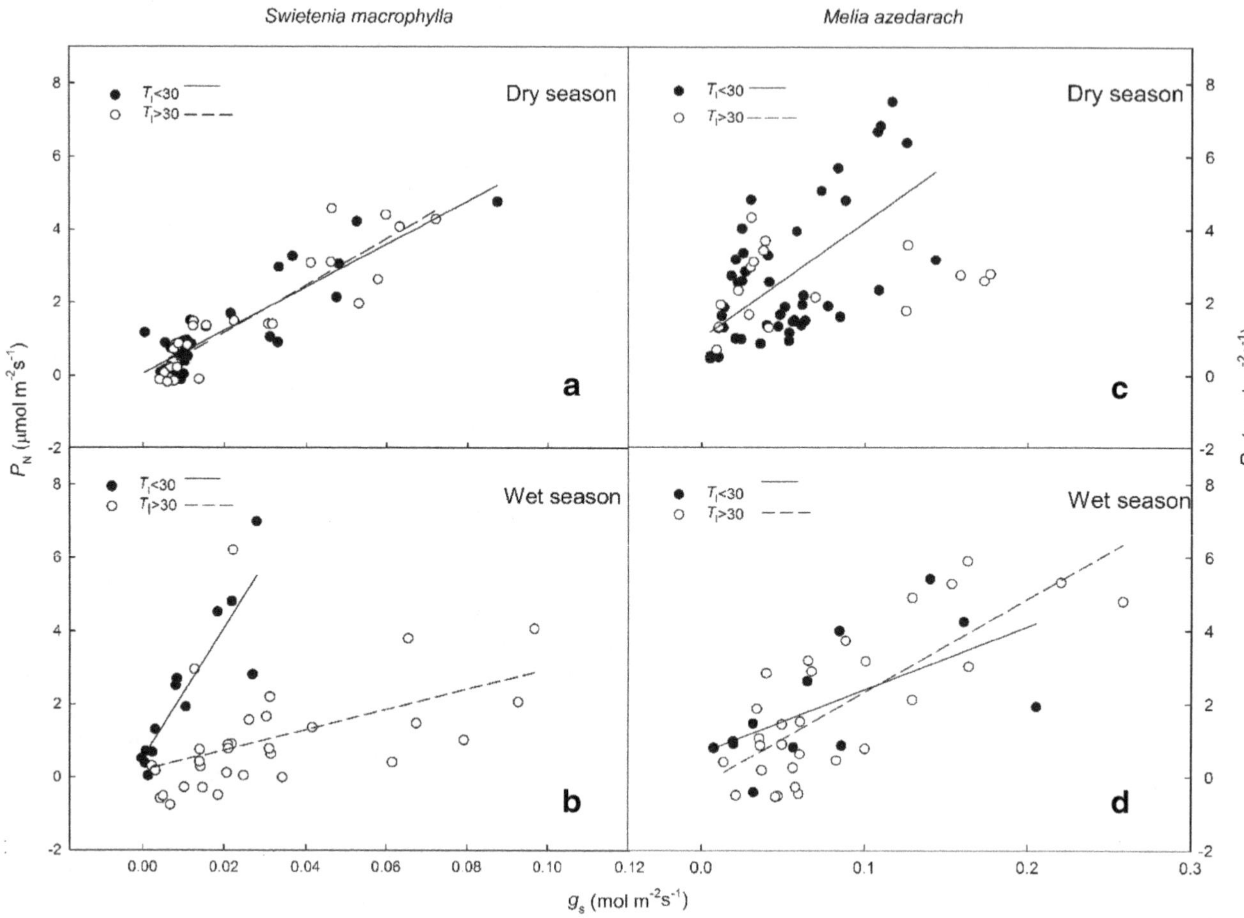


**Fig. 5** Relationship between *Swietenia macrophylla* and *Melia azedarach* stomatal conductance ($g_s$) and net photosynthetic rate ($P_N$) in dry (**a, c**) and wet (**b, d**) seasons at different leaf temperatures ($T_l$)

VPD is an important factor in plant gas exchange, because stomatal opening and closing are strongly dependent on VPD conditions (Cordeiro et al. 2009; Li and Chen 2009; Tucci et al. 2010; Niinemets and Keenan 2014; Chen et al. 2016b). For example, in *Fritillaria cirrhosa*, VPD increased sharply after 10:00, while $g_s$ dropped rapidly (Li and Chen 2009), suggesting that $g_s$ in this species is influenced by the environment. A study on *Dialium pachyphyllum* in Cameroon demonstrated similar results in that species (Koch et al. 1994). Here, we found a significant negative relationship between $g_s$ and $VPD_l$ ($P < 0.01$) in *M. azedarach*, which might suggest a mechanism for water loss prevention. We also observed a negative relationship between $g_s$ and $T_l$ in *S. macrophylla* during the wet season. Together, these results show that stomata are sensitive to environmental conditions and they close at $VPD_l$ above 3 kPa, even in the wet season, in *M. azedarach*. Furthermore, a higher $T_l$ may be the cause of increased $VPD_l$ that is accompanied by lower $g_s$ and $E$ in *S. macrophylla*.

The variation of the leaf area not only reflects the phenology of plantations but also the effects of environmental stress. Thus, there is a strong relationship between LAI and diurnal climate (Hardwick et al. 2015). In *S. macrophylla*, the LAI was reduced only slightly during the dry season, but in *M. azedarach*, more than half of the total leaves fell. Therefore, the different drought tolerance strategies in these species represent the different ways in which they prevent water loss. Arndt et al. (2015) showed that compared to eucalypt savanna, the African mahogany (*Khaya senegalensis*) significantly reduced the total leaf area to moderate water loss in the dry season. Thus, the coordination between plant physiology and leaf area is important for drought adaptation.

Ngugi et al. (2004) reported a non-linear relationship between leaf photosynthesis and $g_s$ in *Eucalyptus argophloia* and *E. cloeziana*. However, Sun et al. (1995) found a significant linear relationship between leaf photosynthesis and $g_s$ in drought-resistant *Nothofagus solandri*, and less so in drought-susceptible *N. menziesii*. In our study, $P_N$ and $g_s$ in *M. azedarach* exhibited a curvilinear relationship during the wet season, whereas under drought ($g_s$ < 0.05 mol/m$^2$/s), the relationship became linear with a steeper slope than that measured in *S. macrophylla*. This steeper slope demonstrates that *M. azedarach* is more drought-resistant than *S. macrophylla*. Generally, the optimum temperature for photosynthesis ranges from 25 to 30 °C for tropical and subtropical broadleaved trees (Larch 1995). The higher leaf temperatures in *S. macrophylla* compared with those in *M. azedarach* may be the cause of the lower $g_s$ in warm–wet seasons, leading to lower $P_N$ (Fig. 5). Additionally, the curvilinear relationship between $C_i$ and $g_s$ in *M. azedarach* indicates that $C_i$ increased with $g_s$ until a certain $CO_2$ concentration was reached, at which point $C_i$ stabilized whenever $g_s$ increased. Colom and Vazzana (2003) found similar evidence for water-stressed *Eragrostis curvula*, and studies on *Quercus pubescens* during a dry summer also demonstrated that $C_i$ remains constant due to metabolic limitations or some degree of mesophyll resistance (Haldimann et al. 2008). In *S. macrophylla*, $C_i$ decreased with increasing $g_s$ during the dry season, and then increased with $g_s$ during the wet season, demonstrating the effects of non-stomatal limitations on photosynthesis. Similarly, Koller et al. (2013) found that high $C_i$ at low $g_s$ indicates an effect of non-stomatal limitation on photosynthesis in three *Quercus* species. In *M. azedarach*, the negative linear relationship between $g_s$ and $VPD_l$ during the dry season ($VPD_l$ > 1.5 kPa) indicates the effects of stomatal limitation on photosynthesis. The same conclusion was made by Zhang et al. (2009) who showed that in dipterocarp species $g_s$ at a given VPD is significantly reduced during the dry season. Photosynthesis is one of the most temperature-sensitive processes in plants (Yamori et al. 2014). In the wet season, a significant linear relationship between $P_N$ and $g_s$ ($P < 0.01$) with different $T_l$ demonstrated that high temperature in summer months might suppress photosynthesis in *S. macrophylla*. The same conclusion was reached by Slot and Winter (2017), who suggested that the negative effect of rising temperatures on photosynthesis in tropical canopy leaves is almost entirely driven by stomatal processes.

In general, elevated WUE helps plants to adapt to water deficiency in arid regions (Aranda et al. 2007). However, plants with higher WUE demonstrate increased growth during the dry but not the wet season (Picotte et al. 2007). A study conducted in Mediterranean climates (Ogaya and Peñuelas 2003) demonstrated that, compared with mesic species (e.g., *Quercus ilex*), drought-resistant species such as *Phillyrea latifolia* typically experience lower photosynthetic rates and higher transpiration rates, which consequently result in lower WUE. However, during drought conditions, $P_N$ and WUE in *P. latifolia* can equal or exceed those of *Q. ilex*. During the wet season in the present study, *S. macrophylla* had relatively higher $P_N$ and WUE values but lower *E* values compared with those in *M. azedarach*. During the dry season, *E* values were very low in both species, but *M. azedarach* had higher WUE than did *S. macrophylla*. This outcome again suggests that *M. azedarach* is a more drought-resistant species when compared to *S. macrophylla*.

Seasonally dry ecosystems such as the Mediterranean or tropical monsoon regions present a challenge for plant growth and survival (Arndt et al. 2015). Noormets et al. (2008) reported that a spring drought might suppress forest canopy development such as leaf sprout and expansion and have long-lasting effects on the forest ecosystem carbon balance. This imbalance remains even when soil water availability improves. In this study, the prolonged drought from October 2009 to April 2010 might have not only decreased leaf photosynthesis but also impaired regional plant productivity. The impact of such stress is represented in the growth performances of *M. azedarach* and *S. macrophylla* (Table 1), where both species showed lower DBH and height compared to those in reported in a previous study in Taiwan (Liao et al. 2011).

## Conclusions

*Melia azedarach* adapted to drought by reducing total leaf area and maintaining higher $P_N$, $g_s$, *E*, and WUE compared with those measured in *S. macrophylla* during the dry season. In contrast, *S. macrophylla* had a higher $P_N$ compared with *M. azedarach* during the wet season and is therefore considered drought-susceptible. In both species, photosynthetic rates during the dry season were affected by multiple, complex environmental factors, but in *S. macrophylla* the temperature had the largest effect on photosynthetic rate during the wet season. However, based on the recovery of $P_N$ patterns in both species during the morning hours, we propose that both *M. azedarach* and *S. macrophylla* acclimatized to the study site, although they differed in diurnal gas-exchange patterns and drought adaptation strategies. *M. azedarach* has better growth performance under prolonged drought compared with *S. macrophylla*. These findings provide information about photosynthesis under drought conditions that support constructive plantation strategies in tropical areas. More attention should be paid to studying the differences in the photosynthetic responses that are dependent on species and the environment because of the changing climate.

**Authors' contributions**

CIC and YNW conceived the study, participated in its design and coordination, and helped draft the manuscript. HCJ and JCY collected, analyzed, and interpreted the data. CIC and HCJ wrote the manuscript. YNW and CSW advised throughout the entire process of the experiment and manuscript preparation. All authors read and approved the final manuscript.

**Author details**

[1] The Experimental Forest, National Taiwan University, 55750 Nantou, Taiwan. [2] Department of Forestry and Resource Conservation, National Taiwan University, 10617 Taipei, Taiwan. [3] Biodiversity Research Center, Academia Sinica, 11529 Taipei, Taiwan.

**Acknowledgements**

We thank the researchers at The Experimental Forest, National Taiwan University, for their help with data collection and field measurements. Many thanks also to the members of the Silviculture laboratory, School of Forestry and Resource Conservation, and to the National Taiwan University for their assistance and cooperation.

**Competing interests**

The authors declare that they have no competing interests.

**Funding**

The author(s) received no specific funding for this work.

**References**

Aranda I, Pardos M, Puértolas J, Jiménez MD, Pardos JA (2007) Water-use efficiency in cork oak (*Quercus suber*) is modified by the interaction of water and light availabilities. Tree Physiol 27:671–677

Arndt SK, Sanders GJ, Bristow M, Hutley LB, Beringer J, Livesley SJ (2015) Vulnerability of native savanna trees and exotic *Khaya senegalensis* to seasonal drought. Tree Physiol 35:783–791

Chen CI, Wang YN, Lih XW, Liao YJ, Yu JC, Wu CS (2013) Growth and $CO_2$ sequestration of *Melia azedarach* in Wan-long farm of Pingtung. Q J Chin For 46:125–134

Chen CI, Yu JC, Wang YN (2016a) Growth and $CO_2$ sequestration of *Swietenia macrophylla* in the Wan-long farm, Pingtung. Q J Chin For 48:331–342

Chen CI, Wang YN, Lih HW, Yu JC (2016b) Three-year study on diurnal and seasonal $CO_2$ sequestration of a young *Fraxinus griffithii* plantation in southern Taiwan. Forests 7:230

Cheng CH, Huang YH, Menyailo OV, Chen CT (2016) Stand development and aboveground biomass carbon accumulation with cropland afforestation in Taiwan. Taiwan J For Sci 31(2):105–118

Colom MR, Vazzana C (2003) Photosynthesis and PSII functionality of drought-resistant and drought-sensitive weeping lovegrass plant. Environ Exp Bot 49:135–144

Cordeiro YEM, Pinheiro HA, Santos Filho BG, Corrêa SS, Silva JRR, Dias-Filho MB (2009) Physiological and morphological responses of young mahogany (*Swietenia macrophylla* King) plants to drought. For Ecol Manage 258:1449–1455

Ding L, Wang KJ, Jiang GM, Li YG, Jiang CD, Liu MZ, Niu SL, Peng YM (2006) Diurnal variation of gas exchange, chlorophyll fluorescence, and xanthophyll cycle components of maize hybrids released in different years. Photosynthetica 44:26–31

Dünisch O, Montóia VR, Bauch J (2003) Dendroecological investigations on *Swietenia macrophylla* King and *Cedrela odorata* L. (Meliaceae) in the central Amazon. Trees Struct Funct 17:244–250

Engelbrecht BMJ, Dalling JW, Pearson TRH, Wolf RL, Gálvez DA, Koehler T, Tyree MT, Kursar TA (2006) Short dry spells in the wet season increase mortality of tropical pioneer seedlings. Oecologia 148:258–269

Gao ZJ, Xu BC, Wang J, Huo LJ, Li S (2015) Diurnal and seasonal variations in photosynthetic characteristics of switchgrass in semiarid region on the Loess Plateau of China. Photosynthetica 53:489–498

Grogan J, Landis RM, Ashton MS, Galvão J (2005) Growth response by big-leaf mahogany (*Swietenia macrophylla*) advance seedling regeneration to overhead canopy release in southeast Pará, Brazil. For Ecol Manage 204:399–412

Guerfel M, Ouni Y, Boujnah D, Zarrouk M (2009) Photosynthesis parameters and activities of enzymes of oxidative stress in two young 'Chemlali' and 'Chetoui' olive trees under water deficit. Photosynthetica 47:340–346

Haldimann P, Gallé A, Feller U (2008) Impact of an exceptionally hot dry summer on photosynthetic traits in oak (*Quercus pubescens*) leaves. Tree Physiol 28:785–795

Hardwick SR, Toumi R, Pfeifer M, Turner EC, Nilus R, Ewers RM (2015) The relationship between leaf area index and microclimate in tropical forest and oil palm plantation: forest disturbance drives changes in microclimate. Agric For Meteorol 201:187–195

IPCC (2015) Climate change 2014: synthesis report. Cambridge University Press, New York

Ishida A, Toma T, Matsumoto Y, Yap SK, Maruyama Y (1996) Diurnal changes in leaf gas exchange characteristics in the uppermost canopy of a rain forest tree, Dryobalanops aromatica Gaertn. f. Tree Physiol 16:779–785

Koch GW, Amthor JS, Goulden ML (1994) Diurnal patterns of leaf photosynthesis, conductance and water potential at the top of a lowland rain forest canopy in Cameroon: measurements from the Radeau des Cimes. Tree Physiol 14:347–360

Koller S, Holland V, Brüggemann W (2013) Effects of drought stress on the evergreen *Quercus ilex* L., the deciduous *Q. robur* L. and their hybrid *Q.* × *turneri* Willd. Photosynthetica 51:574–582

Kuo YL, Fan KS, Hwang CW, Lee YP, Wu HL, Tsay RF (2004) Gas exchange potential in sun-exposed leaves of 30 broadleaf tree species in Taiwan. Taiwan J For Sci 19:375–386

Larch W (1995) Physiological plant ecology, 3rd edn. Springer, Berlin

Leuzinger S, Zotz G, Asshoff R, Körner C (2005) Responses of deciduous forest trees to severe drought in Central Europe. Tree Physiol 25:641–650

Li XW, Chen SL (2009) Diurnal changes in gas exchange and chlorophyll fluorescence parameters of *Fritillaria cirrhosa* and *F. delavayi* under field conditions. Photosynthetica 47:191–198

Li S, Feifel M, Karimi Z, Schuldt B, Choat B, Jansen S (2016) Leaf gas exchange performance and the lethal water potential of five European species during drought. Tree Physiol 36:179–192

Liao WW, Lee MJ, Su PH (2004) Enhancing effects of chinaberry water extract treatment on anti-termite wood property of teak and Taiwan red pine. Q J Chin For 37:453–460

Liao YW, Chen MK, Chen YK, Chung YK, Wu ST (2011) Afforestation carbon storage of Taiwan sugar corporation in Ping-tung County. Q J Chin For 44:373–384

Ngugi MR, Doley D, Hunt MA, Ryan P, Dart P (2004) Physiological response to water stress in *Eucalyptus cloeziana* and *E. argophloia* seedlings. Trees-Struct Funct 18:381–389

Niinemets Ü, Keenan T (2014) Photosynthetic responses to stress in Mediterranean evergreens: mechanisms and models. Environ Exp Bot 103:24–41

Noormets A, McNulty SG, DeForest JL, Sun G, Li Q, Chen J (2008) Drought during canopy development has lasting effect on annual carbon balance in a deciduous temperate forest. New Phytol 179:818–828

Ogaya R, Peñuelas J (2003) Comparative field study of *Quercus ilex* and *Phillyrea latifolia*: photosynthetic response to experimental drought conditions. Environ Exp Bot 50:137–148

Picotte JJ, Rosenthal DM, Rhode JM, Cruzan MB (2007) Plastic responses to temporal variation in moisture availability: consequences for water use efficiency and plant performance. Oecologia 153:821–832

Slot M, Winter K (2017) In situ temperature response of photosynthesis of 42 tree and liana species in the canopy of two Panamanian lowland tropical forests with contrasting rainfall regimes. New Phytol 214:1103–1117

Sun OJ, Sweet GB, Whitehead D, Buchan GD (1995) Physiological responses to water stress and waterlogging in *Nothofagus* species. Tree Physiol 15:629–638

Sun SJ, Meng P, Zhang JS, Wan X (2011) Variation in soil water uptake and its effect on plant water status in *Juglans regia* L. during dry and wet seasons. Tree Physiol 31:1375–1389

Taiz L, Zeiger E (2002) Plant physiology, 3rd edn. Sinauer Associates, Sunderland

Toky OP, Bisht RP (1992) Observations on the rooting patterns of some agroforestry trees in an arid region of north-western India. Agrofor Syst 18:245–263

Tucci MLS, Erismann NM, Machado EC, Ribeiro RV (2010) Diurnal and seasonal variation in photosynthesis of peach palms grown under subtropical conditions. Photosynthetica 48:421–429

Yamori W, Hikosaka K, Way DA (2014) Temperature response of photosynthesis in $C_3$, $C_4$, and CAM plants: temperature acclimation and temperature adaptation. Photosynth Res 119:101–117

Yin CY, Berninger F, Li CY (2006) Photosynthetic responses of *Populus przewalski* subjected to drought stress. Photosynthetica 44:62–68

Zhang JL, Meng LZ, Cao KF (2009) Sustained diurnal photosynthetic depression in uppermost-canopy leaves of four dipterocarp species in the rainy and dry seasons: does photorespiration play a role in photoprotection? Tree Physiol 29:217–228

8

# A method for facilitating the seed germination of a mycoheterotrophic orchid, *Gastrodia pubilabiata*, using decomposed leaf litter harboring a basidiomycete fungus, *Mycena* sp.

Kana Higaki[1], Kento Rammitsu[1], Yumi Yamashita[2], Tomohisa Yukawa[2] and Yuki Ogura-Tsujita[1*]

**Abstract**

**Background:** Mycoheterotrophic plants are one of the most difficult plant groups to conserve because they are entirely dependent on symbiotic fungi. Establishment of viable culture systems would greatly aid their conservation. We describe a simple culture system for the mycoheterotrophic orchid, *Gastrodia pubilabiata*, that does not require laboratory facilities. The orchid is symbiotic with leaf-litter-decomposing fungi.

**Results:** *Gastrodia pubilabiata* seeds were incubated in plastic boxes or glass bottles filled with leaf litter collected from the natural habitat of the species. Seed germination was observed after 35 days and seedling development followed. Fungal isolates from seedlings were identified as Mycenaceae (Basidiomycota), a leaf-litter-decomposing mycorrhizal fungus of *Gastrodia* species.

**Conclusion:** Our method can be used to conserve endangered mycoheterotrophic plants associated with leaf litter-decomposing fungi efficiently, and can also serve as a model system for physiological and molecular studies of such plants.

**Keywords:** Conservation, *Gastrodia*, Leaf litter-decomposing fungi, Mycoheterotrophic plants

## Background

Mycoheterotrophic plants depend on mycorrhizal fungi for their carbon supply throughout their life cycle (Leake 1994). Most such plants are rather rare and seriously endangered, principally because of habitat destruction (Merckx et al. 2013). However, they are also one of the most difficult plant groups to conserve both ex situ and in situ because they are strongly dependent on specific symbiotic fungi. Development of a viable culture system would be of great assistance in terms of efficient plant propagation and subsequent reintroduction into natural habitats. In addition, mycoheterotrophic plants have dust-like seeds with minimal nutrient reserves, and germination occurs mostly underground, rendering it difficult to investigate seed germination ecology in their habitat. Seedling growth in culture would greatly aid studies of seed dormancy, seedling germination, and development. As many biological characteristics of mycoheterotrophic plants remain unknown, a simple culture method would serve as a useful model system for elucidating the physiological and molecular aspects of mycoheterotrophy.

Successful flowering of mycoheterotrophic plants co-cultured with symbionts has been reported for several orchids associated with leaf-litter or wood-decomposing fungi, including *Didymoplexis minor* (Burgeff 1936), *Gastrodia verrucosa* (Tashima et al. 1978), *Gastrodia elata*,

*Correspondence: ytsujita@cc.saga-u.ac.jp
[1] Faculty of Agriculture, Saga University, 1 Honjo-machi, Saga 840-8502, Japan
Full list of author information is available at the end of the article

(Xu and Guo 2000), and *Epipogium roseum* (Yagame et al. 2007). However, these studies used fungal isolates from mycorrhizal tissues and required laboratory facilities. As a wide range of people, from professional scientists to amateur volunteers, are involved in orchid conservation programs, a versatile method without expensive culture facilities is required. An easy symbiotic culture system without fungal isolation has been successfully achieved for Australian green orchids by incubating orchid seeds and soil from their natural habitat (Brundrett and Ager 2011). Seedlings of *Gastrodia nipponica*, a mycoheterotrophic orchid that is associated with litter-decomposing fungi (Kinoshita et al. 2016), were successfully established by incubating the seeds in a plastic box filled with leaf litter from their habitat (Umata and Nishi 2010).

The mycoheterotrophic orchid *Gastrodia pubilabiata* is principally associated with litter-decomposing fungi in the families Mycenaceae and Marasmiaceae (Kinoshita et al. 2016); both the plant and the fungi can be cultured in vitro. In a previous study, seeds of *G. pubilabiata* were successfully germinated and developed into plantlets using in vitro symbiotic culture with a fungus from the mycorrhizal roots of *G. verrucosa* (Umata et al. 2000). Asymbiotic seed germination (Godo et al. 2007) and flowering under symbiotic cultivation (Shimaoka et al. 2017; S. Inagaki personal communication) have also been achieved. Seventeen prefectures in Japan list *G. pubilabiata* as an endangered plant (Association of Wildlife Research and EnVision 2013). Furthermore, this orchid is < 30 cm in height and easy to handle, making it a suitable model plant for physiological and molecular studies.

This paper describes a simple method for germinating seeds of *G. pubilabiata* using decomposing leaf letter, circumventing the need for laboratory facilities. We also describe the basidiomycete fungus implicated in the germination process. This paper presents a new model system that can be applied to other mycoheterotrophic orchids for conservation purposes.

## Methods

### Sample collection and seed germination

Mature seeds from naturally pollinated *G. pubilabiata* plants were collected from two sites in Shizuoka Prefecture, Japan, on 22 October 2016. Three and five capsules from different individuals were collected at Sites 1 and 2, respectively. The collected seeds were kept in paper bags for each individual and stored in glass jars with silica gel at ambient temperature for 19 days until use. To verify the difference in leaf litter components, litter was collected from two different habitat types. Site 1 was a broad-leaved forest dominated by *Quercus serrata* mixed with *Zelkova serrata*, *Cinnamomum yabunikkei*, and *Cleyera japonica*, and Site 2 was a monoculture plantation of *Cryptomeria japonica*. Leaf litter was collected at an approximate depth of 5 cm within 2 m of fruiting *Gastrodia* plants at each site and placed in plastic zipper bags (480 × 340 mm). The bags were then carefully packed into a cardboard box for transport by airplane, transported back to the laboratory within 3 days of collection, and kept at ambient temperature in the dark for 19 days until use. Large and small plastic boxes (with lids) and 900- and 450-mL glass bottles with screw caps were sterilized with 70% ethanol and filled with leaf litter from each site (Table 1, Fig. 1). The inner walls of each box were covered with cardboard, and pumice was used to form layers 2 cm thick at the bottom of the bottles. On 9 November 2016, seeds from three and five capsules were bulked and sprinkled over the boxes and bottles, respectively. The vessels were maintained at 25 °C in a growth chamber in the dark. The screw caps of the glass bottles were loosened to allow gas exchange. Non-sterilized tap water was hand-sprayed as a mist when the surface of the leaf litter became dry. Seed germination was evaluated after 35, 65, 70, and 120 days of culture. Seedlings longer than 1 cm were randomly chosen and used for fungal isolation.

**Table 1 The culture vessels and leaf litter used in this study and the results of seed germination**

| Vessel number | Size of the vessels | Location of leaf litter collected | Seed germination[b] |
|---|---|---|---|
| Box 1 | L260 × W190 × D100[a] | Site 1 | ++ |
| Box 2 | L411 × W337 × D139 | Site 1 | + |
| Bottle 3 | 450 mL | Site 2 | + |
| Bottle 4 | 450 mL | Site 2 | − |
| Bottle 5 | 900 mL | Site 2 | + |

[a] Length (L) × width (W) × depth (D) (mm)

[b] ++, a large number of protocorms and seedlings were observed; +, germination was observed; –, no germination

**Fig. 1** The culture vessels used in this study and flower morphology of *Gastrodia pubilabiata*. Plastic boxes (**a**) and glass bottles (**b**) were filled with leaf litter collected from the natural habitat. **c** A flowering plant in its habitat (photo by T. Yamashita)

### Fungal isolation

Seedlings were washed three times with sterile water and pelotons in mycorrhizal tissue were teased with needles as described by Rasmussen (1995); these pelotons were cultured on potato dextrose agar (PDA) plates with 50 ppm of streptomycin and tetracycline each. Fungal colonies from single pelotons were transferred to fresh PDA plates for subculture. The fungal isolates obtained in this study were preserved in the Biological Resource Center of the National Institute of Technology and Evaluation (NBRC).

### Molecular identification

DNA was extracted from fungal isolates as described by Izumitsu et al. (2012). PCR and sequencing were performed as described by Ogura-Tsujita and Yukawa (2008). To identify the mycorrhizal fungi, fungal nuclear ribosomal internal transcribed spacer (ITS) regions were amplified using the ITS1F/ITS4 primer pair (White et al. 1990; Gardes and Bruns 1993). As the collected leaf litter might have been contaminated with seeds from other orchid species, the plant species of the seedlings were also identified molecularly using the 17SE/26SE primer pair (Sun et al. 1994), and we confirmed whether the germinated seedlings were consistent with the seeds used in this study. An entire seedling from Box 1 and a tuber larger than 1 cm from Bottle 3 were used for plant identification according to the above method.

## Results

Seed germination was observed after 35, 65, 70, and 120 days of culture and protocorms were found in all culture vessels, except Bottle 4 (Table 1, Fig. 2). The leaf litter from each of the two sites with different leaf components induced seed germination. Fungal hyphae and rhizomorphs grew around the protocorms (Fig. 2a, b). All developmental stages, from the protocorm to the seedling stage with roots, were observed (Fig. 2c). Each seedling developed two lateral roots, followed by root elongation and tuber development. After 65 days of culture, the tubers developed soft woolly hairs; the longest root was 14 cm (the root of the largest seedling in Box 1; Fig. 2d). Fungal rhizomorphs originating from dead branches (in the leaf litter) were often connected to the roots (Fig. 2e). After 5 months of culture, seedling growth declined as the litter decomposed.

Four and one fungal isolate(s) from a single peloton were obtained from seedlings in Box 1 and Bottle 3, respectively (Table 2). All Box 1 isolates (Isolate ID. F68) were identical in sequence, being 100% homologous to the ITS sequence of the litter-decomposing fungus *Mycena* cf. *abramsii* (KR673481) and 99% homologous to the symbiotic fungus of *G. nipponica* (LC013372). The isolate from Bottle 3 (Isolate ID. F69) was 99% homologous to the ITS sequence of the symbiotic fungus of *G. pubilabiata* (LC013346), *Mycena plumbea* (JN198391), and a *Mycena* species that induces seed germination in *G. elata* (FJ785523). The rhizomorph connecting the mycorrhizal roots of a seedling in Bottle 3 had a sequence identical to that of F69. Isolates F68 and F69 shared similar morphological characteristics and grew slowly on PDA (Additional file 1: Figure S1). The colonies were white and relatively flat, with diffuse to filamentous edges, and their diameters were approximately 5 cm after 2 weeks.

## Discussion

Seeds of *G. pubilabiata* germinated in four of the five culture vessels used in this study (Table 1), and germination was achieved within 35 days after sowing. There were no differences in seedling establishment among the four types of culture vessel, although the culture in Box 1 resulted in the large number of protocorms (> 50 individuals), while the largest tuber was observed in Bottle 3. The leaf litter components also did not affect seed germination. Fungal rhizomorphs growing from dead leaves and tree branches were often connected to protocorms and roots (Fig. 2a, b, e), indicating that litter-decomposing fungal rhizomorphs associate with *G. pubilabiata* to transfer carbon from leaf litter to the seedlings. Long-term seedling culture was not achieved in this study. As the leaf litter in the box was broken down into small pieces and white vigorous rhizomorphs turned yellowish, it seems likely that the fungal growth resources became depleted over time. Continuous replenishment of leaf litter may be required for long-term seedling culture. Successful fungal isolation from seedlings indicates that the culture system established in this study could be useful as a seed baiting technique to detect symbiotic fungi from orchid habitats, as shown for terrestrial green orchids (Brundrett et al. 2003). In addition to the leaf litter culture, 70 packets containing orchid seeds were buried for 1 year in the same habitats, but only five seedlings (< 5 mm) were obtained from these packets (data not shown). It seems likely that the leaf litter culture may serve as a propagation method for *G. pubilabiata*, although additional replicates are required.

The molecular data imply that the Mycenaceae promote seed germination and seedling growth. The Mycenaceae family constitutes prominent fungal partners of *G. pubilabiata* and two other *Gastrodia* species, *G. confusa* and *G. nipponica* (Ogura-Tsujita et al. 2009; Kinoshita et al. 2016). *Mycena osmundicola* has been shown to induce germination of *G. elata* seeds (Xu and Guo 1989). In our study, the fungal isolates were highly homologous to the symbiotic fungi of *G. nipponica* and *G. elata*, indicating that the symbiotic *Mycena* species that we detected may

**Fig. 2** Seed germination and seedling growth of *Gastrodia pubilabiata*. **a**, **b** Protocorms after 35 days of culture. Fungal hyphae and white rhizomorphs are evident around the protocorms; **c** Seedlings after 35 days of culture. Roots (arrows) and tubers (T) are evident; **d** A Box 1 seedling after 65 days of culture. Elongated roots (arrows) and hairy tubers (T); **e** A seedling root with attached fungal rhizomorphs after 120 days of culture. Rhizomorphs (Rz) growing from a dead branch are connected to a mycorrhizal root (Ro)

be associated with several *Gastrodia* species. Furthermore, the F69 isolate was 99% homologous to a symbiotic fungus from adult *G. pubilabiata* plants, suggesting that a single *Mycena* species may be present throughout the entire life cycle of *G. pubilabiata*, in contrast to the situation in *G. elata*, in which the fungal partner shifts from *Mycena* to *Armillaria* as seedling development proceeds (Xu and Guo 2000).

**Table 2 The results of molecular identification of the fungal isolates from *G. pubilabiata* seedlings**

| Isolate | NBRC no.[a] | Accession no. of ITS sequence | Closest mach in GenBank | Identities |
|---|---|---|---|---|
| F68 | NBRC113007 | LC314114 | *Mycena abramsii* voucher KA12-0434 (KR673481) | 607/607 (100%) |
| | | | Mycorrhizal fungus of *G. nipponica* (LC013372) | 650/659 (99%) |
| F69 | NBRC113008 | LC314115 | Mycorrhizal fungus of *G. pubilabiata* (LC013346) | 642/646 (99%) |
| | | | *Mycena plumbea* (JN198391) | 641/645 (99%) |
| | | | *Mycena* sp. MKACC51882 (FJ785523) | 640/645 (99%) |

[a] All isolates were deposited to the Biological Resource Center of the National Institute of Technology and Evaluation (NBRC)

## Conclusions

We found that *G. pubilabiata* seeds germinated in culture vessels filled with leaf litter collected from the natural habitat; fungal isolation was unnecessary. Our system may assist ex situ conservation programs, allowing seedling culture, propagation, and reintroduction of *Gastrodia* and other mycoheterotrophic orchids associated with litter-decomposing fungi such as *Didymoplexis* (Burgeff 1932). Furthermore, our method may serve as a model system for observing seedling development and physiological and molecular studies of mycoheterotrophic plants.

### Authors' contributions

KH performed the plant cultivation. KR performed the fungal isolation and identification. YY contributed with the sample collection and field survey. TY provides advice and guidance. YO designed the experiments, analyzed the data and wrote the paper. All authors read and approved the final manuscript.


### Author details

[1] Faculty of Agriculture, Saga University, 1 Honjo-machi, Saga 840-8502, Japan. [2] Tsukuba Botanical Garden, National Museum of Nature and Science, 4-1-1 Amakubo, Tsukuba, Ibaraki 305-0005, Japan.



### Acknowledgements

The authors thank M. Goto, A. Kinoshita and T. Tanaka for collecting samples, and S. Inagaki for valuable advices for culture methods.


### Competing interests

The authors declare that they have no competing interests.


### Funding

This study was supported by JSPS KAKENHI (Grant Nos. 17K07536 and 15H04417), Research Grant for Plant Science from The New Technology Development Foundation and a National Museum of Nature and Science research grant entitled "Integrated analysis of natural history collections for conservation of highly endangered species".


### References


Association of Wildlife Research and EnVision (2013) Search system of Japanese Red Data. http://www.jpnrdb.com/

Brundrett M, Ager E (2011) Wheatbelt orchid rescue project final report 7. Seed collecting, soil baiting and propagation of orchids. In: Wheatbelt orchid rescue project, University of western Australia. Available via DIALOG. http://repository.uwa.edu.au/R/-?func=dbin-jump-full&object_id=29545&current_base=GEN01-INS01

Brundrett M, Scade A, Batty AL, Dixon KW, Sivasithamparam K (2003) Development of in situ and ex situ seed baiting techniques to detect mycorrhizal fungi from terrestrial orchid habitats. Mycol Res 107:1210–1220

Burgeff H (1932) Saprophytismus und symbiose. Studien an tropischen orchideen. Gustav Fischer, Jena

Burgeff H (1936) Samenkeimung der orchideen und entwicklung ihrer keimpflanzen. Gustav Fischer, Jena

Gardes M, Bruns TD (1993) ITS primers with enhanced specificity for basidiomycetes-application to the identification of mycorrhizae and rusts. Mol Ecol 2:113–118

Godo T, Hashimoto T, Nakata S, Miyoshi K (2007) Aseptic seed germination of rare mycoheterotrophic orchid, *Gastrodia pubilabiata*. In: Abstracts of the symposium of Japanese society for plant cell and molecular biology, vol 25. pp 100 (In Japanese)

Izumitsu K, Hatoh K, Sumita T, Kitade Y, Morita A, Gafur A, Ohta A, Kawai M, Yamanaka T, Neda H, Ota Y, Tanaka C (2012) Rapid and simple preparation of mushroom DNA directly from colonies and fruiting bodies for PCR. Mycoscience 53:396–401

Kinoshita A, Ogura-Tsujita Y, Umata H, Sato H, Hashimoto T, Yukawa T (2016) How do fungal partners affect the evolution and habitat preferences of mycoheterotrophic plants? A case study in *Gastrodia*. Am J Bot 103:207–220

Leake JR (1994) The biology of myco-heterotrophic ('saprophytic') plants. New Phytol 127:171–216

Merckx VS, Smets EF, Specht CD (2013) Biogeography and conservation. In: Merckx VS (ed) Mycoheterotrophy. Springer, New York

Ogura-Tsujita Y, Yukawa T (2008) High mycorrhizal specificity in a widespread mycoheterotrophic plant, *Eulophia zollingeri* (Orchidaceae). Am J Bot 95:93–97

Ogura-Tsujita Y, Gebauer G, Hashimoto T, Umata H, Yukawa T (2009) Evidence for novel and specialized mycorrhizal parasitism: the orchid *Gastrodia confusa* gains carbon from saprotrophic *Mycena*. Proc R Soc Lond B Biol Sci 276:761–767

Rasmussen H (1995) Terrestrial orchids: from seed to mycotrophic plant. Cambridge University Press, Cembridge

Shimaoka C, Fukunaga Y, Inagaki S, Sawa S (2017) Artificial cultivation of *Gastrodia* species and identification of mycorrhizal fungi. In: Abstracts of the 16th annual meeting of the Japanese society for plant systematics, University of Kyoto, Kyoto, 9–12 March 2017 **(In Japanese)**

Sun Y, Skinner DZ, Liang GH, Hulbert SH (1994) Phylogenetic analysis of *Sorghum* and related taxa using internal transcribed spacers of nuclear ribosomal DNA. TAG Theor Appl Genet 89:26–32

Tashima Y, Terashita T, Umata T, Matsumoto M (1978) In vitro development from seed to flower in *Gastrodia verrucosa* under fungal symbiosis. Trans Mycol Soc Jpn 19:449–453

Umata H, Nishi M (2010) Ecological characteristics and water supply effect on in situ conservation in *Gastrodia nipponica* and *G. pubilabiata*, the endangered mycho-heterotrophic orchids. Res Bull Kagoshima Univ For 37:137–149 **(In Japanese with English abstract)**

Umata H, Yamauchi H, Hashimoto T (2000) In vitro culture of a myco-heterotrphic orchid *Gastrodia pubilabiata* with a mycorrhizal symbiont of *G. verrucosa*. Res Bull Kagoshima Univ For 28:27–30 **(In Japanese with English Abstract)**

White TJ, Bruns T, Lee S, Taylor J (1990) Amplification and direct sequencing of fungal ribosomal RNA genes for phylogenetics. In: Innis MA, Gelfand DH, Sninsky JJ, White TJ (eds) PCR protocols: a guide to methods and applications. Academic Press, California, pp 315–322

Xu J, Guo S (1989) Fungus associated with nutrition of seed germination of *Gastrodia elata–Mycena osmundicola* Lange. Acta Mycol Sin 8:221–226

Xu J, Guo S (2000) Retrospect on the research of the cultivation of *Gastrodia elata* Bl, a rare traditional Chinese medicine. Chin Med J 113:686–692

Yagame T, Yamato M, Mii M, Suzuki A, Iwase K (2007) Developmental processes of achlorophyllous orchid, *Epipogium roseum*: from seed germination to flowering under symbiotic cultivation with mycorrhizal fungus. J Plant Res 120:229–236

9

# Phalaenopsis pollinia storage at sub-zero temperature and its pollen viability assessment

Shih-Chang Yuan, Shih-Wen Chin, Chen-Yu Lee and Fure-Chyi Chen*

**Abstract**

**Background:** In a breeding program, usually only superior parents are chosen for cross hybridization. Pollens of elite cultivars may not be available at hand. Properly stored pollens provide an opportunity for cross hybridization at unavailable time.

**Results:** Pollen of a *Phalaenopsis* hybrid was evaluated for the storage ability at different temperatures, including room temperature, 4, − 20, and − 80 °C for up to 96 weeks. The viability of pollen was assessed by TTC staining, in vitro germination and hand pollination during and after storage. Pollen stored at all temperatures for 4 weeks remained viable and capable of successful pollination. Pollen lost its viability after 4 weeks at room temperature. Pollen remains viable after 40 weeks at 4 °C, and after 96 weeks at both − 20 and − 80 °C of storage. Viable pollen could be successfully pollinated to the female parent at all effective storage conditions and produced seeds.

**Conclusions:** Our results indicate that *Phalaenopsis* pollen can be stored at 4 °C up to 40 weeks for short-term purpose. For long-term storage, pollen can be kept at both − 20 and −80 °C.

**Keywords:** Low temperature storage, Pollen viability, In vitro germination, Hand pollination

## Background

Phalaenopsis is one of the economically important orchids for potted plant and cut flower production worldwide. There are about 60 species of the *Phalaenopsis* genus which are native to tropical Asia and the larger islands of the Pacific Ocean, where they ranged from Sri Lanka and South India in western to Papua New Guinea in eastern and extended to southern China, Taiwan, and the north Philippines (Sweet 1980; Christenson 2001). In Taiwan, the potted plants of *Phalaenopsis* have been exported in large quantity annually during 2001–2016 (Council of Agriculture 2016), partly due to the continuous development of novel varieties. In order to continuously supply enough quantity of young plants to the market, micropropagation of newly selected cultivars from cross-hybridization program is demanded at all times. In a breeding program, usually only superior parents are chosen for cross-hybridization (Ajeeshkumar and Decruse 2013). However, pollens of elite cultivars may not be available at hand, therefore properly stored pollen becomes an alternative to keep the desirable germplasm for cross-pollination at need in plant improvement programs (Metz et al. 2000; Martínez-Gómez et al. 2002; Deng and Harbaugh 2004; Lora et al. 2006; Masum Akond et al. 2012; de Souza et al. 2014; Peng et al. 2015; Wang et al. 2015).

Pollen longevity or viability is defined as the stored pollen retaining its viability after long-term preservation (Vaknin and Disikowitch 2000). Several reports have shown that pollen stored at low temperatures was effective for long-term preservation, such as almond (Martínez-Gómez et al. 2002), cherimoya (Lora et al. 2006), crape myrtle (Masum Akond et al. 2012), Arabidopsis (Bou Daher et al. 2009), bromeliads (Parton et al. 2002; de Souza et al. 2014), pecan (Peng et al. 2015), caladium (Deng and Harbaugh 2004), jojoba (Vaknin et al. 2003), mango (Dutta et al. 2013) and lily (Wang et al. 2004). Pritchard and Prendergast (1989) used several orchid species to evaluate the pollen storage ability, including *Anacamptis pyramidalis*, *Cymbidium elegans*, *C.*

*Correspondence: fchen@mail.npust.edu.tw
Department of Plant Industry, National Pingtung University of Science & Technology, Pingtung 91201, Taiwan

*tracyanum, Dactylorhiza fuchsii, D. maculata, Epipactis purpurata, Gymnadenia conopsea, Listera ovata, Orchis mascula, O. morio,* and *Spiranthes spiralis.* The result indicated that pollen stored at both − 20 and − 196 °C offer the potential to extend its longevity up to at least 1 year. The storage conditions, including temperatures and relative humidity (RH) affect the viability of pollen (van der Walt and Littlejohn 1996; Deng and Harbaugh 2004). Besides, pollen age, flower physiological state as well as pollen moisture content also affect pollen viability during storage (Rosell et al. 2006; Soares et al. 2008).

The evaluation of pollen viability can be performed by use of fluorescein diacetate (FDA) (Heslop-Harrison and Heslop-Harrison 1970), fluorochromatic dye (FCR) (van der Walt and Littlejohn 1996), lactophenol cotton blue (LPCB) (Bellusci et al. 2010), 2,3,5-triphenyltetrazolium chloride (TTC) (Matison et al. 1947; Khatun and Flowers 1995; Sorkheh et al. 2011; Abdelgadir et al. 2012), or thiazolyl blue (MTT) staining (Khatun and Flowers 1995). Another way to examine the pollen viability was by in vitro germination using Brewbaker and Kwack medium (Brewbaker and Kwack 1963) and subsequently stained by Alexander's dye (Alexander 1969; Galleta 1983). Although in vitro pollen germination is handy for determining its viability, it may not reveal the actual potential of pollination ability. Hand pollination probably is the best and easier way to estimate the effectiveness of pollen viability after storage (Lyakh et al. 1998).

Currently, no report is available on the cold storage of *Phalaenopsis* pollen. The objective of this study was to investigate appropriate temperature parameters for both short-term and long-term storage.

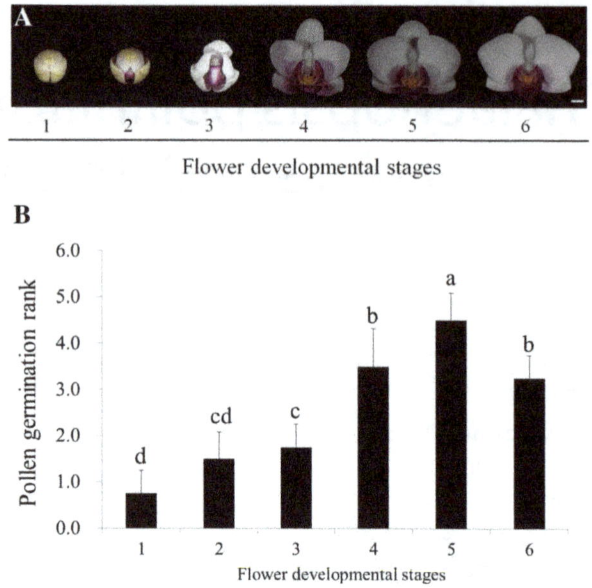


**Fig. 1** Flower developmental stages (**A**) and their corresponding pollen germination rate (**B**) in *Phalaenopsis* Little Gem Stripes. **A** Flower developmental stages are defined as: (1) tight floral bud; (2) sepal half-open; (3) flower half-open; (4) prior to anthesis; (5) flower fully open; (6) 1 day after fully open. **B** In vitro pollen germination of *P.* Little Gem Stripes was evaluated by incubating in Brewbaker and Kwack medium for 7 days at 25 ± 1 °C and then by Alexander staining. Bars represent mean ± SE, n = 4. Pollen germination rank was represented by ranks with area of pollinia stained: (0) no-germination; (1) 1–20%; (2) 21–40%; (3) 41–60%; (4) 61–80%; (5) 81–100%

## Materials and methods

### Pollen source and storage condition

Pollinia of *Phalaenopsis* Little Gem Stripes at 6 stages of flower development were harvested for comparison of in vitro pollen germination ability (Fig. 1A) and a proper stage was chosen for storage experiment. Flower development stages were defined as: (1) floral bud; (2) sepal half-open; (3) flower half-open; (4) prior to anthesis; (5) flower fully open; (6) 1 day after fully open.

Before storage, pollinia were kept at room temperature for 11 days until constant weight. Air-dried pollinia from flower stage 5 which revealed highest in vitro germination rate were wrapped with aluminum foil, sealed and packed into zip-lock bags, and stored at room temperature (RT), 4, − 20, and − 80 °C, respectively. Pollinia were sampled for both viability assessment and pollination tests after 0, 4, 8, 16, 24, 32, 40, 48, 56, 64, 72, 80, 88, and 96 weeks of storage. The stored pollinia were pre-hydrated in reverse osmosis (RO) water at room temperature for at least 30 min before in vitro pollen germination assay.

### Pollen viability evaluated by staining

The viability of stored pollinia was assessed by staining with 2,3,5-triphenyltetrazolium chloride (TTC, Sigma Aldrich, USA) (Matison et al. 1947; Khatun and Flowers 1995; Abdelgadir et al. 2012). The pollinia were placed in a 1.5 ml centrifuge tube containing 0.5% TTC aqueous solution and incubated for 24 h at 25 ± 1 °C in darkness. The 0.5% TTC aqueous solution was prepared by dissolving 0.5 g TTC in 50 mM sodium phosphate buffer solution, pH 7.4. The stained pollinia were rinsed three times with RO water to remove extra stain. The stained pollinia were examined under a stereo microscope (Stemi SV6 Zeiss, Germany). Area of pollinia stained red was considered viable, while the unstained pollinia portion were considered non-viable.

### In vitro pollen germination

Pollens taken from part of pollinia were dipped in 70% alcohol for surface sterilization about 10 s and cultured in a germination medium containing 100 mg $l^{-1}$ boric acid, 300 mg $l^{-1}$ calcium nitrate, 200 mg $l^{-1}$ magnesium sulphate, and 100 mg $l^{-1}$ potassium nitrate, pH 5.6, with 5% sucrose on the slides (Brewbaker and Kwack 1963). Pollinia was incubated at 25 ± 1 °C with three replications for each storage condition, and kept in the dark for 7 days. The Alexander's stain (Alexander 1969) was prepared by mixing 10 ml 95% ethanol, 10 ml Malachite green (1% solution in 95% ethanol), 50 ml distilled water, 25 ml glycerol, 5 ml acid fuchsin (1% solution in water), 0.5 ml orange G (1% solution in water), 4 ml glacial acetic acid and volume adjusted to 100 ml by RO water. Pollens were carefully squashed in the dye solution as evenly as possible, and pollen tube was visualized under a microscope at 100× magnification and about 200 pollen grains of each replication were observed. Pollen grains were considered to be germinated when the pollen tube elongated to twice the pollen grain size. Level of pollen germination was represented by ranking: 0, no-germination; 1, 1–20%; 2, 21–40%; 3, 41–60%; 4, 61–80%; 5, 81–100% germination.

### Hand pollination and observation of in vivo pollen tubes

The stored pollinia at different period were removed and hand pollinated to the female parent *Phalaenopsis* Sogo Vivien 'F858' (Sogo Orchid Nursery, Pingtung, Taiwan). The female plants were grown in the Pad and Fan greenhouse at National Pingtung University Science and Technology (Pingtung, Taiwan), which received a photosynthetic photon flux of 346.4 μmol $m^{-2}$ $s^{-1}$ in summer at noon. Plants were fertilized weekly with 0.5 g $l^{-1}$ of Peters® 20 N-20$P_2O_5$-20$K_2O$ solution (Scotts, USA). Four flowers were pollinated per storage temperature condition. The capsule setting was recorded 4 months after successful pollination. To examine pollen tube growth, parts of the capsule (ovary) were excised at 7, 35, and 70 days after pollination. The ovary tissues were fixed in FAA (formalin: acetic acid: 95% ethanol = 1:1:18) overnight, and washed three times by distilled water, and stored in 70% ethanol. The ovaries were softened in a 4 N sodium hydroxide solution at 60 °C and stained by use of a buffer solution containing 0.1% water-soluble aniline blue dye dissolved in 0.1 N $K_3PO_4$. The ovary tissue was carefully squashed on a slide and observed under a microscope supplied by ultraviolet light of a wavelength at 356 nm (Martin 1959). Finally, the mature capsules were harvested after 4.5 months and the amount of seed was estimated visually. Level of seed amount was represented by ranking: lacking or no any seed (−), low abundance (+), abundant (++), and very abundant (+++), respectively. The presence of embryos in the seeds was examined under a light microscope (Olympus BX50F-3, Japan) by staining with 1% acetocarmine before observation. The seeds were then sterilized with 0.6% sodium hypochlorite solution for 10 min then washed 3 times with sterile water. Later the sterilized seeds were sown on a germination medium containing Hyponex (7N-6$P_2O_5$-19$K_2O$) plus 1 g $l^{-1}$ Bacto-tryptone, 50 g $l^{-1}$ potato homogenate, 50 g $l^{-1}$ banana homogenate, 30 g $l^{-1}$ sucrose, 2 g $l^{-1}$ activated charcoal, and 7.5 g $l^{-1}$ agar as reported previously (Huang et al. 2014).

### Statistical analysis

The experimental design for pollen germination was completely randomized design (CRD) with three replications for each storage condition. Data were analyzed by using analysis of variance (ANOVA) offered in SAS version 9.0 (SAS Institute Inc., Cary, NC, USA) and mean separation was compared by using Least Significance Difference test (LSD) ($p \leq 0.05$).

## Results

### Germination of pollen from different stages of flower development

Pollinia were taken from 6 flower development stages. The in vitro germination rate increased gradually along with the flower stages (Fig. 1A, B). The pollen of first flower stage already showed germination ability. Pollen at fifth flower stage showed highest germination rate followed by the sixth stage. The observation here indicated that fifth-stage probably was the best for pollen storage.

### Pollen storage capability at different temperatures

Freshly-collected pollinia from stage 5 flower showed prominent viability as revealed by heavy TTC staining (Fig. 2). The stained area on the pollinia represented different degree of viability, with yellow or un-stained area being considered non-viable or low viability (Fig. 2).

Next, we examined the in vitro and in vivo germination of stored pollen and the control, the freshly-collected pollen of *P.* Little Gem Stripes. The germination rate of fresh pollen was significantly higher and those stored at all temperatures for 4 weeks were also similar (Fig. 3). Pollen stored at room temperature beyond 4 weeks drastically lost its viability. Pollen germination at lower storage temperatures remained at the reasonable rate but gradually declined with increasing storage period. At 4 °C for 32–40 weeks, germination rate was reduced drastically and showed much lower viability. While the germination rate still remained about the same level as at 4 °C for 40 weeks when stored for 96 weeks at both − 20 or − 80 °C.

To further examine the effectiveness of stored pollen at all period, we pollinated it to a female parent *P.* Sogo Vivien 'F858' and then checked pollen tube growth after 7–70 days (Fig. 4). No pollen tube growth was observed when stored at room temperature and

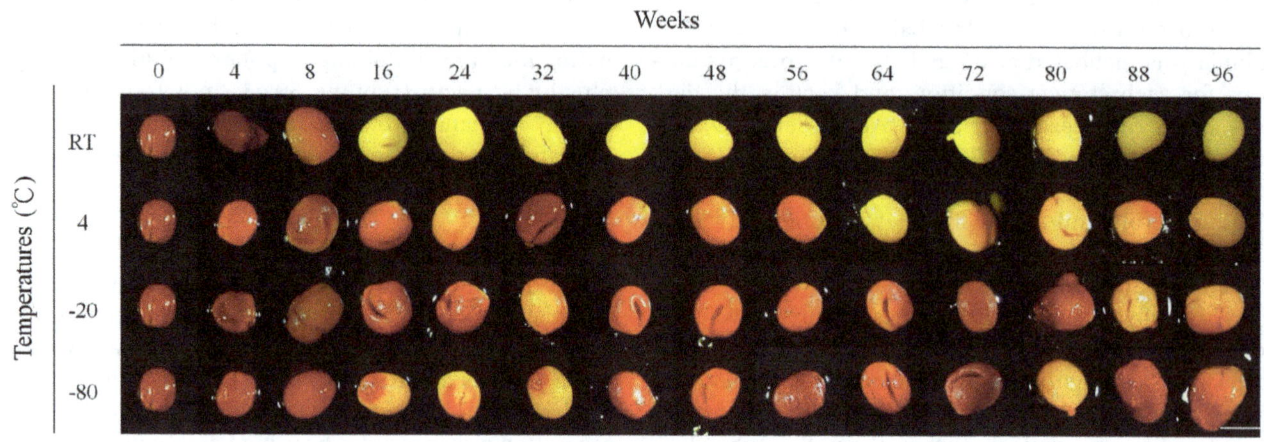


**Fig. 2** The pollen viability of *Phalaenopsis* Little Gem Stripes stored at different temperatures as evaluated by TTC staining. RT, room temperature. Scale bar = 5 mm

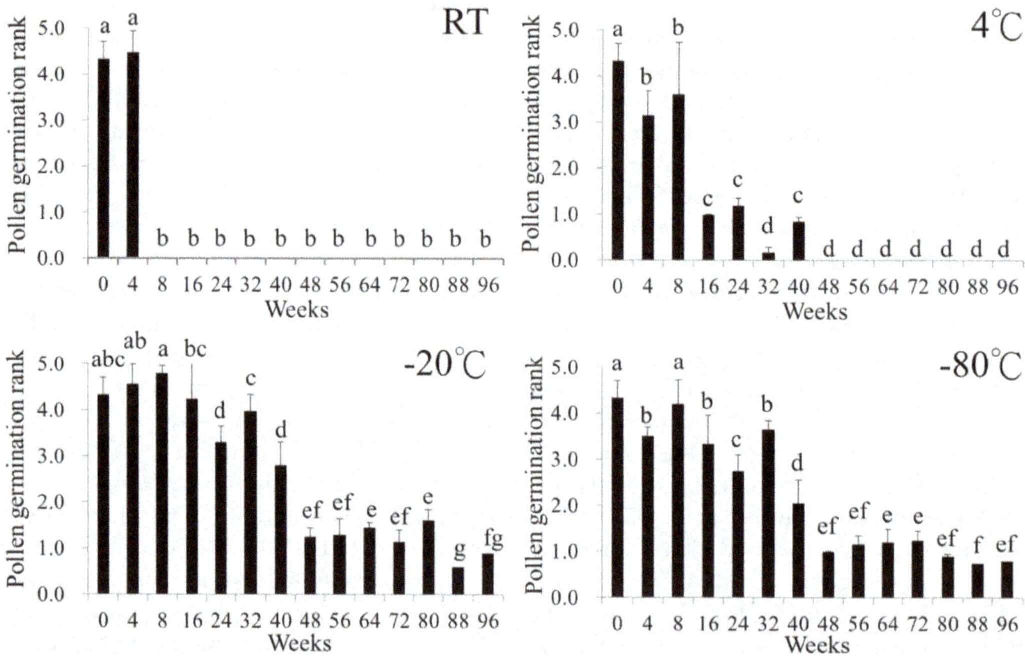


**Fig. 3** Pollen germination rate of *Phalaenopsis* Little Gem Stripes after storage at four temperatures. In vitro pollen germination of *P.* Little Gem Stripes was evaluated by incubating pollinia in Brewbaker and Kwack medium after 7 days and then stained. Bars represent mean ± SE, n = 4. Pollen germination rank was represented by ranks: (0) no-germination; (1) 1–20%; (2) 21–40%; (3) 41–60%; (4) 61–80%; (5) 81–100%

at 4 °C after 96 weeks (Fig. 4a, b). Pollen stored at both − 20 and −80 °C after 96 weeks exhibited pollen tube growth into stigma at 7 days (Fig. 4c, d) and at 35 days (Fig. 4e) and 70 days (Fig. 4f). Immature seeds were observed after 70 days (Fig. 4g) and viable seeds were obtained 4.5 months after pollination. The seeds could be germinated in vitro readily in a germination medium as reported previously (Huang et al. 2014). Since the germination of all examined capsules was all evenly and later developed into seedlings, therefore we did not perform statistical analysis.

The pollen stored at room temperature beyond 4 weeks did not show any viability (Fig. 3) and failed to produce any capsule after pollination, while those stored at lower temperatures resulted in successful fruit set (Table 1). At 4 °C for 56 weeks, 25% capsule set was obtained. When

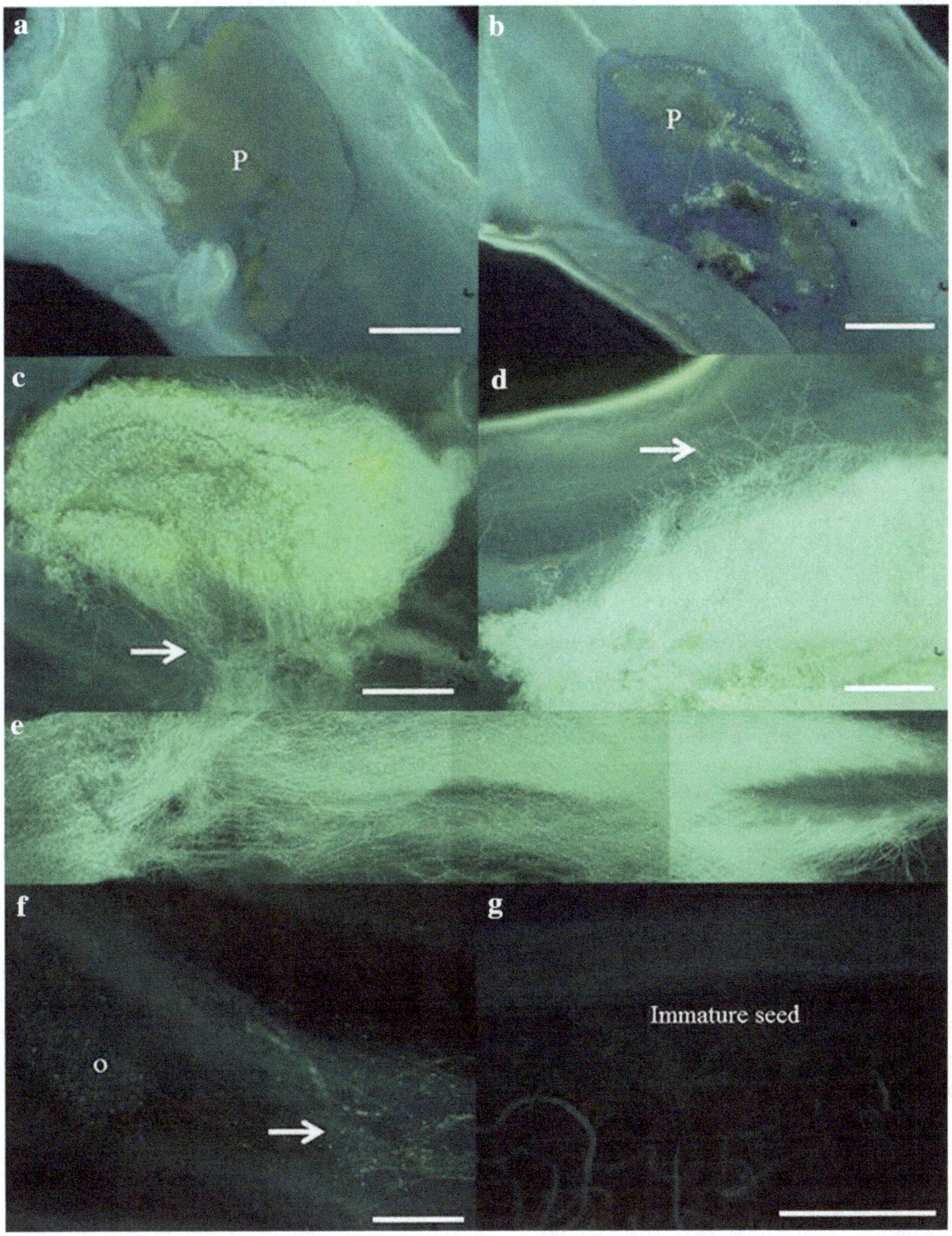


**Fig. 4** Pollen tubes of *Phalaenopsis* Little Gem Stripes grown in stigma and ovary cavity of *P.* Sogo Vivien 'F858' after storage. **a–d** The pollen taken from storage for 96 weeks at four temperatures was pollinated for 7 days. **a** room temperature storage; **b** 4 °C; **c** − 20 °C; **d** − 80 °C. **e** Pollen from storage at − 20 °C was pollinated for 35 days, and **f** for 70 days, **g** partially magnified from **e**. Arrow indicates pollen tube (**c**, **d**, and **f**). P, pollinia; O, ovules. Scale bars = 50 µm

pollen stored at both −20 and − 80 °C after 96 weeks, 50% fruit set was still recorded (Table 1). The mature capsules were harvested after 4.5 months and the relative amount of seeds decreased when the pollen was stored for a longer period (Table 2).

## Discussion

Fresh pollen at different stages of flower development showed germination ability with various degrees. Our previous study on *Doritaenopsis* meiosis revealed matured pollens finished meiosis and were kept at tetrad

**Table 1 Percentage of capsule set after pollination of stored pollinia at four temperatures**

| | Storage weeks | | | | | | | | | | | | | |
|---|---|---|---|---|---|---|---|---|---|---|---|---|---|---|
| | 0 | 4 | 8 | 16 | 24 | 32 | 40 | 48 | 56 | 64 | 72 | 80 | 88 | 96 |
| RT | 100[a] | 100 | 0 | 0 | 0 | 0 | 0 | 0 | 0 | 0 | 0 | 0 | 0 | 0 |
| 4 °C | 100 | 100 | 100 | 100 | 100 | 75 | 50 | 50 | 25 | 0 | 0 | 0 | 0 | 0 |
| − 20 °C | 100 | 100 | 100 | 100 | 100 | 100 | 100 | 100 | 100 | 50 | 25 | 25 | 25 | 50 |
| − 80 °C | 100 | 100 | 100 | 100 | 100 | 100 | 75 | 100 | 100 | 75 | 100 | 100 | 75 | 75 |

RT, room temperature

[a] *P.* Sogo Vivien 'F858' as female parent. Successful pollination percentages recoded after 4 months. n = 4

**Table 2 The amount of seeds in mature capsule after hand pollination of stored pollinia at four temperatures**

| | Storage weeks | | | | | | | | | | | | | |
|---|---|---|---|---|---|---|---|---|---|---|---|---|---|---|
| | 0 | 4 | 8 | 16 | 24 | 32 | 40 | 48 | 56 | 64 | 72 | 80 | 88 | 96 |
| RT | +++[a] | ++ | ND | ND | ND | ND | ND | ND | ND | ND | ND | ND | ND | ND |
| 4 °C | +++ | +++ | +++ | +++ | ++ | ++ | + | − | − | ND | ND | ND | ND | ND |
| − 20 °C | +++ | +++ | ++ | +++ | +++ | +++ | ++ | + | ++ | + | + | + | − | + |
| − 80 °C | +++ | +++ | ++ | ++ | ++ | ++ | ++ | + | + | + | + | + | + | + |

RT, room temperature

[a] ND, not determined (the capsules failed to develop for 7 days after pollination); seeds content per capsule: lacking (−), low (+), abundant (++), and very abundant (+++), respectively

stage that was ready for germination after pollination (Bolaños-Villegas et al. 2008). In cherimoya, pollen exhibited the highest germination rate at anther dehiscence than either at 30 and 5 h before dehiscence or at 20 h after dehiscence (Rosell et al. 2006). Wang et al. (2015) observed that the pollen in litchi showed the better quantity in the filaments extended farther before the anther cracked. Germination increased to 80% in *Anacamptis morio* at the beginning of anthesis (Marks et al. 2014), at which time the moisture of fresh pollen remains higher. The pollen germinability was affected not only by the anther dehiscence but also the pollen age (Bellusci et al. 2010; Marks et al. 2014). Our results showed the pollen of *P.* Little Gem Stripes showed the highest germinability at fifth flower development stage (Fig. 1).

The moisture content at reduced level is important in pollen longevity for long-term preservation. Pollinia weight was reduced to about 50% after 11 days of drying in our study (data not shown). In *Dactylorhiza maculate* pollen moisture of 10% showed the highest germination, however, the pollen germination decreased by increasing the moisture contents (Marks et al. 2014). Our work and others revealed that moisture content of pollen was critical before sub-zero storage. Previous reports have confirmed pollen viability when stored at lower temperature. Our result showed that partially stained pollinia were eminent beginning from 8 to 16 weeks of storage at low temperature. After 96 weeks of storage at 4, − 20 and − 80 °C, pollinia showed light-red to red stain, indicating partial viability remained by TTC staining (Fig. 2). Litchi pollen stored at 4 °C remained viable after 52 days (Wang et al. 2015). Martínez-Gómez et al. (2000) reported almond pollen could be stored at 4 °C for 8 weeks. In this study, the pollen of *P.* Little Gem Stripes stored at 4 °C for at least 40 weeks remained viable (Fig. 3). The pollen germinability stored at lower temperatures varies among plant cultivars. Pollen of four California almond cultivars, 'Nonpareil', 'Ne Plus Ultra', and 'Sonora', stored at 4 °C for 12 months retained 8–50% pollen germination (Martínez-Gómez et al. 2002). On the other hand, pollen of *Protea repens*, *P. magnifica*, *P. eximia*, and *P. aristata* stored at 4 °C could be germinated after storage for 270, 90, 90, and 30 days, respectively, and retained over 60% germination after 360 days (van der Walt and Littlejohn 1996). Furthermore, the pollen of *Dactylorhiza fuchsii* stored at − 20 °C for 6 years still had 64% germination (Marks et al. 2014). Pollen kept at a sub-freezing temperature was used for long-term storage of litchi (Wang et al. 2015).

The fertilizing ability rather than the viability of pollen stored at low temperature was affected (Lyakh et al. 1998; Marks et al. 2014; El-Homosany and Sayed 2015). Metz et al. (2000) observed that pollen of *Hylocereus* stored at 4 °C for 3 or 9 months exhibited only 60–70% fruit set after pollination, but storage at sub-freezing temperature for 3 or 9 months still had 100% fruit set. The pollen of *D. fuchsii* stored at − 20 °C for 6 years showed reduced viability and thus influenced seed yield after pollination and fruit setting (Marks et al. 2014). However, when the pollen of *Brassica napus* L. was stored at 3 or 10 °C, the seed number was decreased with increasing storage period (Lyakh et al. 1998). In our study, when pollen of *P.* Little Gem Stripes stored at both − 20 and − 80 °C after 96 weeks remained viable (Figs. 3, 4), and the pollination capability and amount of seeds were gradually reduced following by increasing storage period (Tables 1, 2) but was still better than those at 4 °C. Although seed yield after pollination of stored pollens of *D. fuchsii* were reduced with increasing storage time, the germination of seed was not affected (Marks et al. 2014). In our study, the mature seeds were successfully obtained 4.5 months after pollination of pollinia stored at subzero temperatures to a receptive parent *P.* Sogo Vivien 'F858' (data not shown).

## Conclusion

This study of subfreezing storage of *Phalaenopsis* pollinia demonstrated the possibility of long-term preservation of at least 96 weeks. For simplicity of viability determination after subfreezing storage, hand-pollination and seed set probably is the best measure in *Phalaenopsis* breeding program for easy operation.

#### Authors' contributions

SWC, CYL and FCC conceived the concept and designed the experiment. SCY carried out the experiment, analyzed data and drafted the manuscript. SWC, CYL and FCC read and commented the manuscript. All authors read and approved the final manuscript.

#### Acknowledgements

This work was supported by grants from the Council of Agriculture, Agriculture and Food Agency, Taiwan, under the contract numbers 105AS-9.6.1-FD-Z1 (12) and 106AS-8.5.3-FD-Z1 (11) to FCC. We thank Dr. Kotapati Kasi Viswanath (Pondicherry Central University, India) for careful reading and suggestions of the manuscript.

#### Competing interests

The authors declare that they have no competing interests.

#### Funding

The Council of Agriculture (COA), Agriculture and Food Agency, Taiwan. 105AS-9.6.1-FD-Z1 (12), 106AS-8.5.3-FD-Z1 (11).

#### References

Abdelgadir HA, Johnson SD, Van Staden J (2012) Pollen viability, pollen germination and pollen tube growth in the biofuel seed crop *Jatropha curcas* (Euphorbiaceae). S Afr J Bot 79:132–139

Ajeeshkumar S, Decruse SW (2013) Fertilizing ability of cryopreserved pollinia of *Luisia macroantha*, an endemic orchid of Western Ghast. Cryo Lett 34:20–29

Alexander MP (1969) Differential staining of aborted and nonaborted pollen. Stain Technol 44:117–122

Bellusci F, Musacchio A, Stabile R, Pellegrino G (2010) Differences in pollen viability in relation to different deceptive pollination strategies in Mediterranean orchids. Ann Bot 106:769–774

Bolaños-Villegas P, Chin SW, Chen FC (2008) Meiotic chromosome behavior and capsule setting in *Doritaenopsis* hybrids. J Amer Soc Hort Sci 133:107–116

Bou Daher F, Chebli Y, Geitmann A (2009) Optimization of conditions for germination of cold-stored *Arabidopsis thaliana* pollen. Plant Cell Rep 28:347–357

Brewbaker JL, Kwack BH (1963) The essential role of calcium ion in pollen germination and pollen tube growth. Am J Bot 50:859–865

Christenson EA (2001) Phalaenopsis: a monograph. Timber Press, Portland, pp 24–25

Council of Agriculture, Executive Yuan (2016) Agricultural statistic information (2016). http://agrstat.coa.gov.tw/sdweb/public/trade/tradereport.aspx. Accessed 29 Mar 2017

de Souza EH, Souza FVD, Rossi ML, Brancalleão N, da Silva Ledo CA, Martinelli AP (2014) Viability, storage and ultrastructure analysis of *Aechmea bicolor* (Bromeliaceae) pollen grains, an endemic species to the Atlantic forest. Euphytica 204:13–28

Deng Z, Harbaugh BK (2004) Technique for in vitro pollen germination and short-term pollen storage in Caladium. HortScience 39:365–367

Dutta SK, Srivastav M, Chaudhary R, Lal K, Patil P, Singh SK, Singh AK (2013) Low temperature storage of mango (*Mangifera inidica* L.) pollen. Sci Hort 161:193–197

El-Homosany AA, Sayed HA (2015) Effect of low temperature and cryopreservation on in vitro pollen germination of some olive cultivars. Am Euras J Agric Environ Sci 15:1803–1808

Galleta G (1983) Pollen and seed management. In: Moore J, Janick J (eds) Methods in fruit breeding. Purdue University Press, West Lafayette, pp 23–47

Heslop-Harrison J, Heslop-Harrison Y (1970) Evaluation of pollen viability by enzymatically induced fluorescence; intracellular hydrolysis of fluorescein diacetate. Biotechnol Histochem 45:115–120

Huang YW, Tsai YJ, Cheng TC, Chen JJ, Chen FC (2014) Physical wounding and ethylene-stimulated embryogenic stem cell proliferation and plantlet regeneration in protocorm-like bodies of *Phalaenopsis* orchids. Genet Mol Res 13:9543–9557

Khatun S, Flowers TJ (1995) The estimation of pollen viability in rice. J Exp Bot 46:151–154

Lora J, Pérez de Oteyza MA, Fuentetaja P, Hormaza JI (2006) Low temperature storage and in vitro germination of cherimoya (*Annona cherimola* Mill.) pollen. Sci Hort 108:91–94

Lyakh VA, Soroka AI, Kalinova MG (1998) Pollen storage at low temperature as a procedure for the improvement of cold tolerance in spring rape, *Brassica napus* L. Plant Breed 117:389–391

Marks TR, Seaton PT, Pritchard HW (2014) Desiccation tolerance, longevity and seed-siring ability of entomophilous pollen from UK native orchid species. Ann Bot 114:561–569

Martin FW (1959) Staining and observing pollen tubes in the style by means of fluorescence. Biotechnol Histochem 34:125–128

Martínez-Gómez P, Gradziel TM, Ortega E, Dicenta F (2000) Short-term storage of almond pollen. HortScience 35:1151–1152

Martínez-Gómez P, Gradziel TM, Ortega E, Dicenta F (2002) Low temperature storage of almond pollen. HortScience 37:691–692

Masum-Akond ASMG, Pounders CT, Blythe EK, Wang XW (2012) Longevity of crapemyrtle pollen stored at different temperatures. Sci Hort 139:53–57

Matison AM, Jensen CO, Dutcher RA (1947) Triphenyltetrazolium chloride as a dye for vital tissues. Science 106:294–295

Metz C, Nerd A, Mizrahi Y (2000) Viability of pollen of two fruit crop cacti of the genus *Hylocereus* is affected by temperature and duration of storage. HortScience 35:22–24

Parton E, Vervaeke I, Delen R, Vandenbussche B, Deroose R, De Proft M (2002) Viability and storage of bromeliad pollen. Euphytica 125:155–161

Peng H-Z, Jin Q-Y, Ye H-L, Zhu T-J (2015) A novel in vitro germination method revealed the influence of environmental variance on the pecan pollen viability. Sci Hort 181:43–51

Pritchard HW, Prendergast FG (1989) Factors influencing the germination and storage characteristics of orchid pollen. In: Pritchard HW (ed) Modern methods in orchid conservation: the role of physiology, ecology, and management. Cambridge Univ. Press, England, pp 1–16

Rosell P, Saúco VG, Herrero M (2006) Pollen germination as affected by pollen age in cherimoya. Sci Hort 109:97–100

Soares TL, Silva SO, Costa MAPC, Santos-Serejo JA, Souza AS, Lino LSM, Souza EH, Jesus ON (2008) In vitro germination and viability of pollen grains of banana diploids. Crop Breed Appl Biotechnol 8:111–118

Sorkheh K, Shiran B, Rouhi V, Khodambashi M (2011) Influence of temperature on the in vitro pollen germination and pollen tube growth of various native Iranian almonds (*Prunus* L. spp.) species. Trees 25:809–822

Sweet HR (1980) The genus *Phalenopsis*. The Orchid Digest, Day Printing Corp., Pomona

Vaknin Y, Disikowitch D (2000) Effects of short-term storage on germinability of pistachio pollen. Plant Breed 119:347–350

Vaknin Y, Mills D, Benzioni A (2003) Pollen production and pollen viability in male jojoba plants. Ind Crop Prod 18:117–123

van der Walt ID, Littlejohn GM (1996) Storage and viability testing of *Protea* pollen. J Amer Soc Hort Sci 121:804–809

Wang M-L, Hsu C-M, Chang L-C, Wang C-S, Su T-H, Huang Y-JJ, Jiang L, Jauh G-Y (2004) Gene expression profiles of cold-stored and fresh pollen to investigate pollen germination and growth. Plant Cell Physiol 45:1519–1528

Wang L-M, Wu J-F, Chen J-Z, Fu D-W, Zhang C-Y, Cai C-H, Ou L-G (2015) A simple pollen collection, dehydration, and long-term storage method for litchi (*Litchi chinensis* Sonn.). Sci Hort 188:78–83

# 10

# Effects of different growth temperatures on growth, development, and plastid pigments metabolism of tobacco (*Nicotiana tabacum* L.) plants

Li Yun Yang[1], Shuang Long Yang[1], Jun Ying Li[2], Jun Hong Ma[2], Tao Pang[2], Cong Ming Zou[2], Bin He[3] and Ming Gong[1*]

**Abstract**

**Background:** Temperature remarkably affects the growth and metabolism of plants. Tobacco is an important cash crop, and the long-term effects of different growth temperatures (18.5, 23.5 and 28.5 °C, daily average) on growth, development and plastid pigments metabolism of tobacco plants were investigated in this study.

**Results:** Compared with tobacco plants grown under 23.5 °C, treatments with 18.5 and 28.5 °C inhibited the expansion of leaves. The contents of superoxide anion ($O_2^{\cdot-}$), hydrogen peroxide ($H_2O_2$) and malonaldehyde (MDA) in the leaves were significantly increased under 28.5 °C from 0 to 60 days, which in turn accelerated the flowering and senescence of tobacco plants. By contrast, the treatment with 18.5 °C remarkably decreased $O_2^{\cdot-}$, $H_2O_2$ and MDA, and delayed the flowering and senescence. Furthermore, treatment with 18.5 °C significantly up-regulated the expression of glutamyl-tRNA reductase (Glu-TR) and magnesium chelatase (MgCH), and down-regulated the ferri chelatase (FeCH), protochlorophyllide oxidoreductase, chlorophyllase (CHLase), phaeophorbide a monooxygenase (PaO) and phytoene synthase (PSY), which further promoted the accumulation of chlorophyll (Chls) and reduced the carotenoids (Cars) in leaves. On the contrary, exposing to 28.5 °C remarkably down-regulated the Glu-TR and MgCH, and up-regulated the FeCH, CHLase, PaO and PSY, which in turn decreased the Chls and increased the Cars in tobacco leaves.

**Conclusion:** As compared with the plants grown under 23.5 °C, lower (18.5 °C) and higher (28.5 °C) growth temperature inhibited the growth of tobacco plants. In general, treatment with 28.5 °C accelerated the flowering and senescence of tobacco plants by enhancing the accumulation of $O_2^{\cdot-}$ and $H_2O_2$ in leaves, while exposing to 18.5 °C had the opposite effects. Treatment with 18.5 °C increased the content of Chls and reduced the Cars in leaves. In contrast, Treatment with 28.5 °C decreased the Chls and increased the Cars. Moreover, both $O_2^{\cdot-}$ and $H_2O_2$ took part in the breakdown of Chls in tobacco leaves to some extent. The results suggest that growth temperature could regulate growth, development, and plastid pigments metabolism, and 23.5 °C could be an optimal temperature for growth, development and metabolism of plastid pigments of tobacco plants under the experimental conditions.

**Keywords:** Growth temperature, Tobacco plants, Growth and development, Plastid pigments metabolism, Regulation

*Correspondence: gongming6307@163.com
[1] School of Life Sciences, Engineering Research Center of Sustainable Development and Utilization of Biomass Energy, Ministry of Education, Key Laboratory of Biomass Energy and Environmental Biotechnology of Yunnan Province, Yunnan Normal University, Kunming 650500, People's Republic of China
Full list of author information is available at the end of the article

## Introduction

Environmental temperature is one of the key factors that affect the yield and quality of crop plants by influencing their distribution, major physiological and biochemical processes (Jumrani and Bhatia 2014; Pham et al. 2015; Souza Lucéia et al. 2016; Wang et al. 2016; Sunoj et al. 2016; Yang et al. 2016; Xu et al. 2016).

In nature, leaf senescence is a process of programmed cell death induced in age-dependent manner and by various environmental cues, such as senescence associated genes (SAGs), genotype, hormones, light and temperature, etc. (Yoshida 2003; Sarwat et al. 2013; Gao et al. 2016). Senescence is the final step in leaf development, which accompanied with the turn yellow of leaf, photosynthesis decline and biomacromolecules degradation, including nucleic acids, proteins, plastid pigments, and fatty acids, and so on. Finally, senescence is usually terminated with the death and abscission of leaf (Ougham et al. 2005; Lim et al. 2007; Zhang et al. 2011; Kim et al. 2016). Leaf is not only the important photosynthetic apparatus for plants but also the crucial edible part of many vegetables for human being. Both the pre-mature and late-mature of plants will greatly affect the yield and quality of many crop plants (Chéour et al. 1992; Navabpour et al. 2003; Kim et al. 2011).

Reactive oxygen species (ROS) can be continuously generated in plant cell as a kind of byproduct of aerobic metabolism (Møller and Sweetlove 2010). Generally, the production of ROS is maintained at lower levels. When the environment changes, such as temperature, go beyond the range required for normal growth and development of plants, excessive generation and accumulation of ROS will induce the oxidative stress, which in turn damage the cellular structures and many bioactive molecules, thereby accelerating the senescence and death of plants (Gill and Tuteja 2010; Suzuki et al. 2012). Previous studies suggested that the metabolism disturbance resulted in a rapid accumulation of ROS in cell is one of the predominant factors that inducing senescence of plants (Mccord and Fridovich 1969; Pastori and Rio 1994). Moreover, the peak of $O_2^{\cdot -}$ generation was coincident with the initial of phase transformation of the membrane, which is caused by peroxidation of membrane lipids in senescing plant (Prochazkova and Wilhelmova 2007; Cheng et al. 2016). Additionally, it has been documented that both $O_2^{\cdot -}$ and $H_2O_2$ take part in the destroy of chloroplast and the breakdown of plastid pigments (Brennan and Frenkel 1977; Adachi and Shimokawa 1995; Zhao et al. 2011).

Chlorophylls (Chls) and carotenoids (Cars) are the predominant plastid pigments in higher plants. Previous study has shown that Chls plays a central and indispensable role in the processes of photosynthetic light-harvesting and energy transduction of photosynthesis (Bollivar 2006). There are about $10^9$ tons of Chls biosynthesis per year, which closely associated with the growth, development, yield and quality of many crop plants (Moser et al. 2009). Therefore, the metabolism of Chls is considered to be one of the most important biochemical pathways on earth (Eckhardt et al. 2004).

In recent years, the synthesis and breakdown of Chls in higher plants have been extensively studied (Gossauer and Engel 1996; Hörtensteine and Kräutler 2011; Steccanella et al. 2015). At the first step of Chls biosynthesis, the formation of aminolevulinate acid (ALA) is the rate-controlling point of whole pathway. Thereby, glutamyl-tRNA reductase (Glu-TR) is the key enzyme for metabolic and environmental control (Kumar and Soll 2000; Gough et al. 2003). At the branch point of Chls biosynthetic pathway, magnesium chelatase (MgCH) and ferri chelatase (FeCH) compete for protoporphyrin (Proto) to synthesis of heme and Mg-Proto, respectively, which also is an important regulating step in Chls biosynthesis. In addition, the higher level accumulation of heme can lead to inhibition of the synthesis of Chls in plants by feedback regulation (Guo et al. 1998; Terry and Kendrick 1999; Brumann et al. 2014). At the later steps of Chls biosynthesis, the light-dependent protochlorophyllide oxidoreductase (POR) catalyzes the reduction of protochlorophyllide (Pchlide) to chlorophyllide (Chlide). In the processes, the tripolymer of the POR-NADPH-Pchlide was disaggregated, and the Chlide was released accompany with the breakdown of POR (Apel et al. 1980; Armstrong et al. 1995; Holtorf and Apel 1996; Reinbothe et al. 2010; Kim and Apel 2012; Talaat 2013; Menon et al. 2016). Finally, Chlide was catalyzed by chlorophyll synthetase (CHL) and chlorophyllide a oxygenase (CAO), respectively, and then turned into Chl a and Chl b, respectively (Beale 1999, 2005).

The Chls breakdown is an important process that influences the quality of many crop plants and agricultural products, during the leaf senescence. In the initial steps of Chls breakdown, Chl b is converted into *Chl a* through the Chls cycle. First, the phytol chain was cleaved from Chls molecular, which was catalyzed by chlorophyllase (CHLase) and generated chlorophyllide a (Chlide a). Secondly, the central $Mg^{2+}$ was removed from Chlide by Mg-dechelatase (MDCase), and pheophorbide was subsequently catalyzed by phaeophorbide a monooxygenase (PaO) and turned into Red Chls catabolite (RCC). Finally, RCC was transformed into non-flourescent Chls catabolite (NCC). This whole process of Chls breakdown was usually named as 'PaO pathway' (Matile and Hörtensteinera 1999; Hörtensteiner 2006; Barry et al. 2008). Previous studies also have demonstrated that Chls catabolism is remarkably regulated by CHLase, MDCase and PaO in plants (Tanaka and Tanaka 2006; Ougham et al. 2008).

Carotenoids (Cars) are the second most abundant naturally occurring pigments on earth, with more than

750 members, which participate in various biological processes in plants, including photosynthesis, photoprotection, oxidation resistance, growth and development (Demming-Adams 1990; Bartley and Scolnik 1995; Stahl and Sies 2005; Shumskaya and Wurtzel 2013; Dias et al. 2014; Esteban et al. 2015; Liu et al. 2015). The first committed step in the process of Cars biosynthesis is the condensation of two geranylgeranyl diphosphate (GGPP) molecules to produce 15-cis phytoene, which is catalyzed by phytoene synthase (PSY) (Liu et al. 2015; Nisar et al. 2015).

Tobacco is one of the most important cash crop plants that widely cultivated worldwide. Since tobacco plants are a thermophilic crop plant originating from tropical region, they are sensitive to the change of temperature (Yamori et al. 2010; Popov et al. 2013; Zhang et al. 2013). It had been reported that the composition and contents of Chls and Cars in tobacco leaves are closely related to the quality and flavor of tobacco products (Kaneko and Harada 1972; Shigenage et al. 1994). Additionally, global temperature is predicted to increased 1.5–4.5 °C by the end of 21st century, and the growth, development, yield and quality of many crop plants will be negatively impacted by both sub- or sup-optimal temperatures (Porter 2005; Sunoj et al. 2016).

Numerous studies had been done to investigate the effects of different temperatures on plants growth and contents of Chls and Cars. However, in these studies, plants were usually exposed to cold or heat stress environment, and these researches were usually conducted under field condition. They were vulnerable to effect of variation of surrounding environment, such as light intensity, rainfall, humidity and temperature. Moreover, only the effects of the short-term or constant temperature on growth and Chls or Cars contents of plants had been examined (Tewari and Tripathy 1998; Zhao et al. 2007, 2011; Liu et al. 2012; Kalisz et al. 2016; Zhou et al. 2016). Until now, it is still unclear about the effects of long-term and non-stress temperatures on growth, development, senescence, and plastid pigments metabolism of plants. Therefore, the present study was conducted in climate chambers to evaluate the long-term effects of different and non-stress growth temperatures on growth, development and plastid pigments metabolism of tobacco plants.

## Materials and methods

### Plant material and experimental design

The experiment was conducted in climate chambers at Yunnan Academy of Tobacco Agricultural Sciences (24°34′N, 102°54′E) from March to August, during 2015 and 2016. Tobacco seeds (*Nicotiana tabacum* L. cv. Yunyan No. 87) were germinated in trays filled with a mixture of peat and vermiculite (2:1), and seedlings were grown under natural light in a greenhouse. The seedlings were transplanted into plastic buckets (volume: 3.14 × 0.04 × 0.5 $m^3$) after 30 days. One seedling per bucket, and cultivated under field condition, till the 12th leaf expanded and leaf length reached 17 cm (marked as 0 day). These plants were then transferred into the climate chambers for the following cultivation under different growth temperatures (Fig. 1).

Three climate chambers (*Kulan Technology Co Ltd., Beijing, China*) were used in the study. Each chamber contained six lamp-supporting brackets with six high pressure sodium lamps (HPSL). Plants were grown under same light intensity (PPFD: 350 μmol $m^{-2}$ $s^{-1}$), relative humidity at 65%/50% (day/night), $CO_2$ concentration at 400 ppm, and photoperiod 14 h/10 h (day: 06:00–19:59, night: 20:00–05:59). Eighty plants were grown in each chamber set to different temperature regimes of 18.5, 23.5 and 28.5 °C (daily average). Sampling of 12th leaf was carried out at 0, 15, 30, 45 and 60 days, and 12 plants were used each time, respectively, after exposed to different growth temperatures.

### Measurement of leaf length and leaf width

Length and width of 12th leaves were measured with a ruler.

### Assay of MDA and soluble protein content in tobacco leaves

MDA was extracted by 10% trichloroacetic acid (TCA), and mixed with 0.6% thiobarbituric acid (TBA). The MDA content was determined by UV/VIS spectrophotometer at 450, 532 and 600 nm, and expressed as nmol $g^{-1}$ FW (Heath and Packer 1968).

Soluble protein content was assayed by coomassie brilliant blue method and determined at 595 nm by spectrophotometry, and expressed as mg $g^{-1}$ FW (Braford 1976).

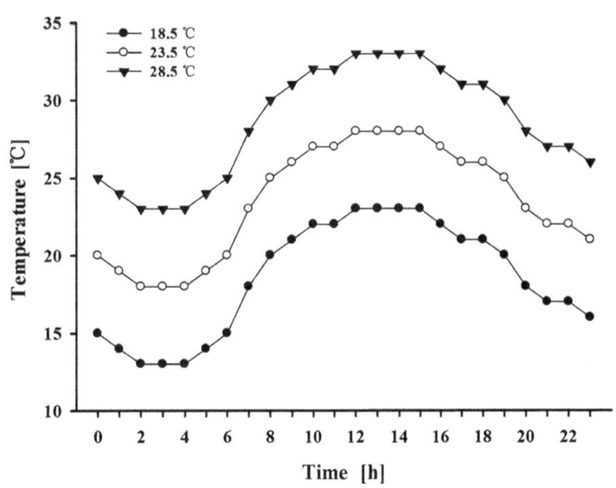


**Fig. 1** Different growth temperature settings in experiments

### Analysis of superoxide and hydrogen peroxide in tobacco leaves

Superoxide was extracted by Tris–Hcl buffer (50 mmol, pH 7.5) containing 1% polyvinylpyrrolidone (PVP) and EDTA (1 mmol $L^{-1}$). The supernatant was mixed with Tris–Hcl buffer (50 mmol, pH 7.5) and 300 μL XTT (100 μmol $L^{-1}$). The superoxide was analyzed by UV/VIS spectrophotometer (DU800, Beckman, USA) at 470 nm, and expressed as nmol $g^{-1}$ FW (Frahy and Schopfer 2001).

Hydrogen peroxide was extracted by phosphate buffer (50 mM, pH 6.5) containing 1 mM hydroxylamine. And the extracted solution was mixed with 0.1% titanium sulfate in 20% (v/v) $H_2SO_4$. The $H_2O_2$ content was determined by UV/VIS spectrophotometer at 410 nm, and expressed as μmol $g^{-1}$ FW (Li et al. 2012).

### Determination of Chls and Cars contents in tobacco leaves

Chls and Cars were extracted by 80% cold acetone, and determined at 663 (Chl a), 646 (Chl b) and 470 nm (Car) by UV/VIS spectrophotometer, and expressed as mg $g^{-1}$ FW (Dere et al. 1998).

### Assay of relative mRNA abundance of genes by real-time PCR

Fresh 12th leaf tissue was sampled at 0, 15, 30, 45 and 60 days, respectively, and reserved in liquid nitrogen. 0.1 g leaf tissue was used for extraction of total RNA according to the instruction of RNAiso plus reagent (TakaRa, Japan). Total RNA was dissolved with RNase free $H_2O$, and the cDNA was synthesis by PrimeScript RT reagent Kit with gDNA Eraser (Perfect Real Time,Takara, Japan). Sequence of genes, which were responsible for the biosynthesis and degradation of Chls and Cars was derived from the result of transcriptome analysis of our research team. The expression of genes in Chls and Cars metabolism pathway was detected according the method of SYBR Premix Ex Taq (Perfect Real Time, Takara, Japan). Primers for qPCR analysis were showed in Table 1.

### Statistical analysis

All statistical analyses were conducted using *SPSS 11.5* (NY, USA) with the *Turkey's text* ($P < 0.05$), and the means and standard deviation (SD) were performed by analysis of variance (ANOVA) procedure using multiple comparisons. The figures were drawn with *Sigmaplot* 12.0 (*Systa software inc.*, Chicago, IL, USA).

## Results

### Leaf length and width

The length and width of tobacco leaves were rapidly increased during 0–30 days, and reached the maximum size from 30 to 60 days. Treatment with 18.5 and 28.5 °C decreased their length and width of leaves, compared with plants under 23.5 °C (Fig. 2a, b).

### Contents of MDA and soluble protein in tobacco leaves

The content of MDA was increased, while the soluble protein was decreased in tobacco leaves during 0–60 days. Compared with plants grown under 23.5 °C, treatment with 28.5 °C induced the accumulation of MDA in leaves, while exposing to 18.5 °C reduced the MDA content (Fig. 3a). Moreover, the reduction of the soluble protein in leaves was accelerated under 28.5 °C. On the contrary, the breakdown of soluble protein was slowed down under 18.5 °C, compared with plants grown under 23.5 °C (Fig. 3b).

**Table 1 Primer sequences used in real-time PCR analysis**

| Gene and abbreviation | Forward primer (5′–3′) | Reverse primer |
|---|---|---|
| β-Actin | CTGAGGTCCTTTTCCAACCA | TACCCGGGAACATGGTAGAG |
| CL10247<br>Glutamyl-tRNA reductase (Glu-TR) | TGGTGTCCGTTTCGCTGT | GCTGTGGTATCTTCTGGCTTTTC |
| CL941. Contig 3<br>Ferri chelatase (FeCH) | GGACGAGAAATGAATCCTACCG | GCCCCGACAGAACAAAACA |
| Gi2318136<br>Magnesium chelatase (MgCH) | TGATGCCACATTCCAGAACC | GGCTTCTTCCCGTCTTTCCT |
| CL1870. Contig 2<br>Protochlorophyllide oxidoreductase (POR) | CCACGAAGAAACTGGCATTACA | CTGGAATGGAGGGAAAAGGAG |
| CL4058. Contig 1<br>Chlorophyllase (CHLase) | CTCTCTTCCCAGCTTGTGCTC | CCTTGCCTCTAATCCCTTTTGTT |
| Unigene40510_Rwg20A<br>Mg-dechelatase (MDcase) | TCTATCCATCATCTGGCGTTTG | GGCTTCTTCCCGTCTTTCCT |
| Unigene38644_Rwg20A<br>Phaeophorbide a monooxygenase (PaO) | ATCGCTTCGTCTTGGCATTT | TCTATCCATCATCTGGCGTTTG |
| CL4383.Contig 1<br>Phytoene synthase (PSY) | GGAATTTGGGCTTGTTGAGTG | TTGCGGAGTTATGTGGGATG |

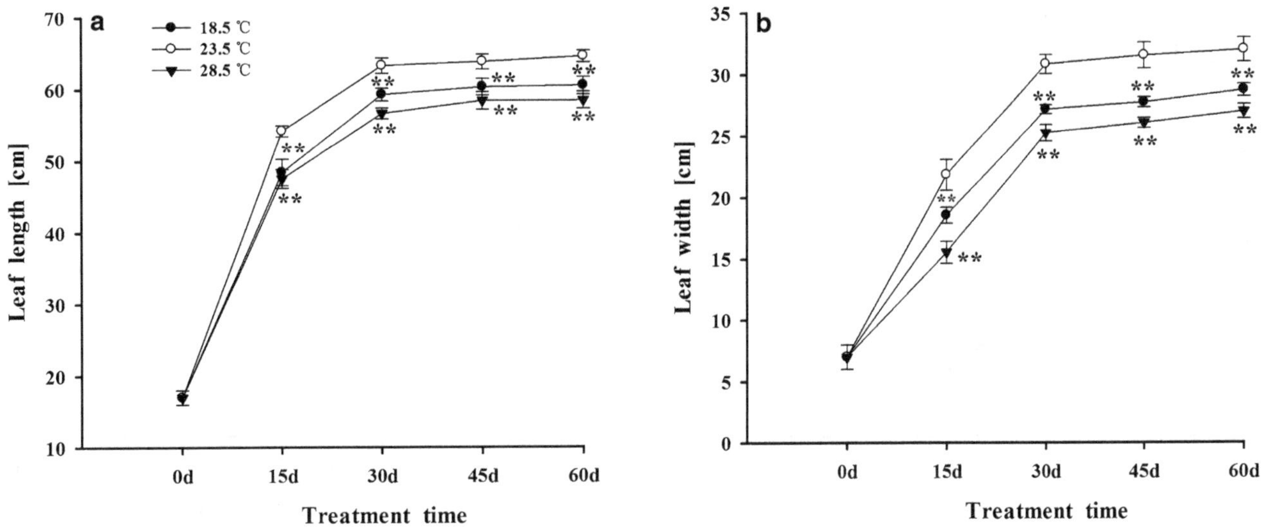


**Fig. 2** Effects of different growth temperatures on leaf length (**a**) and width (**b**) of tobacco leaves. Level of significant difference: ** and * indicate significant difference at $P \leq 0.01$ and $P \leq 0.05$, respectively, same as the following figures

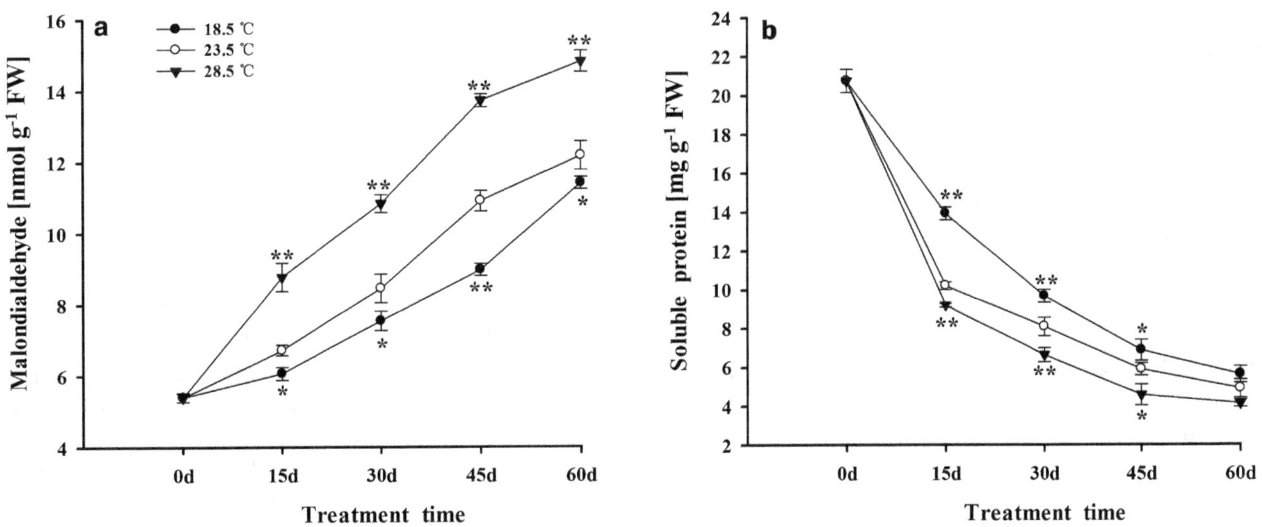


**Fig. 3** Effects of different growth temperatures on contents of MDA (**a**) and soluble protein (**b**) in tobacco leaves

### Levels of $O_2^{\cdot-}$ and $H_2O_2$ in tobacco leaves

The levels of $O_2^{\cdot-}$ and $H_2O_2$ were gradually increased in tobacco leaves during 0–60 days. Compared with plants grown under 23.5 °C, treatment with 28.5 °C enhanced the accumulation of $O_2^{\cdot-}$ and $H_2O_2$ in leaves, while treatment with 18.5 °C decreased $O_2^{\cdot-}$ and $H_2O_2$ levels (Fig. 4a, b).

### Accumulation of Chls in tobacco leaves

The contents of Chla, Chlb and total Chls were gradually increased in tobacco leaves during 0–15 days, and then declined from 15 to 60 days (Fig. 5a–c). Compared with plants grown under 23.5 °C, the content of Chla, Chlb and total Chls in leaves increased under 18.5 °C, while the Chla and total Chls were decreased under 28.5 °C.

### Expression of Chls biosynthesis related genes in tobacco leaves

The expression of Glu-TR and MgCH were increased in tobacco leaves from 0 to 15 days, and then they decreased during 15–60 days. Compared with plants under 23.5 °C, Treatment with 18.5 °C up-regulated the expression of Glu-TR and MgCH but decreased the expression of POR and FeCH in leaves. Treatment with 28.5 °C

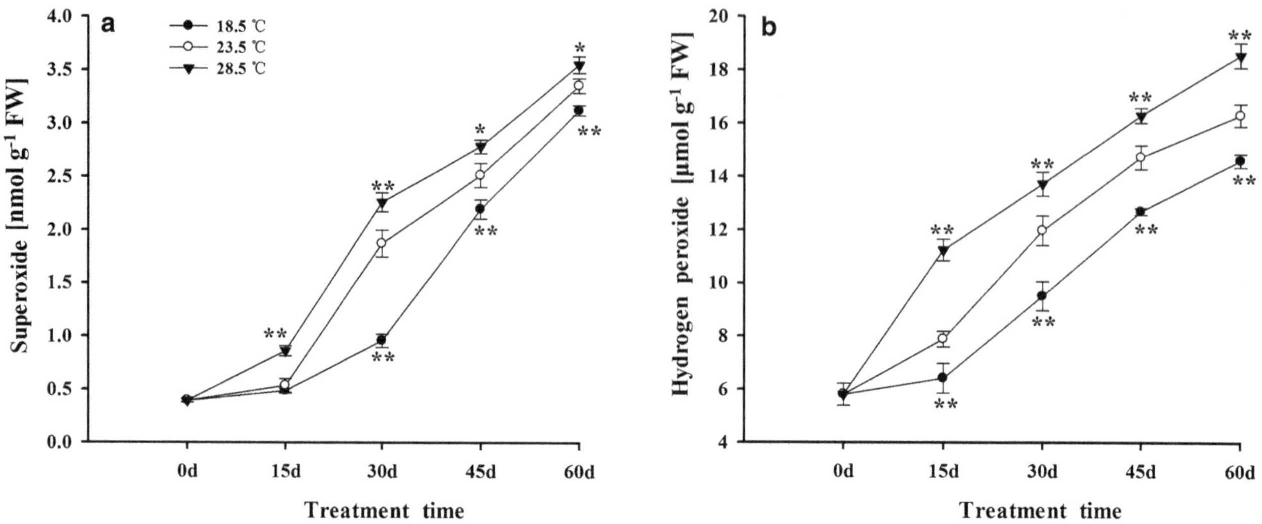


**Fig. 4** Effects of different growth temperatures on levels of superoxide (**a**) and hydrogen peroxide (**b**) in tobacco leaves

down-regulated the expression of Glu-TR and MgCH but increased the expression of FeCH (Fig. 6).

### Expression of Chls degradation related-genes in tobacco leaves

In general, the expression of CHLase and PaO increased, while the MDCase was decreased in tobacco leaves during 0–60 days. Treatment with 28.5 °C significantly up-regulated the expression of CHLase and PaO in leaves. On the contrary, treatment with 18.5 °C down-regulated the expression of CHLase and PaO, compared with plants grown under 23.5 °C (Fig. 7a, c).

### Content of Cars and expression of PSY gene in tobacco leaves

The content of Cars and the expression of PSY gene in tobacco leaves were gradually decreased from 0 to 60 days. Compared with plants grown under 23.5 °C, the content of Cars and the expression of PSY gene increased when grown under 28.5 °C, and decreased under 18.5 °C (Fig. 8).

## Discussion

### Temperature affected the growth and development of tobacco plants

Temperature influences the distribution, growth, development and metabolism of plants (Sunoj et al. 2016; Yang et al. 2016; Xu et al. 2016). Our previous study also found that treatment with 16.5 and 30.5 °C remarkably inhibited the growth and photosynthesis of tobacco plants, compared with plants under 23.5 °C (Yang et al. 2017). In the present work, it was observed that tobacco plants showed distinct growth responses to different temperatures. Compared with plants grown under 23.5 °C, treatment with 18.5 and 28.5 °C significantly inhibited the growth and expansion of tobacco leaves (Fig. 2a, b). It was shown that both the relative lower temperature (16–18 °C) and the relative higher temperature (35 °C) remarkably inhibited the growth of leaves and reduced the leaf area of tomato, cherry and chicory plants, compared with the plants grown under medium ones (22–25 °C) (Venema et al. 1999; Mathieu et al. 2014). Plant morphogenesis was regulated by various environmental factors including light, temperature and so on (Berleth and Sachs 2001; Frankin 2009). In the present study, it was also observed that different temperatures significantly affected the morphogenesis of tobacco plants. Exposing to 18.5 °C promoted the elongation of petiole and stem, reduced leaf area and increased the angle between leaf and stem of tobacco plants. Whereas, treatment with 28.5 °C reduced plant height, leaf area, and the angle between leaf and stem of plants, compared with plants grown under 23.5 °C. Similar finding has also been observed when *Arabidopsis thaliana* plants were grown at 16, 22 and 28 °C (Frankin 2009).

Flowering was controlled by plant genotype, hormone and environmental factors (Cerdán and Chory 2003; Shalit et al. 2009; Song et al. 2013). This study observed that treatment with 28.5 °C significantly accelerated the flowering and senescence of tobacco plants; in contrast, treatment with 18.5 °C greatly delayed the flowering and senescence, as compared with those plants grown under 23.5 °C. Other study also demonstrated that 16 °C delayed the flowering time of *Arabidopsis thaliana* than those under 23 °C (Blázquez et al. 2003). Additionally, treatment of witloof plants with 35/28 °C accelerated the flowering than those plants under 17 °C (Mathieu et al. 2014).

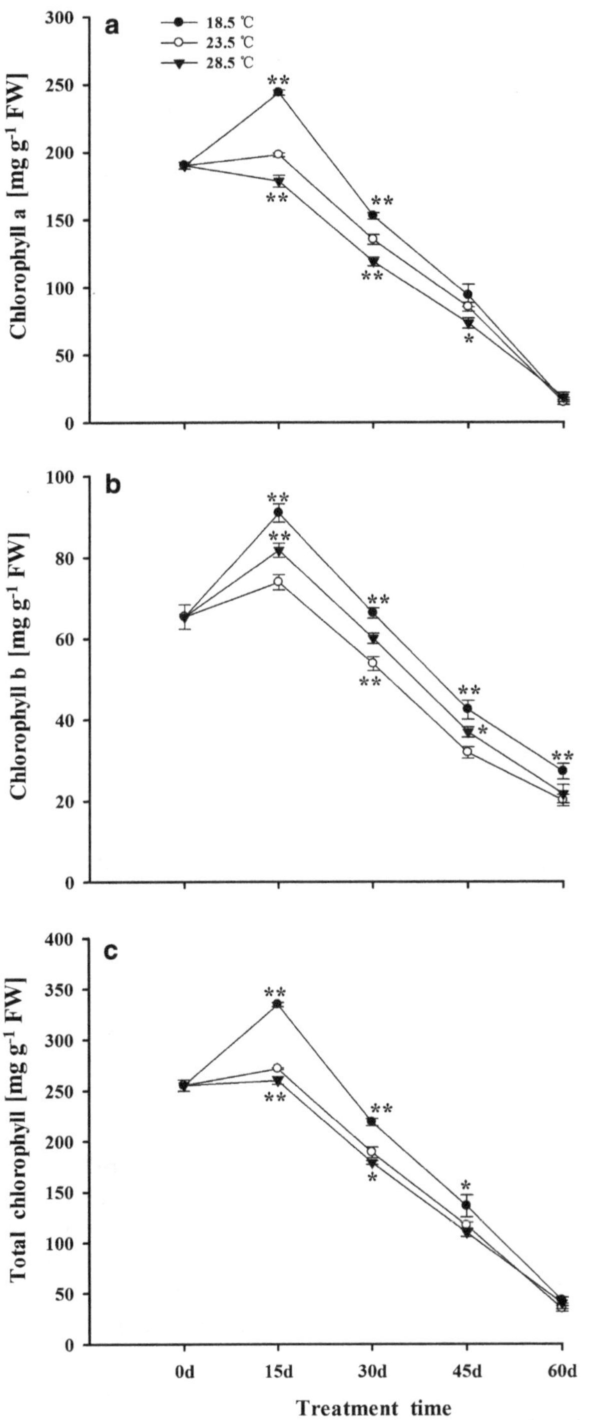


**Fig. 5** Effects of different growth temperatures on contents of Chla (**a**), Chlb (**b**) and total Chl (**c**) in tobacco leaves

## Temperature regulated the senescence of tobacco plants

Excessive accumulation of ROS in cell is one of the predominant factors inducing the senescence of plants. Moreover, the increase in MDA and the decrease in soluble protein in plants were normally considered the indicator of senescence of plants (Pastori and Rio 1994; Cheng et al. 2016). It was shown that the activities of proteolytic enzymes and the accumulation of ROS are increased during senescence of plants, accompanying with the rapid breakdown of proteins, especially the soluble proteins, and the increase in MDA in plants (Belknap and Garbarino 1996; Yoshida 2003; Lim et al. 2007). In the present study, treatment with 28.5 °C remarkably speeded up the degradation of soluble protein and increased the content of MDA in tobacco leaves, which in turn accelerated the senescence of tobacco plants. On the contrary, exposing to 18.5 °C slowed down the breakdown of soluble protein and reduced the content of MDA in leaves, followed by delaying of senescence of tobacco plants, when compared with plants grown under 23.5 °C (Fig. 3a, b). The finding had also been observed when wheat, sunflower and cucumber plants were grown at relative higher temperature of 33–44 °C (Ferguson et al. 1993; Zhao et al. 2011; Haba et al. 2014). Similarly, treatment with 15/13 °C (day/night) delayed the senescence of leaves in soybean, when compared with those under 30/25 °C (Kao 1980).

In the present work, the level of ROS was gradually increased in tobacco leaves during 0–60 days. In addition, treatment with 28.5 °C remarkably increased the levels of $O_2^{\cdot-}$ and $H_2O_2$ in tobacco leaves, which in turn accelerated the process of senescence of tobacco plants. On the contrary, exposing to 18.5 °C significantly decreased the $O_2^{\cdot-}$ and $H_2O_2$ levels in leaves, followed by delaying the senescence of tobacco leaves, when compared with plants grown under 23.5 °C (Fig. 4a, b). The above-mentioned results could be ascribed to the primary factors affecting senescence of tobacco plants, which speeded up by 28.5 °C, while delayed by 18.5 °C in the study. It has also been reported that high temperatures decreased the photosynthesis and increased the contents of $O_2^-$ and $H_2O_2$ in citrus, reed and tomato plants (Guo et al. 2006; Song et al. 2008; Ogweno et al. 2009). Additionally, it was demonstrated that higher temperature environment enhanced the generation and accumulation of ethylene, and then accelerated the senescence of soybean plants (Djanaguiraman and Prasad 2010), which can also be considered as a potential factor accelerating senescence of tobacco plants by 28.5 °C in this study.

## Temperature controlled metabolism of Chls in tobacco leaves

It had been shown that contents of Chls in plants are influenced by environmental temperature. It was observed that 18.5 °C significantly increased the content of total Chls in tobacco leaves, whereas exposing to 28.5 °C decreased Chls, when compared with plants

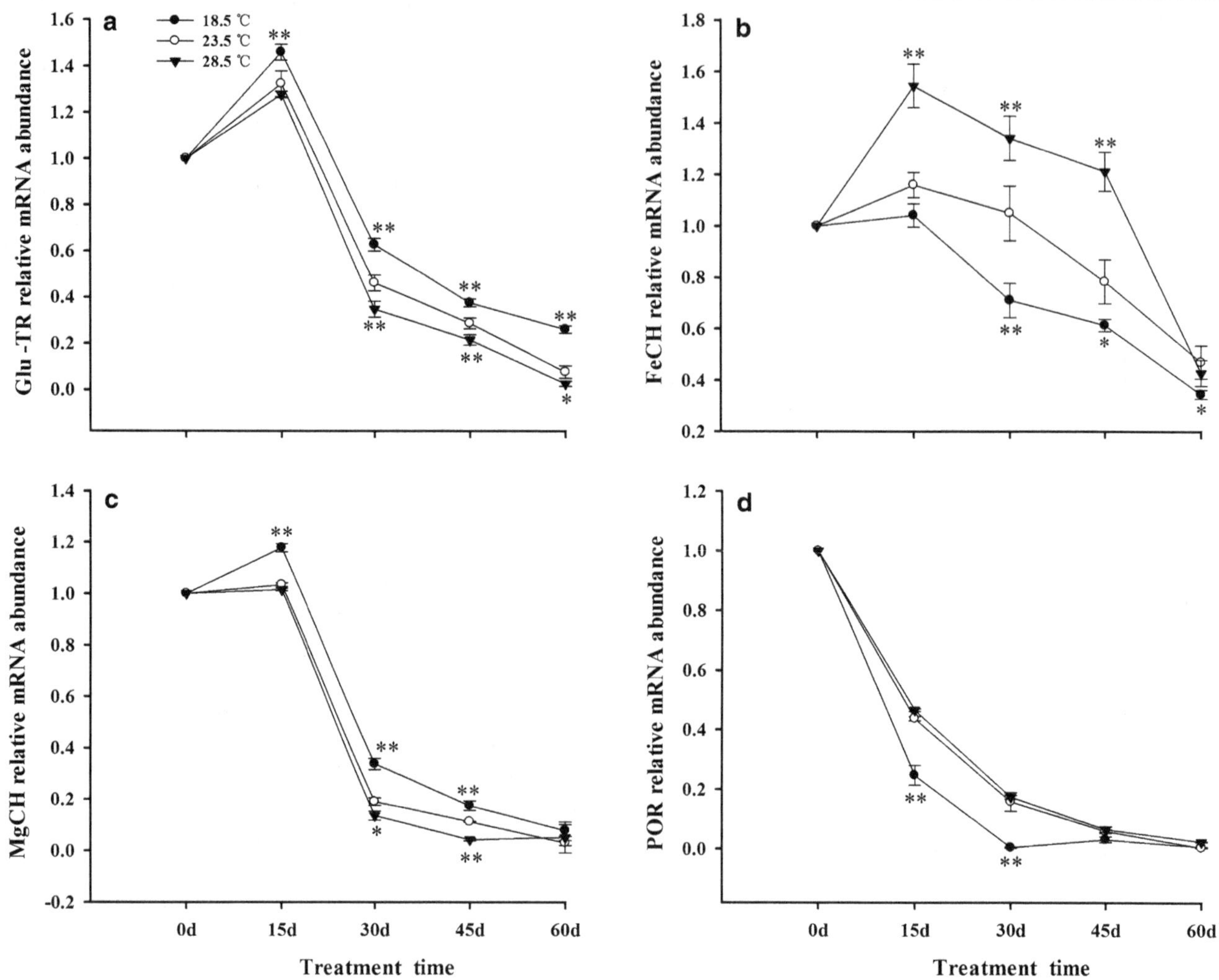


**Fig. 6** Effects of different growth temperatures on expression of Chls biosynthesis related-genes in tobacco leaves. Glu-TR (**a**); FeCH (**b**); MgCH (**c**); and POR (**d**), respectively

grown under 23.5 °C (Fig. 5). Zhao et al. (2011) reported that 36 °C reduced the content of Chls in cucumber leaves. On the other hand, it was found that treatment with 16/14 °C (day/night) increased the level of Chls in lycopersicon leaves (Venema et al. 1999). It was also observed that the degradation rate of Chls is more rapidly than that of Cars in tobacco leaves, and this finding was also reported in other study (Whitfield and Rowan 1974). Additionally, Grover et al. (1986) have demonstrated that the senescence-induced degradation of Chls was faster at 35 °C than that at 25 °C in detached leaves of wheat, and Cars degradation was faster at 25 °C than that at 35 °C.

The accumulation of plastid pigments in plants is determined by both the processes of their synthesis and degradation. Glu-TR catalyzes the synthesis of ALA, which is a key step of biosynthesis of Chls in plants (Cornah et al. 2003). It was found that 18.5 °C increased the expression of Glu-TR and MgCH, but it decreased FeCH in tobacco leaves (Fig. 6), which in turn promoted the synthesis of Chls in leaves (Fig. 5). On the contrary, treatment with 28.5 °C significantly reduced the expression of Glu-TR and MgCH, but increased FeCH (Fig. 6), followed by reducing the content of Chls in leaves (Fig. 5), as compared with the plants grown under 23.5 °C. In addition, it had been reported that the enzyme activity of Glu-TR was also affected by accumulation of heme, suggesting that high level of heme would inhibit the activity of Glu-TR by feedback regulation and reduce the synthesis of Chls in plants (Terry and Kendrick 1999; Armas-Ricard et al. 2011). The higher expression of FeCH in leaves also was observed in the study, and this could be ascribed to the reduction of Chls in tobacco leaves under 28.5 °C. At the later steps of Chls biosynthesis in plants, POR catalyzes the conversion of Pchlide into

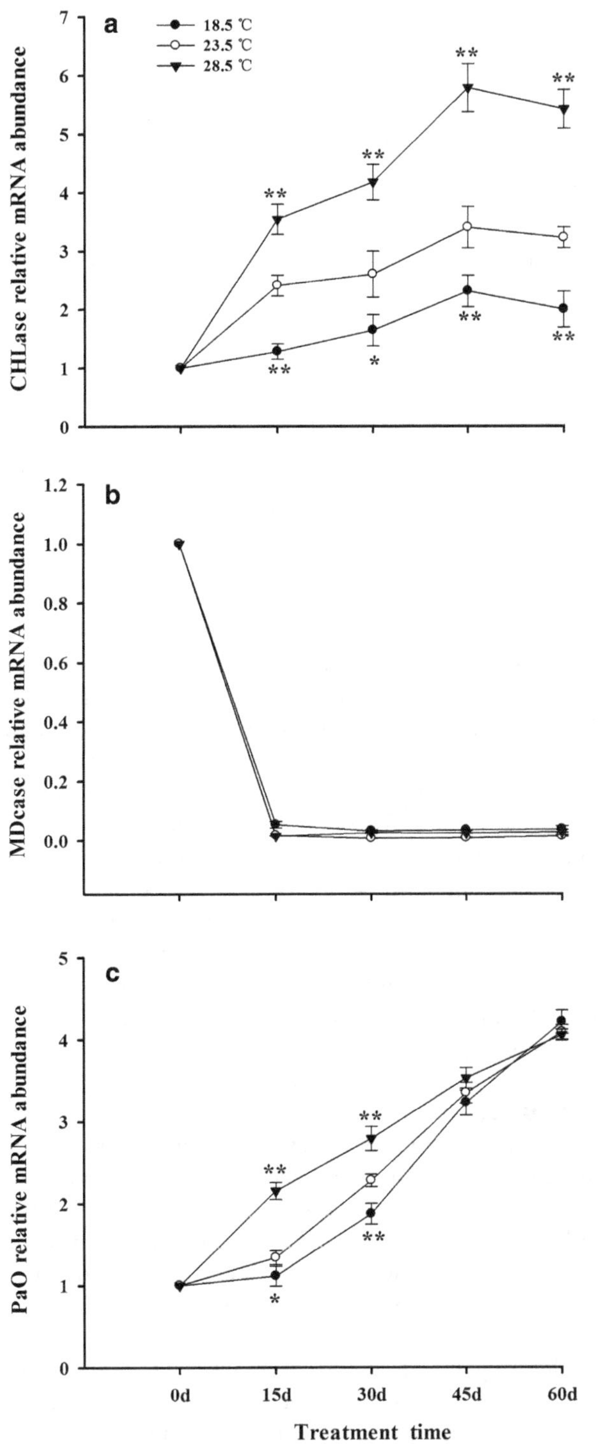


**Fig. 7** Effects of different growth temperatures on expression of Chls degradation related-genes in tobacco leaves. CHLase (**a**); MDcase (**b**); and PaO (**c**), respectively

Chlide, accompanying with the degradation of POR and the synthesis of Chls (Beale 1999, 2005; Kim and Apel 2012; Talaat 2013; Menon et al. 2016). In the present work, it was observed that 18.5 °C significantly reduced the expression of POR gene in tobacco leaves during 15–30 days (Fig. 6d), indicating that 18.5 °C accelerated the degradation of POR and the release of Chlide, which in turn promoted the synthesis of Chls in leaves, when compared with plants grown under 23.5 °C.

CHLase is considered as the first enzyme in the pathway of Chls breakdown, which is intimately concerned with the stability and degradation of Chls molecules in the chloroplast (Moll et al. 1978; Matile et al. 1997; Todorov et al. 2003; Harpaz-Saad et al. 2007). Generally, the activity of CHLase is decreased progressively during the senescence of plants, indicating that Chls degradation was also regulated by other factors (Yamauchi and Watada 1991; Minguez-Mosquera and Gallardo-Guerrero 1996). Results showed that treatment with different temperatures increased the expression of CHLase gene in tobacco leaves from 0 to 45 days, then decreased the expression from 45 to 60 days (Fig. 7a). Such a phenomenon has also been observed in leaves of *Phaseolus vulgaris* (Fang et al. 1998).

It was demonstrated that the level of endogenous abscisic acid (ABA) was increased and the leaf senescence was accelerated during the process of growth and development of plant (Zhang et al. 2012). Moreover, the accumulation of ABA was elevated when plants were grown under stress conditions, such as drought, chilling and extreme temperature, followed by up-regulating the expression of CHLase of plants (Guo and Gan 2005; Gupta et al. 2012). Additionally, it was reported that high temperature enhanced the de novo synthesis of CHLase in plants (Trebitsh et al. 1993; Gong and Mattheis 2003). These could be considered to be the primary factors for the increase in expression of CHLase gene in tobacco leaves under 28.5 °C, which in turn reduced the content of Chls in leaves in this study.

Removal of $Mg^{2+}$ from the Chlide is an important step in Chls breakdown pathway and leaf senescence (Costa et al. 2002). In the present work, expression of MDcase gene was drastically decreased in tobacco leaves from 0 to 15 days, then maintained at a lower level during 15–60 days. No significant difference of MDcase was observed in leaves among different growth temperature treatments (Fig. 7b). Previous study demonstrated that the stable activity of MDCase depends on the continuously synthesis of soluble protein of plants (Langmeier et al. 1993). However, it was observed that content of soluble protein was decreased during the process of senescence of tobacco leaves, which could be responsible for the decrease of expression of MDCase gene in tobacco leaves.

Finally, PaO converts the pheophorbide into RCC, which is induced by senescence of plants (Smart 1994;

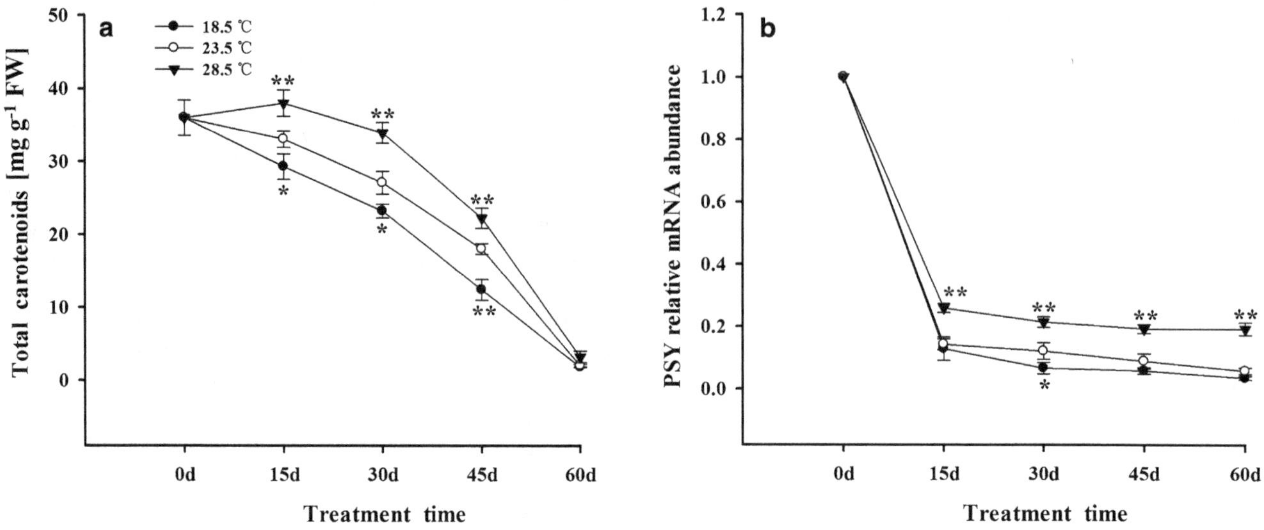


**Fig. 8** Effects of different growth temperatures on content of Cars (**a**) and expression of PSY gene (**b**) in tobacco leaves

Hörtensteiner et al. 1995; Matile and Hörtensteinera 1999; Hörtensteiner 2006). In this work, treatment of plants with 28.5 °C up-regulated the expression of PaO gene, followed by accelerating the breakdown of Chls in tobacco leaves. On the contrary, exposing to 18.5 °C significantly down-regulated the PaO and slowed down the degradation of Chls in leaves, when compared with the plants grown under 23.5 °C (Fig. 7c). Previous study has also found that treatment with heat shock increased the activity of PaO, followed by reducing the content of Chls and accelerating the senescence of barley leaves (Rodoni et al. 1998). Our results suggest that treatment with 28.5 °C speeded up the breakdown of Chls by up-regulating the expression of CHLase and PaO genes in tobacco leaves, and exposing to 18.5 °C slowed down the degradation of Chls by down-regulating the expression of CHLase and PaO genes in the leaves.

Furthermore, the change in level of ROS was coincident with the variation of content of Chls in tobacco leaves under different growth temperatures (Figs. 4, 7), which indicated that ROS takes part in the breakdown of Chls in leaves to some extent. Similar results also had been reported in the previous studies (Brennan and Frenkel 1977; Adachi and Shimokawa 1995; Zhao et al. 2011).

### Temperature influenced metabolism of Cars in tobacco leaves

Previous studies showed that PSY is the key enzyme for synthesis of Cars in plants, and overexpression of PSY gene remarkably increased the content of Cars in *Brassica napus* (Shewmaker et al. 1999; Shumskaya and Wurtzel 2013). In the present work, it was observed that 28.5 °C significantly increased the expression of PSY gene in tobacco leaves, which in turn increased the accumulation of Cars in leaves, when compared with these plants grown under 23.5 and 18.5 °C (Fig. 8a, b). Lefsrud and Kopsell (2005) have reported that the content of Cars was the highest in kale under 30 °C than those at 15, 20 and 25 °C. Moreover, Cars content was increased linearly with increasing air temperature. Additionally, Ikoma et al. (2001) have also showed that the PSY transcript in mature leaves of Satsuma mandarin fruit was higher than in young ones. These could be ascribed to the increase in content of Cars in tobacco leaves under 28.5 °C in the present study.

## Conclusion

Different temperatures affected the growth, development and morphogenesis of tobacco plants. Compared with tobacco plants grown under 23.5 °C, treatment with 18.5 and 28.5 °C significantly inhibited the growth of tobacco plants. Treatment with 28.5 °C increased the contents of $O_2^{\cdot-}$, $H_2O_2$, MDA and Cars and reduced the contents of soluble protein and Chls in tobacco leaves, followed by accelerating the flowering, maturity and senescence of tobacco plants. On the other hand, exposing to 18.5 °C significantly decreased the contents of $O_2^{\cdot-}$, $H_2O_2$, MDA and Cars and increased the contents of soluble protein and Chls in tobacco leaves, which in turn slowed down the flowering, maturity and senescence, as compared with the plants grown under 23.5 °C. Moreover, both the $O_2^{\cdot-}$ and $H_2O_2$ participated in the degradation of Chls in tobacco leaves to some extent. As a result temperature of 23.5 °C is more beneficial for the growth, development

**and metabolism of plastid pigments of tobacco plants under the experimental conditions of this study.**

### Abbreviations

Chl: chlorophyll; Car: carotenoid; CHLase: chlorophyllase; FeCH: ferri chelatase; Glu-TR: glutamyl-tRNA reductase; $H_2O_2$: hydrogen peroxide; MDA: malondialdehyde; MDCase: Mg-dechelatase; MgCH: H subunit of magnesium chelatase; $O_2^-$: superoxide; PaO: phaeophorbide a monooxygenase; POR: protochlorophyllide oxidoreductase; PSY: phytoene synthase.

### Authors' contributions

LYY and MG conceived the experiment, analyzed the data and drafted the manuscript. LYY, SLY, JHM, BH, JYL performed the experiment. CMZ and MG revised the manuscript. All authors read and approved the final manuscript.

### Author details

[1] School of Life Sciences, Engineering Research Center of Sustainable Development and Utilization of Biomass Energy, Ministry of Education, Key Laboratory of Biomass Energy and Environmental Biotechnology of Yunnan Province, Yunnan Normal University, Kunming 650500, People's Republic of China. [2] Yunnan Academy of Tobacco Agricultural Sciences, Kunming 650031, People's Republic of China. [3] Yunnan Tobacco Leaf Company, Kunming 650218, People's Republic of China.

### Acknowledgements

This study was supported by several grants from the National Natural Science Foundation of China (Nos. 31260064, 31460059), the special project of science and technology (110201101003 TS03), State Tobacco Monopoly Bureau, China, and the financial support by foundations (2014YN16) from CNTC (China National Tobacco Corporation). The New Ph.D. Candidate Academician Scholarship of Yunnan Province of 2016 (01700205020516002).

### Competing interests

The authors declare that they have no competing interests.

### References

Adachi M, Shimokawa K (1995) Evidence for the involvement of superoxide anion in the ethylene-inducted chlorophyll a catabolism of *Raphanus sativus* cotyledons. Phytochemistry 3:527–530

Apel K, Santel HJ, Redlinger TE, Falk H (1980) The protochlorophyllide holochrome of barley (*Hordeum Vulgare* L.). Isolation and characterization of the NADPH: protochlorophyllide oxidoreductase. FEBS J 111:251–258

Armas-Ricard Md, Levicán G, Katz A, Moser J, Jahn D, Orellana O (2011) Cellular levels of heme affect the activity of dimeric glutamyl-tRNA reductase. Biochem Biophys Res Commun 405:134–139

Armstrong GA, Runge S, Frick G, Sperling U, Apel K (1995) Identification of NADPH: protochlorophyllide oxidoreductases A and B: a branched pathway for light-dependent chlorophyll biosynthesis in *Arabidopsis thaliana*. Plant Physiol 108:1505–1517

Barry CS, McQuinn RP, Chung MY, Besuden A, Giovannoni JJ (2008) Amino acid substitutions in homologs of the STAY-GREEN protein are responsible for the green-fl esh and chlorophyll retainer mutations of tomato and pepper. Plant Physiol 147:179–187

Bartley GE, Scolnik PA (1995) Plant cartenoids: pigment for photoprotection visual attraction, and human health. Plant Cell 7:1027–1038

Beale SI (1999) Enzymes of chlorophyll biosynthesis. Photosynth Res 60:43–73

Beale SI (2005) Green genes gleaned. Trends Plant Sci 10(7):309–312

Belknap WR, Garbarino JE (1996) The role of ubiquitin in plant senescence and stress response. Trends Plant Sci 1:331–335

Berleth T, Sachs T (2001) Plant morphogenesis: long-distance coordination and local patterning. Curr Opin Plant Biol 1:57

Blázquez MA, Ahn JH, Weigel DA (2003) Thermosensory pathway controlling flowering time in *Arabidopsis thaliana*. Nat Genet 33:168–171

Bollivar DW (2006) Recent advances in chlorophyll biosynthesis. Photosynth Res 90:173–194

Braford MM (1976) A rapid and sensitive method for the quantification of microgram quantities of protein utilizing the principle of protein-dye binding. Anal Biochem 72:248–254

Brennan T, Frenkel C (1977) Involvement of hydrogen peroxide in regulation of senescence in pear. Plant Physiol 59:411–416

Brumann I, Stein N, Hansson M (2014) Reduced chlorophyll biosynthesis in heterozygous barely magnesium chelatase mutants. Plant Physiol Biochem 8:10–14

Cerdán PD, Chory J (2003) Regulation of flowering time by light quality. Nature 6942:881

Cheng XX, Yu M, Zhang N, Zhou ZQ, Xu QT (2016) Reactive oxygen species regulate programmed cell death progress of endosperm in winter wheat (*Triticum aestivum* L.) under waterlogging. Protoplasma 253:311–327

Chéour F, Arul J, Makhlouf J, Willemot C (1992) Delay of membrane lipid degradation by calcium treatment during cabbage leaf senescence. Plant Physiol 4:1656–1660

Cornah JE, Terry MJ, Smith AG (2003) Green or red: what stops the traffic in the tetrapyrrole pathway? Trends Plant Sci 5:224–230

Costa M, Civello PM, Chaves AR, Martínez GA (2002) Characterization of Mg-dechelatase activity obtained from *Fragaria* × *ananassa* fruit. Plant Physiol Biochem 40:111–118

Demming-Adams B (1990) Carotenoids and photoprotection in plants: a role for xanthophylls zeaxanthin. Biochem Biophys Acta 1020:1–24

Dere S, Günes T, Sivaci R (1998) Spectrophotometric determination of chlorophyll-*a*, *b* and total carotenoid of some algae species using different solvents. Turk J Bot 22:13–17

Dias MG, Camoes MF, Oliveira L (2014) Carotenoid stability in fruits, vegetables and working standards—effect of storage temperature and time. Food Chem 2:37–41

Djanaguiraman M, Prasad PVV (2010) Ethylene production under high temperature stress cause premature leaf senescence in soybean. Funct Plant Biol 37:1071–1084

Eckhardt U, Grimm B, Hörtensteiner S (2004) Recent advances in chlorophyll biosynthesis and breakdown in higher plants. Plant Mol Biol 56:1–14

Esteban R, Moran JF, Becerril JM, García-Plazaola JI (2015) Versatility of carotenoids: an integrated view on diversity, evolution, functional roles and environmental interactions. Environ Exp Bot 119:63–75

Fang ZY, Bouwkamp JC, Solomos T (1998) Chlorophyllase activities and chlorophyll degradation during leaf senescence in non-yellowing mutant and wild types of *Phaseolus vulgaris* L. J Exp Bot 320:503–510

Ferguson DL, Ai-Khatib K, Guikema JA, Paulsen GM (1993) Degradation of proteins from thylakoid membranes in senescing wheat leaves at high temperature. Plant Cell Environ 4:421–442

Frahy G, Schopfer P (2001) NADH-stimulated, cyanide-resistant superoxide production in maize coleoptiles analyzed with a tetrazolium-based assay. Planta 212:175–183

Frankin KA (2009) Light and temperature signal crosstalk in plant development. Curr Opin Plant Biol 12:63–68

Gao S, Gao J, Zhu XY, Song Y, Li ZP, Ren GD, Zhou X, Kuai BK (2016) ABF2, ABF3, and ABF4 promote ABA-mediated chlorophyll degradation and leaf senescence by transcriptional activation of chlorophyll catabolic genes and senescence-associated genes in Arabidopsis. Mol Plant 9:1272–1285

Gill SS, Tuteja N (2010) Reactive oxygen species and antioxidant machinery in abiotic stress tolerance in crop plants. Plant Physiol Biochem 48:909–930

Gong YP, Mattheis JP (2003) Effect of ethylene and 1-methylcyclopropene on chlorophyll catabolism of broccoli florets. Plant Growth Regul 40:33–38

Gossauer A, Engel N (1996) Chlorophyll catabolism—structures, mechanism, conversions. J Photochem Photobiol B 32:141–151

Gough SP, Westergen T, Hansson M (2003) Chlorophyll biosynthesis in higher plants. Regulatory aspects of 5-aminolevulinate formation. J Plant Biol 3:135–160

Grover A, Sabat SC, Mohanty P (1986) Effect of temperature on photosynthetic activities of senescing detached wheat leaves. Plant Cell Physiol 1:117–126

Guo Y, Gan S (2005) Leaf senescence: signals, execution, and regulation. Curr Top Dev Biol 71(71):83–112

Guo R, Luo M, Weinstein JD (1998) Magnesium-chelatase from developing pea leaves. Characterization of a soluble extract from chloroplasts and resolution into tree required proteins fractions. Plant Physiol 116:605–615

Guo YP, Zhou HF, Zhang LC (2006) Photosynthetic characteristics and protective mechanism against photooxidation during high temperature stress in two citrus species. Sci Hortic 108:260–267

Gupta S, Gupta SM, Sane AP, Kumar NC (2012) Chlorophyllase in *Piper betle* L. has a role in chlorophyll homeostasis and senescence dependent chlorophyll breakdown. Mol Biol Rep 39:7133–7142

Haba PDI, Mata DI, Molina E, Agüera E (2014) High temperature promotes early senescence in primary leaves of sunflower (*Helianthus annuus* L.) plants. Can J Plant Sci 94:659–669

Harpaz-Saad S, Azoulay T, Arazi T, Ben-Yaakov E, Mett A, Shiboleth YM, Hörtensteiner S, Gidoni D, Gal-On A, Goldschmidt EE, Eyal Y (2007) Chlorophyllase is a rate-limiting enzyme in chlorophyll catabolism and is posttranslationally regulated. Plant Cell 3:1007–1022

Heath RL, Packer L (1968) Photoperoxidation in isolated chloroplasts. 1. Kinetics and stoichiometry of fatty acid peroxidation. Arch Biochem Biophys 25:189–198

Holtorf H, Apel K (1996) Transcripts of two NADPH protochlorophyllide oxidereductase genes PorA and PorB are differentially degraded in etiolated barley seedlings. Plant Mol Biol 31:387–392

Hörtensteine S, Kräutler B (2011) Chlorophyll breakdown in higher plants. BBA Bioenerg 1807:977–988

Hörtensteiner S (2006) Chlorophyll degradation during senescence. Annu Rev Plant Biol 57:55–77

Hörtensteiner S, Vicentini F, Matile P (1995) Chlorophyll breakdown in senescent cotyledons of rape, *Brassica napus* L.: enzymatic cleavage of phaeophorbide a in vitro. New Phytol 2:237–246

Ikoma Y, Komatsu A, Kita M, Ogawa K, Omura M, Yano M, Moriguchi T (2001) Expression of a phytoene synthase and characteristic carotenoid accumulation during citrus fruit development. Physiol Plant 111:232–238

Jumrani K, Bhatia VS (2014) Impacts of elevate temperature on growth and yield of chickpea (*Cicer arietinum* L.). Field Crop Res 164:90–97

Kalisz A, Jezdinský A, Pokluda R, Sękara A, Grabowska A, Gil J (2016) Impacts of chilling on photosynthesis and chlorophyll pigment content in juvenile basil cultivars. Hortic Environ Biotechnol 4:330–339

Kaneko H, Harada M (1972) 4-Hydroxy-β-damascone and 4-hydroxy-dihydro-β-damascone from cigar tobacco. Agric Biol Chem 1:168–171

Kao CH (1980) Retardation of senescence by low temperature and benzyladenine in intact primary leaves of soybean. Plant Cell Physiol 2:339–344

Kim C, Apel K (2012) Arabidopsis light-dependent NADPH: protochlorophyllide oxidoreductase A (PORA) is essential for normal plant growth and development: an addendum. Plant Mol Biol 80:237–240

Kim JW, Shon JY, Lee CK, Yang WH, Yoon YW, Yang WH, Kim YG, Lee BW (2011) Relationship between grain filling and leaf senescence of temperate rice under high temperature. Field Crop Res 122:207–213

Kim J, Woo HR, Nam HG (2016) Toward systems understanding of leaf senescence: an integrated multi-omics perspective on leaf senescence research. Mol Plant 9:813–825

Kumar AM, Soll D (2000) Antisense HEMA1 RNA expression inhibits heme and chlorophyll biosynthesis in Arabidopsis. Plant Physiol 122(1):49–56

Langmeier M, Ginsburg S, Matile P (1993) Chlorophyll breakdown in senescent leaves: demonstration of Mg-dechelatase activity. Physiol Plant 2:347–353

Lefsrud MG, Kopsell D (2005) Air temperature affects biomass and carotenoid pigment accumulation in kale and spinach grown in a controlled environment. HortScience 7:2026–2030

Li ZG, Gong M, Liu P (2012) Hydrogen sulfide is a mediator in $H_2O_2$-induced seed germination in *Jatropha curcas*. Acta Physiol Plant 34:2207–2213

Lim PO, Kim HJ, Nam HG (2007) Leaf senescence. Annu Rev Plant Biol 1:115–136

Liu XG, Xu H, Zhang JY, Liang GW, Liu YT, Guo AG (2012) Effect of low temperature on chlorophyll biosynthesis in albinism line of wheat (*Triticum aestivum*) FA85. Physiol Plant 145:384–394

Liu LH, Shao ZY, Zhang M, Wang QM (2015) Regulation of carotenoid metabolism in tomato. Mol Plant 8:28–39

Mathieu AS, Lutts S, Vandoorne B, Descamps C, Périlleux C, Dielen V, Herck JCV, Quinet M (2014) High temperature limit plant growth but hasten flowing in root chicory (*Cichorium intybus*) independently of vernalisation. J Plant Physiol 171:109–118

Matile P, Hörtensteinera S (1999) Chlorophyll degradation. Plant Biol 50:67–95

Matile P, Schellenberg M, Vicentini F (1997) Location of chlorophyllase in the chloroplast envelope. Planta 201:96–99

Mccord JM, Fridovich I (1969) Superoxide dismutase an enzymic function for erythrocuprein (hemocuprein). J Biol Chem 22:6049–6055

Menon BR, Hardman SJ, Scrutton NS, Heyes DJ (2016) Multiple active site residues are important for photochemical efficiency in the light-activated enzyme protochlorophyllide oxidoreductase (POR). J Photochem Photobiol B 161:236–243

Minguez-Mosquera MI, Gallardo-Guerrero L (1996) Role of chlorophyllase in chlorophyll metabolism in olives cv. Grodal. Phytochemistry 41:691–697

Moll WAW, de Wit B, Lutter R (1978) Chlorophyllase activity in developing leaves of *Phaseolus vulgaris* L. Planta 139:79–83

Møller IM, Sweetlove LJ (2010) ROS signaling-specificity is required. Trends Plant Sci 15:370–374

Moser S, Müller T, Oberhuber M, Kräutler B (2009) Chlorophyll catabolites-chemical and structural footprints of a fascinating biological phenomenon. Eur J Org Chem 1:21–31

Navabpour S, Morris K, Allen R, Harrison E, A-H-Mackerness S, Buchanan-Wollaston V (2003) Expression of senescence-enhanced genes in response to oxidative stress. J Exp Bot 54:2285–2292

Nisar N, Li L, Lu S, Khin NC, Pogson BJ (2015) Carotenoid metabolism in plants. Mol Plant 8:68–82

Ogweno JO, Song XS, Hu WH, Shi K, Zhou YH, Yu JQ (2009) Detached leaves of tomato differ in their photosynthetic physiological response to moderate high and low temperature stress. Sci Hortic 123:17–22

Ougham HJ, Morris P, Thomas H (2005) The colors of autumn leaves as symptoms of cellular recycling and defenses against environmental stresses. Curr Top Dev Biol 66:135–160

Ougham H, Hörtensteiner S, Armstead I, Donnison I, King I, Thomas H, Mur L (2008) The control of chlorophyll catabolism and the status of yellowing as a biomarker of leaf senescence. Plant Biol 10:4–14

Pastori GM, Rio LA (1994) An activated-oxygen-mediated role for peroxisomes in the mechanism of senescence of *Pisum sativum* L. leaves. Planta 193:385–391

Pham VT, Herero M, Hormaza JI (2015) Effect of temperature on pollen germination and pollen tube growth in longan (*Dimocarpus longan* Lour.). Sci Hortic 197:470–475

Popov VN, Antipina OV, Burakhanova EA (2013) Involvement of cell-wall invertase in low-temperature hardening of tobacco plants. Russ J Plant Physiol 2:221–226

Porter JR (2005) Rising temperatures are likely to reduced crop yields. Nature 436:174–175

Prochazkova D, Wilhelmova N (2007) Leaf senescence and activities of the antioxidant enzymes. Biol Plant 3:401–406

Reinbothe C, Bakkouri MEI, Buhr F, Muraki N, Nomata J, Kurisu G, Fujita Y, Reinbothe S (2010) Chlorophyll biosynthesis: spotlight on protochlorophyllide reduction. Cell 11:614–624

Rodoni S, Schelienberg M, Matile P (1998) Chlorophyll breakdown in senescing barley leaves as correlated with phaeophorbide a oxygenase activity. J Plant Physiol 152:139–144

Sarwat M, Naqvi AR, Ahmad P, Ashraf M, Akram NA (2013) Phytohormones and microRNAs as sensors and regulators of leaf senescence: assigning macro roles to small molecules. Biotechnol Adv 31:1153–1171

Shalit A, Rozman A, Goldshmidt A, Alvarez JP, Bowman JL, Eshed Y, Lischitz E (2009) The flowering hormone florigen functions as a general systemic regulator of growth and termination. PNAs USA 20:8392–8397

Shewmaker CK, Sheehy JA, Daley M, Colburn S, Ke DY (1999) Seed-specific overexpression of phytoene synthase: increase in carotenoids and other metabolic effects. Plant J 20(4):401–412

Shigenage MK, Hagen TM, Ames BN (1994) Oxidative damage and mitochondrial decay in aging. Proc Natl Acad Sci USA 91:10771–10778

Shumskaya M, Wurtzel ET (2013) The carotenoid biosynthetic pathway: thinking in all dimensions. Plant Sci 208:58–63

Smart CM (1994) Gene expression during leaf senescence. New Phytol 3:419–448

Song LL, Ding W, Shen J, Zhang ZG, Bi YR, Zhang LX (2008) Nitric oxide mediate abscisic acid and induced thermotolerance in the calluses from two ecotypes of reed under heat stress. Plant Sci 175:826–832

Song YH, Ito S, Imaizumi T (2013) Flowering time regulation: photoperiod- and temperature-sensing in leaves. Trends Plant Sci 10:575–583

Souza Lucéia F, Gasparetto Bruno F, Lopes Rodrigo R, Brrros Ingrid BI (2016) Temperature requirements for seed germination of *Pereskia aculeata* and *Pereskia grandifolia*. J Therm Biol 57:6–10

Stahl W, Sies H (2005) Bioactivity and protective effects of natural carotenoids. BBA Mol Basis Dis 1740:101–107

Steccanella V, Hansson M, Jensen PE (2015) Linking chlorophyll biosynthesis to a dynamic plastoquinone pool. Plant Physiol Biochem 97:207–216

Sunoj VSJ, Shroyer KJ, Jagadish SVK, Prasad PV (2016) Diurnal temperature amplitude alters physiological and growth response of maize (*Zea mays* L.) during the vegetative stage. Environ Exp Bot 130:113–121

Suzuki N, Koussevitzky S, Mittler R, Miller G (2012) ROS and redox signaling in the responses of plants to abiotic stress. Plant Cell Environ 35:259–270

Talaat NB (2013) RNAi based simultaneous silencing of all forms of light-dependent NADPH: protochlorophyllide oxidoreductase genes result in the accumulation of protochlorophyllide in tobacco (*Nicotiana tabacum*). Plant Physiol Biochem 71:31–36

Tanaka A, Tanaka R (2006) Chlorophyll metabolism. Curr Opin Plant Biol 9:248–255

Terry MJ, Kendrick RE (1999) Feedback inhibition of chlorophyll synthesis in the phytochrome-deficient aurea and yellowgreen-2 mutants of tomato. Plant Physiol 119:143–152

Tewari AK, Tripathy BC (1998) Temperature-stress-induced impairment of chlorophyll biosynthetic reactions in cucumber and wheat. Plant Physiol 117:851–858

Todorov DT, Laranov EN, Smith AR, Hall MA (2003) Chlorophyllase activity and chlorophyll content in wild type and *eti* 5 mutant of *Arabidopsis thaliana* subjected to low and high temperature. Biol Plant 4:633–636

Trebitsh T, Goldschmidt EE, Riov J (1993) Ethylene induces de novo synthesis of chlorophyllase, a chlorophyll degrading enzyme, in citrus fruit peel. Proc Natl Acad Sci USA 90:9441–9445

Venema JH, Posthumus F, Hasselt PRV (1999) Impact of suboptimal temperature on growth, photosynthesis, leaf pigments and carbohydrates of domestic and high-altitude wild *Lycopersicon* species. J Plant Physiol 155:711–718

Wang PC, Mo BT, Long ZF, Fan SQ, Wang HH (2016) Factors affecting seed germination and emergence of *Sophora davidii*. Ind Crop Prod 87:261–265

Whitfield DM, Rowan KS (1974) Changes in the chlorophylls and carotenoids of leaves of *Nicotiana tabacum* during senescence. Phytochemistry 13:77–83

Xu GL, Singh SK, Reddy VR, Barnaby JY (2016) Soybean grown under elevated $CO_2$ benefits more under low temperature than high temperature stress: varying response of photosynthetic limitations, leaf metabolites, growth, and seed yield. J Plant Physiol 205:20–32

Yamauchi N, Watada AE (1991) Regulated chlorophyll degradation in spinach leaves during storage. J Am Soc Hortic Sci 116:58–62

Yamori W, Evans JR, Von CS (2010) Effects of growth and measurement light intensities on temperature dependence of $CO_2$ assimilation rate in tobacco leaves. Plant Cell Environ 3:332

Yang DL, Sun P, Li MF (2016) Chilling temperature stimulates growth, gene over-expression and podophyllotoxin biosynthesis in *Podophyllum hexandrum* royle. Plant Physiol Biochem 107:197–203

Yang LY, Duan SZ, Li JY, Pang T, Yang SL, Wang SS, Gong M (2017) Effects of different temperatures on growth, development and photosynthesis of tobacco (*Nicotiana tabacum* L.) plants. Acta Bot Boreal Occident Sin 2:0330–0338 **(in Chinese with English abstract)**

Yoshida S (2003) Molecular regulation of leaf senescence. Curr Opin Plant Biol 1:79–84

Zhang XL, Zhang ZQ, Li J, Wu LJ, Guo JY, Ouyang LQ, Xia YY, Huang XM, Pang XQ (2011) Correlation of leaf senescence and gene expression/activities of chlorophyll degradation enzymes in harvested Chinese flowering cabbage (*Brassica rapa var. parchinensis*). J Plant Physiol 168:2081–2087

Zhang KW, Xia XY, Zhang YY, Gan SS (2012) An ABA-regulated and golgi-localizaed protein phosphatase controls water loss during leaf senescence in Arabidopsis. Plant J 69:667–678

Zhang W, Huang W, Yang Q, Zhang S, Hu H (2013) Effects of growth temperature on the electron flow for photorespiration in leaves of tobacco grown in the field. Physiol Plant 1:141–150

Zhao H, Dai T, Jing Q, Jiang Q, Jiang D, Cao W (2007) Leaf senescence and grain filling affected by post-anthesis high temperature in two different wheat cultivars. Plant Growth Regul 2:149–158

Zhao XH, Nishimura Y, Fukumoto Y, Li JC (2011) Effects of high temperature on active oxygen species, senescence and photosynthetic properties in cucumber leaves. Environ Exp Bot 70:212–216

Zhou H, Guo SR, An YH, Shan X, Wang Y (2016) Exogenous spermidine delays chlorophyll metabolism in cucumber leaves (*Cucumis sativus* L.) under high temperature stress. Acta Physiol Plant 8:224

# 11

# Acetylsalicylic acid enhance tolerance of *Phaseolus vulgaris* L. to chilling stress, improving photosynthesis, antioxidants and expression of cold stress responsive genes

Mona H. Soliman[1,2], Aisha A. M. Alayafi[3], Amr A. El Kelish[4*] and Abdelghafar M. Abu-Elsaoud[4]

**Abstract**

**Background:** High and low temperatures constitute the most damaging type of abiotic stress and limit the survival, and productivity of plants. The present study aimed to evaluate the role of exogenous applications of acetylsalicylic acid (ASA) in reducing the deleterious effects of cold stress. *Phaseolus vulgaris* L. seedlings were treated with foliar-sprayed ASA at concentrations of 0–3 mM and then subjected to chilling stress at 4 °C for 2 or 4 days.

**Results:** Growth, photosynthesis, biochemical alterations, oxidative damage and antioxidant enzyme activities as well as the expression of cold-responsive genes (CBF3–COR47), were monitored during the experiment. ASA applications substantially improved several growth and photosynthetic parameters, including shoot biomass, dry weight, and photosynthetic pigments, of *P. vulgaris* seedlings exposed to different durations of chilling stresses. The ASA foliar spray treatments significantly ($p < 0.05$) rescued the growth and photosynthetic pigments of *P. vulgaris* seedlings under different chilling stresses. The total soluble sugars markedly increased during 0–4 days of chilling stress following ASA foliar spraying. The exogenous application of ASA significantly ($p < 0.05$) increased the accumulation of proline in *P. vulgaris* seedlings under chilling stress. At the gene expression level, ASA significantly ($p < 0.05$) upregulated the cold-responsive genes CBF3 and COR47.

**Conclusions:** As a result, we speculate that, the application of exogenous ASA alleviated the adverse effects of chilling stress on all measured parameters, and 1 and 2 mM ASA exhibited the greatest effects.

**Keywords:** Chilling stress, Acetylsalicylic acid, ASA, Antioxidants enzymes, Superoxide dismutase, Catalase, Peroxidase

## Background

Climate change and global warming generate different kinds of biotic and abiotic stresses that in turn alter plant responses at the transcriptomic, proteomic and metabolomic levels (Chartzoulakis and Psarras 2005; Khan and Khan 2013; El Kelish et al. 2014). Moreover, climate change severely disturbs the biochemistry, quantity, and quality of crop yields (Jaleel et al. 2009; Miller et al. 2010; Cramer et al. 2011; Pereira 2016). Environmental challenges lead to a wide range of responses in plants, such as morphological, physiological, and molecular changes (Zandalinas et al. 2017). According to the Food and Agriculture Organization (FAO), the population is increasing rapidly and will reach approximately 10 billion in 2050. Moreover, food production is decreasing because of various abiotic stresses. As such, the earth will require 70% more food by 2050 (Mahajan and Tuteja 2005; Gill and Tuteja 2010). Therefore, it is necessary to obtain crops that display both enhanced vigor and tolerance to various environmental factors to cope with the future problem of food security (Bita and Gerats 2013).

*Correspondence: amr.elkelish@science.suez.edu.eg
[4] Botany Department, Faculty of Science, Suez Canal University, Ismailia, Egypt
Full list of author information is available at the end of the article

Chilling injury can occur at temperatures between 0 and 15 °C (Kratsch and Wise 2000; Allen and Ort 2001). Various phenotypic symptoms of plants in response to chilling stress are clear, including reductions in leaf chlorophyll content, wilting, chlorosis, and necrosis. Moreover, different metabolic modifications are induced, such as reductions in unsaturated fatty acid contents and increases in the permeability of cell membranes, which collectively reduce plant performance (Bracale and Coraggio 2003; Tong et al. 2005; Miura and Furumoto 2013). One of the essential cellular response throughout a freeze–thaw cycle is the maintenance of plasma membrane integrity. This maintenance reduces the efficiency of photosystem II (PSII); damages photosystem I (PSI); alters the carbon reduction cycle, $CO_2$ assimilation, and photosynthetic pigment complex systems; and accelerates reactive oxygen species (ROS) formation (Guy 1990; Thomashow 2001; Allen and Ort 2001; Saibo et al. 2009).

Beans (*Phaseolus lunatus*), corn (*Zea mays*) and tomato (*Solanum lycopersicum*) are considered relatively sensitive to environmental stress. Many studies have reported the impact of chilling injury and how short exposure to temperatures below freezing can destroy whole crops (Kratsch and Wise 2000; Chinnusamy et al. 2007). Most plants have developed numerous systems to manage performance during environmental stress. For instance, to minimize chilling-induced injury, plants can upregulate different scavenging systems, such as enzymatic antioxidants and non-enzymatic metabolites (Gill and Tuteja 2010). Moreover, plants can synthesize plant growth regulators [salicylic acid (SA)] and osmoprotectants (proline) (Gautam and Singh 2009; AbdElgawad et al. 2016; Tabassum et al. 2017). Most of these compounds shield membranes and the photosynthetic apparatus from the harmful effects of environmental stress (Foyer and Noctor 2003).

Among the best-recognized gene family is the CBF/DREB1 family, whose members play a major role in chilling tolerance in Arabidopsis (Thomashow 1998, 2001). CBF transcription factors act as a regulatory hub for low-temperature acclimation, controlling the expression of cold-regulated (COR) target genes that include CRT/DRE *cis*-elements in their promoters (Thomashow 2010). The overexpression of CBF leads to an intense activation of COR genes and successively enhances chilling and dehydration tolerance in sensitive plants (Kasuga et al. 1999; Jaglo et al. 2001). Furthermore, the constitutive overexpression of Arabidopsis CBF1/DREB1b confers chilling stress tolerance to cucumber and potato (Gupta et al. 2012; Movahedi et al. 2012; Caffagni et al. 2014). Thus, CBF/DREB1 is the primary and most critical regulatory gene associated with chilling stress in plants.

SA and ortho-hydroxybenzoic acid are essential signaling phenolic compounds associated with plant development and tolerance to abiotic and biotic stresses (Raskin 1992; Khan et al. 2003, 2012). SA is involved in the adjustment of vital plant physiochemical activities, such as the light and dark reactions, carbon–nitrogen metabolism, proline metabolism, and ROS scavenger systems, and therefore offers protection against abiotic stresses in plants (Khan et al. 2003; Saleh et al. 2007; Simaei et al. 2012; Miura and Tada 2014; Ruelland 2017). Aspirin, a trade name for acetylsalicylic acid (ASA), can be obtained by the spontaneous hydrolysis of SA (Senaratna et al. 2000). Similarities between the chemical, physical, and physiological characteristics of ASA and SA have encouraged plant scientists to use the former in biological experiments (Kupferwasser et al. 1999; Canakci and Munzuroğlu 2007).

The objective of this study was to evaluate the role of exogenous ASA in rescuing *Phaseolus vulgaris* from the deleterious effects of chilling stress. The exogenous application of ASA may represent a strategy to increase plant tolerance to chilling stress by regulating antioxidant defense systems and increasing the levels of key metabolites and genes involved in chilling stress tolerance.

## Methods

### Plant materials and ASA treatment

A greenhouse experiment was conducted at the Department of Botany of the Faculty of Science, Suez Canal University, Ismailia, Egypt, during November and December 2016. Seeds of common white bean (*P. vulgaris* L.) were purchased from the Department of Vegetable Crop Research of the Agricultural Research Center, Giza, Egypt. Before they were sown, *P. vulgaris* seeds were superficially sterilized with 5% sodium hypochlorite, rinsed thoroughly with distilled water and then dried on filter paper for 30 min. Pure crystals of ASA (99.5%) (Sigma Chemical Company, USA) was prepared at five concentrations (0.1, 0.5, 1, 2 and 3 mM) in addition to a non-ASA control. The ASA crystals were dissolved in 0.1 mL of ethanol (95%). NaOH and HCl were used for neutral pH adjustments. Five sterilized *P. vulgaris* seeds were planted in plastic pots (20 cm in length × 15 cm in diameter) filled with a clay: sand mixture (1:1, w/w). Germination was carried out for 30 days under normal greenhouse conditions of 25 ± 4.0 °C and a 16-h photoperiod. Once they had emerged, the seedlings were irrigated weekly with tap water to field capacity. Thirty-day-old seedlings were subjected to foliar sprays of ASA at concentrations of 0.1, 0.5, 1, 2 or 3 mM concentrations. The foliar sprays were applied using a hand sprayer until the droplets began to run off, after which the seedlings were allowed to grow for 24 h. The non-ASA control treatment was sprayed with distilled water only, and there were three replicates of each treatment as well as the control.

### Chilling stress experiments

For chilling stress treatments, 24 h after ASA spraying, the seedlings were thinned to three seedlings pot$^{-1}$, after which the pots were transferred to a 2 °C cold room that contained metal halide lamps that generated 250 μmol m$^{-2}$ s$^{-1}$ of photosynthetic photon flux density (PPFD) under a 12-h photoperiod. The seedlings remained under cold-shock conditions at 4 °C for 2 or 4 days under 16-h photoperiod and a PPFD of 300 μmol m$^{-2}$ s$^{-1}$. Each treatment consisted of three replicates in a randomized complete block design. Fresh samples were taken after 0, 2 and 4 days of chilling for morphological and biochemical analyses. The remaining samples were collected and stored at − 80 °C until use.

### Determination of photosynthetic pigments and soluble sugars

The photosynthetic pigments [chlorophyll a (chl. a) and chlorophyll b (chl. b)] were determined in accordance with the spectrophotometric method (Jenway UV/Vis spectrophotometer, UK) recommended by Holder (1965). The absorbance was measured against a blank sample of pure acetone (85%) at 2 wavelengths: A645 and A663 (nm). The pigment fractions (chl. a and chl. b) were expressed as milligrams per gram of dry weight (DW) (Strain and Svec 1966).

The procedures for the extraction of the total soluble sugars from *P. vulgaris* leaves were described by Ciha and Brun (1978). The total soluble sugars were quantified using a modified anthrone acid assay (Irigoyen et al. 1992).

### Determination of protein, amino acid, and proline contents

The total soluble protein content in the fresh leaves of *P. vulgaris* was determined as described by Bradford (1976) using bovine serum albumin (BSA) as a standard. The total free amino acid content was determined as described by Dubey and Rani (1989), with modifications. A 0.2-g sample of dried leaves was grinded in 10 mL ethanol (80%). After filtration, one hundred microliters of the extract and 5 mL of ninhydrin reagent were mixed, after which the mixture was shaken gently and then heated for 10 min in a boiling water bath. Subsequently, the mixture cooled, its absorbance was measured spectrophotometrically at 570 nm. The extraction of proline was carried out with aqueous sulfosalicylic acid, and the proline contents were determined spectrophotometrically in accordance with the methods of Sadasivam (1992). The proline content in each sample was measured in milligrams per gram of fresh weight (FW) and determined using a standard curve of analytical-grade proline.

### Cellular lipid peroxidation

Lipid peroxidation was investigated via the malondialdehyde (MDA) content, which was estimated spectrophotometrically using thiobarbituric acid (TBA)-MDA (TBA-MDA) assays (De Vos et al. 1991). The extraction of lipid peroxides was carried out using 500 mg of fresh shoot tissue, 0.3 mL of 10% trichloroacetic acid, and 1 mL of 0.5% TBA. The TBA-chromogen color was measured spectrophotometrically at 532 nm.

### Determination of ascorbic acid (AA) content

The AA content was measured by redox titration using iodine solution in accordance with the methods of McHenry and Graham (1935).

### Determination of antioxidant enzymes

Enzyme extracts were prepared by homogenizing 1 g of fresh leaves of *P. vulgaris* with 5 mL of cold phosphate buffer (pH = 7.0). The extracts were then centrifuged at 18,000*g* for 30 min at 4 °C, after which the supernatant was filtered and stored at − 20 °C for further enzyme assays.

The ascorbate peroxidase (APX) activity was determined as described by Rao et al. (2007). The APX activity was recorded by following the decrease in absorbance at 290 nm for 3 min in 1 mL of a reaction mixture that contained 100 mM phosphate buffer (pH 7.5), 0.5 mM ascorbate, 0.2 mM $H_2O_2$ and 30 μL of enzyme extract. The enzyme activity was expressed as moles of oxidized ascorbate per minute per mg of protein.

The catalase (CAT) activity was determined by recording at 240 nm the consumption of $H_2O_2$ for 30 s in 3 mL of a reaction mixture that consisted of 100 mM phosphate buffer (pH 7.0), 20 μL of 30% $H_2O_2$ and 30 μL of enzyme extract (Aebi 1984).

The activity of peroxidase (POD) was determined as described by Jiang et al. (2002). The reaction mixture consisted of 1 mL of enzyme extract and guaiacol as a substrate. A 3-mL reaction mixture consisted of 100 mM sodium phosphate buffer (pH 7.0) and 20 mM guaiacol. The increase in absorbance at 470 nm during a 3-min period was measured spectrophotometrically after 20 μL of $H_2O_2$ was added to the mixture. The enzyme activity was defined as the change in the optical density per mg of proteins per minute.

The activity of superoxide dismutase (SOD) was determined photochemically in accordance with the methods of Giannopolitis and Ries (1977). The assays were carried out under illumination. One unit of SOD activity was defined as the amount of enzyme required to inhibit

50% of a nitro blue tetrazolium (NBT) chloride reaction, which was recorded at 560 nm.

### Gene expression of cold-responsive genes (CBF3 and COR47)

#### *RNA extraction and first-strand cDNA synthesis*

RNA extraction was carried out by a QIAGEN RNA extraction kit (QIAGEN, Germany). DNA digestion was performed with DNaseI (QIAGEN, Germany). The RNA yield and quality were determined using a NanoDrop system (ND-1000; Thermo Fisher Scientific). First-strand cDNA was synthesized from 1 µg of high-quality total RNA using SuperScript II Reverse Transcriptase in accordance with the manufacturer's instructions (QIAGEN, Omniscript RT).

#### *Real-time PCR analysis*

The specific primer sequences of two cold-responsive genes, CBF3 and COR47, were used (Table 1), and the β-actin gene was used as an internal standard. The obtained cDNA was used in qRT-PCR, which was performed in a 20-µL reaction mixture that contained one unit of Taq polymerase (Fermentas, Canada), 3.5 mM of $MgCl_2$, 1 unit of PCR buffer, 0.5 mM of each primer, 10 mM dNTPs, 0.5 units of SYBR Green I, 0.6 mL of DMSO (Sigma-Aldrich), and 2 µL of template cDNA, using a Roter-Gene Real-Time thermocycler (QIAGEN, USA). The PCR conditions were as follows: 95 °C for 1 min, followed by 40 cycles of 95 °C for 15 s, 58–60 °C for 30 s and 72 °C for 30 s. The relative expression of the cold-responsive genes (CBF3 and COR47) was calculated using the $2^{-\Delta\Delta Ct}$ comparative CT method (Livak and Schmittgen 2001). Differences in the CBF3 and COR47 transcriptional patterns in response to different ASA concentrations were analyzed using SPSS 16.0 software. Statistical significances were determined using one-way analysis of variance (ANOVA) and post hoc Duncan multiple range tests (DMRTs). Significance was established at $p < 0.05$. The means and standard deviations were calculated from experiments performed in triplicate and are presented as n-fold differences in expression.

### Statistical analysis

The data obtained from various growth, photosynthetic, and biochemical parameters were analyzed statistically Using IBM-SPSS 23.0 statistical software for Mac OS. Two-way ANOVA was performed to assess differences between various chilling stress levels and ASA concentrations. All comparisons were made at a probability level of 95% ($p \leq 0.05$). The means are the average of three replicates, and DMRTs were used for comparing means following the 2-way ANOVA.

## Results

The seedlings of *P. vulgaris* grown under normal conditions (non-stressed) and treated with foliar sprays of different concentrations of ASA (0 to 3 mM ASA) exhibited no significant differences in various growth parameters ($p > 0.05$), especially shoot biomass (g plant$^{-1}$), root biomass (g plant$^{-1}$), the shoot:root ratio (g g$^{-1}$), chl. a (mg g$^{-1}$ FW), chl. b (mg g$^{-1}$ FW), and total chlorophyll (mg g$^{-1}$ FW) (Table 2). However, *P. vulgaris* seedlings exposed to chilling stress at two levels (2 or 4 days) and seedlings not treated with ASA foliar sprays showed general deleterious effects on growth parameters (Table 2). The shoot biomass decreased significantly ($p < 0.05$) from 2.85 ± 0.002 g before chilling stress to 1.67 ± 0.03 and 0.094 g plant$^{-1}$ after 2 and 4 days of chilling stress, respectively. Likewise, the shoot DW significantly decreased from 1.65 ± 0.001 g plant$^{-1}$ in the non-stressed control plants to levels of 0.064 ± 0.01 and 0.55 ± 0.00 g plant$^{-1}$ in *P. vulgaris* seedlings exposed to chilling stress for 2 and 4 days, respectively. The photosynthetic pigments showed a similar damaging pattern under chilling stress, e.g., the chl. a content significantly decreased from 6.95 ± 0.01 mg g$^{-1}$ FW in the non-stressed control seedlings to 4.86 ± 0.00 and 4.75 ± 0.02 mg g$^{-1}$ FW in the seedlings exposed to 2 and 4 days of chilling stress (Table 2).

The application of ASA foliar sprays significantly improved various growth parameters of *P. vulgaris* seedlings exposed to different durations of chilling stress (Table 2). The shoot biomass of *P. vulgaris* seedlings subjected to 2 days of chilling stress significantly ($p < 0.05$) increased from 1.67 ± 0.03 g in the non-ASA-treated

**Table 1 The sequences of the primers used in qRT-PCR**

| Primer name | | Primer sequence (5_–3_) | Annealing temperature (°C) |
|---|---|---|---|
| CBF3 | F | 5′-TTCCGTCCGTACAGTGGAAT-3′ | 58 |
| | R | 5′-AACTCCATAACGATACGTCGTC-3′ | |
| COR47 | F | 5′-CGGTACCAGTGTCGGAGAGT-3′ | 58 |
| | R | 5′-ACAGCTGGTGAATCCTCTGC-3′ | |
| β-Actin | F | 5′-CGCGACCTCACAGACTACCTG-3′ | 58 |
| | R | 5′-CGTAGGACTTCTCCAGGGAGC-3′ | |

**Table 2 Shoot fresh weight (g plant$^{-1}$), dry weight (g plant$^{-1}$), shoot: root ratio (g g$^{-1}$), chlorophyll-a (mg g$^{-1}$ FW), chlorophyll-b (mg g$^{-1}$ FW), chlorophyll a/b, and total chlorophyll contents (mg g$^{-1}$ FW) of *P. vulgaris* L. after foliar application of different concentration of acetylsalicylic acid (ASA; 0, 0.1, 0.5, 1, 2, 3 mM) and exposed to different level of chilling stress (0, 2, 4 days)**

| Chilling (d) | ASA foliar spray (mM) | Shoot FW (g plant$^{-1}$) | Shoot DW (g plant$^{-1}$) | Shoot: root ratio (g g$^{-1}$) | Chl-a (mg g$^{-1}$ FW) | Chl-b (mg g$^{-1}$ FW) | Chl a/b | Total chlorophyll (mg g$^{-1}$ FW) |
|---|---|---|---|---|---|---|---|---|
| 0 | 0 | 2.85 ± 0.002 g | 1.65 ± 0.00ij | 0.78 ± 0.09i | 6.95 ± 0.01 h | 3.75 ± 0.03de | 1.85 ± 0.01i | 10.70 ± 0.04a |
| | 0.1 | 2.87 ± 0.02 g | 1.64 ± 0.00i | 0.69 ± 0.01i | 6.92 ± 0.00 h | 3.74 ± 0.02de | 1.85 ± 0.01i | 10.67 ± 0.02a |
| | 0.5 | 2.85 ± 0.01 g | 1.66 ± 0.02ij | 0.68 ± 0.00i | 6.95 ± 0.01 h | 3.76 ± 0.02de | 1.85 ± 0.00i | 10.71 ± 0.03a |
| | 1 | 2.85 ± 0.01 g | 1.66 ± 0.00ij | 0.70 ± 0.01i | 6.93 ± 0.00 h | 3.71 ± 0.00def | 1.87 ± 0.00i | 10.64 ± 0.00a |
| | 2 | 2.85 ± 0.00 g | 1.64 ± 0.02ij | 0.68 ± 0.01i | 6.96 ± 0.03 h | 3.72 ± 0.00ef | 1.87 ± 0.01i | 10.68 ± 0.03a |
| | 3 | 2.85 ± 0.00 g | 1.72 ± 0.05j | 0.70 ± 0.01i | 6.96 ± 0.02 h | 3.72 ± 0.00ef | 1.87 ± 0.01i | 10.68 ± 0.02ab |
| 2 | 0 | 1.67 ± 0.03bc | 0.64 ± 0.01b | 2.25 ± 0.05b | 4.86 ± 0.00b | 2.68 ± 0.00d | 1.82 ± 0.00b | 7.54 ± 0.00bc |
| | 0.1 | 1.81 ± 0.04bcd | 0.68 ± 0.00bc | 1.75 ± 0.35bc | 4.92 ± 0.00c | 2.97 ± 0.00b | 1.66 ± 0.00c | 7.88 ± 0.00c |
| | 0.5 | 1.92 ± 0.01d | 0.86 ± 0.00d | 1.85 ± 0.28ef | 5.13 ± 0.00 fg | 3.34 ± 0.00a | 1.54 ± 0.00 fg | 8.48 ± 0.00c |
| | 1 | 2.11 ± 0.08e | 1.25 ± 0.01 g | 1.39 ± 0.01f | 5.22 ± 0.01de | 3.18 ± 0.00b | 1.64 ± 0.00ef | 8.40 ± 0.01a |
| | 2 | 1.82 ± 0.03 cd | 0.73 ± 0.01c | 1.48 ± 0.00f | 5.18 ± 0.00d | 3.12 ± 0.10b | 1.66 ± 0.05de | 8.29 ± 0.10a |
| | 3 | 1.73 ± 0.05bc | 0.83 ± 0.01d | 1.86 ± 0.02de | 5.07 ± 0.05c | 2.95 ± 0.02c | 1.72 ± 0.03c | 8.02 ± 0.03a |
| 4 | 0 | 0.94 ± 0.03a | 0.55 ± 0.00a | 2.56 ± 0.01a | 4.75 ± 0.02a | 2.37 ± 0.00 g | 2.01 ± 0.01a | 7.12 ± 0.01a |
| | 0.1 | 1.79 ± 0.01bcd | 0.71 ± 0.00bc | 1.69 ± 0.01 cd | 4.98 ± 0.01def | 3.22 ± 0.01a | 1.55 ± 0.00d | 8.21 ± 0.02a |
| | 0.5 | 2.17 ± 0.01e | 1.16 ± 0.01f | 1.59 ± 0.04 h | 5.53 ± 0.01 g | 3.41 ± 0.01b | 1.62 ± 0.01 h | 8.94 ± 0.00ab |
| | 1 | 2.48 ± 0.02f | 1.72 ± 0.01j | 0.88 ± 0.01 h | 5.74 ± 0.02efg | 3.28 ± 0.00c | 1.75 ± 0.01 h | 9.02 ± 0.01bc |
| | 2 | 2.25 ± 0.00e | 1.44 ± 0.01 h | 1.48 ± 0.01 h | 5.70 ± 0.01c | 2.86 ± 0.00 g | 2.00 ± 0.00 fg | 8.56 ± 0.01c |
| | 3 | 1.66 ± 0.01b | 0.97 ± 0.02e | 1.77 ± 0.06 h | 5.64 ± 0.00c | 2.95 ± 0.01f | 1.91 ± 0.00 g | 8.59 ± 0.00c |
| Univariate; two-way ANOVA | | | | | | | | |
| $F_{(corrected)}$ | | 423.8*** | 1013.3*** | 2559.6*** | 263.5*** | 85.6*** | 1768.3*** | 34.23*** |
| $F_{(intercept)}$ | | 111,905*** | 131,774*** | 2,011,183*** | 286,459*** | 239,993*** | 1,652,469*** | 622.28*** |
| $F_{(ASA)}$ | | 2540.9*** | 5482.6*** | 20,123*** | 1506.7*** | 232.4*** | 13,321.5*** | 68.25*** |
| $F_{(chilling)}$ | | 203.1*** | 686.9*** | 353.3*** | 169.7*** | 104.2*** | 395.9*** | 4.87** |
| $F_{(ASA*chilling)}$ | | 110.8*** | 282.7*** | 149.9*** | 61.9*** | 46.95*** | 143.9*** | 42.11*** |

Data represented are mean of three points ± standard deviation

* Significant at p < 0.05; ** p < 0.01; *** *p* value < 0.001, Variations between different chilling stresses and foliar ASA concentration were assessed by univariate analyses followed by Duncan's test statistic. Means with the same letters are not significantly different according to Duncan's multiple comparisons

seedlings to 2.11 ± 0.08 g after foliar treatment with 1 mM ASA (Table 2). The differences among treatments were assessed statistically using two-way ANOVA followed by DMRT comparisons; the means with different letters significantly differ at the $p < 0.05$ level. The shoot dry weight also increased from 0.64 ± 0.01 g in the non-ASA-treated *P. vulgaris* seedlings to 1.25 ± 0.01 g after foliar treatment with 1 mM ASA (Table 2). The ASA foliar sprays significantly improved the growth of *P. vulgaris* seedlings subjected to chilling stress for 4 days. The shoot FW and DW significantly ($p < 0.05$) increased. Maximum shoot FW and DW values of 2.48 ± 0.002 and 1.72 ± 0.01 g, respectively, were recorded following foliar treatment with 1 mM ASA. The ASA foliar sprays significantly rescued the growth and photosynthetic pigments of *P. vulgaris* seedlings under different chilling stress levels (2 and 4 days).

The carbohydrate metabolism in *P. vulgaris* seedlings was assessed with respect to the total soluble sugar content (mg g$^{-1}$ FW). The total soluble sugars significantly ($F_{(chilling)} = 50{,}361.9$; $p < 0.05$) increased between 0 and 4 days of chilling stress in seedlings treated with all concentrations of foliar-sprayed ASA (Fig. 1). The total soluble sugars in the non-stressed *P. vulgaris* seedlings did not significantly ($p > 0.05$) change after being treated with different concentrations of foliar-sprayed ASA (Fig. 1). However, compared with that in non-stressed *P. vulgaris* control seedlings, the total soluble sugar content in *P. vulgaris* seedlings subjected to 2 or 4 days of chilling stress significantly increased in response to all concentrations of ASA foliar sprays. Foliar treatment with ASA

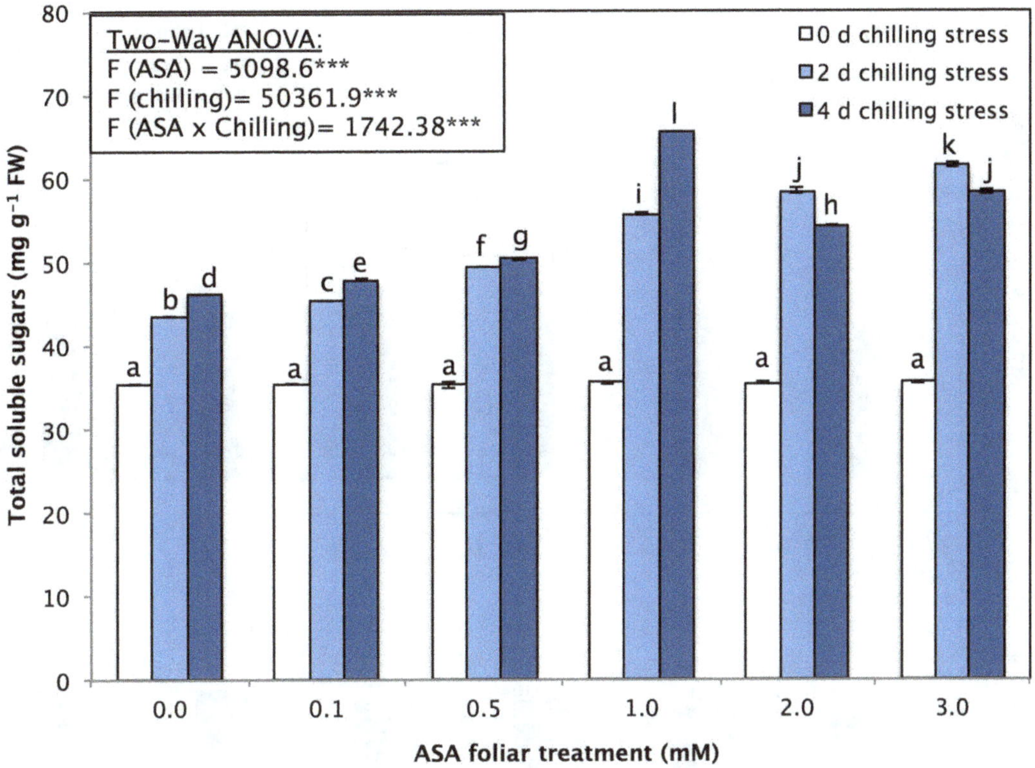


**Fig. 1** Total soluble sugars (mg $g^{-1}$ FW) of *P. vulgaris* L. after foliar application of different concentration of acetylsalicylic acid (ASA; 0, 0.1, 0.5, 1, 2, 3 mM) and exposed to different level of chilling stress (0, 2, 4 days). Data represented are mean of two replicates ± standard deviation. Variations between different chilling stresses and foliar ASA concentration were assessed by univariate analyses followed by post hoc analysis. Means with the same letters are not significantly different according to Duncan's multiple comparisons

significantly increased the total soluble sugars; the 1 mM concentration especially increased the soluble sugars from 35.55 ± 0.18 in non-stressed plants to 55.67 ± 0.073 or 65.47 ± 0.0028 mg $g^{-1}$ FW after 2 or 4 days of chilling stresses (Fig. 1).

Chilling stress induced a significant ($p < 0.05$) decrease in total protein and free amino acid contents in *P. vulgaris* seedlings (Fig. 2a). The total protein content was 9.69 ± 0.007, 6.76 ± 0.006 and 4.67 ± 0.019 mg $g^{-1}$ FW in non-ASA-sprayed seedlings subjected to chilling stress for 0, 2 and 4 days, respectively; the total free amino acids in the non-ASA-sprayed seedlings reached levels of 15.25 ± 0.014, 13.15 ± 0.028 and 12.95 ± 0.014 under the same chilling stress durations; respectively. The treatment of *P. vulgaris* with different concentrations of foliar-sprayed ASA did not induce significant changes in either total protein content or total free amino acids (Fig. 2a, b). However, the protein content of seedlings slightly increased after they were treated with foliar sprays with certain concentrations of ASA. The maximum increase in protein content was recorded in the seedlings treated with 0.5 mM ASA; the protein contents reached 7.13 ± 0.013 and 5.27 ± 0.001 mg $g^{-1}$ FW in the seedlings subjected to chilling stress for 2 and 4 days, respectively. However, the maximum levels of free amino acids were 13.93 ± 0.00 and 13.93 ± 0.02 mg $g^{-1}$ FW in the seedlings subjected to chilling stress for 2 and 4 days, respectively, following treatment with 3 mM ASA (Fig. 2b).

Proline metabolism, in terms of proline content (mg $g^{-1}$ FW), was monitored in the *P. vulgaris* seedlings (Fig. 2c). Proline accumulated significantly ($p < 0.05$) in the cells of *P. vulgaris* seedlings not treated with ASA in response to chilling stress. The proline accumulation increased significantly ($p < 0.05$) from 24.36 ± 0.011 mg $g^{-1}$ FW in the non-stressed seedlings to 28.40 ± 0.20 and 68.51 ± 0.068 mg $g^{-1}$ FW in those subjected to chilling stress for 2 and 4 days (Fig. 2c). However, treatment with ASA significantly ($p < 0.05$) increased the accumulation of proline in *P. vulgaris*, e.g., treatment with 3 mM ASA increased the accumulation from 24.35 ± 0.005 mg $g^{-1}$ FW in the non-stressed seedlings to 41.16 ± 0.44 and 91.35 ± 0.007 mg $g^{-1}$ FW in the seedlings subjected to chilling stress for 2 and 4 days (Fig. 2c).

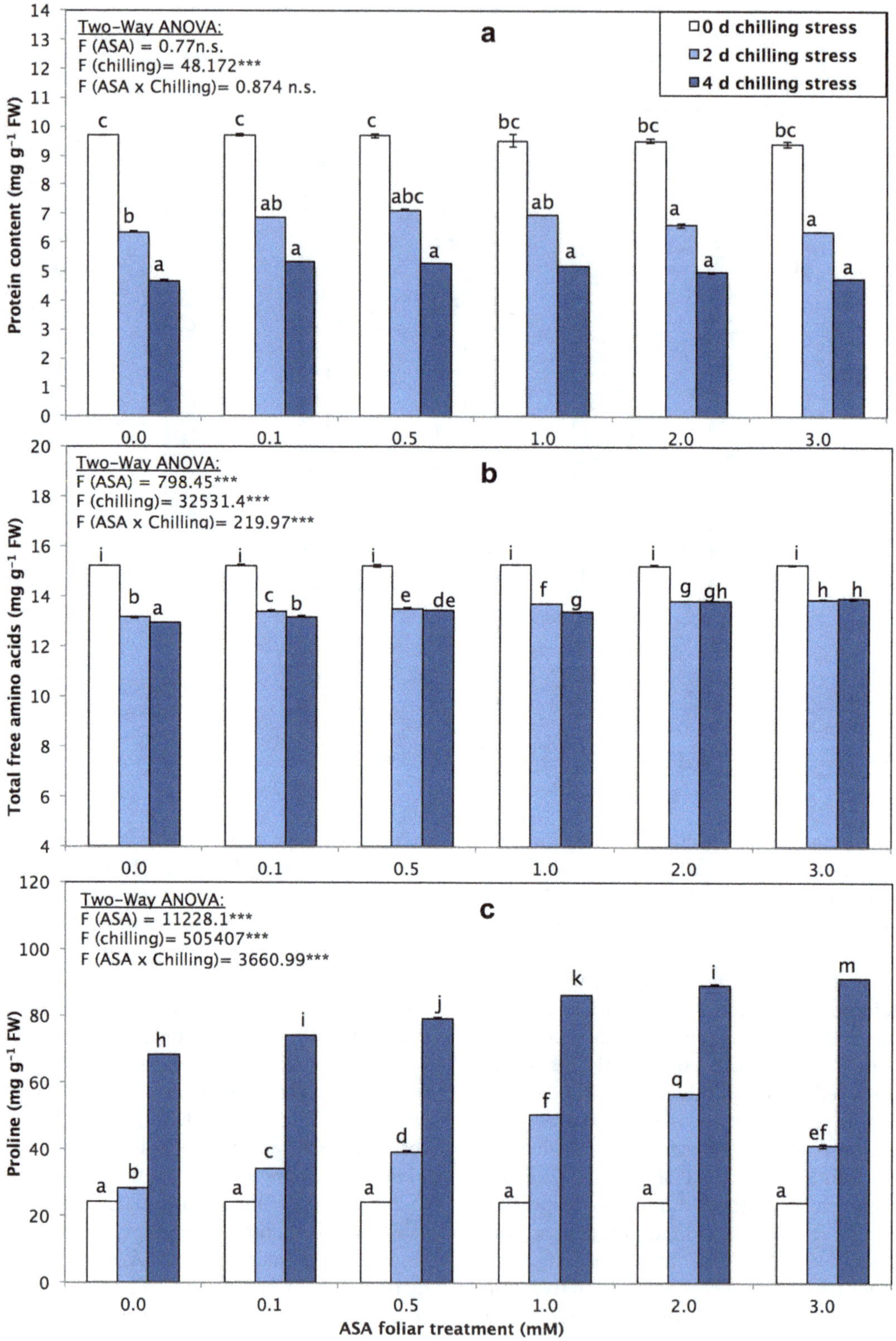


**Fig. 2** **a** Protein content (mg $g^{-1}$ FW), **b** total free amino acids (mg $g^{-1}$ FW), **c** proline (mg $g^{-1}$ FW) of *P. vulgaris* L. after foliar application of different concentration of acetylsalicylic acid (ASA; 0, 0.1, 0.5, 1, 2, 3 mM) and exposed to different level of chilling stress (0, 2, 4 days). Data represented are mean of two replicates ± standard deviation. Variations between different chilling stresses and foliar ASA concentration were assessed by univariate analyses followed by post hoc analysis. Means with the same letters are not significantly different according to Duncan's multiple comparisons

Cellular lipid peroxidation in terms of MDA was monitored. This method is well recognized to reflect oxidative damage caused by chilling stress. Chilling stress significantly increased cellular lipid peroxidation both in seedlings treated and not treated with ASA (Fig. 3). Compared with the non-stressed control *P. vulgaris* seedlings, the non-stressed ASA-treated *P. vulgaris* seedlings had significantly higher MDA levels under chilling stress and displayed levels of 30.35 ± 0.07 and 39.05 ± 0.070 µmol $g^{-1}$ FW under chilling stress of 2 and 4 days, respectively, following treatment with 3 mM ASA (Fig. 3). The MDA concentrations increased as the duration of low-temperature stress increased during the experiment (Fig. 3). Moreover, the MDA contents in non-stressed seedlings were slightly affected by ASA applications. However, these contents significantly increased in seedlings exposed to different durations of chilling stress (Fig. 3).

The content of AA, a non-enzymatic antioxidant, increased significantly ($p < 0.0.5$) during the experiment and peaked at 39.33 ± 0.14 mg $g^{-1}$ FW after 2 mM ASA and 2 days of chilling stress. The highest level of increase in AA at 4 days of chilling stress was 35.90 ± 0.11 mg $g^{-1}$ FW, recorded after 3 mM ASA treatment (Fig. 4a).

The enzymatic activities of four antioxidant enzymes, namely, APX, SOD, CAT, and POD, were also monitored in all the experimental variants (Figs. 4b, 5, 6). The activities of the antioxidant enzymes APX, POD, and SOD significantly ($p < 0.05$) increased in response to the different levels of chilling stress; however, CAT activity decreased (Figs. 4b, 5, 6). The activities of the antioxidant enzymes APX, SOD, POD, and CAT significantly increased in response to the application of exogenous ASA, alleviating the adverse effects of chilling stress.

With respect to the qRT-PCR analysis results, the fluorescence intensity of each clone was divided by its corresponding control, after which the relative mRNA levels of CBF3 and COR47 after 0, 2 and 4 days of chilling stress were quantified using RT-qPCR for the seedlings in each of the ASA concentration treatments (0.1, 0.5, 1, 2 and 3 mM). The CBF3 gene expression results presented in Fig. 7a that, compared with the CBF3 gene expression in the control treatment at 0 days (0.81-fold), the maximum expression of the CBF3 gene of 7.72 ± 0.10 was recorded in response to the 2 mM ASA concentration after 2 days

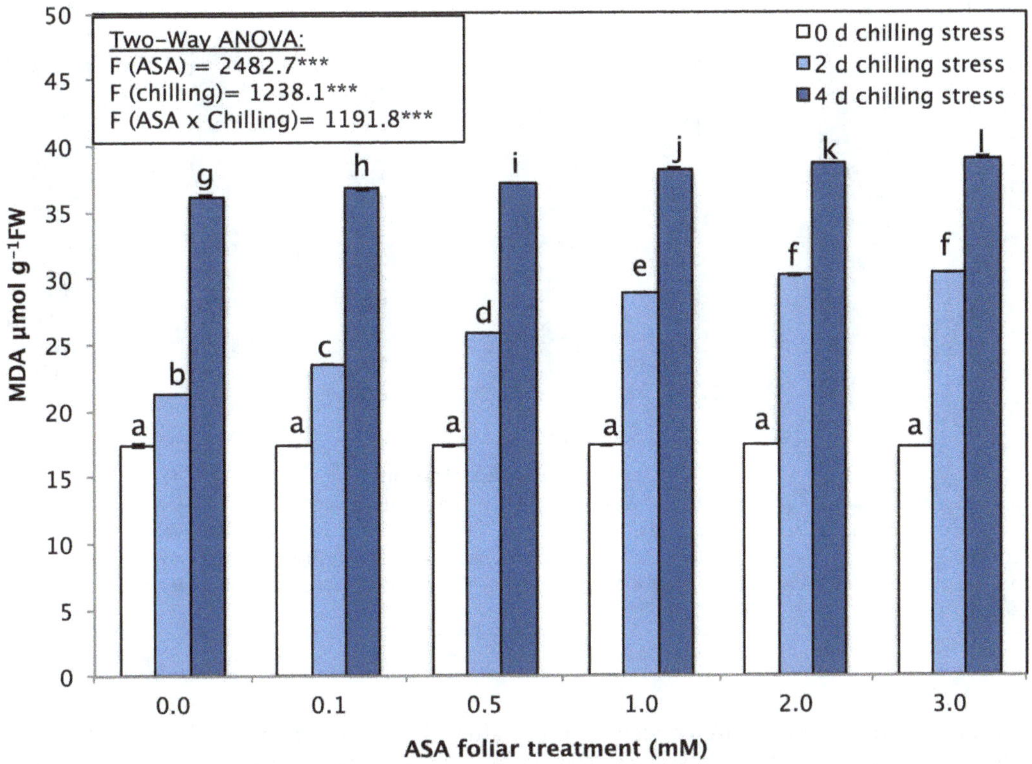


**Fig. 3** Cellular lipid peroxidation in terms of MDA (µmol $g^{-1}$ FW) of *P. vulgaris* L. after foliar application of different concentration of acetylsalicylic acid (ASA; 0, 0.1, 0.5, 1, 2, 3 mM) and exposed to different level of chilling stress (0, 2, 4 days). Data represented are mean of two replicates ± standard deviation. Variations between different chilling stresses and foliar ASA concentration were assessed by univariate analyses followed by post hoc analysis. Means with the same letters are not significantly different according to Duncan's multiple comparisons

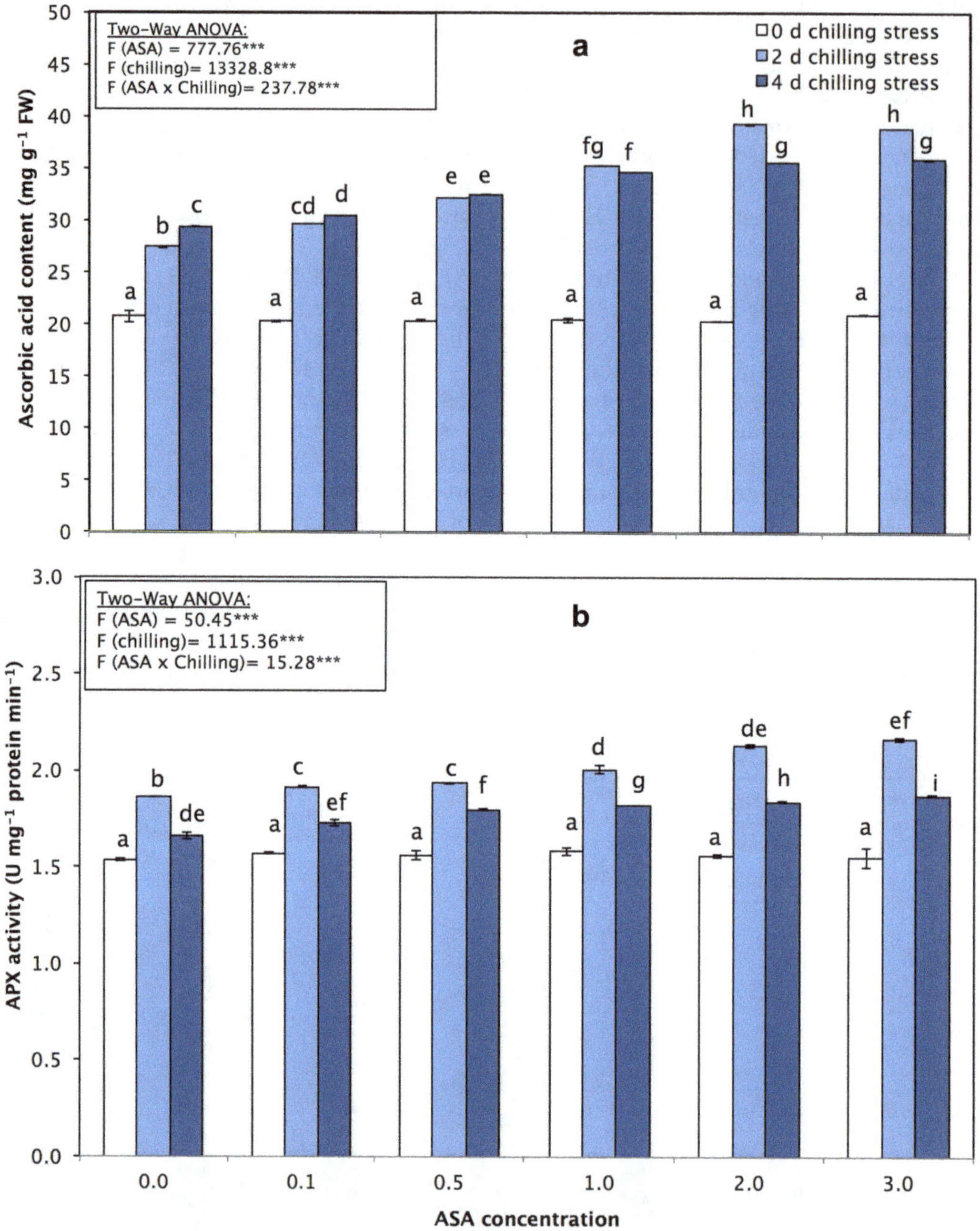


**Fig. 4** **a** Ascorbic acid content (mg $g^{-1}$ FW) and **b** ascorbic peroxidase (APX; mmol ascorbate $min^{-1}$ $g^{-1}$ FW) in leaves of *P. vulgaris* L. after foliar application of different concentration of acetylsalicylic acid (ASA; 0, 0.1, 0.5, 1, 2, 3 mM) and exposed to different level of chilling stress (0, 2, 4 days). Data represented are mean of two replicates ± standard deviation. Variations between different chilling stresses and foliar ASA concentration were assessed by univariate analyses followed by post hoc analysis. Means with the same letters are not significantly different according to Duncan's multiple comparisons

(7.72-fold), followed by the 1 mM ASA concentration after 4 days (7.47-fold), the 1 mM ASA concentration after 2 days (7.46-fold), the 0.5 mM ASA concentration after 2 days (6.72-fold), the 0.5 mM ASA concentration after 4 days (6.54-fold) and the 3 mM ASA concentration after 2 days (6.35-fold). The COR47 gene expression results (Fig. 7b) revealed that, compared with the COR47 gene expression levels of the control treatment after 0 days (0.69-fold), the maximum expression levels of the COR47 gene were recorded in response to the 2 mM

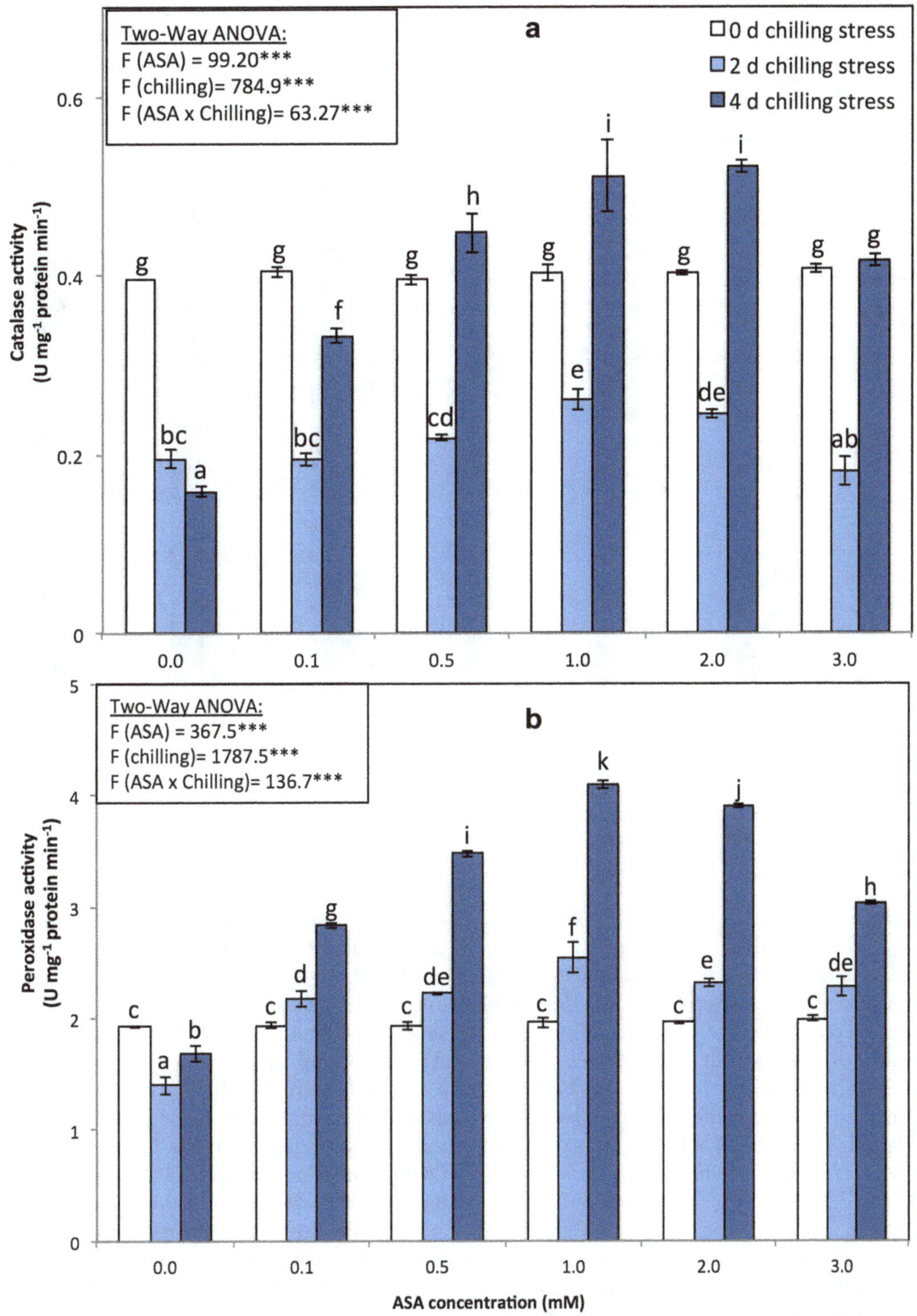


**Fig. 5** **a** Catalase (unit mg$^{-1}$ protein min$^{-1}$) and **b** peroxidase activities (unit mg$^{-1}$ protein min$^{-1}$) of *P. vulgaris* L. after foliar application of different concentration of acetylsalicylic acid (ASA; 0, 0.1, 0.5, 1, 2, 3 mM) and exposed to different level of chilling stress (0, 2, 4 days). Data represented are mean of two replicates ± standard deviation. Variations between different chilling stresses and foliar ASA concentration were assessed by univariate analyses followed by post hoc analysis. Means with the same letters are not significantly different according to Duncan's multiple comparisons

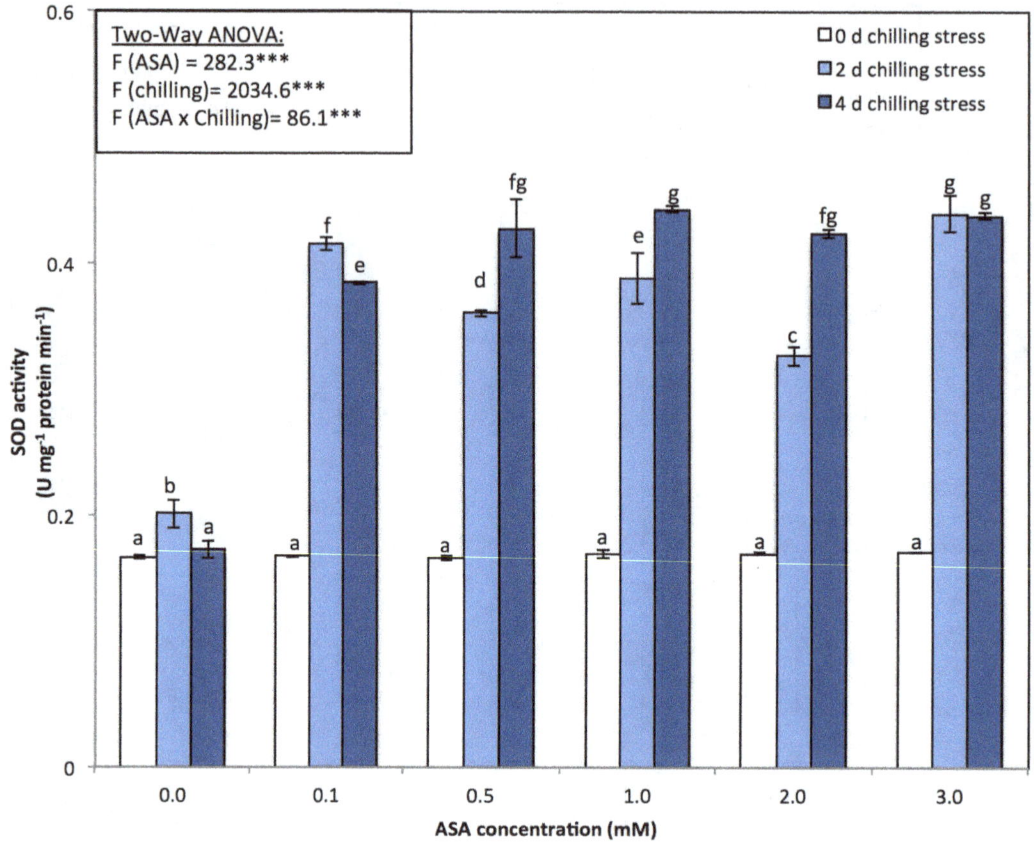


**Fig. 6** Superoxide dismutase activities (unit $mg^{-1}$ protein $min^{-1}$) of *P. vulgaris* L. after foliar application of different concentration of acetylsalicylic acid (ASA; 0, 0.1, 0.5, 1, 2, 3 mM) and exposed to different level of chilling stress (0, 2, 4 days). Data represented are mean of two replicates ± standard deviation. Variations between different chilling stresses and foliar ASA concentration were assessed by univariate analyses followed by post hoc analysis. Means with the same letters are not significantly different according to Duncan's multiple comparisons

ASA concentration after 2 days (2.72-fold), followed by the 1 mM ASA concentration after 2 days (2.72-fold), the 1 mM ASA concentration after 4 days (2.69-fold), the 2 mM ASA concentration after 4 days and the 3 mM ASA concentration after 2 days.

The application of exogenous ASA alleviated the adverse effects of chilling stresses for all measured parameters, and the strongest effects were observed in response to 1 and 2 mM ASA. As such, pretreatment with ASA appeared to induce chilling tolerance.

## Discussion

Low-temperature or chilling stress tolerance is a critical feature for economic crops that originate in tropical or temperate regions worldwide. ASA and other phenolic compounds (e.g., SA and benzoic acid) can enhance the chilling tolerance of plants (Kang and Saltveit 2001; Kang et al. 2003; Chinnusamy et al. 2007).

The current study revealed that all growth characteristics and photosynthetic pigments of *P. vulgaris* increased as ASA concentrations increased to a certain level during the chilling stress conditions and that the most significant reduction was observed in the control plants. The most adverse effects of chilling stress observed in the untreated seedlings included reductions in FW, DW, the shoot: root ratio and chlorophyll content (Table 2). The ASA treatment improved the growth characteristics to a certain level under chilling stress; ASA, therefore, acts as a growth stimulant. These effects of ASA application were in agreement with the findings of Senaratna et al. (2000), who reported that 100% of tomato and bean seedlings treated with 0.5 mM ASA survived under chilling stress, while the control plants did not. Moreover, Gharib and Hegazi (2010) reported that soaking *P. vulgaris* seeds in 0.1 mM SA solution also significantly enhance the germination rate, germination percentage and seedling trait under standard and low-temperature stress conditions. Similarly, different morphological and growth criteria of tomato plants treated with foliar applications of salicylaldehyde were enhanced (Kord and Hathout 1992).

Low temperatures cause an imbalance between light harvest and energy use; in turn, the superoxide anion is

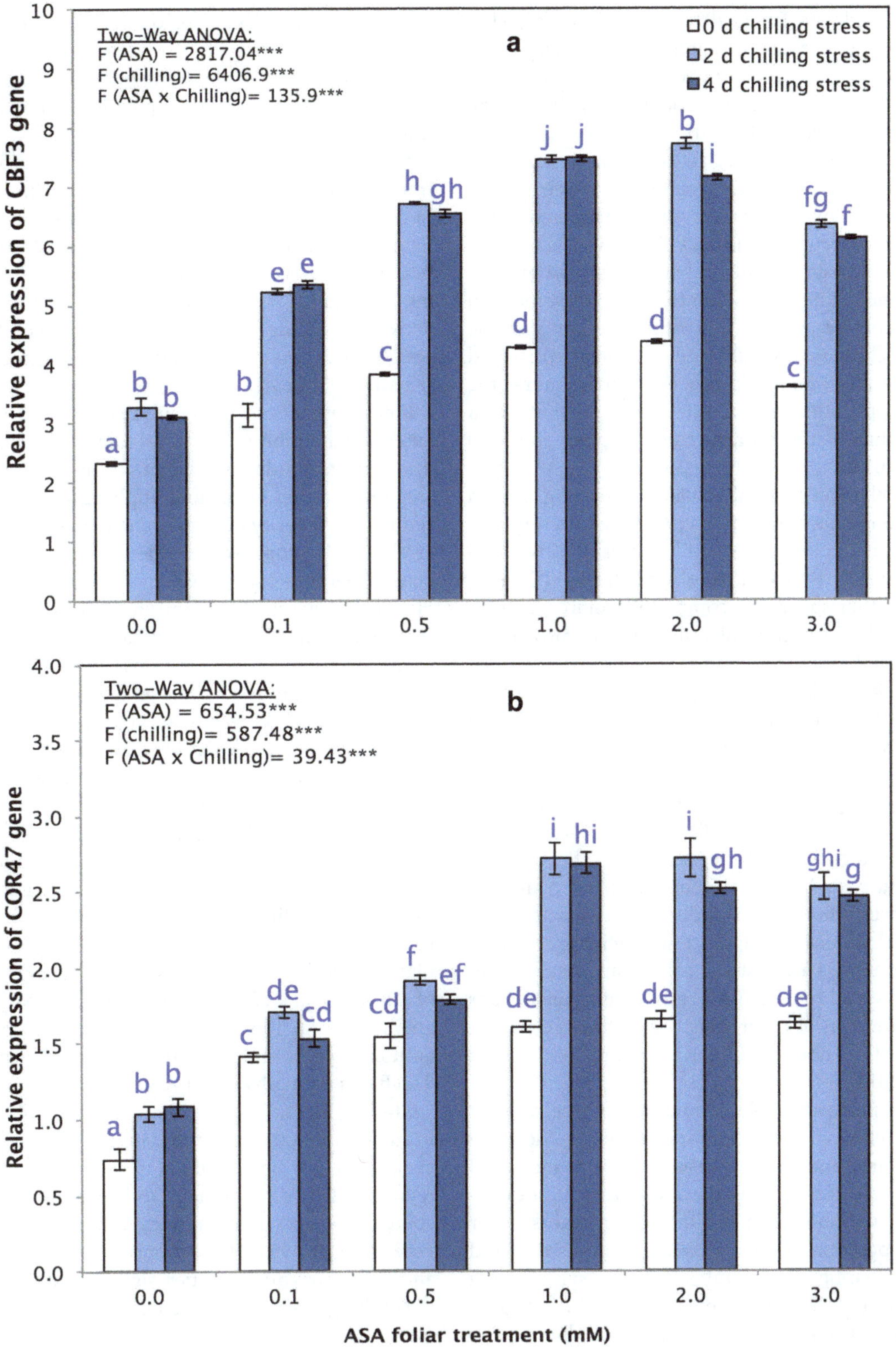


**Fig. 7** Relative expression of genes **a** CBF3, and **b** COR47 of *P. vulgaris* L. after foliar application of different concentration of acetylsalicylic acid (ASA; 0, 0.1, 0.5, 1, 2, 3 mM) and exposed to a different level of chilling stress (0, 2, 4 days). Data represented are mean of two replicates ± standard deviation. Variations between different chilling stresses and foliar ASA concentration were assessed by univariate analyses followed by post hoc analysis. Means with the same letters are not significantly different according to Duncan's multiple comparisons

generated, inducing oxidative stress in chloroplasts (Einset et al. 2007; Gill and Tuteja 2010). ROS can damage the photosynthetic apparatus through the disarrangement of thylakoid structures, suppression of chloroplastic enzymes, and inhibition of D1 protein biosynthesis; the D1 protein is needed for the PSII repair process (Gururani et al. 2015; Choudhury et al. 2017). The exogenous application of ASA most likely significantly increases the activity of the antioxidant system, directly impacting the intensity of photosynthesis. Li and his co-workers (2014) reported that SA significantly enhances *Torreya grandis* biomass under salinity stress due to the improvement in chlorophyll content and the flow of antioxidant enzyme activity that consequently alleviates oxidative stress and enhances the photosynthetic process. Furthermore, the exogenous application of SA effectively prevents iron deficiency, which is the primary cause of plant chlorosis (Kong et al. 2014).

Moreover, changes in total soluble sugars (Fig. 1) and carbohydrate metabolism act as energy factors required for driving acclimation under chilling stress (Gusta and Wisniewski 2013). By interacting with the lipid bilayer, sucrose and other sugars (as compatible solutes) significantly affect plant freezing tolerance and the protection of cellular membranes from damage (Hansen et al. 1997; Liu et al. 1998; Shalaev and Steponkus 2001; Frankow-Lindberg 2001; Arroyo et al. 2003; Shao et al. 2007). The accumulation of soluble sugars contributes to increased membrane cryostability, which is a prerequisite for freezing tolerance in plants (Arroyo et al. 2003; Shao et al. 2007; Gusta and Wisniewski 2013). In agreement with the findings of Nafeh and Hegazi (2009), the present study showed that ASA treatment improved plant tolerance against chilling stress and, compared with those under normal conditions, significantly increases in soluble sugars content. Under stress conditions, increases in soluble sugars lead to enhanced resistance against water loss, the protection of chloroplasts and the acceleration of plant growth (Khodary 2004; Kader et al. 2011; Fayez and Bazaid 2014).

ASA has a protective role. It neutralizes free radicals induced by chilling stress and prevents the destruction of proteins, and in the present study, it increased the amount of protein in treated seedlings (Shinozaki et al. 2003). In response to chilling stress, plants lower the osmotic potential of the cytosol by synthesizing and accumulating compatible solutes and by synthesizing cold acclimation-induced proteins, all of which trigger crop tolerance to chilling stress (Frankow-Lindberg 2001; Murakeözy et al. 2003; Yancey 2005).

Free proline is involved in plant resistance and accumulates under several stress conditions (Parvanova et al. 2004; Demiral and Turkan 2005; Awasthi et al. 2015). In our findings, the proline content significantly increased in the chilling-stressed seedlings of *P. vulgaris* treated with different levels of ASA (Fig. 2). Our data suggest that proline plays a role in the protection against chilling stress by assisting osmotic regulation, removing hydroxyl free radicals and avoiding the destruction of enzymes (Kuznetsov and Shevyakova 1999). The accumulation of free proline also constitutes a vital element of tolerance in other species (Barka et al. 2006; Apostolova et al. 2008; Esra et al. 2010). Treatment with 0.5 mM SA mitigates heat stress by increasing proline production by increasing γ-glutamyl kinase and decreasing proline oxidase activity, resulting in the promotion of osmotic potential necessary for maintaining crucial physiological processes (Khan et al. 2013).

Measurements of MDA are usually used to assess the extent of membrane damage caused by chilling stress. The present study found that, compared with the control applications, exogenous applications of ASA failed to reduce the severity of membrane damage (Fig. 3). Mittler et al. (2004) proposed that membrane damage might be caused by high $H_2O_2$ levels, which could accelerate the Haber–Weiss reaction, resulting in hydroxyl radical formation and thus lipid peroxidation.

The total AA concentrations increased markedly in ASA-treated common bean seedlings as well as in those subjected to chilling stress (Fig. 4). This increase may occur because the AA proved to be a ubiquitous compound that effectively boosts stress tolerance in plants (Miguel et al. 2006; Khan et al. 2012; Naz et al. 2016). AA have a major role during the minimization of ROS activity by enzymatic and non-enzymatic detoxification (Mittler et al. 2004; Borland et al. 2006; Shao et al. 2007). Our findings are in line with those of Kader et al. (2011), who reported that the ascorbate content increased in wheat plants under chilling stress. Our results agree with those of Tirani et al. (2013), who confirmed that SA induces an increase in the AA content in canola plants; those authors suggested that the increase in AA occurs because SA affects the pathways of AA biosynthesis.

Chilling stress triggers the formation of ROS, which in turn leads to stronger oxidative stress in plants. However, ROS can be classified as signaling molecules that regulate plant development and responses to biotic and abiotic stresses (Apel and Hirt 2004; Mittler et al. 2004; Perez and Brown 2014). In our research, the SOD, POD and APX enzyme activities significantly increased during chilling stress and in response to foliar applications of ASA, while the CAT enzyme activity decreased (Figs. 4, 5). The enhancement of antioxidant enzyme activity under chilling stress has been reported in many plant species (Hodges et al. 1996; Kratsch and Wise 2000; Bracale and Coraggio 2003; Einset et al. 2007; Zhao et al.

2013; Ruelland 2017). When applied at low concentrations, SA causes transient oxidative stress in plants, which acts as a hardening process, increasing the antioxidant capacity of plants (Wang et al. 2003; Horváth et al. 2007). Exogenous applications of SA can prevent isozyme activity, including that of CAT-1 and CAT-2, which in turn can mediate responses to low-temperature stress in *Z. mays* plants (Horváth et al. 2007). In the present study, by inducing the activity of APX, POX, and SOD, ASA strongly reduced the effects of chilling stress on all parameters measured in the plants. Similar results were also reported by Mutlu et al. (2013) and He and Zhu (2008) in wheat and tomato, respectively. Positive correlations among SOD, POD and APX activities suggest that increased SOD activity is accompanied by increases in POD and APX activities because of the high demand from $H_2O_2$ quenching (Senaratna et al. 2000).

Low-temperature stress triggers the expression of chilling stress-associated genes. Among several cold signaling pathways, the involvement of *CBF3* and *COR47* is critical in plant chilling tolerance and cold acclimation (Miura and Furumoto 2013). Our results demonstrate that the CBF3 and COR47 genes exhibit a significant increase in their relative expression level (fold change) in accordance with the application of different ASA concentrations to plants under chilling stress (Fig. 7a, b). Chilling stress amplified the expression of the tested CBF3 genes in the *P. vulgaris* seedlings (Fig. 7a), suggesting a functional similarity of CBFs in *P. vulgaris* as well as Arabidopsis in response to low-temperature stress (Gilmour et al. 2004; Thomashow 2010). Application of ASA under chilling stress enhanced the expression of CBF3 together with the COR47 gene (Fig. 7b). In Arabidopsis, three CBF/DREB1 proteins take part in the control of COR gene expression and chilling tolerance (Gilmour et al. 2000, 2004). A calmodulin-binding transcription activator, CAMTA3/AtSR1, identifies the promoter of CBF2/DREB1C to upregulate genes associated with chilling tolerance, indicating that SA signaling and low-temperature signaling are interconnected (Du et al. 2009; Doherty et al. 2009).

Overexpression of *CfCBF3* increases chilling tolerance and causes no dwarf phenotype (Hanin et al. 2011). This overexpression also leads to multiple biochemical and physiological changes associated with chilling stress. Higher contents of proline and soluble sugars and lower contents of ROS have been observed in transgenic plants (Yang et al. 2011; Miura and Furumoto 2013).

## Conclusions

Chilling temperatures are responsible for a range of physiological disturbances in chilling-sensitive plants and can cause chilling injury and death of many horticulture plants such as *P. vulgaris*. In this study, the chilling tolerance of *P. vulgaris* could be significantly improved by the exogenous application of ASA. The physiological and molecular data ultimately revealed that ASA could mutually induce and maintain homeostasis to exert synergistic effects on common bean plant chilling stress. The present study confirmed that the optimum concentrations of ASA for alleviating the effects of chilling stress proved to be 1 and 2 mM ASA, as maximum stimulation of the antioxidant enzyme system occurred in response to these concentrations. These results indicate that ASA can effectively be used to protect *P. vulgaris* from the damaging effects of chilling stress during the early stages of growth.

### Abbreviations

AA: ascorbic acid; ASA: acetylsalicylic acid; APX: ascorbate peroxidase; BSA: bovine serum albumin; CAT: catalase; DAC: days after chilling; DMRTs: Duncan multiple range tests; FW: fresh weight; MDA: malondialdehyde; PPFD: photosynthetic photon flux density; POD: peroxidase; ROS: reactive oxygen species; SOD: superoxide dismutase.

### Authors' contributions

MS considered the experiments, and wrote the manuscript. AE participated in experimental design, wrote and proofreading. AAA revised the manuscript and made some laboratory work. AMA contributed in conducting the experiments and performed the statistical data analysis. All authors read and approved the final manuscript.

### Author details

[1] Biology Department, Faculty of Science, Taibah University, Al-Sharm, Yanbu El-Bahr 46429, Kingdom of Saudi Arabia. [2] Botany and Microbiology Department, Faculty of Science, Cairo University, 12613 Giza, Egypt. [3] Biological Sciences Department, Faculty of Science, University of Jeddah, Jeddah, Kingdom of Saudi Arabia. [4] Botany Department, Faculty of Science, Suez Canal University, Ismailia, Egypt.

### Acknowledgements

Not applicable.

### Competing interests

The authors declare that they have no competing interests.

### Funding

Not applicable.

### References

AbdElgawad H, Zinta G, Hegab MM et al (2016) High salinity induces different oxidative stress and antioxidant responses in maize seedlings organs. Front Plant Sci. https://doi.org/10.3389/fpls.2016.00276

Aebi H (1984) [13] Catalase in vitro. In: Packer L (ed) Methods in enzymology. Academic Press, New York, pp 121–126

Allen DJ, Ort DR (2001) Impacts of chilling temperatures on photosynthesis in warm-climate plants. Trends Plant Sci 6:36–42. https://doi.org/10.1016/S1360-1385(00)01808-2

Apel K, Hirt H (2004) Reactive oxygen species: metabolism, oxidative stress, and signal transduction. Annu Rev Plant Biol 55:373–399. https://doi.org/10.1146/annurev.arplant.55.031903.141701

Apostolova P, Yordanova R, Popova L (2008) Response of antioxidative defence system to low temperature stress in two wheat cultivars. Gen Appl Plant Physiol 34:281–294

Arroyo A, Bossi F, Finkelstein RR, León P (2003) Three genes that affect sugar sensing (abscisic acid insensitive 4, abscisic acid insensitive 5, and constitutive triple response 1) are differentially regulated by glucose in Arabidopsis. Plant Physiol 133:231–242

Awasthi R, Bhandari K, Nayyar H (2015) Temperature stress and redox homeostasis in agricultural crops. Front Environ Sci. https://doi.org/10.3389/fenvs.2015.00011

Barka EA, Nowak J, Clement C (2006) Enhancement of chilling resistance of inoculated grapevine plantlets with a plant growth-promoting rhizobacterium, *Burkholderia phytofirmans* strain PsJN. Appl Environ Microbiol 72:7246–7252. https://doi.org/10.1128/AEM.01047-06

Bita CE, Gerats T (2013) Plant tolerance to high temperature in a changing environment: scientific fundamentals and production of heat stress-tolerant crops. Front Plant Sci. https://doi.org/10.3389/fpls.2013.00273

Borland A, Elliott S, Patterson S et al (2006) Are the metabolic components of crassulacean acid metabolism up-regulated in response to an increase in oxidative burden? J Exp Bot 57:319–328. https://doi.org/10.1093/jxb/erj028

Bracale M, Coraggio I (2003) Chilling and freezing stresses in plants: cellular responses and molecular strategies for adaptation. In: di Toppi LS, Pawlik-Skowrońska B (eds) Abiotic stresses in plants. Springer, Dordrecht, pp 23–51

Bradford MM (1976) A rapid and sensitive method for the quantitation of microgram quantities of protein utilizing the principle of protein-dye binding. Anal Biochem 72:248–254. https://doi.org/10.1016/0003-2697(76)90527-3

Caffagni A, Pecchioni N, Francia E et al (2014) Candidate gene expression profiling in two contrasting tomato cultivars under chilling stress. Biol Plant 58:283–295. https://doi.org/10.1007/s10535-014-0403-z

Canakci S, Munzuroğlu O (2007) Effects of acetylsalicylic acid on germination, growth and chlorophyll amounts of cucumber (*Cucumis sativus* L.) seeds. Pak J Biol Sci PJBS 10:2930–2934

Chartzoulakis K, Psarras G (2005) Global change effects on crop photosynthesis and production in Mediterranean: the case of Crete, Greece. Agric Ecosyst Environ 106:147–157. https://doi.org/10.1016/j.agee.2004.10.004

Chinnusamy V, Zhu J, Zhu JK (2007) Cold stress regulation of gene expression in plants. Trends Plant Sci 12:444–451. https://doi.org/10.1016/j.tplants.2007.07.002

Choudhury FK, Rivero RM, Blumwald E, Mittler R (2017) Reactive oxygen species, abiotic stress and stress combination. Plant J 90:856–867. https://doi.org/10.1111/tpj.13299

Ciha AJ, Brun WA (1978) Effect of pod removal on nonstructural carbohydrate concentration in soybean tissue. Crop Sci 18:773–776. https://doi.org/10.2135/cropsci1978.0011183X001800050020x

Cramer GR, Urano K, Delrot S et al (2011) Effects of abiotic stress on plants: a systems biology perspective. BMC Plant Biol 11:163. https://doi.org/10.1186/1471-2229-11-163

De Vos CHR, Schat H, De Waal MAM et al (1991) Increased resistance to copper-induced damage of the root cell plasmalemma in copper tolerant *Silene cucubalus*. Physiol Plant 82:523–528. https://doi.org/10.1111/j.1399-3054.1991.tb02942.x

Demiral T, Turkan I (2005) Comparative lipid peroxidation, antioxidant defense systems and proline content in roots of two rice cultivars differing in salt tolerance. Environ Exp Bot 53:247–257. https://doi.org/10.1016/j.envexpbot.2004.03.017

Doherty CJ, Van Buskirk HA, Myers SJ, Thomashow MF (2009) Roles for Arabidopsis CAMTA transcription factors in cold-regulated gene expression and freezing tolerance. Plant Cell 21:972–984. https://doi.org/10.1105/tpc.108.063958

Du L, Ali GS, Simons KA et al (2009) Ca(2+)/calmodulin regulates salicylic-acid-mediated plant immunity. Nature 457:1154–1158. https://doi.org/10.1038/nature07612

Dubey RS, Rani M (1989) Influence of NaCl salinity on growth and metabolic status of protein and amino acids in rice seedlings. J Agron Crop Sci 162:97–106. https://doi.org/10.1111/j.1439-037X.1989.tb00694.x

Einset J, Winge P, Bones A (2007) ROS signaling pathways in chilling stress. Plant Signal Behav 2:365–367

El Kelish A, Zhao F, Heller W et al (2014) Ragweed (*Ambrosia artemisiifolia*) pollen allergenicity: SuperSAGE transcriptomic analysis upon elevated $CO_2$ and drought stress. BMC Plant Biol 14:176. https://doi.org/10.1186/1471-2229-14-176

Esra K, İŞLEK C, Üstün AS (2010) Effect of cold on protein, proline, phenolic compounds and chlorophyll content of two pepper (*Capsicum annuum* L.) varieties. Gazi Univ J Sci 23:1–6

Fayez KA, Bazaid SA (2014) Improving drought and salinity tolerance in barley by application of salicylic acid and potassium nitrate. J Saudi Soc Agric Sci 13:45–55. https://doi.org/10.1016/j.jssas.2013.01.001

Foyer CH, Noctor G (2003) Redox sensing and signalling associated with reactive oxygen in chloroplasts, peroxisomes and mitochondria. Physiol Plant 119:355–364. https://doi.org/10.1034/j.1399-3054.2003.00223.x

Frankow-Lindberg BE (2001) Adaptation to winter stress in nine white clover populations: changes in non-structural carbohydrates during exposure to simulated winter conditions and "spring" regrowth potential. Ann Bot 88:745–751. https://doi.org/10.1006/anbo.2001.1432

Gautam S, Singh PK (2009) Salicylic acid-induced salinity tolerance in corn grown under NaCl stress. Acta Physiol Plant 31:1185. https://doi.org/10.1007/s11738-009-0338-8

Gharib FA, Hegazi AZ (2010) Salicylic acid ameliorates germination, seedling growth, phytohormone and enzymes activity in bean (*Phaseolus vulgaris* L.) under cold stress. J Am Sci 6:675–683

Giannopolitis CN, Ries SK (1977) Superoxide dismutases: I. Occurrence in higher plants. Plant Physiol 59:309–314

Gill SS, Tuteja N (2010) Reactive oxygen species and antioxidant machinery in abiotic stress tolerance in crop plants. Plant Physiol Biochem 48:909–930. https://doi.org/10.1016/j.plaphy.2010.08.016

Gilmour SJ, Sebolt AM, Salazar MP et al (2000) Overexpression of the Arabidopsis CBF3 transcriptional activator mimics multiple biochemical changes associated with cold acclimation. Plant Physiol 124:1854–1865

Gilmour SJ, Fowler SG, Thomashow MF (2004) Arabidopsis transcriptional activators CBF1, CBF2, and CBF3 have matching functional activities. Plant Mol Biol 54:767–781. https://doi.org/10.1023/B:PLAN.0000040902.06881.d4

Gupta N, Rathore M, Goyary D et al (2012) Marker-free transgenic cucumber expressing Arabidopsis cbf1 gene confers chilling stress tolerance. Biol Plant 56:57–63. https://doi.org/10.1007/s10535-012-0016-3

Gururani MA, Venkatesh J, Tran LSP (2015) Regulation of photosynthesis during abiotic stress-induced photoinhibition. Mol Plant 8:1304–1320. https://doi.org/10.1016/j.molp.2015.05.005

Gusta LV, Wisniewski M (2013) Understanding plant cold hardiness: an opinion. Physiol Plant 147:4–14. https://doi.org/10.1111/j.1399-3054.2012.01611.x

Guy CL (1990) Cold acclimation and freezing stress tolerance: role of protein metabolism. Annu Rev Plant Physiol Plant Mol Biol 41:187–223. https://doi.org/10.1146/annurev.pp.41.060190.001155

Hanin M, Brini F, Ebel C et al (2011) Plant dehydrins and stress tolerance. Plant Signal Behav 6:1503–1509. https://doi.org/10.4161/psb.6.10.17088

Hansen J, Turkington R, Vogg G et al (1997) Conifer carbohydrate physiology: updating classical views. In: Rennenberg H, Eschrich W, Ziegler H (eds.) Trees—contributions to modern tree physiology. Backhuys, Leiden, pp 97–108

He Y, Zhu ZJ (2008) Exogenous salicylic acid alleviates NaCl toxicity and increases antioxidative enzyme activity in *Lycopersicon esculentum*. Biol Plant 52:792–795. https://doi.org/10.1007/s10535-008-0155-8

Hodges DM, Andrews CJ, Johnson DA, Hamilton RI (1996) Antioxidant compound responses to chilling stress in differentially sensitive inbred maize lines. Physiol Plant 98:685–692. https://doi.org/10.1034/j.1399-3054.1996.980402.x

Holder M (1965) Chlorophylls: chemistry and biochemistry of plant pigments. In: Goodwin TW (ed) Chemistry and biochemistry of plant pigments. Academic Press, London

Horváth E, Szalai G, Janda T (2007) Induction of abiotic stress tolerance by salicylic acid signaling. J Plant Growth Regul 26:290–300. https://doi.org/10.1007/s00344-007-9017-4

Irigoyen JJ, Einerich DW, Sanchez-Diaz M (1992) Water stress induced changes in concentrations of proline and total soluble sugars in nodulated alfalfa (*Medicago sativa*) plants. Physiol Plant 84:55–60. https://doi.org/10.1111/j.1399-3054.1992.tb08764.x

Jaglo KR, Kleff S, Amundsen KL et al (2001) Components of the Arabidopsis C-repeat/dehydration-responsive element binding factor cold-response pathway are conserved in *Brassica napus* and other plant species. Plant Physiol 127:910–917. https://doi.org/10.1104/pp.010548

Jaleel CA, Manivannan P, Wahid A et al (2009) Drought stress in plants: a review on morphological characteristics and pigments composition. Int J Agric Biol 11:100–105

Jiang AL, Tian S, Xu Y (2002) Effect of controlled atmospheres with high $O_2$ or high-$CO_2$ concentrations on postharvest physiology and storability of "Napoleon" sweet cherry. Acta Bot Sin 44:925–930

Kader DZA, Saleh AAH, Elmeleigy SA, Dosoky NS (2011) Chilling-induced oxidative stress and polyamines regulatory role in two wheat varieties. J Taibah Univ Sci 5:14–24. https://doi.org/10.1016/S1658-3655(12)60034-X

Kang H-M, Saltveit ME (2001) Activity of enzymatic antioxidant defense systems in chilled and heat shocked cucumber seedling radicles. Physiol Plant 113:548–556. https://doi.org/10.1034/j.1399-3054.2001.1130414.x

Kang G, Wang C, Sun G, Wang Z (2003) Salicylic acid changes activities of $H_2O_2$-metabolizing enzymes and increases the chilling tolerance of banana seedlings. Environ Exp Bot 50:9–15. https://doi.org/10.1016/S0098-8472(02)00109-0

Kasuga M, Liu Q, Miura S et al (1999) Improving plant drought, salt, and freezing tolerance by gene transfer of a single stress-inducible transcription factor. Nat Biotechnol 17:287–291. https://doi.org/10.1038/7036

Khan MIR, Khan NA (2013) Salicylic acid and jasmonates: approaches in abiotic stress tolerance. J Plant Biochem Physiol. https://doi.org/10.4172/2329-9029.1000e113

Khan W, Prithiviraj B, Smith DL (2003) Photosynthetic responses of corn and soybean to foliar application of salicylates. J Plant Physiol 160:485–492. https://doi.org/10.1078/0176-1617-00865

Khan NA, Nazar R, Iqbal N, Anjum NA (eds) (2012) Phytohormones and abiotic stress tolerance in plants. Springer, Berlin

Khan MIR, Iqbal N, Masood A et al (2013) Salicylic acid alleviates adverse effects of heat stress on photosynthesis through changes in proline production and ethylene formation. Plant Signal Behav 8:e26374. https://doi.org/10.4161/psb.26374

Khodary SEA (2004) Effect of salicylic acid on the growth, photosynthesis and carbohydrate metabolism in salt-stressed maize plants. Int J Agric Biol 6:5–8

Kong J, Dong Y, Xu L et al (2014) Effects of foliar application of salicylic acid and nitric oxide in alleviating iron deficiency induced chlorosis of *Arachis hypogaea* L. Bot Stud 55:9. https://doi.org/10.1186/1999-3110-55-9

Kord M, Hathout T (1992) Changes in some growth criteria, metabolic activities and endogenous hormones in tomato plants consequent to spraying with different concentrations of salicylaldehyde. Egypt J Physiol Sci. https://eurekamag.com/research/002/318/002318832.php. Accessed 20 Sep 2017

Kratsch HA, Wise RR (2000) The ultrastructure of chilling stress. Plant Cell Environ 23:337–350. https://doi.org/10.1046/j.1365-3040.2000.00560.x

Kupferwasser LI, Yeaman MR, Shapiro SM et al (1999) Acetylsalicylic acid reduces vegetation bacterial density, hematogenous bacterial dissemination, and frequency of embolic events in experimental Staphylococcus aureus endocarditis through antiplatelet and antibacterial effects. Circulation 99:2791–2797

Kuznetsov VV, Shevyakova NI (1999) Proline under stress: biological role, metabolism, and regulation. Russ J Plant Physiol 46:274–287

Li YC, Jiang XX, Long XJ (2014) Effects and action mechanisms of sodium fluoride (NaF) on the growth and cephalotaxine production of *Cephalotaxus mannii* suspension cells. Enzyme Microb Technol 67:77–81. https://doi.org/10.1016/j.enzmictec.2014.09.006

Liu Q, Kasuga M, Sakuma Y et al (1998) Two transcription factors, DREB1 and DREB2, with an EREBP/AP2 DNA binding domain separate two cellular signal transduction pathways in drought- and low-temperature-responsive gene expression, respectively, in Arabidopsis. Plant Cell 10:1391–1406

Livak KJ, Schmittgen TD (2001) Analysis of relative gene expression data using real-time quantitative PCR and the 2(-Delta Delta C(T)) method. Methods San Diego Calif 25:402–408. https://doi.org/10.1006/meth.2001.1262

Mahajan S, Tuteja N (2005) Cold, salinity and drought stresses: an overview. Arch Biochem Biophys 444:139–158. https://doi.org/10.1016/j.abb.2005.10.018

McHenry EW, Graham M (1935) Observations on the estimation of ascorbic acid by titration. Biochem J 29:2013–2019

Miguel G, Fontes C, Martins D et al (2006) Effects of post-harvest treatment and storage time on the organic acid content in Assaria and Mollar pomegranate (*Punica granatum* L.) fruit. Ital J Food Sci 18:317–322

Miller G, Suzuki N, Ciftci-Yilmaz S, Mittler R (2010) Reactive oxygen species homeostasis and signalling during drought and salinity stresses. Plant Cell Environ 33:453–467. https://doi.org/10.1111/j.1365-3040.2009.02041.x

Mittler R, Vanderauwera S, Gollery M, Van Breusegem F (2004) Reactive oxygen gene network of plants. Trends Plant Sci 9:490–498. https://doi.org/10.1016/j.tplants.2004.08.009

Miura K, Furumoto T (2013) Cold signaling and cold response in plants. Int J Mol Sci 14:5312–5337. https://doi.org/10.3390/ijms14035312

Miura K, Tada Y (2014) Regulation of water, salinity, and cold stress responses by salicylic acid. Front Plant Sci 5:4. https://doi.org/10.3389/fpls.2014.00004

Movahedi S, Tabatabaei BES, Alizade H et al (2012) Constitutive expression of Arabidopsis DREB1B in transgenic potato enhances drought and freezing tolerance. Biol Plant 56:37–42. https://doi.org/10.1007/s10535-012-0013-6

Murakeözy ÉP, Nagy Z, Duhazé C et al (2003) Seasonal changes in the levels of compatible osmolytes in three halophytic species of inland saline vegetation in Hungary. J Plant Physiol 160:395–401. https://doi.org/10.1078/0176-1617-00790

Mutlu S, Karadağoğlu Ö, Atici Ö, Nalbantoğlu B (2013) Protective role of salicylic acid applied before cold stress on antioxidative system and protein patterns in barley apoplast. Biol Plant 57:507–513. https://doi.org/10.1007/s10535-013-0322-4

Nafeh A, Hegazi A (2009) Effect of acetylsalicylic acid, indole-3-bytric acid and gibberellic acid on plant growth and yield of pea (*Pisum sativum* L.). Aust J Basic Appl Sci 3:3514–3523

Naz H, Aisha N, Ashraf M (2016) Impact of ascorbic acid on growth and some physiological attributes of cucumber (Cucumis sativus) plants under water-deficit conditions. Pak J Bot 48:877–883

Parvanova D, Ivanov S, Konstantinova T et al (2004) Transgenic tobacco plants accumulating osmolytes show reduced oxidative damage under freezing stress. Plant Physiol Biochem PPB 42:57–63. https://doi.org/10.1016/j.plaphy.2003.10.007

Pereira A (2016) Plant abiotic stress challenges from the changing environment. Front Plant Sci 7. https://doi.org/10.3389/fpls.2016.01123

Perez IB, Brown PJ (2014) The role of ROS signaling in cross-tolerance: from model to crop. Front Plant Sci. https://doi.org/10.3389/fpls.2014.00754

Rao RP, Yuan C, Allegood JC et al (2007) Ceramide transfer protein function is essential for normal oxidative stress response and lifespan. Proc Natl Acad Sci 104:11364–11369. https://doi.org/10.1073/pnas.0705049104

Raskin I (1992) Role of salicylic acid in plants. Annu Rev Plant Physiol Plant Mol Biol 43:439–463. https://doi.org/10.1146/annurev.pp.43.060192.002255

Ruelland E (2017) Plant responses to chilling temperatures. In: Shabala S (ed) Plant stress physiology, 2nd edn. CABI, Wallingford, pp 97–137

Sadasivam S (1992) Biochemical methods for agricultural sciences. New Age International Pub. (P) Limited

Saibo NJM, Lourenço T, Oliveira MM (2009) Transcription factors and regulation of photosynthetic and related metabolism under environmental stresses. Ann Bot 103:609–623. https://doi.org/10.1093/aob/mcn227

Saleh AAH, Abdel-Kader D, El Kelish A (2007) Role of heat shock and salicylic acid in antioxidant homeostasis in mungbean (*Vigna radiata* L.) plant subjected to heat stress. Am J Plant Physiol 2:344–355. https://doi.org/10.3923/ajpp.2007.344.355

Senaratna T, Touchell D, Bunn E, Dixon K (2000) Acetyl salicylic acid (Aspirin) and salicylic acid induce multiple stress tolerance in bean and tomato plants. Plant Growth Regul 30:157–161. https://doi.org/10.1023/A:1006386800974

Shalaev EY, Steponkus PL (2001) Phase behavior and glass transition of 1,2-dioleoylphosphatidylethanolamine (DOPE) dehydrated in the presence of sucrose. Biochim Biophys Acta 1514:100–116

Shao HB, Chu LY, Lu ZH, Kang CM (2007) Primary antioxidant free radical scavenging and redox signaling pathways in higher plant cells. Int J Biol Sci 4:8–14

Shinozaki K, Yamaguchi-Shinozaki K, Seki M (2003) Regulatory network of gene expression in the drought and cold stress responses. Curr Opin Plant Biol 6:410–417

Simaei M, Khavari-Nejad RA, Bernard F (2012) Exogenous application of salicylic acid and nitric oxide on the ionic contents and enzymatic activities in NaCl-stressed soybean plants. Am J Plant Sci 3:1495–1503. https://doi.org/10.4236/ajps.2012.310180

Strain H, Svec (1966) Extraction, separation, estimation and isolation of the chlorophylls. In: Vernon LP, Seely GR (eds) The chlorophylls. Academic Press, New York, pp 21–65

Tabassum T, Farooq M, Ahmad R et al (2017) Seed priming and transgenerational drought memory improves tolerance against salt stress in bread wheat. Plant Physiol Biochem 118:362–369. https://doi.org/10.1016/j.plaphy.2017.07.007

Thomashow MF (1998) Role of cold-responsive genes in plant freezing tolerance. Plant Physiol 118:1–8. https://doi.org/10.1104/pp.118.1.1

Thomashow MF (2001) So what's new in the field of plant cold acclimation? Lots! Plant Physiol 125:89–93

Thomashow MF (2010) Molecular basis of plant cold acclimation: insights gained from studying the CBF cold response pathway. Plant Physiol 154:571–577. https://doi.org/10.1104/pp.110.161794

Tirani MM, Nasibi F, Kalantari KM (2013) Interaction of salicylic acid and ethylene and their effects on some physiological and biochemical parameters in canola plants (*Brassica napus* L.). Photosynthetica 51:411–418. https://doi.org/10.1007/s11099-013-0041-2

Tong L, DanWei M, WuYuan D, Fang C (2005) Effects of low temperature on physiological indices of Jatropha curcas. Chin J Oil Crop Sci 27:50–58

Wang W, Vinocur B, Altman A (2003) Plant responses to drought, salinity and extreme temperatures: towards genetic engineering for stress tolerance. Planta 218:1–14. https://doi.org/10.1007/s00425-003-1105-5

Yancey PH (2005) Organic osmolytes as compatible, metabolic and counteracting cytoprotectants in high osmolarity and other stresses. J Exp Biol 208:2819–2830. https://doi.org/10.1242/jeb.01730

Yang S, Tang XF, Ma NN et al (2011) Heterology expression of the sweet pepper CBF3 gene confers elevated tolerance to chilling stress in transgenic tobacco. J Plant Physiol 168:1804–1812. https://doi.org/10.1016/j.jplph.2011.05.017

Zandalinas SI, Mittler R, Balfagón D et al (2017) Plant adaptations to the combination of drought and high temperatures. Physiol Plant. https://doi.org/10.1111/ppl.12540

Zhao XQ, Wang WS, Zhang F et al (2013) Temporal profiling of primary metabolites under chilling stress and its association with seedling chilling tolerance of rice (*Oryza sativa* L.). Rice. https://doi.org/10.1186/1939-8433-6-23

# 12

# Revealing proteins associated with symbiotic germination of *Gastrodia elata* by proteomic analysis

Xu Zeng†, Yuanyuan Li†, Hong Ling, Juan Chen* and Shunxing Guo*

**Abstract**

**Background:** *Gastrodia elata*, a mycoheterotrophic orchid, is a well-known medicinal herb. In nature, the seed germination of *G. elata* requires proper fungal association, because of the absence of endosperm. To germinate successfully, *G. elata* obtains nutrition from mycorrhizal fungi such as *Mycena*. However, *Mycena* is not able to supply nutrition for the further development and enlargement of protocorms into tubers, flowering and fruit setting of *G. elata*. To date, current genomic studies on this topic are limited. Here we used the proteomic approach to explore changes in *G. elata* at different stages of symbiotic germination.

**Results:** Using mass spectrometry, 3787 unique proteins were identified, of which 599 were classified as differentially accumulated proteins. Most of these differentially accumulated proteins were putatively involved in energy metabolism, plant defense, molecular signaling, and secondary metabolism. Among them, the defense genes (e.g., pathogenesis-/wound-related proteins, peroxidases, and serine/threonine-protein kinase) were highly expressed in late-stage protocorms, suggesting that fungal colonization triggered the significant defense responses of *G. elata*.

**Conclusions:** The present study indicated the metabolic change and defensive reaction could disrupt the balance between *Mycena* and *G. elata* during mycorrhizal symbiotic germination.

**Keywords:** *Gastrodia elata*, Mycorrhizal fungi, Symbiotic germination, Plant defense

## Background

In nature, orchid seeds possess no endosperm therefore are devoid of nutrient supply. Mycorrhizal fungi provide the orchid seeds with signals and nutrients for germination, a mechanism named symbiotic germination. After germination, the orchid seeds give rise to protocorms. The protocorm is a post-embryonic structure from which both shoot and root systems subsequently differentiate. After the differentiation of green leaves, most orchid seedlings acquire autotrophy, while some orchids are achlorophyllous and obtain their entire carbon source from their mycorrhizal fungi. These orchids are known as fully mycoheterotrophic plants (Leake 2004; Dearnaley 2007).

*Correspondence: kibchenjuan@126.com; sxguo2015@163.com
†Xu Zeng and Yuanyuan Li contributed equally to this work
Institute of Medicinal Plant Development, Chinese Academy of Medical Sciences, Beijing 100193, People's Republic of China

*Gastrodia elata*, a fully mycoheterotrophic orchid, associates with two groups of fungal partners, *Mycena* and *Armillaria*. The ontogenesis of *G. elata* has four stages: seed germination, tuber formation, flowering and fruiting. *Mycena* species acts as its first symbiont during the early-stage of seed germination and protocorm development (Kim et al. 2006). However, *Mycena* cannot provide a regular supply of nutrients for further development of *G. elata* during the late-stage of protocorm development. Once a protocorm has been formed, *G. elata* switches its association to the second symbiont *Arminaria*, which subsequently invades the adult rhizome rapidly. For the enlargement of tubers, flowering, and fruit setting, the *Armillaria* becomes essential for the nutrient supply (Tsai et al. 2016). For more than 30 years, our research group only identified some *Mycena* species (e.g. *M. dendrobii*, *M. orchidicola*, *M. anoectochila*, and *M. osmundicola*) which were able to promote the germination of *G. elata*. In addition, *G. elata* is a well-known Chinese medicinal plant. Recent studies have indicated that it had strong potential to combat Alzheimer's disease, Parkinson's disease, and other neurodegenerative diseases (Manavalan et al. 2012).

The researches of the plant-fungus interactions in orchid mycorrhiza are still limited (Dearnaley 2007). Most studies agreed that the orchid mycorrhiza deserved more attention because arbuscular mycorrhiza (Hogekamp et al. 2011) and ectomycorrhiza (Plett and Martin 2011) had been profoundly studied. Recently, Perotto et al. (2014) investigated gene expression in mycorrhizal orchid protocorms (*Serapias vomeracea* colonized by *Tulasnella calospora*) to understand the molecular bases of plant-fungus interactions. Furthermore, Valadares et al. (2014) performed 2D-LC–MS/MS coupled to isobaric tagging for relative and absolute quantification and identify proteins with differential accumulation in *Oncidium sphacelatum* at different stages of mycorrhizal protocorm development. These studies suggested that the protocorm was a relatively constant period in all stage of orchid ontogenesis. It could represent a more feasible experimental system to analyze molecular and cellular aspects of the early plant-fungus interactions of orchid mycorrhiza. By contrast, orchids at maturity have a complicated symbiotic relationship, depending largely upon the trophic strategy of the host plant species and on the environmental conditions.

In this study, we investigated the changes of proteomic profiles during the symbiotic seed germination of *G. elata* inoculated with *M. dendrobii*. Our results may be useful for elucidating the reasons for partnership change during *G. elata* seed germination.

## Methods

### Plant materials

The flowers of *G. elata* were pollinated by hand in the production base at Shaanxi, China. After 16 days of pollination, capsules were collected just prior to dehiscence. The mycorrhizal fungal isolate (*M. dendrobii*) was incubated on fresh Potato Dextrose Agar (PDA: potato infusion, boiling 200 g potato in 1000 ml water; 20 g dextrose; 15 g agar powder) in the dark at 25 °C. The procedure for symbiotic germination of *G. elata* was performed according to the description in a previous report (Kim et al. 2006). Briefly, the culture media for *M. dendrobii* was prepared by fallen leaves of *Quercus* and rice bran at the ratio of 8:2 (V/V). Culturing in dark was benefit to the growth of *M. dendrobii* for 4 weeks at 25 °C. After 4 weeks of incubation, the culture media were fully colonized by *M. dendrobii*. Leaves infested with *M. dendrobii* were placed on fresh PDA and spread on several hundred seeds of *G. elata*, and then dark-cultured for 4–8 weeks at 25 °C. Germination performance was evaluated weekly under a stereo-microscope for 2 months. The early-stage of protocorms (EP) and the late-stage of protocorms (LP) were carefully collected under a stereo-microscope (Fig. 1). For further proteomic study, samples of equal fresh weight of EP and LP of *G. elata* were collected and immediately stored at − 80 °C.

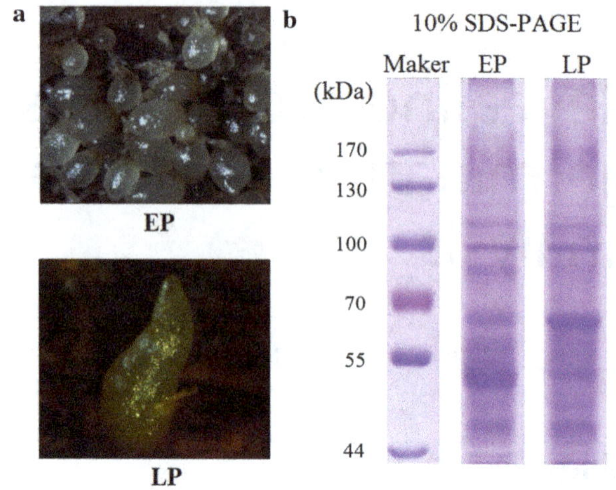


**Fig. 1** **a** The early-stage and the late-stage protocorms (EP and LP) of *G. elata*; **b** 10% SDS-PAGE of total protein

### Protein extraction, digestion, iTRAQ labeling

Protein extraction, digestion, iTRAQ labeling, and mass spectrometry were conducted using protocols from previous paper (Xing et al. 2014). Briefly, total protein was extracted from each sample using Plant Total Protein Extraction Kit PE0230 (Sigma-Aldrich, USA) according to the manufacturer's instructions. Bradford method was selected to determine the protein content. Proteins of each sample were detected by 10% SDS-PAGE (Fig. 1). Each of the samples (75 μg in total) was deoxidized with 20 mM DTT, alkylated with 50 mM IAA and digested with trypsin.

The peptide samples were labeled using the iTRAQ 4-plex kit (AB sciex, USA) according to the manufacturer's protocol. Then, EP and LP were labeled by 116 and 117 iTRAQ, respectively. After labeling, the equal amounts of each sample were mixed together and lyophilized.

### Mass spectrometry

The pooled mixture from labeled samples was dissolved in mobile phases A and fractioned by Durashell RP column (5 μm, 150 Å, 250 mm × 4.6 mm, Agela) from L-3000 HPLC system (Rigol, China). Eluent was collected every minute, pooled into 12 samples and dried under vacuum. Peptides were eluted from the C18 analytical column with a 40 min gradient at a speed of 350 nl/min on an Eksigent Ultra HPLC (AB sciex, USA). The mass spectrum conditions for Triple TOF 5600 was set as follow: The spray voltage was set at 2.5 kV and the temperature of heater was 150 °C. The mass spectrum scan range was *m/z* 350–1250 and the tandem mass spectrometry (MS/MS) scan range was *m/z* 100–1500.

### Database search and functional annotation

Raw proteome data were searched using the Proteinpilot™ v4.0 search engine with percolator against the *G. elata* unigene translation database, including 9908 sequences. Unigene translation sequences were obtained from *G. elata* protocorm transcriptome sequences (Zeng et al. 2017). Based on RNA-seq, the clean reads from protocorm libraries were pooled together and *denovo* assembled into 139,756 unigenes, including 42,140 well-annotated unigenes. Finally, 9908 unigenes were able to translate into protein sequences and establish the self-built database (unigene translation database). The average iTRAQ ratios and standard deviations were calculated for each protein using all of the available treatment control iTRAQ pairs. A 1.5-fold cutoff was used to determine up-accumulated or down-accumulated proteins, with a $P$ value of $< 0.05$.

Functional annotation of the proteins was performed using GO and KEGG annotation. Gene Ontology (GO) analyses were performed by WEGO (Web Gene Ontology Annotation Plot, http://wego.genomics.org.cn/) for plotting GO annotation results (Ye et al. 2006; Zeng et al. 2017). KEGG is a database for recoding the collection of high-level functions and the utility of the biological system. Here, KOBAS software was used for the statistical of DAPs in KEGG pathways (Kanehisa et al. 2008).

### Quantitative PCR

Total RNA of seed and protocorm samples was extracted using RNeasy® Plant Mini Kit (QIAGEN, Germany) according to the manufacturer's instructions. Primers designed with Primer Premier 5.0 are shown in Table 1. A PrimerScoript™ RT reagent Kit (TaKaRa, Japan) was used for reverse transcription. First, 1 μl RT product diluted with 20 μl $ddH_2O$ was used as a template. Then, qPCR was performed in 15 μl reaction mixture containing 7.5 μl of 2× SYBR® Premix Ex TaqTM II (TaKaRa, Japan), 1.5 μl of cDNA template, and 0.3 μl of each gene specific primers. Overall, we preformed three biological replicates and three technical replicates using the LightCycler® 480 II RT-PCR System (Roche, Switzerland). The parameters for the reactions were: 95 °C for 30 s, 40 cycles of 95 °C for 5 s, and 60 °C for 30 s. The cDNA libraries were standardized to housekeeping gene 18S rRNA. The $2^{-\Delta\Delta Ct}$ method was used for evaluating gene expression.

**Table 1 The quantitative PCR primers of putative genes**

| Gene ID | Forward primer (5′–3′) | Reverse primer (5′–3′) |
|---|---|---|
| 18S rRNA | CCAGGTCCAGACATAGTAAG | GTACAAAGGGCAGGGACGTA |
| c48836 | AACCTCTTCCGCCATACCTGC | GCTCTCCGCTTCAACTGACCA |
| c54199 | GGCGTTGTGGAGAGCATTG GA | TTTGTCCGTGCCATGCCTTTCA |
| c81881 | ATGCCGCCTCGTGGAAGACA | GTTGAGACCGCTGCCCGTTTAG |
| c51606 | CCAATCGCAGTGCCAGTT CTTC | AGAGCATCCTGTGTTCCGTTGT |
| c81941 | GATGCGGCACAAGGAGACC AA | TGAGTCGTCGTCAGCACTACCT |
| c47345 | GTCTCAGCGAGCAACAGAT GGT | GCGAGCAAAGAGCAAGCACAT |

## Results and discussion

### Seed germination

The mature seed of *G. elata* contains a globular-shaped embryo covered by a thin layer of seed coat. After 4 weeks of inoculation (Fig. 1), the seed had been infected by fungal hyphae and germinated. The embryo enlarged further and resulted in the formation of mycorrhizal protocorms (early-satge, EP). Afterward, the protocorm elongated further and the shoot tip became visible (late-stage, LP).

### Proteome profiles

Unfortunately, it was technically impossible for us to remove the intracellular fungal hyphae (*Mycena*) from the protocorms. This meant that symbiotic cells contained transcripts produced by both partners (*G. elata* and *M. dendrobii*). Therefore, RNA from the *Mycena* library was used for establishing *Mycena* reference transcriptome. All of *M. dendrobii* reads derived from protocorm were removed by mapping all reads against the *Mycena* reference transcriptome. The clean reads from *G. elata* were *denovo* assembled into transcripts. We identified 139,756 unigenes. Among them, 9908 unigenes were able to translate into protein sequences and use for establishing the *G. elata* unigene translation database. The bioinformatics analyses were selected as reported in the previous studies (Liu et al. 2015a, b; Wang et al. 2016; Zeng et al. 2017). In our results, 3787 proteins were identified and quantified from our self-built database (unigene translation database) at a false discovery rate (FDR) of 1%. By analyzing, our self-built database was obviously suited for proteomic analyses.

Based on GO annotation (Fig. 2), the proteins were functionally classified according to the BP, CC, and MF

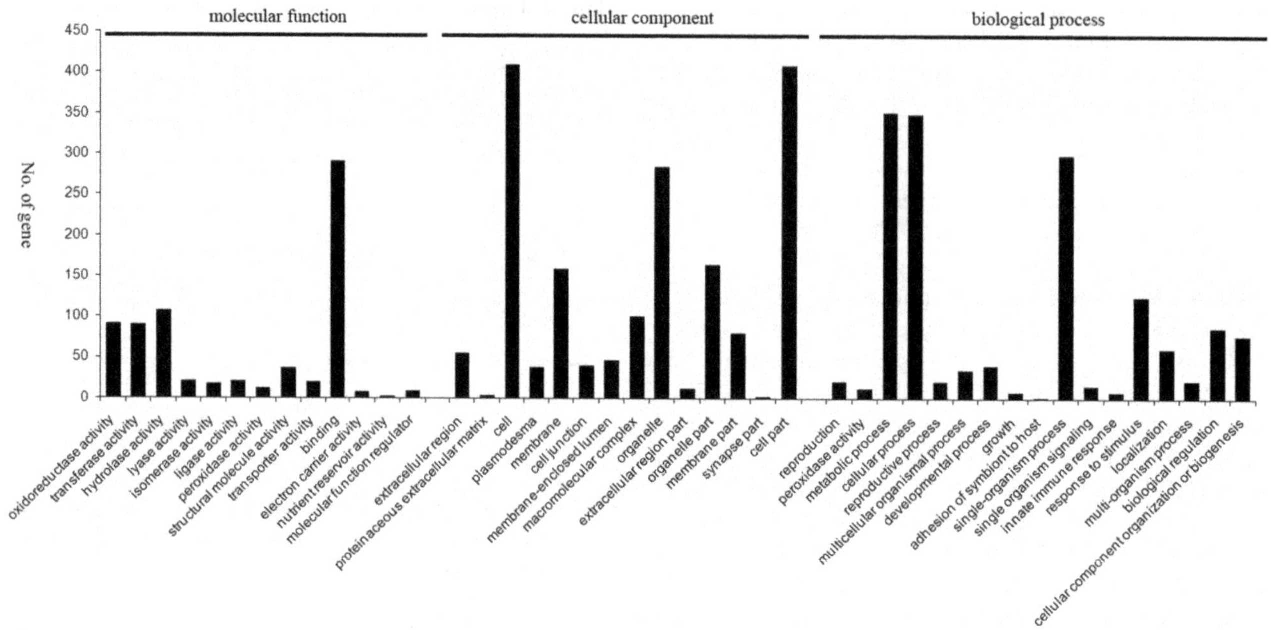


**Fig. 2** Go categorization of the total proteins

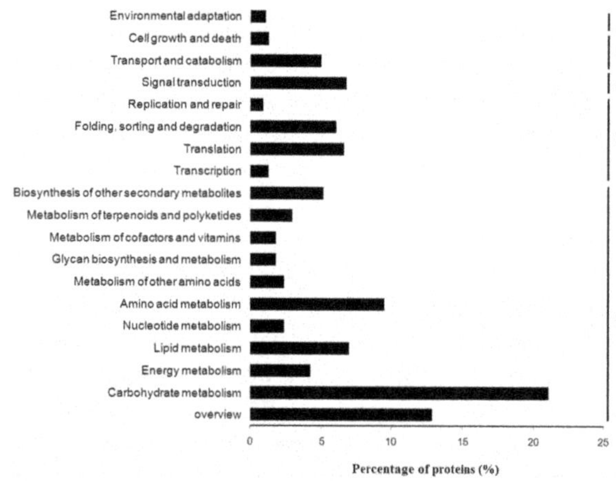


**Fig. 3** KEGG annotation of the total proteins. *A* Cellular processes; *B* Environmental information processing; *C* Genetic information processing; *D* Metabolism; *E* Organismal systems

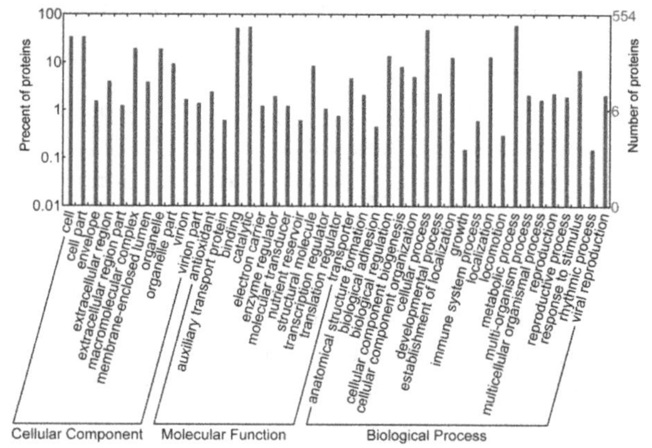


**Fig. 4** Go categorization of the differentially accumulated proteins

categories and their subcategories. The largest subcategories for each functional group were as follows: "binding", "hydrolase activity", "transferase activity" and "oxidoreductase activity" for MF; "cell", "cell part" and "organelle" for CC; and "cellular process", "metabolic process", "single-organism signaling" and "response to stimulus" for BP. According to KEGG (Fig. 3), most of the proteins were assigned to "signal transduction", "translation", "Amino acid metabolism" and "carbohydrate metabolism" pathways.

### Differentially accumulated proteins (DAPs)

Compared to EP, 599 proteins were differentially accumulated in LP. Among these proteins, 321 proteins were significantly up-regulated, and 278 proteins were significantly down-regulated in protocorm. Based on GO analyses (Fig. 4), DAPs were functionally classified according to the BP, CC, and MF categories and their subcategories. The largest subcategories for each functional group were as follows: "cellular process", "metabolic process", and "response to stimulus" for BP; "cell", "cell part" and "organelle" for CC; "binding", "catalytic", and "structural molecule" for MF. According to KEGG (Fig. 5), most of DAPs were assigned to "signal transduction", "translation",

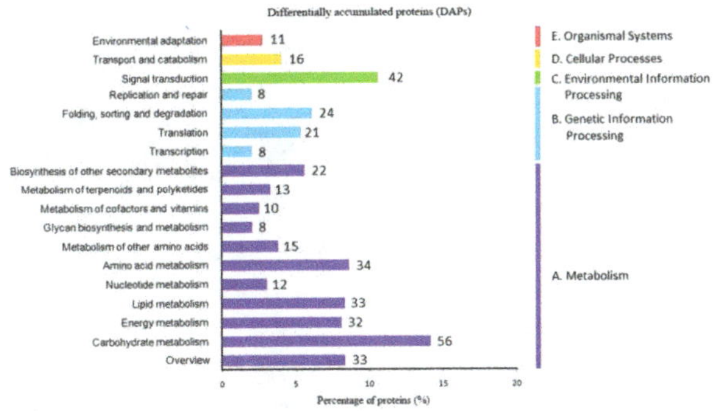


**Fig. 5** KEGG annotation of the differentially accumulated proteins

and "carbohydrate metabolism" pathways. Our results indicated that fungal colonization altered the metabolic processes of *G. elata* and could disrupt the balance between *Mycena* and *G. elata* during mycorrhizal symbiotic germination.

### Putative genes involved in plant defense

In general, the defense mechanism of the host plant is established after fungi infection, including the synthesis and accumulation of reactive oxygen species (ROS), phytoalexins and pathogenesis-related proteins. Recent studies have suggested that activation of defense mechanisms of rice plants by mycorrhizal fungi (*Glomus intraradices*) did not affect arbuscular mycorrhizal symbiosis (Campos-Soriano et al. 2010). Mycorrhizal fungi have evolved the capacity to circumvent defense mechanisms that are controlled by the plant's immune system. However, our results indicated that defensive reaction could disrupt the balance the mycorrhizal symbiosis between *Mycena* and *G. elata* during germination. Here, we identified a number of proteins involved in plant defense (Table 2). Most of them were significantly up-regulated in the late-stage of protocorm. Therefore, *G. elata* defended itself against the colonization of *Mycena* in protocorm and interrupted the mycorrhizal symbiosis. In addition, *Mycena* fungi do not have the ability to suppress, neutralize, or evade the plant defense response.

Pathogenesis-related (PR) proteins are proteins produced in plants in the event of a pathogen attack. In our result (Table 2), a pathogenesis-related protein (PR1) was significantly up-regulated in LP compared with EP. In general, PR genes were strongly induced by fungal inoculation. For instance, after infection by fungi, PR1 proteins accumulate in maize seedlings that are primarily in contact with the pathogen and, as a second barrier, in papillae in the inner parts of the infected tissue. Several studies that have overexpressed PR genes have demonstrated that the role of PR proteins in plant-pathogen interactions was to enhance resistance to fungi (Maschietto et al. 2016). Moreover, PR proteins showed a broad-spectrum resistance to infection by bacterial and fungal pathogens. They displayed a basal expression level of endogenous defense genes and stronger and quicker defense responses during fungal infection (Ozgonen et al. 2009).

**Table 2 Defense genes involved in symbiotic germination of *G. elata***

| Protein ID | Ratio change | P value | Uniprot ID | Reference gene |
|---|---|---|---|---|
| c54199 | 6.36 | 2.47E−02 | Q9ZT66 | Endoglucanase |
| c51606 | 3.76 | 2.52E−02 | P00434 | Peroxidase |
| c48836 | 3.34 | 3.16E−02 | Q05736 | Pathogenesis-related protein 1 |
| c47606 | 2.96 | 3.81E−02 | Q9FJZ9 | Peroxidase |
| c81881 | 2.38 | 1.44E−05 | Q9SW70 | Stress-related protein |
| c81784 | 2.09 | 2.11E−05 | O23814 | Phospholipid hydroperoxide glutathione peroxidase |
| c41136 | 2.04 | 1.41E−04 | Q6ZJJ1 | L-Ascorbate peroxidase |
| c29958 | 1.77 | 2.10E−03 | F4IXW2 | Brefeldin A-inhibited guanine nucleotide-exchange protein |
| c47345 | 1.61 | 2.00E−03 | Q10716 | Cysteine proteinase |
| c81941 | 1.59 | 2.36E−03 | Q7XQP4 | Serine/threonine-protein kinase |

A recent molecular study by Perotto et al. (2014) indicated that none of the wound/stress-related genes were significantly up-regulated in mycorrhizal tissues (*Serapias vomeracea* infected with *Tulasnella calospora*). Meanwhile, Girlanda et al. (2011) demonstrated that *S. vomeracea* was a typical terrestrial orchid in the Mediterranean (partial mycoheterotrophic). In contrast, one stress-related protein was remarkably up-accumulated in the LP of *G. elata*. Our investigation from *G. elata* (fully mycoheterotrophic) suggested that fungal colonization triggered plant defense responses.

Previous studies have suggested that infection stress is accompanied by the production of ROS ($H_2O_2$, superoxide anion, etc.) in organisms (Liu et al. 2015a; Maschietto et al. 2016). The induction of enzymes, such as superoxide dismutase, peroxidases and catalases, could be involved in the protection of tissues against oxidative damage under infection conditions. The major functions of peroxidase include removal of $H_2O_2$, oxidation of toxic reductants, and response to stress, such as wounding, pathogen attack and oxidative stress. We identified that four peroxidases were up-accumulated in protocorms of *G. elata*. Among them, L-ascorbate peroxidase plays a key role in $H_2O_2$ removal (Teixeira et al. 2004). In our results, the LP of *G. elata* showed a constitutive higher level of L-ascorbate peroxidase, which was significantly increased after *Mycena* inoculation, contributing to efficient $H_2O_2$ scavenging. In addition, one up-regulated protein, phospholipid hydroperoxide glutathione peroxidase, protects cells and enzymes against oxidative damage, by catalyzing the reduction of $H_2O_2$, lipid peroxides, and organic hydroperoxide, by glutathione (Sugimoto et al. 1997).

Interestingly, one serine/threonine-protein kinase was found to be highly expressed in the LP tissues. MAPKKK serine/threonine-protein kinase confers sensitivity to various pathogens. This is required for resistance to some hemibiotrophic/necrotrophic fungal pathogens through the induction of defensin expression, probably by repressing MYC2, an inhibitor of defensin genes. Together with KEEP ON GOING protein, MAPKKK serine/threonine-protein kinase may regulate endocytic trafficking and/or the formation of signaling complexes on trans-Golgi network early endosome vesicles during stress responses (Gu and Innes 2011; Hiruma et al. 2011).

We also found two proteins up-accumulated in late-stage protocorms from subcategory "response to biotic stimulus" based on GO analyses. One protein (Brefeldin A-inhibited guanine nucleotide-exchange protein) plays a broad role in PAMP-triggered immunity, effector-triggered immunity, and salicylic acid-regulated immunity (Nomura et al. 2011). The other protein (cysteine protease) plays a role in immunity, senescence, and biotic and abiotic stress and may be involved in immunity against the necrotrophic fungal pathogen (Shindo et al. 2012).

The study of glucanase is hot topic in plant genetic engineering of disease resistance, and great progress on this subject has been made in the past few years (Day and Graham 2007). Glucans are major components of fungal cell wall. We analyzed hydrolases involved in the degradation of glucans. According to the results, one endoglucanase from *G. elata* protocorm was expressed to a great extent after *Mycena* infection. This protein could be involved in fungal cell wall hydrolysis.

### qPCR analysis of putative genes

We performed quantitative PCR analysis on 6 selected genes putatively involved in plant defense. The results of qPCR related to expression changes of these genes are shown in Fig. 6. In our results, most of the genes showed high gene expression in LP, and low gene expression in EP during the developmental process of seed germination of *G. elata*. In future, the expression of putative proteins in mycorrhizal symbiotic germination will be confirmed by the western blot.

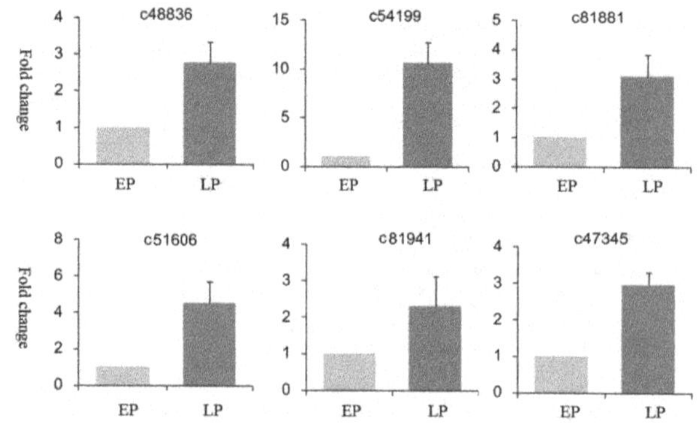


**Fig. 6** Quantitative PCR validations of six genes that were differentially expressed between the early-stage protocorms (EP) and the late-stage protocorms (LP)

## Conclusion

In conclusion, analysis of differentially accumulated proteins based on LC–MS/MS platform was a powerful method for investigating putative proteins involved in plant-fungus interactions. In the study, fungal colonization altered the metabolic processes of *G. elata*. We also analyzed pathogenesis-/stress-related proteins, peroxidases, and serine/threonine-protein kinase produced in the process of plant defense. These results indicated that the metabolic change and defense response of *G. elata* and could disrupt the balance between *Mycena* and *G. elata* during mycorrhizal symbiotic germination.

**Authors' contributions**

Conceived and designed the experiments: XZ, YYL and SXG. Performed the experiments: XZ, YYL and JC. Analyzed the data: XZ and HL. Contributed reagents/materials/analysis tools: HL. Wrote the paper: XZ and YYL. All authors read and approved the final manuscript.

**Acknowledgements**

This study was financially supported by CAMS basic research fund for central public research institutes (2016ZX350062) and the Program for Innovative Research Team in IMPLAD (PIRTI-IT 1302).

**Competing interests**

The authors declare that they have no competing interests.

**Funding**

This study was financially supported by CAMS basic research fund for central public research institutes (2016ZX350062) and the Program for Innovative Research Team in IMPLAD (PIRTI-IT 1302).

**References**

Campos-Soriano L, Garcia-Garrido JM, Segundo BS (2010) Activation of basal defense mechanisms of rice plants by Glomus intraradices does not affect the arbuscular mycorrhizal symbiosis. N Phytol 188:597–614

Day B, Graham T (2007) The plant host pathogen interface. Ann N Y Acad Sci 1113:123–134

Dearnaley JD (2007) Further advances in orchid mycorrhizal research. Mycorrhiza 17:475–486

Girlanda M, Segreto R, Cafasso D, Liebel HT, Rodda M, Ercole E, Cozzolino S, Gebauer G, Perotto S (2011) Photosynthetic mediterranean meadow orchids feature partial mycoheterotrophy and specific mycorrhizal associations. Am J Bot 98:1148–1163

Gu Y, Innes RW (2011) The KEEP ON GOING protein of *Arabidopsis* recruits the ENHANCED DISEASE RESISTANCE1 protein to trans-Golgi network/early endosome vesicles. Plant Physiol 155:1827–1838

Hiruma K, Nishiuchi T, Kato T, Bednarek P, Okuno T, Schulze-Lefert P, Takano Y (2011) Arabidopsis ENHANCED DISEASE RESISTANCE 1 is required for pathogen-induced expression of plant defensins in nonhost resistance, and acts through interference of MYC2-mediated repressor function. Plant J 67:980–992

Hogekamp C, Arndt D, Pereira PA, Becker JD, Hohnjec N, Kuster H (2011) Laser microdissection unravels cell-type-specific transcription in arbuscular mycorrhizal roots, including CAAT-box transcription factor gene expression correlating with fungal contact and spread. Plant Physiol 157:2023–2043

Kanehisa M, Araki M, Goto S, Hattori M, Hirakawa M, Itoh M, Katayama T, Kawashima S, Okuda S, Tokimatsu T (2008) KEGG for linking genomes to life and the environment. Nucleic Acids Res 36(Suppl 1):D480–D484

Kim YI, Chang KJ, Ka KH, Hur H, Hong IP, Shim JO, Lee TS, Lee JY, Lee MW (2006) Seed germination of *Gastrodia elata* using symbiotic fungi, *Mycena osmundicola*. Mycobiology 34:79–82

Leake JR (2004) Myco-heterotroph/epiparasitic plant interactions with ectomycorrhizal and arbuscular mycorrhizal fungi. Curr Opin Plant Biol 7:422–428

Liu MM, Xing YM, Zhang DW, Guo SX (2015a) Transcriptome analysis of genes involved in defence response in *Polyporus umbellatus* with *Armillaria mellea* infection. Sci Rep 5(3):16075

Liu SS, Chen J, Li SC, Zeng X, Meng ZX, Guo SX (2015b) Comparative transcriptome analysis of genes involved in GA-GID1-DELLA regulatory module in symbiotic and asymbiotic seed germination of *Anoectochilus roxburghii* (Wall.) Lindl. (Orchidaceae). Int J Mol Sci 16:30190–30203

Manavalan A, Feng L, Sze SK, Hu JM, Heese K (2012) New insights into the brain protein metabolism of *Gastrodia elata*-treated rats by quantitative proteomics. J Proteom 75(8):2468–2479

Maschietto V, Lanubile A, Leonardis SD, Marocco A, Paciolla C (2016) Constitutive expression of pathogenesis-related proteins and antioxydant enzyme activities triggers maize resistance towards *Fusarium verticillioides*. J Plant Physiol 200:53–61

Nomura K, Mecey C, Lee YN, Imboden LA, Chang JH, He SY (2011) Effector-triggered immunity blocks pathogen degradation of an immunity-associated vesicle traffic regulator in *Arabidopsis*. Proc Natl Acad Sci 108:10774–10779

Ozgonen H, Yardimci N, Kilic HC (2009) Induction of phenolic compounds and pathogenesis-related proteins by mycorrhizal fungal inoculations against *Phytophthora capsici* Leonian in pepper Pakistan. J Biol Sci 12:1181

Perotto S, Rodda M, Benetti A, Sillo F, Ercole E, Rodda M, Girlanda M, Murat C, Balestrini R (2014) Gene expression in mycorrhizal orchid protocorms suggests a friendly plant-fungus relationship. Planta 239:1337–1349

Plett JM, Martin F (2011) Blurred boundaries: lifestyle lessons from ectomycorrhizal fungal genomes. Trends Genet 27:14–22

Shindo T, Misas-Villamil JC, Hörger AC, Song J, Hoorn RA (2012) A role in immunity for *Arabidopsis* cysteine protease RD21, the ortholog of the tomato immune protease C14. PLoS ONE 7:e29317

Sugimoto M, Furui S, Suzuki Y (1997) Molecular cloning and characterization of a cDNA encoding putative phospholipid hydroperoxide glutathione peroxidase from spinach. Biosci Biotechnol Biochem 61:1379–1381

Teixeira FK, Menezes-Benavente L, Margis R, Margis-Pinheiro M (2004) Analysis of the molecular evolutionary history of the ascorbate peroxidase gene family: inferences from the rice genome. J Mol Evol 59:761–770

Tsai CC, Wu KM, Chiang TY, Huang CY, Chou CH, Li SJ, Chiang YC (2016) Comparative transcriptome analysis of *Gastrodia elata* (Orchidaceae) in response to fungus symbiosis to identify gastrodin biosynthesis-related genes. BMC Genom 17:1–11

Valadares RB, Perotto S, Santos EC, Lambais MR (2014) Proteome changes in *Oncidium sphacelatum* (Orchidaceae) at different trophic stages of symbiotic germination. Mycorrhiza 24(5):349–360

Wang L, Cao H, Chen C, Yue C, Hao X, Yang Y, Wang X (2016) Complementary transcriptomic and proteomic analyses of a chlorophyll-deficient tea plant cultivar reveal multiple metabolic pathway changes. J Proteom 130:160–169

Xing X, Zhang C, Li N, Zhai L, Zhu Y, Yang X, Xu P (2014) Qualitative and quantitative analysis of the adult *Drosophila melanogaster* proteome. Proteomics 14:286–290

Ye J, Fang L, Zheng H, Zhang Y, Chen J, Zhang Z, Wang J, Li S, Li R, Bolund L (2006) WEGO: a web tool for plotting GO annotations. Nucleic Acids Res 34(suppl 2):W293–W297

Zeng X, Li YY, Ling H, Liu S, Liu M, Chen J, Guo S (2017) Transcriptomic analyses reveal clathrin-mediated endocytosis involved in symbiotic seed germination of *Gastrodia elata*. Bot Stud 58(1):31

# 13

# *Phyllanthus rufuschaneyi*: a new nickel hyperaccumulator from Sabah (Borneo Island) with potential for tropical agromining

Roderick Bouman[1,2,3], Peter van Welzen[1,3], Sukaibin Sumail[4], Guillaume Echevarria[5], Peter D. Erskine[6] and Antony van der Ent[5,6*]

**Abstract**

**Background:** Nickel hyperaccumulator plants are of much interest for their evolution and unique ecophysiology, and also for potential applications in agromining—a novel technology that uses plants to extract valuable metals from soil. The majority of nickel hyperaccumulators are known from ultramafic soils in tropical regions (Cuba, New Caledonia and Southeast Asia), and one genus, *Phyllanthus* (Phyllanthaceae), is globally the most represented taxonomic entity. A number of tropical *Phyllanthus*-species have the potential to be used as 'metal crops' in agromining operations mainly because of their ease in cultivation and their ability to attain high nickel concentrations and biomass yields.

**Results:** One of the most promising species globally for agromining, is the here newly described species *Phyllanthus rufuschaneyi*. This species can be classified in subgenus *Gomphidium* on account of its staminate nectar disc and pistillate entire style and represents the most western species of this diverse group. The flower structure indicates that this species is probably pollinated by *Epicephala* moths.

**Conclusions:** *Phyllanthus rufuschaneyi* is an extremely rare taxon in the wild, restricted to Lompoyou Hill near Kinabalu Park in Sabah, Malaysia. Its utilization in agromining will be a mechanism for conservation of the taxon, and highlights the importance of habitat and germplasm preservation if rare species are to be used in novel green technologies.

**Keywords:** *Epicephala* pollination, Nickel hyperaccumulation, Phyllanthaceae, *Phyllanthus* subgenus *Gomphidium*, Sabah

## Background

Whereas the great majority of plants growing on naturally nickel (Ni) rich ultramafic soils exclude it from uptake and translocation, a minority of plants display a highly unusual response with enhanced uptake and transfer to the shoots (Reeves 2003; Van der Ent et al. 2013a). These plants are called 'hyperaccumulators' and they have the ability to accumulate trace elements to extreme concentrations in their living tissues (Jaffré et al. 1976; Van der Ent et al. 2013a). The Ni concentrations in some species can reach up to 16.9 Wt% in the phloem sap (Van der Ent and Mulligan 2015). Although there are over 400 known Ni hyperaccumulators species (>0.1 Wt% shoot dry weight), there are just ca. 50 *hypernickelophores* (e.g. hyperaccumulator species with >1 Wt% Ni shoot dry weight) known globally (Reeves 2003; Reeves et al. 2017). Hyperaccumulator plants can be used as 'metal crops' in agromining (phytomining) operations to generate metal-rich biomass for commercial gain (Chaney et al. 1998; Van der Ent et al. 2015a). This innovative approach enables access to resources not accessible by conventional mining techniques such as abundant low-grade sources of valuable elements (Li et al. 2003; Van

*Correspondence: a.vanderent@uq.edu.au
[6] Centre for Mined Land Rehabilitation, Sustainable Minerals Institute, The University of Queensland, St Lucia, QLD 4072, Australia
Full list of author information is available at the end of the article

der Ent et al. 2015a). Agromining can also benefit local communities, by providing new income opportunities for farmers in developing countries (Bani et al. 2015; Chaney et al. 2018). The greatest potential for agromining is in tropical regions (Cuba, New Caledonia and Southeast Asia) where some of the world's largest low-grade nickel sources are located (Van der Ent et al. 2013b).

On a global scale, Ni hyperaccumulation occurs most frequently in the order Malpighiales, particularly in the families Dichapetalaceae, Phyllanthaceae, Salicaceae and Violaceae.

The Phyllanthaceae has the greatest numbers of hyperaccumulators with representatives in the genera *Actephila* Blume, *Antidesma* L., *Breynia* J.R.Forst. & G.Forst., *Cleistanthus* Hook.f. ex Planch., *Glochidion* J.R.Forst. & G.Forst. and *Phyllanthus* L. The latter is pantropical and the most speciose genus of the family with over 800 species globally (Govaerts et al. 2000; Kathriarachchi et al. 2006; Bouman et al. under review). Due to its great diversity in morphology, *Phyllanthus* is currently classified in many subgenera and (sub)sections, which were often former separate genera. The genus is characterized by its unisexual flowers, the absence of petals and a characteristic branching system called phyllanthoid branching (see Fig. 2a; Webster 1956). Species with this particular type of branching have deciduous, plagiotropic branchlets that are subtended by reduced scale-like leaves (cataphylls) (Webster 1956). Normal leaves and flowers are only found on the plagiotropic branchlets. This branching system has been lost several times (Kathriarachchi et al. 2006) and in species with non-phyllanthoid branching, leaves can be found on all axes and the branchlets are not deciduous and subtended by normal leaves. The genus *Phyllanthus* is currently paraphyletic with the genera *Breynia, Glochidion and Synostemon* F.Muell. nested within (Kathriarachchi et al. 2006; Pruesapan et al. 2012). Discussions on how to resolve the paraphyly of the genus *Phyllanthus* are still ongoing (see Hoffmann et al. 2006; Van Welzen et al. 2014). One of the most speciose subgenera is subgenus *Gomphidium* (Baill.) G.L.Webster. Species in the subgenus are characterised by phyllanthoid branching in combination with flowers with biseriate sepals, stamens with free filaments (but connate in one section) and three duplex nectar glands (sometimes divided and then 6 or absent). The phylogenetic position of subgenus *Gomphidium* was not consistent between the markers used in the study of Kathriarachchi et al. (2006) and requires further study.

Major centres of diversity for *Phyllanthus* are in New Caledonia with over 100 species of which 15 species are Ni hyperaccumulators (Kersten et al. 1979; Schmid 1991; Jaffré et al. 2013), in Cuba with at least 40 species of which 17 species are Ni hyperaccumulators (Leon and Alain 1953; Reeves et al 1996), and in the Malesian Region about 100 species of which 5 species are Ni hyperaccumulators (Van der Ent et al. 2015a, b, c; Wu et al. 2016; Galey et al. 2017).

A number of taxa in the genus *Phyllanthus* are among the most promising 'metal crops' due to their fast growth and other favourable growth characteristics, including easy propagation and pest resistance (Nkrumah et al. 2016). Some *Phyllanthus*-species also reach some of the highest Ni concentrations known in any hyperaccumulator plants, with 3.8 Wt% in the leaves of *P. serpentinus* S.Moore from New Caledonia (Jaffré 1977; Kersten et al. 1979; as *P. favieri* M.Schmid), 3.9 Wt% in the leaves of *Phyllanthus insulae-japen* Airy Shaw from Indonesia (Reeves 2003), and 6 Wt% in the leaves *P.* × *pallidus* C.Wright ex Griseb. from Cuba (Reeves et al. 1996).

Further study of other genera within the Phyllanthaceae continues to yield new Ni hyperaccumulator records, such as in the genus *Antidesma* (Nkrumah et al. 2018), and even new species that are hyperaccumulators, such as the recently described *Actephila alanbakeri* Welzen & Ent (Van der Ent et al. 2016a). Kinabalu Park is the world's most species-rich hotspot with over 5000 species in 1000 genera and 200 families recorded to date (Beaman and Beaman 1990; Beaman 2005) of which 2542 plant species have been found on the ultramafic soils inside the Park (Van der Ent et al. 2015c). In Sabah, a total of 8 species of *Phyllanthus* occur, of which two are known Ni hyperaccumulators: *P. balgooyi* Petra Hoffm. & A.J.M.Baker which can accumulate up to 1.6 Wt% Ni in the leaves and up to 16.9 Wt% Ni in the phloem sap and the here newly described *Phyllanthus rufuschaneyi* (Hoffmann et al. 2008; Van der Ent and Mulligan 2015; Mesjasz-Przybylowicz et al. 2016). *Phyllanthus balgooyi*, also occurs in the Philippines, in addition to *Phyllanthus securinegioides* Merr. (which can accumulate up to 3.5 Wt% in the leaves), and a third Ni hyperaccumulator species from the genus, *P. erythrotrichus* C.B.Rob. which can accumulate up to 1.1 Wt% Ni in the leaves (Baker et al. 1992; Quimado et al. 2015). The extreme levels of Ni accumulation in these species poses important questions about the ways in which these plants take up, transport and store Ni, while avoiding the potential effects of metabolic toxicity. In leaves, Ni appears to be associated mainly with organic acids, such as citrate, malate and malonate (Kersten et al. 1980; Homer 1991; Montargès-Pelletier et al. 2008; Van der Ent et al. 2017).

In early 2013, a hitherto unknown species of *Phyllanthus* was planted in the local garden by staff at the Monggis substation of Kinabalu Park. Spot-testing with dimethyl-glyoxime-impregnated paper revealed it to be a strong Ni hyperaccumulator, which was subsequently confirmed through Inductively Coupled Plasma Atomic

Emission Spectroscopy (ICP-AES) on acid-digested samples of leaves in the laboratory (first reported in Van der Ent et al. 2013c). The plant was collected from an unknown location near the Kinabalu Park boundary and could not be re-collected at the time. In 2015, the taxon was 're-discovered' during fieldwork on Lompoyou Hill, approximately 12 km from Monggis substation, where it was locally abundant (Fig. 1). The taxon is of significant scientific interest because of the extremely high levels of Ni accumulation, reaching up to 2.8 Wt% Ni in leaves and 1.8 Wt% in the phloem tissue (Van der Ent and Mulligan 2015). Following its discovery, the ecophysiology was studied in detail, which revealed that Ni is mainly concentrated in the phloem in roots and stems, and in the leaves in the epidermis (Van der Ent et al. 2017). Apart from scientific interest, the taxon has great potential for agromining, which was studied in a pioneering nursery pot experiment and field trial (Nkrumah et al. 2016). This experimental work demonstrated that *P. rufuschaneyi* has advantageous characteristics for utilization as a 'metal crop' including fast growth rate, easy re-growth after coppicing, tolerance for exposed conditions on eroded soils and high Ni accumulation in its whole biomass. Taken together, *P. rufuschaneyi* has the best combination of characteristics of any tropical 'metal crop' presently known. The bio-ore (e.g. ashed biomass) of *P. rufuschaneyi* contains up to 12.7 Wt% Ni and the extractive hydrometallurgy of this material was studied for producing high-purity Ni salts for the electrochemical industry (Vaughan et al. 2017). At first, this plant was only known as *Phyllanthus* cf. *securinegioides* Merr. (and reported as such in Van der Ent et al. 2015b, 2016b, 2017), because of its superficial resemblance to this species from the Philippines. However, after examination of the literature and other specimens, we conclude that it is a new species that is of great interest for its taxonomic position and its qualities as a hyperaccumulator. Here we describe this taxon and provide information about its taxonomic relations, distribution, ecology, hyperaccumulation properties, and conservation status.

**Fig. 1** Different aspects of the habitat of *P. rufuschaneyi* in Sabah, Malaysia. **a** Lompoyou Hill seen from Nalumad village; **b** Garas—the eastern end of Lompoyou Hill, the outcropping ultramafic (serpentinite) bedrock is clearly visible in the road cuts; **c** *Phyllanthus rufuschaneyi* growing in situ on Lompoyou Hill; **d** the summit of Lompoyou Hill with secondary scrub and dead standing trees after forest fires. Images by A. van der Ent

**Fig. 2** Detail of *P. rufuschaneyi* plants. **a** Inflorescences of *P. rufuschaneyi*, note the difference between main stem and side stem with at the base small structures that signal phyllanthoid branching; **b** fruit capsules of *P. rufuschaneyi*. Images by A. van der Ent

## Methods

### Taxonomical investigation

The material studied comprised of herbarium specimens loaned from the Sabah Parks Herbarium (SNP) in Sabah, Malaysia. Descriptions were made using standard taxonomical techniques and morphological terminology follows Beentje (2016).

The IUCN conservation status was assessed applying the IUCN Red List Categories (IUCN 2001) that considers (i) extent of occurrence (EOO), and (ii) area of occurrence (AOO) in order to generate applicable IUCN threat categories.

### Collection of plant samples for chemical analysis

Plant tissue samples (leaves, wood, bark, flowers) for bulk chemical analysis were collected in the habitat (6°06′29.6″N 116°47′36.7″E) near Kinabalu Park in Sabah, Malaysia. These samples were dried at 70 °C for 5 days in a drying oven and subsequently packed for transport to Australia and gamma irradiated at Steritech Pty. Ltd. in Brisbane following Quarantine Regulations in Australia. The dried plant tissue samples were subsequently ground and digested using 4 mL $HNO_3$ (70%) and 1 mL $H_2O_2$ (30%) in a microwave oven (Milestone Start D) for a 45-min programme and diluted to 30 mL with ultrapure water (Millipore 18.2 MΩ cm at 25 °C) before analysis with ICP-AES (Varian Vista Pro II) (Huang et al. 2004). The elemental concentrations originate from previously reported data (Van der Ent and Mulligan 2015; Van der Ent et al. 2015b, 2017) augmented with new data from plant tissue samples collected for this study.

## Results

### Taxonomic treatment

*Phyllanthus rufuschaneyi* Welzen, R.W.Bouman and Ent, *sp. nov.*—TYPE: MALAYSIA. Sabah, near Kampong Nalumad, eastern boundary Kinabalu Park, Lompoyou Hill, Antony Van der Ent et al. SNP 32987! (holo SNP; iso L). Paratype: SNP 22039!, Lompoyou Hill, Sabah, Malaysia (Figs. 2, 3, 4).

This species is most similar to *P. securinegoides* from the Philippines, from which it can be distinguished by its smaller leaves, staminate flowers with connate filaments and pistillate flowers with connate tubular stigmas.

Shrub to tree(let), up to 6 m high, monoecious, phyllanthoid branching present, main branches often hollow. *Indument* generally absent, on some parts asperities (stiff, short, papillae-like hairs) on orthotropic and plagiotropic branches and sometimes petioles and pedicels. *Plagiotropic branches* 1–1.5 mm wide, not flattened, with 2 narrow longitudinal wings, to 0.3 mm wide, asperities usually present between wings. *Stipules* ovate, 1.5–6 by 0.8–1 mm, persistent, becoming brown when dry, basally eared, ears sometimes elongated. *Leaves* distichous, simple, dimorphous (reduced on orthotropic branch to cataphylls (scale leaves), not reduced on plagiotropic branches). *Cataphylls* on main trunk below branches, stipule-like, ovate to triangular, c. 3 mm long, early caducous. *Leaves on plagiotropic branches*: petiole more or less dorsoventrally flattened, 1–1.5 mm long, glabrous or with asperities, transversely wrinkled when dry; blade ovate to elliptic, 1.6–3.7 (without mucro) by 0.5–1.2 cm, 2.2–3.7 times longer than wide, coriaceous, base attenuate, asymmetric, margin entire, somewhat thickened, slightly revolute, apex rounded to acuminate, mucronate up to 5.2 mm, mucron usually breaking off, upper and lower surface glabrous, smooth, lower surface glaucous; venation pinnate, above hardly visible, slightly raised underneath, secondary veins 6–8 per side, looped and closed near margin, higher order veins reticulate. *Inflorescences* consisting of unisexual or bisexual fascicles, flowers 1–4 per node; staminate flowers single or a few, generally in lower axils but sometimes at end of branches, usually together with a single pistillate flower; latter also single in upper axils. *Staminate flowers* 1.4–2 mm diameter, remaining closed, actinomorphic; pedicel slender,

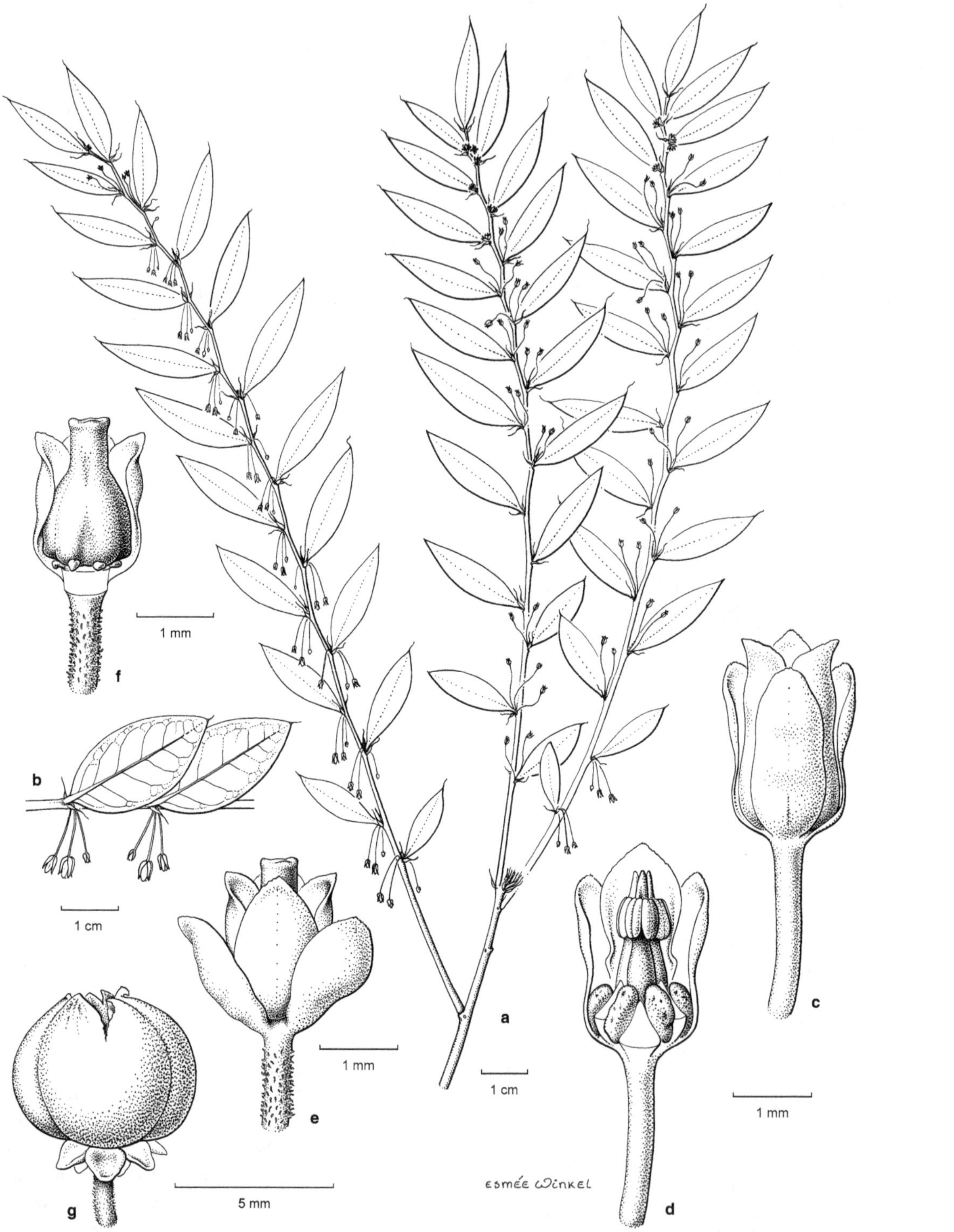


**Fig. 3** *Phyllanthus rufuschaneyi* Welzen, R.W.Bouman & Ent: **a** a branch with only scars of cataphylls and cataphyllary stipules present at the base of branchlets as these are caducous (drawn from herbarium specimen with leaves glued sideways and staminate flowers sometimes upright instead of hanging); **b** detail of sidebranch with leaves and staminate flowers in natural position; **c** staminate flower; **d** staminate flower with part of sepals removed showing disc glands and androecium; **e** pistillate flower; **f** pistillate flower with part of sepals removed showing disc glands and ovary; **g** fruit (**a**, **c**, **d** *Daim Endau 225*; **b** *Lomudin Tadon g257*; **e**, **f** *SNP 32987*; **g** *Lomudin Tadon 257*; all SNP). Drawing by Esmée Winkel (2017)

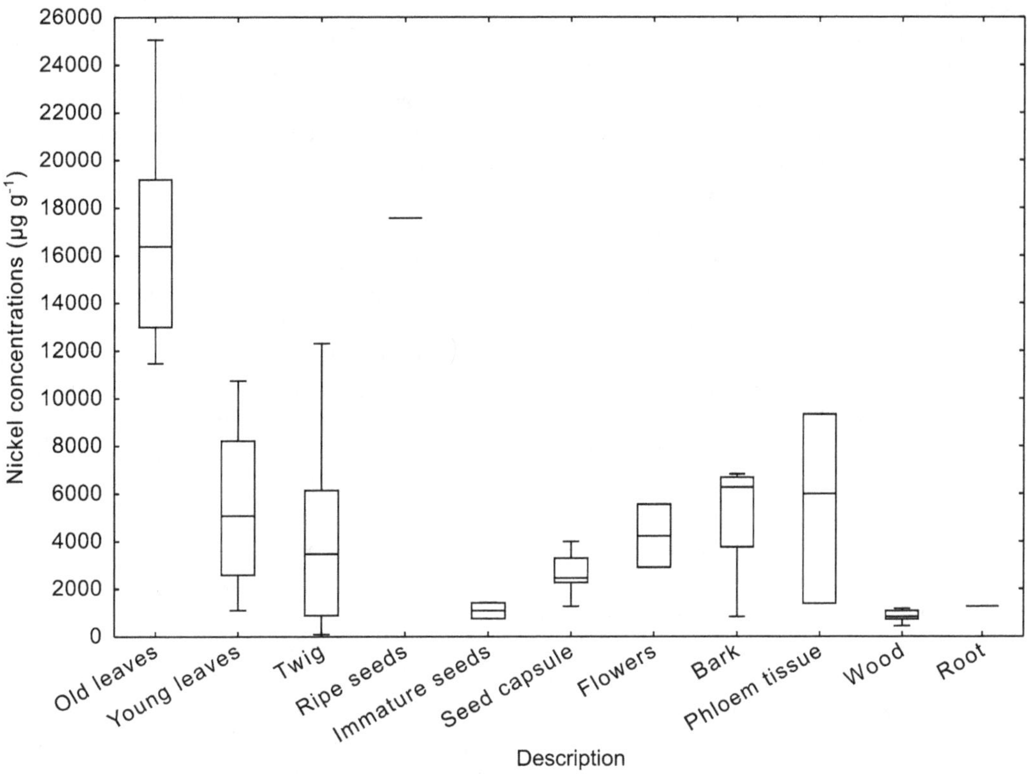


**Fig. 4** Elemental concentration in various plant parts of *P. rufuschaneyi*. Data from Tables 1 and 2

6.7–12 mm long, apically somewhat thickened; sepals 6, upright, tightly packed, margin entire, apex rounded; outer three smaller (when young) to slightly longer than inner ones, 1.9–2 by 0.9–1 mm, central part thickened, pink, margins and upper half thin, white to whitish pink; inner 3 1.4–2.3 by 1.3–1.4 mm, lower part almost nail-like, central midrib area thickened and darker coloured, basal central part inside attached to androphore via a ^-like structure; disc lobes 6, paired, vertical, kidney-like, c. 0.6 mm long, greyish blue when dry, attached to broad part of flower receptacle, smooth; stamens 3, filaments connate, c. 1 mm high, broadly cone-shaped, apically with erect anthers, these elliptic, c. 0.7 by 0.4 mm, opening extrorse via lengthwise slits, apically on connective a c. 0.3 mm long slender appendix. *Pistillate flowers* 1.4–1.6 mm diameter, actinomorphic; pedicel 1–2 mm long, round, glabrous or with asperities; sepals 6, tight to ovary in flower, spreading in fruit, base very thickened, attached to receptacle, margin entire, outer 3 slightly smaller than inner 3, elliptic, 1.7–2 by c. 1 mm, apex truncate to erose, inner 3 elliptic, c. 2 by 1.2–1.6 mm, apex rounded; disc lobes 6, small, globose, less than 0.2 mm in diameter; gynophore c. 0.3 mm high, ovary ovoid, c. 1.3 by 1.1–1.3 mm, 3-locular, 2 ovules per locule, smooth, glabrous, three stigmas united into a cone of 0.8–1 mm high, apically somewhat erose (slightly split stigmas) and hollow inside. *Fruits* c. 7.5 mm in diameter, c. 4.5 mm high, opening completely septicidal and partly to completely loculicidal, exocarp separating from meso- and endocarp, wall thin, woody when dry, smooth, glabrous; columella broadly triangular, 2–2.3 mm high. *Seed* ovoid-triangular, c. 2.8 by 1.8 mm, brown, smooth, not seen mature.

## Etymology

The specific epithet "*rufuschaneyi*" honours Dr. Rufus L. Chaney (b. 1942), an agronomist who is widely credited for inventing phytomining (agromining) (Chaney 1983), leading to the technology being patented (Chaney et al. 1998). Dr. Chaney has worked for 47 years at the USDA Agricultural Research Service (USA) on risk assessment for metals in soils and crops, and the food-chain transfer and bioavailability of soil and crop metals to humans. He published over 490 publications and won the Gordon Award for Lifetime Achievement and Excellence in Phytoremediation Research. The fact that *P. rufuschaneyi* is the most promising tropical Ni 'metal crop' presently known, makes this recognition fitting.

### Phenology and pollination

The species flowers and fruits all year round. Especially young plants growing in open habitats flower profusely on many branches. Flowering is less frequent when plants grow under more shaded conditions in developing forest. Many *Phyllanthus* species are pollinated by moths of the genus *Epicephala* Meyrick. Females of these moths actively pollinate the flowers and then deposit eggs into the floral ovaries, after which the larvae consume some of the developing seeds (Kawakita 2010). *Phyllanthus rufuschaneyi*, like all other species in subgenus *Gomphidium*, is likely pollinated by *Epicephala* moths (Kato and Kawakita 2017) as the staminate flowers are highly closed (inner sepals even grown together with the broad androphore) and possess vertical anthers, the pistillate flowers have grown together, cone-like, non-opening stigmas. Although pollination has not been specifically studied in this species, *Epicephala* larvae were observed in the fruit capsules of *P. rufuschaneyi* in the field. The similarity with *Glochidion* flowers, also pollinated by *Epicephala* moths, is remarkable. The co-evolution and high specificity for *Epicephala*-species for *Phyllanthus*-species has been linked to the extensive diversification of this genus in New Caledonia (Kawakita and Kato 2004).

### Distribution, habitat and ecology of *Phyllanthus rufuschaneyi*

*Phyllanthus rufuschaneyi* is known only from two populations; one (very small) population at the foot of Bukit Hampuan, and another larger population on Lompoyou Hill approximately 5 km from the first population. The habitat in both localities is open secondary scrub that has been affected by recurring forest fires (Fig. 1). Lompoyou Hill is close to the villages of Nalumad and Pahu. The hill (400 m asl) has been burnt at least once as a result of an uncontrolled forest fire in 1998. Prior to burning, the site was already disturbed by logging. The area has a short and open scrub community (dominated by shrubs 1–3 m tall) with pioneer species such as *Macaranga kinabaluensis* Airy Shaw (Euphorbiaceae). In this habitat type several other Ni hyperaccumulator plant species occur, including *Phyllanthus balgooyi*, *Actephila alanbakeri*, *Mischocarpus sundaicus* Blume (Sapindaceae), and *Xylosma luzonensis* Clos (Salicaceae). The local conditions are xeric, and the soils are shallow and heavily eroded with limited amounts of organic matter. In pot experiments *P. rufuschaneyi* responded negatively to increasing organic matter amendments (Nkrumah et al. 2017). *Phyllanthus rufuschaneyi* occurs exclusively on these young eroded soils (hypermagnesian Cambisols) that occur at low elevation (700 m asl) on strongly serpentinised bedrock. These soils have extremely high magnesium (Mg) to calcium (Ca), circum-neutral pH, and high available Ni as a result of the disintegration of phyllosilicates and re-sorption onto secondary iron (Fe)-oxides or high-charge clays (Echevarria 2018). In Sabah, Ni hyperaccumulator plant species are restricted to these soils with a pH > 6.3 and relatively high total soil Ni concentrations > 630 μg g$^{-1}$ (Van der Ent et al. 2016b).

### Elemental concentrations in the plant tissues

Bulk elemental concentrations of macro-elements (mainly essential nutrients) are given in Table 1; Fig. 4. Aluminium (Al) concentrations are uniformly low in all plant parts, but highest in the flowers (up to 370 μg g$^{-1}$). Calcium is high to extremely high in many parts of the plant, especially in the old leaves (up to 1.09 Wt%), in the bark (up to 2.34 Wt%) and in the twigs (up to 2.59 Wt%). Potassium (K) concentrations are remarkably high for a plant species growing on severely K-deficient soils. The highest K concentrations are in the old leaves (up to 1.34 Wt%) and twigs (up to 1.37 Wt%), and the lowest in the roots (53 μg g$^{-1}$). Magnesium concentrations are particularly high in the immature seeds (mean of 4931 μg g$^{-1}$). Sodium (Na) concentrations are unremarkable, but highest in the young leaves with up to 2325 μg g$^{-1}$ (mean is 390 μg g$^{-1}$). Finally, sulfur (S) concentrations vary widely with the lowest concentrations in the roots (81 μg g$^{-1}$) and the highest in the immature seeds (up to 3290 μg g$^{-1}$).

Bulk elemental concentrations of trace-elements (transition group elements) are given in Table 2. Cobalt (Co) concentrations are the highest in the young leaves with up to 198 μg g$^{-1}$, but are comparatively low in relation to Ni concentrations. The mean Co:Ni quotient is 1:532, evidencing highly selective uptake of Ni over Co, compared to soil concentrations of these elements (generally 10:1 Ni to Co). Chromium (Cr) concentrations are universally low, and this element is clearly excluded from uptake. Copper (Cu) concentrations have a narrow range between 0.7–17 μg g$^{-1}$, with mean values of 6.7 and 4.7 μg g$^{-1}$ in young and old leaves respectively. Iron concentrations are more variable, but generally low (maximum of 564 μg g$^{-1}$ in the phloem tissue). Manganese (Mn) concentrations too are low in all plant parts, with the highest concentrations on the old leaves (up to 461 μg g$^{-1}$). Nickel concentrations are the highest in the old leaves with up to 2.50 Wt% (mean of 1.65 Wt%) and somewhat lower in the young leaves (up to 1.08 Wt% and a mean of 5390 μg g$^{-1}$). Nickel concentrations are also high in the twigs (up to 1.23 Wt%) and in the phloem tissue (up to 0.93 Wt%). It is remarkable that the reproductive organs (flowers, fruit capsule and seeds) are highly Ni-enriched. The seeds contain 1.76 Wt% Ni and the

**Table 1 Elemental concentration (macro elements: Al, Ca, K, Mg, Na, P, S) ranges and means in parentheses**

| Plant part | *N* samples | Al | Ca | K | Mg | Na | P | S |
|---|---|---|---|---|---|---|---|---|
| Flowers | 2 | 20–370 [195] | 1020–3330 [2180] | 431–4600 [2520] | 2810–3630 [3220] | 65–301 [183] | 392–1070 [732] | 347–904 [625] |
| Fruit capsule | 5 | 9.0–70 [27] | 672–4160 [2410] | 2700–5920 [4600] | 563–1800 [1080] | 38–305 [123] | 191–1240 [738] | 757–1390 [972] |
| Immature seeds | 2 | 26–296 [161] | 4340–4480 [4410] | 6010–6090 [6050] | 2740–7120 [4930] | 70–661 [365] | 3330–3880 [3610] | 1940–3290 [2610] |
| Ripe seeds | 1 | 85 | 9460 | 4570 | 4100 | 603 | 1856 | 2380 |
| Young leaves | 12 | 1.0–52 [20] | 124–12,200 [4100] | 89–19,500 [7250] | 352–6900 [3020] | 1.9–2320 [390] | 34–5180 [1010] | 152–3610 [1440] |
| Old leaves | 21 | 12–69 [30] | 2416–10900 [4730] | 3101–13,400 [5760] | 1060–5500 [3140] | 30–525 [120] | 430–3220 [789] | 1140–3100 [1660] |
| Bark | 7 | 14–71 [43] | 92–23,400 [10,870] | 45–6240 [3750] | 361–2160 [728] | 19–530 [213] | 16–505 [332] | 108–1127 [749] |
| Phloem tissue | 3 | 29–95 [51] | 129–36,400 [16,500] | 84–8420 [4970] | 13–1360 [975] | 26–384 [222] | 30–339 [180] | 152–1100 [735] |
| Twigs | 14 | 1.0–71 [21] | 130–25,900 [5200] | 246–13,700 [5520] | 109–5280 [966] | 8.8–856 [313] | 16–1730 [539] | 53–1720 [610] |
| Wood | 7 | 2.8–23 [14] | 125–1296 [762] | 78–2725 [1585] | 85–249 [150] | 5.0–483 [200] | 23–791 [359] | 48–506 [306] |
| Roots | 1 | 91 | 125 | 53 | 382 | 31 | 40 | 81 |

All values provided in $\mu g\ g^{-1}$

**Table 2 Elemental concentration (trace elements: Co, Cr, Cu, Fe, Mn, Ni, Zn) ranges and means in parentheses**

| Plant part | *N* samples | Co | Cr | Cu | Fe | Mn | Ni | Zn |
|---|---|---|---|---|---|---|---|---|
| Flowers | 2 | 16–104 [60] | 14–17 [16] | 6.5–9.5 [8.0] | 15–24 [20] | 34–945 [490] | 2905–5560 [4230] | 20–100 [60] |
| Fruit capsule | 5 | 4.3–31 [13] | 2.3–17 [8.4] | 2.2–10 [5.9] | 8.0–30 [20] | 11–83 [35] | 1274–4001 [2660] | 9.0–22 [17] |
| Immature seeds | 2 | 23–89 [56] | 15–28 [22] | 8.5–13 [11] | 30–43 [36] | 58–64 [61] | 754–1421 [1090] | 26–34 [30] |
| Ripe seeds | 1 | 3 | 114 | 14 | 131 | 187 | 17,570 | 182 |
| Young leaves | 12 | 3.0–198 [46] | 1.3–66 [16] | 0.1–17 [6.7] | 9–241 [63] | 21–182 [93] | 1105–10,750 [5388] | 10–244 [52] |
| Old leaves | 21 | 12–75 [31] | 1.2–13 [6.9] | 0.9–11 [4.7] | 22–90 [46] | 95–461 [170] | 11,470–25,060 [16,490] | 45–249 [85] |
| Bark | 7 | 4.4–28 [14] | 2.3–13 [6.3] | 1.0–10 [4.5] | 22–429 [85] | 33–89 [66] | 842–6820 [5260] | 18–83 [69] |
| Phloem tissue | 3 | 12–49 [27] | 6.3–15 [10] | 1.3–6.6 [4.1] | 37–564 [220] | 62–126 [103] | 1390–9340 [5570] | 28–191 [100] |
| Twigs | 14 | 1.9–36 [16] | 2.4–42 [8.0] | 0.5–33 [6.2] | 4.0–40 [17] | 5.0–259 [72] | 99–12,300 [4330] | 5.1–103 [49] |
| Wood | 7 | 0.8–15 [5.0] | 1.9–4.7 [3.4] | 0.7–6.3 [2.5] | 3.7–10 [5.7] | 5.9–18 [11] | 443–1170 [862] | 4.8–14 [10] |
| Roots | 1 | 26 | 6.5 | 0.7 | 512 | 164 | 1272 | 24 |

All values provided in $\mu g\ g^{-1}$

flowers on average 0.43 Wt% Ni. The elemental fractionation of Ni in the different plant parts is depicted in Fig. 5. Finally, zinc (Zn) concentrations are unremarkable with on average 85 $\mu g\ g^{-1}$ in the old leaves.

### Conservation status

The habitat of *P. rufuschaneyi* is outside any protected areas, in patches of remaining scrub in an area devastated by recurring forest fires. The restriction of this species to just one main population (the second population

**Fig. 5** Agromining growth trials using *P. rufuschaneyi* in Sabah, Malaysia. **a** Mass propagation of *P. rufuschaneyi* using cutting grown in perlite; **b** *Phyllanthus rufuschaneyi* shrubs planted out in the 'Hyperaccumulator Botanic Garden' of Sabah Parks; **c** plant nutrition growth experiment using *P. rufuschaneyi* at Monggis substation; **d** full-scale (1.5 ha) agromining field trial using *P. rufuschaneyi* near Pahu village 6 months after planting

is very small) and the small area of occupancy (< 10 $km^2$) means that this species is sensitive to disturbances which could ultimately lead to its extinction. Therefore, the species can be classified as Critically Endangered (CR) on the basis of IUCN Red List Criteria (Version 3.1: IUCN 2001) considering Criterion B. Geographic range; B2. Area of occupancy estimated to be less than 10 $km^2$, and a. Severely fragmented or known to exist at only a single location; b. Continuing decline in (i) extent of occurrence; and (iii) area, extent and/or quality of habitat.

## Discussion

The taxonomic position of *P. rufuschaneyi* is noteworthy as it represents the most western species of its subgenus. It is placed in *Phyllanthus* subgenus *Gomphidium* section *Nymania* (K.Schum) J.J.Sm. on account of its three paired staminate disc glands and its connate stamens (Fig. 3c). Section *Nymania* is mainly distributed in Papua New Guinea, but a few species also occur in the Philippines. *Phyllanthus rufuschaneyi* is most similar to *P. securinegioides*, however, it differs in its significantly smaller leaves, its smaller fascicled inflorescences with pinkish flowers (Fig. 1a; vs larger fascicles with yellow flowers in *P. securinegioides*), the completely fused anther filaments (Fig. 3c) and the fused style (Fig. 3e). The other Philippine species of *Gomphidium*, *P. apiculatus* Merr., *P. ramosii* Quisumb., *P. glochidioides* Elmer and *P. cordatulus* Rob., all have staminate flowers with free stamens. Though *P. cordatulus* has a similarly fused style of c. 1.3 mm long, it is spreading for about the same distance, whereas the style of *P. rufuschaneyi* is only fused for up to 1 mm and slightly erose at the end. The style is quite prominent and emerges out of the pistillate flowers (Fig. 3d, e). The other species of *Phyllanthus* occurring in Borneo are from other subgenera and differ either in their branching system (e.g. phyllanthoid vs non-phyllanthoid in subgenus *Macraea* (Wight) Jean F.Brunel) or in the morphology of their flowers (e.g. no difference between inner and outer sepals in subgenera *Emblica* (Gaertn.) Kurz, *Kirganelia* (A.Juss.) Kurz and *Eriococcus* (Hassk.) Croizat & Metcalf). The structure on top of the anther connective is slightly reminiscent of the apiculate anthers in subgenus *Phyllanthodendron* (Hemsl.) G.L.Webster. However, *P. rufuschaneyi* does not have the characteristic ligulate nectar disc and apiculate anthers that are also found in subgenus *Gomphidium*. Nickel hyperaccumulators are found in several subgenera within the genus *Phyllanthus*, which suggests that it evolved several times.

The restricted distribution of *P. rufuschaneyi* cannot be easily explained. In Sabah, Ni hyperaccumulators are restricted to circum-neutral soils with relatively high phytoavailable calcium, magnesium and Ni concentrations with at least 20 $\mu g\ g^{-1}$ carboxylic acid extractable Ni or 630 $\mu g\ g^{-1}$ total nickel, and a soil pH > 6.3 (Van der Ent et al. 2016b). However, these types of Cambisols derived from serpentinized bedrock are relatively widespread in Sabah, but *P. rufuschaneyi* has not been found elsewhere despite extensive fieldwork on all major ultramafic outcrops in Sabah (Van der Ent et al. 2015b). Although some Ni hyperaccumulators are comparatively common (*P. balgooyi*, *Psychotria sarmentosa* Blume, Rubiaceae, *Xylosma luzonensis*) on several ultramafic outcrops in Sabah, the situation of *P. rufuschaneyi* mirrors that of *Actephila alanbakeri*, which is also restricted to a single site at the same location (Van der Ent et al. 2016a). Experience in cultivation shows that *P. rufuschaneyi* is highly shade-intolerant and requires exposed conditions with minimal light competition (Fig. 5). It suffers from fungal infections when grown under shaded and moist conditions, and stops flowering. The dependence on a specific pollinator that is characteristic for the *Phyllanthus-Epicephala* mutualism and limited dispersal capabilities might be possible explanations why this species is not more widespread in Borneo.

*Phyllanthus rufuschaneyi* has particularly attractive characteristics for cultivation as 'metal crop' in agromining operations. These favourable properties include its multi-stemmed habit, the rapid re-growth after coppicing, and high Ni concentrations in woody parts of the biomass. Other *Phyllanthus*-species, such as *P. balgooyi*, albeit having equally high foliar Ni concentrations, have lower growth rates and do not tolerate open and exposed conditions on bare soils. A major uncertainty currently pertains to the effective mass-propagation of *P. rufuschaneyi*, however, and the specialised pollination strategy of this species presents a challenge for using seed stock.

More than any other species, obligate hyperaccumulator plants that have a restricted distribution on isolated ultramafic outcrops are susceptible to habitat degradation (Galey et al. 2017). *Phyllanthus rufuschaneyi* is known only from a site that has been severely affected by over-logging and (man-made) forest fires for clearing of agricultural land, and neither the area nor the species has statutory protection. The current expansion of oil palm plantations on Lompoyou Hill cast a further shadow over its continued existence in the wild. As such, the case of *P. rufuschaneyi* highlights the importance of habitat and germplasm preservation if rare species are to be used in green technologies such as agromining. Therefore, concerted efforts must be made to screen for hyperaccumulator species (Whiting et al. 2004), followed by appropriate methods for conserving them both in and ex situ (Erskine et al. 2012). The utilization of *P. rufuschaneyi* in agromining operations means that there are now likely to be more plants in cultivation than in the wild, and this will for now safeguard its future survival.

**Authors' contributions**

AvdE, PE and GE conducted the fieldwork and collected the plant specimens in Sabah. RB and PvW carried out the taxonomical study. AvdE undertook the laboratory analyses. All authors read and approved the final manuscript.

**Author details**

[1] Naturalis Biodiversity Center, Botany 2300 RA, The Netherlands. [2] Hortus Botanicus, Leiden University, Leiden 2311 GJ, The Netherlands. [3] Institute of Biology Leiden, Leiden University, Leiden 2300 RA, The Netherlands. [4] Herbarium, Sabah Parks, Kota Kinabalu, Sabah, Malaysia. [5] Laboratoire Sols et Environnement, Université de Lorraine, INRA, Nancy 54000, France. [6] Centre for Mined Land Rehabilitation, Sustainable Minerals Institute, The University of Queensland, St Lucia, QLD 4072, Australia.

**Acknowledgements**

We thank Rimi Repin, Rositti Karim (Sabah Parks) and Postar Miun (Sabah Forestry Department) for their help during the fieldwork in Malaysia.

**Competing interests**

The authors declare that they have no competing interests.

**Funding**

The French National Research Agency through the national "Investissements d'avenir" program (ANR-10-LABX-21, LABEX RESSOURCES21) and through the ANR-14-CE04-0005 Project "Agromine" is acknowledged for funding support to G. Echevarria and A. van der Ent. A. van der Ent is the recipient of a Discovery Early Career Researcher Award (DE160100429) from the Australian Research Council.

**References**

Baker AJM, Proctor J, van Balgooy MMJ, Reeves RD (1992) Hyperaccumulation of nickel by the flora of the ultramafics of Palawan, Republic of the Philippines. In: Baker AJM, Proctor J, Reeves RD (ed) The vegetation of ultramafic (serpentine) soils. Proceedings of the first international conference on serpentine ecology. Intercept, Andover pp 291–304

Bani A, Echevarria G, Zhang X, Benizri E, Laubie B, Morel J-L, Simonnot M-O (2015) The effect of plant density in nickel-phytomining field experiments with Alyssum murale in Albania. Aust J Bot 63:72–77

Beaman JH (2005) Mount Kinabalu: hotspot of plant diversity in Borneo. Biol Skr 55:103–127

Beaman JH, Beaman RS (1990) Diversity and distribution patterns in the flora of Mount Kinabalu. In: Baas P, Kalkman K, Geesink R (eds) The plant diversity of Malesia. Kluwer Academic Publishers, Dordrecht, pp 147–160

Beentje H (2016) The Kew plant glossary: an illustrated dictionary of plant terms. Kew Royal Botanic Gardens, Richmond

Chaney RL (1983) Plant uptake of inorganic waste constituents. In: Parr JF, Marsh PB, Kla JM (eds) Land treatment of hazardous wastes. Noyes Data Corp, Park Ridge, pp 50–76

Chaney RL, Angle JS, Baker AJM, Li YM (1998) Method for phytomining of nickel, cobalt and other metals from soil. US Patent 5: 711–784, 27 January 1998

Chaney RL, Baker AJM, Morel JL (2018) The long road to developing agromining/phytomining. In: Van der Ent A, Echevarria G, Baker AJM, Morel JL (eds) Agromining: farming for metals. Mineral resource reviews. Springer, Cham

Echevarria G (2018) Genesis and behaviour of ultramafic soils and consequences for nickel biogeochemistry. In: Van der Ent A, Echevarria G, Baker AJM, Morel JL (eds) Agromining: farming for metals. Mineral resource reviews. Springer, Cham

Erskine P, van der Ent A, Fletcher A (2012) Sustaining metal-loving plants in mining regions. Science 337:1172–1173

Galey MC, van der Ent A, Iqbal MCM, Rajakaruna N (2017) Ultramafic geoecology of south and southeast asia. Bot Stud 58:18

Govaerts R, Frodin DG, Radcliffe-Smith A (2000) World checklist and bibliography of Euphorbiaceae (with Pandaceae), vol 4. Royal Botanic Gardens, Kew

Hoffmann P, Kathriararachchi H, Wurdack KJ (2006) A phylogenetic classification of Phyllanthaceae (Malpighiales; Euphorbiaceae sensu lato). Kew Bull 61:37–53

Hoffmann P, Baker AJM, Proctor J, Madulid D (2008) *Phyllanthus balgooyi* (Euphorbiaceae s.l.), a new nickel-hyperaccumulating species from Palawan and Sabah. Blumea 48:193–199

Homer F (1991) Chemical studies on some plants that hyperaccumulate nickel. Ph.D. thesis, Massey University, pp 1–285

Huang L, Bell RW, Dell B, Woodward J (2004) Rapid nitric acid digestion of plant material with an open-vessel microwave system. Commun Soil Sci Plant Anal 35:427–440

IUCN (2001) IUCN red list categories and criteria: version 3.1. IUCN Species Survival Commission, Gland

Jaffré T (1977) Composition chimique élémentaire des tissus foliaires des espèces végétales colonisatrices des anciennes mines de nickel en Nouvelle Calédonie. Cah. O.R.S.T.O.M., sér. Biol. XII: 323–330

Jaffré T, Brooks RR, Lee J, Reeves RD (1976) *Sebertia acuminata*: a hyperaccumulator of nickel from New Caledonia. Science 193:579–580

Jaffré T, Pillon Y, Thomine S, Merlot S (2013) The metal hyperaccumulators from New Caledonia can broaden our understanding of nickel accumulation in plants. Front Plant Sci. https://doi.org/10.3389/fpls.2013.00279/abstract

Kathriarachchi H, Samuel R, Hoffmann P, Mlinarec J, Wurdack KJ, Ralimanana H, Stuessy TF, Chase MW (2006) Phylogenetics of tribe Phyllantheae (Phyllanthaceae; Euphorbiaceae *sensu lato*) based on nrITS and plastid matK DNA sequence data. Am J Bot 93:637–655

Kato M, Kawakita A (2017) Obligate pollination mutualism, XII edn. Springer, Tokyo, p 309

Kawakita A (2010) Evolution of obligate pollination mutualism in the tribe Phyllantheae (Phyllanthaceae). Plant Species Biol 25:3–19

Kawakita A, Kato M (2004) Evolution of obligate pollination mutualism in New Caledonian *Phyllanthus* (Euphorbiaceae). Am J Bot 91:410–415

Kersten W, Brooks R, Reeves R, Jaffré T (1979) Nickel uptake by New Caledonian species of *Phyllanthus*. Taxon 28:529–534

Kersten WJ, Brooks RR, Reeves RD, Jaffré A (1980) Nature of nickel complexes in *Psychotria douarrei* and other nickel-accumulating plants. Phytochemistry 19:1963–1965

Leon B, Alain B (1953) Flora de Cuba. III. Euphorbiaceae. Contribuciones Ocasionales del Museo de Historia Natural del Colegio de la Salle 13:44–59

Li Y-M, Chaney R, Brewer E, Roseberg R, Angle JS, Baker A, Reeves R, Nelkin J (2003) Development of a technology for commercial phytoextraction of nickel: economic and technical considerations. Plant Soil 249:107–115

Mesjasz-Przybylowicz J, Przybylowicz W, Barnabas A, van der Ent A (2016) Extreme nickel hyperaccumulation in the vascular tracts of the tree *Phyllanthus balgooyi* from Borneo. New Phytol 209:1513–1526

Montargès-Pelletier E, Chardot V, Echevarria G, Michot LJ, Bauer A, Morel J-L (2008) Identification of nickel chelators in three hyperaccumulating plants: an X-ray spectroscopic study. Phytochemistry 69:1695–1709

Nkrumah PN, Baker AJM, Chaney RL, Erskine PD, Echevarria G, Morel J-L, van der Ent A (2016) Current status and challenges in developing nickel phytomining: an agronomic perspective. Plant Soil 406:55–69

Nkrumah P, Echevarria G, Erskine, PD, van der Ent A, Sumail S (2017) First growth trial of tropical 'metal shrubs' to be used in economic agromining. In: Proceedings of the 9th international conference on Serpentine ecology, 5–9th June 2017. Tirana-Pogradec, Albania, p 169

Nkrumah P, Echevarria G, Erskine PD, van der Ent A (2018) The discovery of nickel hyper-accumulation in *Antidesma montis-silam*: from herbarium identification to field re-discovery. Ecol Res. https://doi.org/10.1007/s11284-017-1542-4

Pruesapan K, Telford IRH, Bruhl JJ, Van Welzen PC (2012) Phylogeny and proposed circumscription of Breynia, Sauropus and Synostemon (Phyllanthaceae), based on chloroplast and nuclear DNA sequences. Aust Syst Bot 25:313–330

Quimado MO, Fernando ES, Trinidad LC, Doronila A (2015) Nickel-hyperaccumulating species of *Phyllanthus* (Phyllanthaceae) from the Philippines. Aus J Bot 63(2):103–110

Reeves RD (2003) Tropical hyperaccumulators of metals and their potential for phytoextraction. Plant Soil 249(1):57–65

Reeves RD, Baker AJM, Borhidi A, Berazaín R (1996) Nickel-accumulating plants from the ancient serpentine soils of Cuba. New Phytol 133:217–224

Reeves RD, Baker AJM, Jaffré T, Erskine PD, Echevarria G, van der Ent A (2017) A global database for hyperaccumulator plants of metal and metalloid trace elements. New Phytol. https://doi.org/10.1111/nph.14907

Schmid M (1991) *Phyllanthus*. In: Morat P, Mackee HS (eds) Flore de la Nouvelle-Calédonie et Dépendances, vol 17. Muséum National d'Histoire Naturelle, Paris, pp 31–320

Van der Ent A, Mulligan D (2015) Multi-element concentrations in plant parts and fluids of Malaysian nickel hyperaccumulator plants and some economic and ecological considerations. J Chem Ecol 41:396–408

Van der Ent A, Baker AJM, Reeves RD, Pollard AJ, Schat H (2013a) Hyperaccumulators of metal and metalloid trace elements: facts and fiction. Plant Soil 362:319–334

Van der Ent A, Baker AJM, van Balgooy MMJ, Tjoa A (2013b) Ultramafic nickel laterites in Indonesia (Sulawesi, Halmahera): mining, nickel hyperaccumulators and opportunities for phytomining. J Geochem Explor 128:72–79

Van der Ent A, Mulligan D, Erskine PD (2013c) Discovery of nickel hyperaccumulators from Kinabalu Park, Sabah (Malaysia) for potential utilization in phytomining. In: Enviromine 2013, Santiago, Chile, 4–6 December 2013. pp 213–221

Van der Ent A, Baker AJM, Reeves RD, Chaney RL, Anderson CWN, Meech JA, Erskine PD, Simonnot M-O, Vaughan J, Morel J-L et al (2015a) Agromining: farming for metals in the future? Environ Sci Technol 49:4773–4780

Van der Ent A, Erskine P, Sumail S (2015b) Ecology of nickel hyperaccumulator plants from ultramafic soils in Sabah (Malaysia). Chemoecology 25:243–259

Van der Ent A, Repin R, Sugau J, Wong KM (2015c) Plant diversity and ecology of ultramafic outcrops in Sabah (Malaysia). Aust J Bot 63:204–215

Van der Ent A, van Balgooy M, van Welzen P (2016a) *Actephila alanbakeri* (Phyllanthaceae): a new nickel hyperaccumulating plant species from localised ultramafic outcrops in Sabah (Malaysia). Bot Stud 57:19

Van der Ent A, Echevarria G, Tibbett M (2016b) Delimiting soil chemistry thresholds for nickel hyperaccumulator plants in Sabah (Malaysia). Chemoecology 26:67–82

Van der Ent A, Damien L, Callahan DL, Noller BN, Mesjasz-Przybylowicz J, Przybylowicz WJ, Barnabas A, Harris HH (2017) Nickel biopathways in tropical nickel hyperaccumulating trees from Sabah (Malaysia). Sci Rep 7:41861

Van Welzen PC, Pruesapan K, Telford IRH, Esser H, Bruhl JJ (2014) Phylogenetic reconstruction prompts taxonomic changes in Sauropus, Synostemon and Breynia (Phyllanthaceae tribe Phyllantheae). Blumea 59(2):77–94

Vaughan J, Riggio J, Chen J, Peng H, Harris HH, van der Ent A (2017) Characterisation and hydrometallurgical processing of nickel from tropical agromined bio-ore. Hydrometallurgy 169:346–355

Webster GL (1956) A monographic study of the West Indian species of *Phyllanthus*. J Arnold Arbor 37: 91–122, 217–268, 340–359
Whiting S, Reeves RD, Richards D, Johnson M, Cooke J, Malaisse F, Paton A, Smith J, Angle J, Chaney R et al (2004) Research priorities for conservation of metallophyte biodiversity and their potential for restoration and site remediation. Restor Ecol 12:106–116
Wu M-J, Huang T-C, Liu C-C, Chen Y-J, Chang Y-S, Hsu C-L, Wu S-Y, Tseng A-Y, Chang Y-C, Liu C-C, Kaewmuan A (2016) Pollen morphology and taxonomy in Malesian *Phyllanthus* (Phyllanthaceae). J Jpn Bot 91:257–292

# Grain width 2 (GW2) and its interacting proteins regulate seed development in rice (*Oryza sativa* L.)

Kyu Ho Lee[1], Sang Woo Park[1], Yeon Jeong Kim[1], Yeon Jong Koo[2], Jong Tae Song[3] and Hak Soo Seo[1*]

**Abstract**

**Background:** Seed size has been extensively studied in crop plants, as it determines crop yield. However, the mechanism of seed development remains elusive. In this study, we explored the mechanism of seed development in rice (*Oryza sativa* L.), and identified proteins affecting seed size.

**Results:** Proteomic analysis showed that glyceraldehyde 3-phosphate dehydrogenase, chitinase 14 (CHT14), and phosphoglycerate kinase (PGK) accumulated to high levels in the seeds of the natural *japonica* rice mutant Oochikara, which carries a loss-of-function mutation in the *grain width 2* (*GW2*) gene; *GW2* encodes a RING-type E3 ubiquitin ligase. In vitro pull-down and ubiquitination assays showed that CHT14 and PGK directly interacted with GW2 but were not ubiquitinated by GW2. Immunoblot analysis revealed that protein disulfide isomerase-like 1-1 accumulated to high levels in young developing seeds of the *gw2* mutant compared with the wild type. Histochemical *β-glucuronidase* staining showed strong expression of *GW2* in leaf and root tissues but weak expression in leaf sheaths and internodes. In addition, transformation of the *green fluorescent protein* (*GFP*) gene under the control of the *GW2* promoter in rice revealed *GFP* expression in the aleurone layer of seeds.

**Conclusions:** Collectively, these results suggest that GW2 regulates seed size through direct interactions with proteins involved in carbohydrate metabolism by modulating their activity or stability and controlling disulfide bond formation in various proteins during seed development. Additionally, GW2 participates in vegetative as well as reproductive growth, and protects the seed from pathogen attack.

**Keywords:** Aleurone layer, CHT14, E3 ubiquitin ligase, GW2, PDIL1-1, PGK, Rice, Seed size

## Background

Post-translational modification of proteins involves the covalent addition of functional molecules of proteins, cleavage of protein subunits, or complete degradation of proteins. Ubiquitination is a type of post-translational modification in which the ubiquitin protein is covalently conjugated to the target protein (Ciechanover 1994). Ubiquitination involves three sequential reactions, each catalyzed by three unique enzymes: an ubiquitin-activating enzyme E1, an ubiquitin-conjugating enzyme E2, and a specific ligase E3 (Vierstra 2009; Bie and Ciechanover 2011; Kelley 2018). Ubiquitination is involved in the regulation of protein localization, stability, and activity (Kelley 2018). Target proteins undergo either monoubiquitination or polyubiquitination. The 26S proteasome complex degrades polyubiquitinated proteins but not monoubiquitinated proteins. Monoubiquitination affects the localization and function of the modified protein (Bie and Ciechanover 2011). Ubiquitination is emerging as a key process in the regulation of seed size in crop plants (Song et al. 2007; Choi et al. 2018).

Seed size is a major factor affecting grain yield in rice (*Oryza sativa* L.) (Fan et al. 2006; Bai et al. 2010; Song et al. 2007) and is controlled by the hull. Rice grain width 2 (GW2) is an E3 ubiquitin ligase, which contains a RING

*Correspondence: seohs@snu.ac.kr
[1] Department of Plant Science, College of Agriculture and Life Sciences, Research Institute of Agriculture and Life Sciences, Seoul National University, Gwanakro 200, Gwanak-gu, Seoul 151-921, South Korea
Full list of author information is available at the end of the article

finger motif and shows both auto-ubiquitination and substrate ubiquitination activities (Song et al. 2007; Choi et al. 2018). GW2 homologs have been identified in wheat (*Triticum aestivum*; TaGW2) and maize (Zea mays; ZmGW2); these are involved in the regulation of kernel size, weight, width, and maturity (Su et al. 2011; Li et al. 2010). Although progress has been made towards the cloning and characterization of seed size-related quantitative trait loci and genes, mechanisms determining seed size remain unclear. Recently, it was reported that GW2 directly interacts with and ubiquitinates EXPANSIN-LIKE 1 (EXPLA1), thus inactivating it.

Here, we showed that GW2 strongly interacted with chitinase 14 (CHT14) and phosphoglycerate kinase (PGK) but did not ubiquitinate these proteins, suggesting that GW2 regulates seed size control via its interacting partners, CHT14 and PGK. Additionally, the level of protein disulfide isomerase-like 1-1 (PDIL1-1) was higher in young developing seeds of the *gw2* mutant than in the wild-type. Moreover, GW2 was specifically expressed in the aleurone layer. These data suggest that GW2 controls the development of the aleurone layer and the activity or stability of various proteins by regulating disulfide bond formation during seed development.

## Results

### Proteomic analysis of gw2 seeds

Because GW2 protein contains a RING finger domain and has auto-ubiquitination activity, we hypothesized that it functions as an E3 ubiquitin ligase of target proteins involved in GW2-mediated seed development. We used a proteomics approach to identify GW2-interacting proteins. Proteomics is a useful method to identify proteins that accumulate in an organ- or development-specific manner (Cho et al. 2006). Soluble proteins were extracted from the seeds of the wild-type rice cultivar Norin 22 and those of the *japonica* rice *gw2* mutant, Oochikara. These proteins were separated in the first dimension via isoelectric focusing (IEF) on a pH gradient ranging from 4 to 7 and in the second dimension by sodium dodecyl sulfate–polyacrylamide gel electrophoresis (SDS-PAGE) using 10% polyacrylamide gels (Fig. 1a). Protein spots in parallel gels were cross-matched, demonstrating reproducibility. After electrophoresis, gels were analyzed using PDQuest software, and two-dimensional (2-D) gels were systematically compared. Three protein spots showing significant accumulation in *gw2* seeds were selected (Fig. 1a, b). Proteins in these spots were identified via Ettan matrix-assisted laser desorption/ionization time-of-flight mass spectrometry (MALDI-TOF MS), followed by a homology search. The three proteins were identified as glyceraldehyde 3-phosphate dehydrogenase (GAPDH), CHT14, and PGK.

### GW2 interacts with but does not ubiquitinate PGK and CHT14

To characterize GAPDH, PGK, and CHT14, we isolated the corresponding cDNAs using reverse transcriptase PCR (RT-PCR). Gene-specific primers were designed on the basis of sequences present in the nucleotide database of the National Center for Biotechnology Information (NCBI). Results of proteomic analysis showed the accumulation of these proteins in *gw2* seeds, implying that GW2 directly interacts with and possibly ubiquitinates these proteins, thus rendering them liable to degradation by the 26S proteasome complex. Next, we examined the possible interaction of GW2 with each of the three proteins using in vitro pull-down assay. Three recombinant plasmids encoding the maltose-binding protein (MBP) fused with GW2 (MBP-GW2) and glutathione *S*-transferase (GST) fused with CHT14 (GST-CHT14) and PGK (GST-PGK) were constructed and expressed in *Escherichia coli* BL21 (DE3) cells. Recombinant proteins were then affinity purified (Fig. 2a), and GST-CHT14 and GST-PGK were pulled down with either MBP or MBP-GW2. In vitro pull-down assay using MBP-GW2 confirmed that GW2 interacted with CHT14 and PGK (Fig. 2b). However, no interaction was detected between GW2 and GAPDH.

Direct interaction of GW2 with CHT14 and PGK raised the possibility that GW2 may ubiquitinate CHT14 and PGK. However, in vitro ubiquitination assays showed that both CHT14 and PGK were not ubiquitinated (Fig. 3a, b), indicating that GW2 does not ubiquitinate CHT14 and PGK.

### PDIL1-1 accumulates during early seed development in gw2 mutants

PDIL1-1 regulates the amounts of various seed proteins, which in turn influence the development of endosperm and the aleurone layer (Kim et al. 2012). In addition, seeds of *PDIL1-1Δ* exhibit a floury phenotype due to the production of irregular starch granules and protein bodies (Kim et al. 2012). Seeds of the *gw2* mutant also display a floury phenotype (Choi et al. 2018). Therefore, we examined the level of PDIL1-1 during seed development in the *gw2* mutant. Results showed that PDIL1-1 accumulated to high levels in *gw2* seeds at 5 and 10 days after flowering (DAF), whereas its level remained very low in wild-type seeds (Fig. 4a, b), suggesting that GW2 affects the level of PDIL1-1 protein.

### GW2 is highly expressed in panicles and anthers

Previous studies showed that *GW2* is expressed in anthers, stamens, pistil, panicles, hull, and endosperm (Song et al. 2007; Choi et al. 2018). To further examine

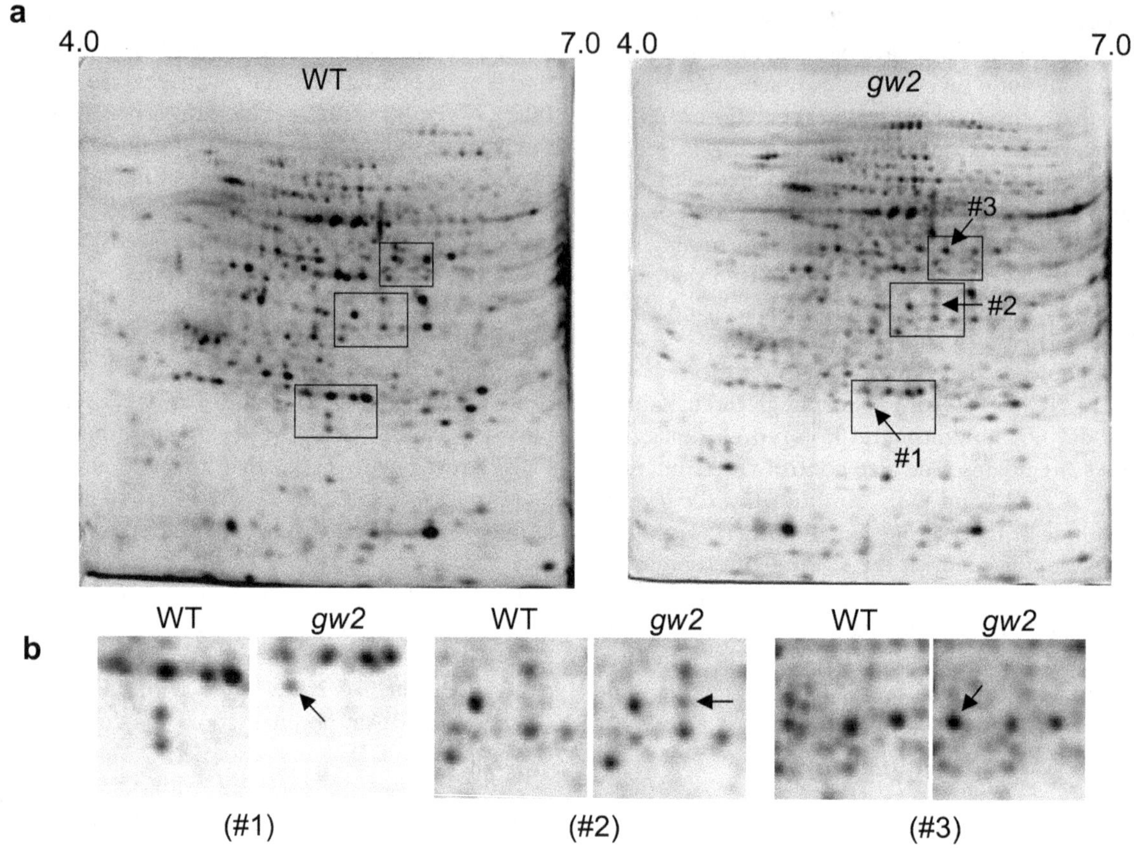


**Fig. 1** Proteomic analysis of seed proteins in the *grain width 2* (*gw2*) mutant of rice, Oochikara. Total proteins were extracted from whole seeds and subjected to isoelectric focusing (IEF) in the first dimension, using immobilized pH gradient dry strips (pH 4–7 nonlinear, 24 cm in length), and SDS-PAGE in the second dimension, using 10–16% gradient polyacrylamide gels. Gels were stained with silver nitrate. Three independent biological replicates were performed. **a** 2D gels showing protein spots in wild-type and *gw2* mutant rice. Arrowheads indicate highly accumulated proteins in the *gw2* mutant compared with the wild type. Boxed regions shown in **a** are magnified in **b**. Spot #1, glyceraldehyde 3-phosphate dehydrogenase (GAPDH); spot #2, chitinase 14 (CHT14); spot #3, phosphoglycerate kinase (PGK)

the expression of *GW2*, we generated transgenic rice lines overexpressing *β-glucuronidase* (*GUS*) and *green fluorescent protein* (*GFP*) reporter genes under the control of the *GW2* promoter. Histochemical staining of $GW2_{pro}$-*GUS* plants showed that *GUS* was strongly expressed in leaves and roots but weakly in leaf sheaths and internodes (Fig. 5a). Additionally, fluorescence detection in $GW2_{pro}$-*GFP* plants revealed a strong GFP signal in the seed aleurone layer (Fig. 5b).

## Discussion

In this study, we show that GW2-interacting proteins, CHT14 and PGK, regulate GW2-mediated seed development. Additionally, GW2 participates in vegetative as well as reproductive growth. Previously, auto-ubiquitination and substrate ubiquitination activities of GW2 have been reported in vitro (Song et al. 2007; Choi et al. 2018), suggesting that the E3 ligase activity of GW2 is required for the regulation of cell size in the spikelet hull. To further study the role of GW2 during seed development, we isolated GW2-interacting proteins whose level and stability were regulated by GW2 activity. Proteomic analysis of *gw2* mutant and wild-type seeds revealed that the levels of several proteins were higher in *gw2* mutant seeds than in wild-type seeds (Fig. 1a, b). Determination of amino acid sequences revealed that GAPDH, CHT14, and PGK proteins accumulated to high levels in *gw2* mutant seeds, suggesting that these proteins are destabilized by the E3 ubiquitin ligase activity of GW2. In vitro pull-down assay showed interaction of GW2 with CHT14 and PGK but not with GAPDH (Fig. 2a, b). Moreover, in vitro ubiquitination assay revealed that GW2 did not ubiquitinate CHT14 and PGK (Fig. 3a, b). CHT enzymes, which are encoded by a multigene family, degrade chitins and repress pathogen growth (Nakahara and Masuta 2014). This suggests that CHTs play a pivotal role in plant immunity. Direct interaction between GW2 and CHT14 suggests that GW2 can be involved

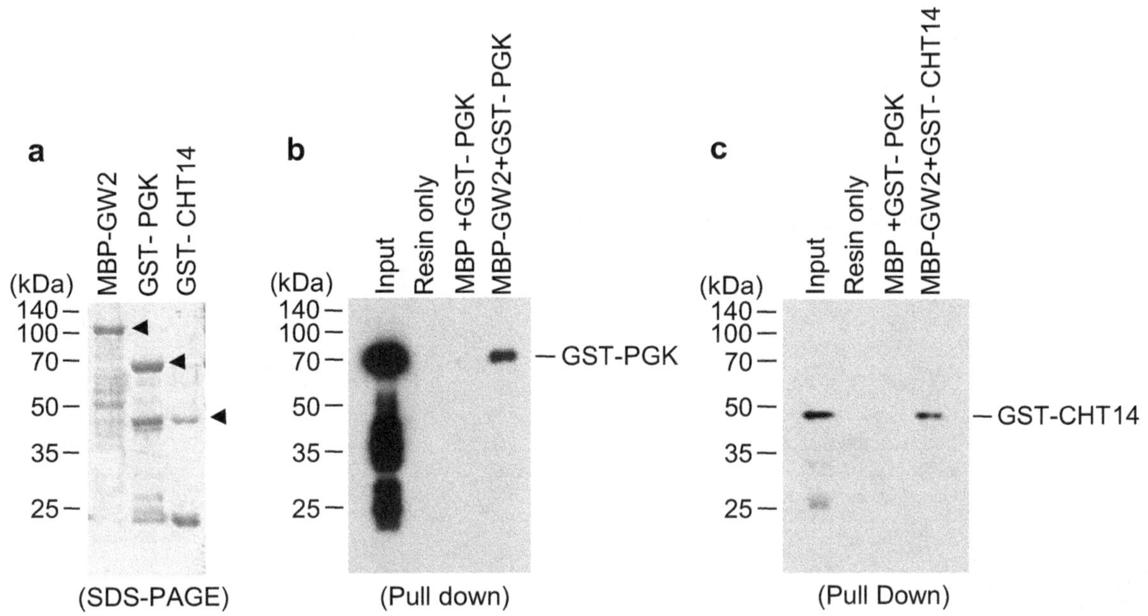


**Fig. 2** GW2 interacts with PGK and CHT14. **a** Recombinant fusion proteins MBP-GW2, GST-PGK, and GST-CHT14 were overexpressed in *E. coli* and purified using amylose or glutathione resins. Arrowheads indicate MBP-GW2, GST-PGK, and GST-CHT14. **b**, **c** In vitro pull-down assay showing the proteins GST-PGK (**b**) and GST-CHT14 (**c**) pulled down with MBP-GW2, separated on 10% SDS–polyacrylamide gels, and detected by western blotting using anti-GST antibody. *GST* glutathione *S*-transferase, *MBP* maltose-binding protein, *GW2* grain width 2, *PGK* phosphoglycerate kinase, *CHT14* chitinase 14

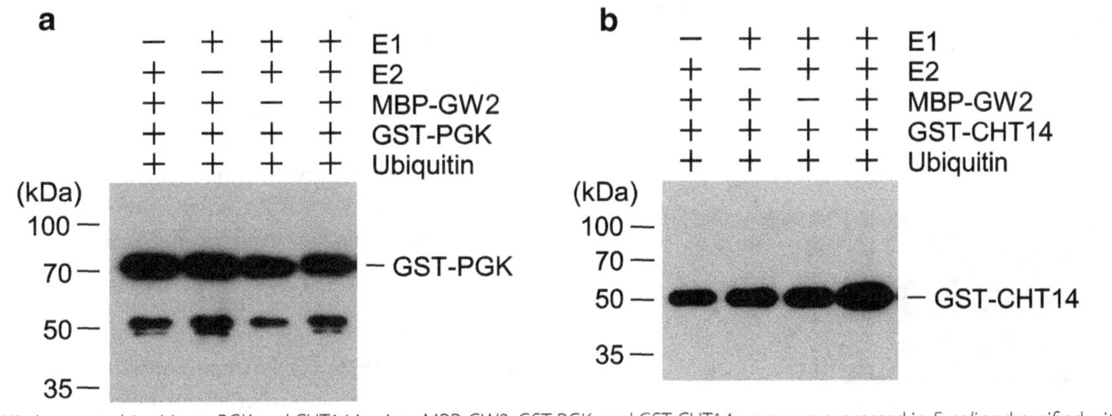


**Fig. 3** GW2 does not ubiquitinate PGK and CHT14 in vitro. MBP-GW2, GST-PGK, and GST-CHT14 were overexpressed in *E. coli* and purified with amylose or glutathione affinity columns. Ubiquitination of GST-PGK (**a**) or GST-CHT14 (**b**) was examined in the presence or absence of rabbit E1, UbcH5b (E2), MBP-GW2 (E3), and $His_6$-ubiquitin. PGK and CHT14 proteins were detected by western blotting using anti-GST antibody

in the protection of the seeds from pathogen attack during development. PGK converts 1,3-bisphosphoglycerate into 3-phosphoglycerate during glycolysis but also participates in the reverse reaction during gluconeogenesis and the Calvin–Benson cycle, indicating that PGK provides the building blocks for carbohydrate biosynthesis (Watson et al. 1982; Brinkmann and Martin 1996). Therefore, direct interaction between GW2 and PGK suggests that GW2 regulates carbohydrate biosynthesis during seed development. Together, these results suggest that the levels of CHT14 and PGK are not directly regulated by the E3 ligase activity of GW2. Nevertheless, direct interactions between GW2 and CHT14 or PGK suggest that GW2 can regulate the activity and level of CHT14 and PGK or, that CHT14 and PGK can control the activity of GW2.

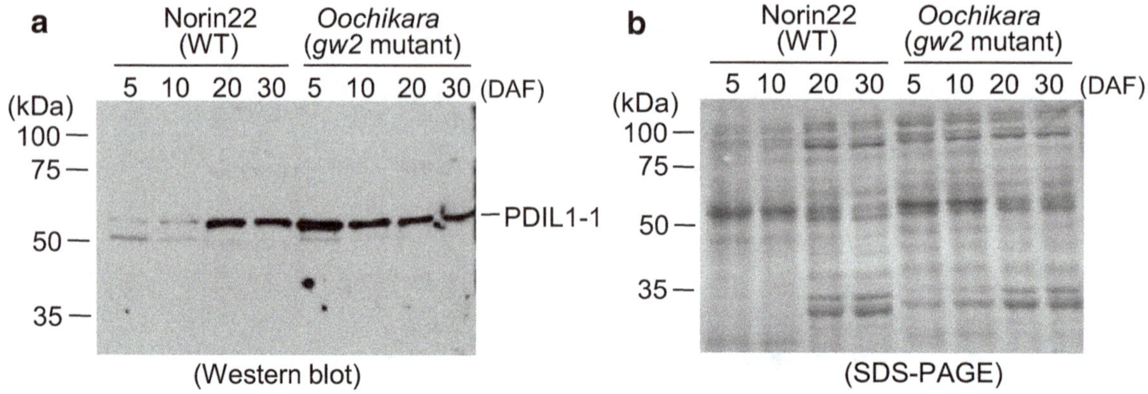


**Fig. 4** Accumulation of PDIL1-1 protein in *gw2* mutant (Oochikara) and wild-type (Norin 22) seeds during development. Total proteins were extracted from Norin 22 and Oochikara seeds at the indicated time points. After SDS-PAGE, proteins were transferred onto a nitrocellulose membrane. **a** Immunodetection of PDIL1-1 protein using anti-PDIL1-1 antibody. **b** SDS–polyacrylamide gel stained with Coomassie brilliant blue after blotting. PDIL1-1, protein disulfide isomerase-like 1-1

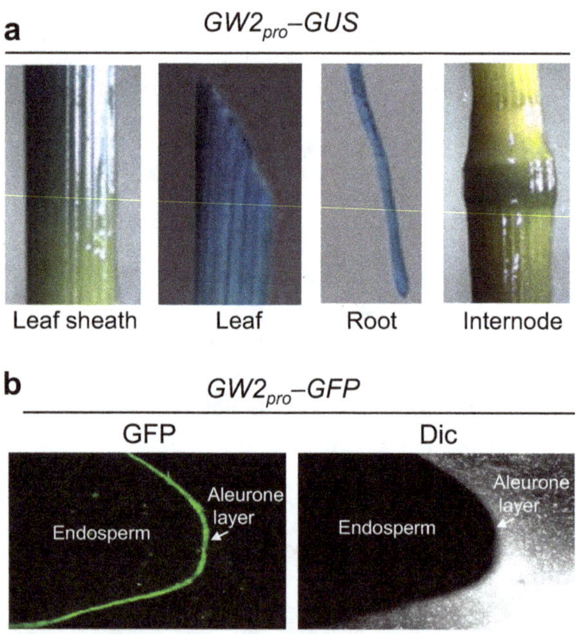


**Fig. 5** Expression of *GUS* and *GFP* reporter genes under the control of the *GW2* promoter. **a** Examination of *GUS* expression in *GW2*$_{pro}$-*GUS* transgenic rice. **b** Detection of GFP fluorescence in *GW2*$_{pro}$-*GFP* transgenic rice. GFP imaging was performed using a LAS 4000 imager. *GW2*$_{pro}$ *GW2* promoter, *GUS* β-glucuronidase, *GFP* green fluorescent protein

The natural *gw2* mutant Oochikara has a floury endosperm due to small, spherical, and loosely packed starch granules (Song et al. 2007; Choi et al. 2018). This endosperm phenotype of Oochikara is similar to that of the *PDIL1-1Δ* mutant; the floury endosperm phenotype of the *pdil1-1* deletion mutant is caused by the production of irregular starch granules and protein bodies (Kim et al. 2012). However, in Oochikara, PDIL1-1 accumulated to high levels during the early stages of seed development (Fig. 4a). PDIL ensures proper protein folding through the formation and breakage of intramolecular disulfide bonds (Wilkinson and Gilbert 2004). How a high level of PDIL1-1 protein leads to the production of a floury endosperm is currently unknown. It is possible that PDIL1-1 stabilizes the enzymes involved in branching and debranching of carbohydrates or those involved in formation of protein bodies during early seed development in Oochikara. Failure to degrade these enzymes in a timely manner may lead to their stabilization, thus increasing their activity. Another possibility is that the loss-of-function of *GW2* changes the stability or activity of target proteins involved in endosperm development, including those involved in the formation of starch granules and protein bodies, leading to a floury endosperm. Further in vitro and in vivo analyses are required to elucidate whether PDIL1-1 level and activity are regulated by GW2 through direct interaction between PDIL1-1 and GW2. In addition, the effect of PDIL1-1 on seed phenotypes using transgenic Norin 22 and Oochikara plants overexpressing *PDIL1-1* also needs to be investigated.

Investigation of gene expression in *GW2*$_{pro}$-*GUS* and *GW2*$_{pro}$-*GFP* transgenic rice plants showed that *GUS* was highly expressed in leaves and roots but weakly in leaf sheaths and internodes (Fig. 5a). Notably, *GFP* was highly expressed in the aleurone layer of seeds (Fig. 5b). Additionally, *GW2* expression has been previously demonstrated in various floral organs and seeds (Song et al. 2007; Choi et al. 2018). All of these data strongly indicate that *GW2* regulates rice growth and development during vegetative as well as reproductive growth.

## Conclusions

Overall, we conclude that GW2 strongly interacts with CHT14 and PGK but does not ubiquitinate these proteins. Our data also suggest that CHT14 and PGK can participate in the regulation of the function of GW2, and that GW2 controls seed size through the regulation of CHT14 and PGK levels or activities as well as its E3 ubiquitin ligase activity. Further analysis of transgenic plants will provide insights into how CHT14, PGK, and PDIL1-1 participate in GW2-mediated regulation of vegetative and reproductive growth of rice plants.

## Methods

### Plant material and growth conditions

The natural *japonica* rice *gw2* mutant Oochikara (Accession No. 54075) was obtained from the National Institute of Agrobiological Sciences (NIAS), Japan, and the wild-type rice cultivar Norin 22 was kindly provided by Dr. Hee Jong Koh, Seoul National University. Rice plants were grown at 26 °C under long day conditions (16 h light/8 h dark) in the greenhouse or field.

### 2D PAGE and MALDI-TOF MS

Total proteins were extracted from the seeds of 'Norin 22' and 'Oochikara' using homogenizing buffer (7 M urea, 2 M thiourea, 4% CHAPS, 1% DTT, 2% pharmalyte, and 1 mM benzamidine). Protein concentrations were determined using the Bradford assay (Bio-Rad, Bradford 1976), and proteins were subjected to 2D PAGE as described previously (Yeu et al. 2007). The signal intensity of each spot was normalized relative to the total spot intensity. Three protein spots, which accumulated to high levels in *gw2* mutant seeds, were subjected to in-gel digestion using modified porcine trypsin. Proteins were identified using an Ettan MALDI-TOF (Amersham Biosciences) as described previously (Yeu et al. 2007).

### Construction of recombinant plasmids

To produce MBP-GW2, GST-CHT14, GST-GAPDH, and GST-PGK fusion proteins, cDNA sequences encoding full-length GW2, CHT14, GAPDH, and PGK were PCR amplified using gene-specific primers and cloned into pMALc2 (New England Biolabs) or pGEX4T-1 (Amersham Biosciences) vectors. Recombinant plasmids were transformed into *E. coli* strain BL21 (DE3) pLysS, and the expression of recombinant proteins was induced with the addition of 5 mM isopropyl-β-D-thiogalactoside (IPTG).

### Purification of recombinant proteins

Recombinant proteins expressed in *E. coli* BL21 cells were purified according to the manufacturer's instructions. Cells expressing MBP or MBP-GW2 were resuspended and sonicated in a buffer containing 20 mM Tris–HCl (pH 7.4), 200 mM NaCl, 1 mM EDTA, 1% Triton X-100, 2 mM PMSF, and proteinase inhibitor cocktail (Roche). Total protein extracts were loaded onto a column packed with amylose resin (New England Biolabs). Columns were washed with buffer containing 20 mM Tris–HCl (pH 7.4), 200 mM NaCl, 1 mM EDTA, and 1% Triton X-100, and MBP and MBP-GW2 recombinant proteins were eluted with 10 mM maltose.

To purify GST fusion proteins, *E. coli* cells expressing GST-CHT14, GST-GAPDH, and GST-PGK were resuspended and sonicated in buffer containing 50 mM Tris, 50 mM NaCl, 1 mM EDTA, 1% Triton X-100, 2 mM PMSF, and proteinase inhibitor cocktail (Roche). Total protein extracts were loaded onto a column packed with glutathione resin (Pharmacia), and columns were washed with buffer containing 50 mM Tris, 50 mM NaCl, 1 mM EDTA, and 1% Triton X-100. The recombinant proteins GST-CHT14, GST-GAPDH, and GST-PGK were eluted using 10 mM glutathione. All protein concentrations were determined using the Bradford assay (Bio-Rad).

### In vitro pull-down assays

Interactions of MBP-GW2 with GST-CHT14, GST-PGK, and GST-GAPDH were examined using in vitro pull-down assay. Two micrograms of MBP-GW2 was mixed with 2 μg of GST-PGK, GST-CHT14, or GST-GAPDH, and this mixture was added to the reaction buffer containing 50 mM Tris–HCl (pH 7.5), 100 mM NaCl, 1% Triton X-100, 0.2% glycerol, and 0.5 mM 2-mercaptoethanol. Reaction mixtures were incubated at 25 °C for 2 h and then purified using the amylose resin. Resins were washed with buffer containing 50 mM Tris–HCl (pH 7.5), 100 mM NaCl, and 1% Triton X-100 five times. Proteins were eluted using 10 mM glutathione and separated by 10% SDS-PAGE. The GST fusion proteins were detected by immunoblotting using anti-GST antibody (0.4 μg $ml^{-1}$; Santa Cruz Biotechnology).

### In vitro ubiquitination assay

In vitro ubiquitination reactions were carried out in reaction buffer containing 50 mM Tris–HCl (pH 7.4), 5 mM $MgCl_2$, 150 mM NaCl, 2 mM ATP, and 1 mM DTT. Each reaction mixture contained 50 ng of E1 (Boston Biochem), 100 ng of E2 (Boston Biochem), 5 μg of $His_6$-ubiquitin (Sigma), 200 ng of MBP-GW2, and 100 ng of GST-CHT14 or GST-PGK. Reactions were incubated at 30 °C for 3 h, and proteins were separated by 10% SDS-PAGE. Ubiquitination of GST-CHT14 and GST-PGK was examined by immunoblotting with anti-GST antibody (0.4 μg $ml^{-1}$; Santa Cruz Biotechnology).

### Promoter activity analysis

Transgenic rice lines overexpressing *$GW2_{pro}$-GUS* and *$GW2_{pro}$-GFP* were generated as described previously (Choi et al. 2018). Histochemical GUS staining assays were carried out as described previously (Shomura et al. 2008). Specimens were photographed using the OLYMPUS SZX7 microscope. GFP fluorescence in flowers and immature seeds of transgenic and wild-type rice plants was observed using a fluorescence microscope (LAS 4000 imager, GE Life Sciences).

### Western blot analysis

The level of PDIL1-1 protein during seed development was examined in wild-type and *gw2* mutant rice seeds harvested at 5, 10, 20, and 30 DAF. Total proteins were extracted from whole grains using buffer containing 50 mM Tris–Cl (pH 7.5), 150 mM NaCl, 1% Triton X-100, and 1 mM PMSF, and then separated by 10% SDS-PAGE. The PDIL1-1 protein was detected by immunoblotting with anti-PDIL1-1 antibody as described previously (Kim et al. 2012).

#### Abbreviations

CHT14: chitinase; *E. coli*: *Escherichia coli*; EXPLA1: EXPANSIN-LIKE 1; GAPDH: glyceraldehyde 3-phosphate dehydrogenase; GFP: green fluorescence protein; GST: glutathione *S*-transferase; GUS: β-glucuronidase; GW2: grain width 2; MALDI-TOF: matrix assisted laser desorption/ionization time-of-flight mass spectrometry; MBP: maltose binding protein; PAGE: polyacrylamide gel electrophoresis; PDIL1-1: protein disulfide isomerase-like 1-1; PGK: phosphoglycerate kinase; RING: real interesting new gene; RT-PCR: real time polymerase chain reaction; SDS: sodium dodecyl sulfate.

#### Authors' contributions

HSS designed the project. YJK and MK carried out experiments. KHL, SWP, YJK, JTS and HSS analyzed and interpreted the data. KHL, SWP and HSS wrote the manuscript. All authors commented on the results and the manuscript. All authors read and approved the final manuscript.

#### Author details

[1] Department of Plant Science, College of Agriculture and Life Sciences, Research Institute of Agriculture and Life Sciences, Seoul National University, Gwanakro 200, Gwanak-gu, Seoul 151-921, South Korea. [2] Department of Biological Chemistry, Chonnam National University, Gwangju 61186, South Korea. [3] School of Applied Biosciences, Kyungpook National University, Daegu 41566, South Korea.

#### Acknowledgements

We appreciate Dr. Hee Jong Koh for providing wild-type Norin 22 rice.

#### Competing interests

The authors declare that they have no competing interests.

#### Funding

This work was supported by a grant from the Next-Generation BioGreen 21 Program (Plant Molecular Breeding Center No. PJ01327601), Rural Development Administration, Republic of Korea. This work was also supported by a grant from the Next-Generation BioGreen 21 Program (Plant Molecular Breeding Center No. PJ01330802), Rural Development Administration, Republic of Korea and supported by Basic Science Research Program through the National Research Foundation of Korea (NRF) funded by the Ministry of Education (Grant Number: 2015057573).

#### References

Bai X, Luo L, Yan W, Kovi MR, Zhan W, Xing Y (2010) Genetic dissection of rice grain shape using a recombinant inbred line population derived from two contrasting parents and fine mapping a pleiotropic quantitative trait locus qGL7. BMC Genet 11:16

Bie P, Ciechanover A (2011) Ubiquitination of E3 ligases: self-regulation of the ubiquitin system via proteolytic and non-proteolytic mechanisms. Cell Death Differ 18:1393–1402

Bradford MM (1976) A rapid and sensitive method for the quantitation of microgram quantities of protein utilizing the principle of protein-dye binding. Anal Biochem 72:248–254

Brinkmann H, Martin W (1996) Higher-plant chloroplast and cytosolic 3-phosphoglycerate kinases: a case of endosymbiotic gene replacement. Plant Mol Biol 30(1):65–75

Cho K, Torres NL, Subbramanyam S, Deepak SA, Sardesai N, Han O, Williams CE, Ishii H, Iwahashi H, Rakwal R (2006) Protein extraction/solubilization protocol for monocot and dicot plant gel-based proteomics. J Plant Biol 49:413–420

Choi BS, Kim YJ, Markkandan K, Koo YJ, Song JT, Seo HS (2018) GW2 functions as an E3 ubiquitin ligase for rice expansin-like 1. Int J Mol Sci 19(7):E1904

Ciechanover A (1994) The ubiquitin-proteasome proteolytic pathway. Cell 79(1):13–21

Fan C, Xing Y, Mao H, Lu T, Han B, Xu C, Li X, Zhang Q (2006) GS3, a major QTL for grain length andweight and minor QTL for grain width and thickness in rice, encodes a putative transmembrane protein. Theor Appl Genet 12:1164–1171

Kelley DR (2018) E3 ubiquitin ligases: key regulators of hormone signaling in plants. Mol Cell Proteomics 17(6):1047–1054

Kim YJ, Choi SH, Han YJ, Jang IC, Song JT, Bhoo SH, Seo HS (2012) Rice protein disulfide isomerase (OsPDI) and rice chaperone binding protein (OsBiP) levels are not directly correlated with the eating quality of rice. J Korean Soc Appl Biol Chem 53:508–511

Li Q, Yang X, Bai G, Warburton ML, Mahuku G, Gore M, Dai J, Li J, Yan J (2010) Cloning and characterization of a putative GS3 ortholog involved in maize kernel development. Theor Appl Genet 120:753–763

Nakahara KS, Masuta C (2014) Interaction between viral RNA silencing suppressors and host factors in plant immunity. Curr Opin Plant Biol 20:88–95

Shomura A, Izawa T, Ebana K, Ebitani T, Kanegae H, Konishi S, Yano M (2008) Deletion in a gene associated with grain size increased yields during rice domestication. Nat Genet 40:1023–1028

Song XJ, Huang W, Shi M, Zhu MZ, Lin HX (2007) A QTL for rice grain width and weight encodes a previously unknown RING-type E3 ubiquitin ligase. Nat Genet 39:623–630

Su Z, Hao C, Wang L, Dong Y, Zhang X (2011) Identification and development of a functional marker of TaGW2 associated with grain weight in bread wheat (*Triticum aestivum* L.). Theor Appl Genet 122(1):211–223

Vierstra RD (2009) The ubiquitin-26S proteasome system at the nexus of plant biology. Nat Rev Mol Cell Biol 10(6):385–397

Watson HC, Walker NP, Shaw PJ, Bryant TN, Wendell PL, Fothergill LA, Perkins RE, Conroy SC, Dobson MJ, Tuite MF (1982) Sequence and structure of yeast phosphoglycerate kinase. EMBO J 1(12):1635–4160

Wilkinson B, Gilbert HF (2004) Protein disulfide isomerase. Biochim Biophys Acta 1699:35–44

Yeu SY, Park BS, Sang WG, Choi YD, Kim MC, Song JT, Paek NC, Koh HJ, Seo HS (2007) The serine proteinase inhibitor OsSerpin is a potent tillering regulator in rice. J Plant Biol 50(5):600–604

# 15

# Age and site-specific pattern on encrustation of charophytes

Anne Herbst* and Hendrik Schubert

**Abstract**

Encrustation and element content (Ca, Fe, K, Mg and P) of charophytes was studied along plant thalli to investigate the dependency of thallus age and site-specificity. Charophytes were collected from five sampling sites (Angersdorfer Teiche, Asche, Bruchwiesen, Krüselinsee and Lützlower See) which were distinct with respect to water chemistry. Furthermore, photosynthesis was measured to identify the physiological state of plants in habitat waters and with the addition of different ion concentrations ($Ca^{2+}$, $K^{+}$, $Mg^{2+}$ and $Na^{+}$). Age pattern on encrustation of charophytes was site-specific: carbonate content increased from the youngest to the oldest part (Angersdorfer Teiche), younger parts were less encrusted than older parts in Asche, Bruchwiesen and Krüselinsee, whereas encrustation in Lützlower See was the same along plants thallus. Charophytes showed species-specific encrustation in investigated sites. Encrustation of *C. hispida* in Angersdorfer Teiche was also as high as of individuals from hard-water lakes irrespective of 10.15 mS $cm^{-1}$ (salinity of 6.3). For species growing in Angersdorfer Teiche, K/Na content and photosynthesis was lowest when compared to other sites. Photosynthesis of charophytes was enhanced after the addition of KCl and adversely affected by $CaCl_2$, $MgCl_2$ and NaCl. In summary, it was shown that encrustation of charophytes in water sites with strong ion anomalies could be as high as in hard-water lakes. It is assumed that ion composition, rather than ion concentration of $Na^{+}$, $Mg^{2+}$ and $SO_4^{2-}$, impact on the encrustation of charophytes. The age pattern on encrustation in this study showed a strong site-specificity, whereas encrustation of charophytes was species-specific. Ion concentrations, either of habitats or actively added in laboratory measurements, impact on encrustation, element content and photosynthesis of charophytes.

**Keywords:** Carbonate precipitation, Age gradient, Plant thallus, Photosynthesis, Potassium

## Background

Charophytes are macrophytes of the order Charales Lindley 1836 belonging to the streptophyta (Jeffrey 1967; Stewart and Mattox 1975). This lineage also includes the embryophyta, thus charophytes are closely related to land plants (Jeffrey 1967; McCourt 1995; Nishiyama et al. 2018). Charophytes can form dense meadows which provide habitats and shelter for epiphytes, invertebrates and fish, indicating their important ecological function in the ecosystem (Kairesalo et al. 1987; Kingsford and Porter 1994; Kuczyńska-Kippen 2007). The presence or absence of charophytes seems to depend on abiotic lake characteristics; water chemistry especially plays an important role (van den Berg et al. 1998a; Blindow et al. 2014). They often occur in waters with low nutrient concentrations. Therefore, some charophytes can serve as bioindicators for nutrient-poor conditions (Krause 1981; Doege et al. 2016). In clear-water ecosystems charophytes are more competitive than other macrophytes but are outcompeted under eutrophic conditions (Ozimek and Kowalczewski 1984; Pieczyńska et al. 1988; Blindow 1992). However, charophytes are not limited to oligotrophic waters, also inhabiting eutrophic waters when niches are available (van den Berg et al. 1998b; Schubert et al. 2018). In hard-water lakes, charophytes encrust on the plant surface (Wahlstedt 1875; Sand-Jensen et al. 2018). Calcium carbonate content can thereby account for up to 80% of plant dry weight (Pukacz et al. 2016a). In the case of heavy encrustation of charophytes, it has been described that supersaturation with calcium carbonate

*Correspondence: anne.herbst2@uni-rostock.de
University of Rostock, Biosciences, Albert-Einstein-Straße 3, 18059 Rostock, Germany

of the water was not required (Nõges et al. 2003; Kufel et al. 2016). Several authors pointed out the importance of site-specificity, and the influence that depth, temperature, pH, and water chemistry parameters have on the encrustation of charophytes (Kufel et al. 2013, 2016; Pukacz et al. 2016a, b). So far age dependent encrustation was analyzed by Kawahata et al. (2013); younger internodes of *C. globularis* Thuill. 1799 were less encrusted than older ones. In situ studies on encrustation regarding age and site-specificity have never been done. Therefore, encrustation was analyzed in parts of corticated thallus for *C. canescens* Loisel. 1810, *C. hispida* L. 1753, *C. subspinosa* Rupr. 1846 and *C. tomentosa* L. 1753 from five areas of water in Germany. We investigated whether carbonate and element (Ca, Fe, K, Mg and P) contents of plant dry weight (DW) are dependent on thallus age and sites sampled. Furthermore, photosynthesis was measured to identify the physiological state of plants in habitat waters and with the addition of different ion concentrations ($Ca^{2+}$, $K^{+}$, $Mg^{2+}$ and $Na^{+}$).

## Materials and methods

### Sampling sites

In July 2016 five areas of water were investigated in the northeast and eastern part of Germany (Fig. 1). These included two residual mining holes: Angersdorfer Teiche, a clay pit filled with rain and groundwater, and Asche in Teutschenthal, which is located close to a potassium mining heap, in the lake area Mansfelder region. Bruchwiesen, a natural spring, in Bad Tennstedt was also sampled, as well as two hard-water lakes, Krüselinsee in Mecklenburg Lake District and Lützlower See in Uckermark.

### Sampling processing

Plant material was collected at a depth of 0.5–1.5 m from the shore, or by snorkeling. Undamaged plants of *C. canescens*, *C. hispida*, *C. subspinosa*, and *C. tomentosa* were picked. In the laboratory, epibionts were removed gently with a brush. For further analyses, plants were analyzed in parts of thalli. First to third, forth to sixth, seventh to ninth whorls and internodes were used for analysis (Fig. 2). Side branches and other parts of thalli were removed.

Carbonate proportion of plant dry weight was determined by the loss of ignition (LOI) method as described by Heiri et al. 2001. Sample size was n = 7 for the LOI analysis. The samples were dried at 60 °C (UM 400, Memmert) overnight. Plant DW was weighed and combusted in crucibles in a two-step process, at 550 °C ($DW_{550}$) and 925 °C ($DW_{925}$) for two hours, respectively (LE 6/11/B 150, Nabertherm). Carbonate content of DW (%) was calculated of the carbonate bounded $CO_2$ loss multiplied by 1.36 (fraction of $CO_2$ in $CO_3^{2-}$) after Pełechaty et al. 2013 (Eq. 1).

$$\text{carbonate content of DW (\%)} = \left(\frac{DW550 - DW925}{DW925} \times 100\right) \times 1.36 \quad (1)$$

Element contents of Ca, Fe, K, Mg and P were measured with inductively coupled plasma-optic emission spectroscopy (ICP-OES). Plant DW was ground and 0.1 g was weighted for the digestion. Samples were extracted in duplicate with 5 ml $HNO_3$ (65%) and 3 ml $H_2O_2$ (30%) for 1.5 h in CEM MARS 6 microwave. After extraction, samples were filled up to 25 ml with ultrapure water and were filtered (601P, Rotilabo). Samples were analyzed spectrometrically for Ca (317.9 nm), Fe (238.2 nm), K (766.5 nm), Mg (285.2 nm) and P (214.9 nm) with Optima 8300 (Perkin Elmer, Singapore).

Photosynthesis of *C. canescens*, *C. hispida* from Angersdorfer Teiche, *C. hispida* from Asche, and *C. tomentosa* from Bruchwiesen was measured by means of $O_2$ evolution with a Clark-type electrode (Microelectrodes, USA). Within a day after sampling, plants were dark adapted at least for 30 min and were set in a temperature-controlled 2.5 ml cuvette (15 °C). The cuvette was filled with habitat medium which was stirred during measurement. Photon flux densities were applied by a light dispenser system (MK2, Illuminova, Sweden) described by Wolfstein and Hartig (1998). Dark respiration was measured for 10 min, followed by increasing irradiation levels of 14, 18, 31, 102, 494, 912, 1462, 1811, 2120 μmol photons $m^{-2}\ s^{-1}$. Each irradiation level was set for 3.3 min; measurement took 40 min in total.

Photosynthesis of *C. tomentosa* from Asche was measured with the addition of $CaCl_2$, KCl, $MgCl_2$ and NaCl. Plants were cultivated in habitat water and at a constant temperature (15 °C) and light (50–80 μmol photons $m^{-2}\ s^{-1}$ on a 12 h/12 h light–dark cycle) to allow consistent conditions for measurements. Light saturation irradiance ($I_s$) of cultivated *C. tomentosa* was calculated in preliminary measurements ($I_s = 120$ μmol $m^{-2}\ s^{-1}$). For 20 min measurements, $1.5 \times I_s$ ($1.5I_s = 180$ μmol $m^{-2}\ s^{-1}$) was used to gain light saturation. After a dark period of 3 min, constant irradiance level of 180 μmol $m^{-2}\ s^{-1}$ was set. Concentration of 1.2 mmol $L^{-1}$ of chloride salt was added at minutes 5, 7, 9, 11, 13, 15, and 17 of measurement. The end concentration was 8.4 mmol $L^{-1}$ added chloride salt in the cuvette. Control measurements were conducted without addition of chloride salt.

Photosynthesis/irradiance (P/I) curve was calculated using the model of Walsby (1997) to determine photosynthesis (P), maximum photosynthesis ($P_m$), dark respiration ($R_d$), $\alpha$ (slope of P/I curve at limiting irradiance), light

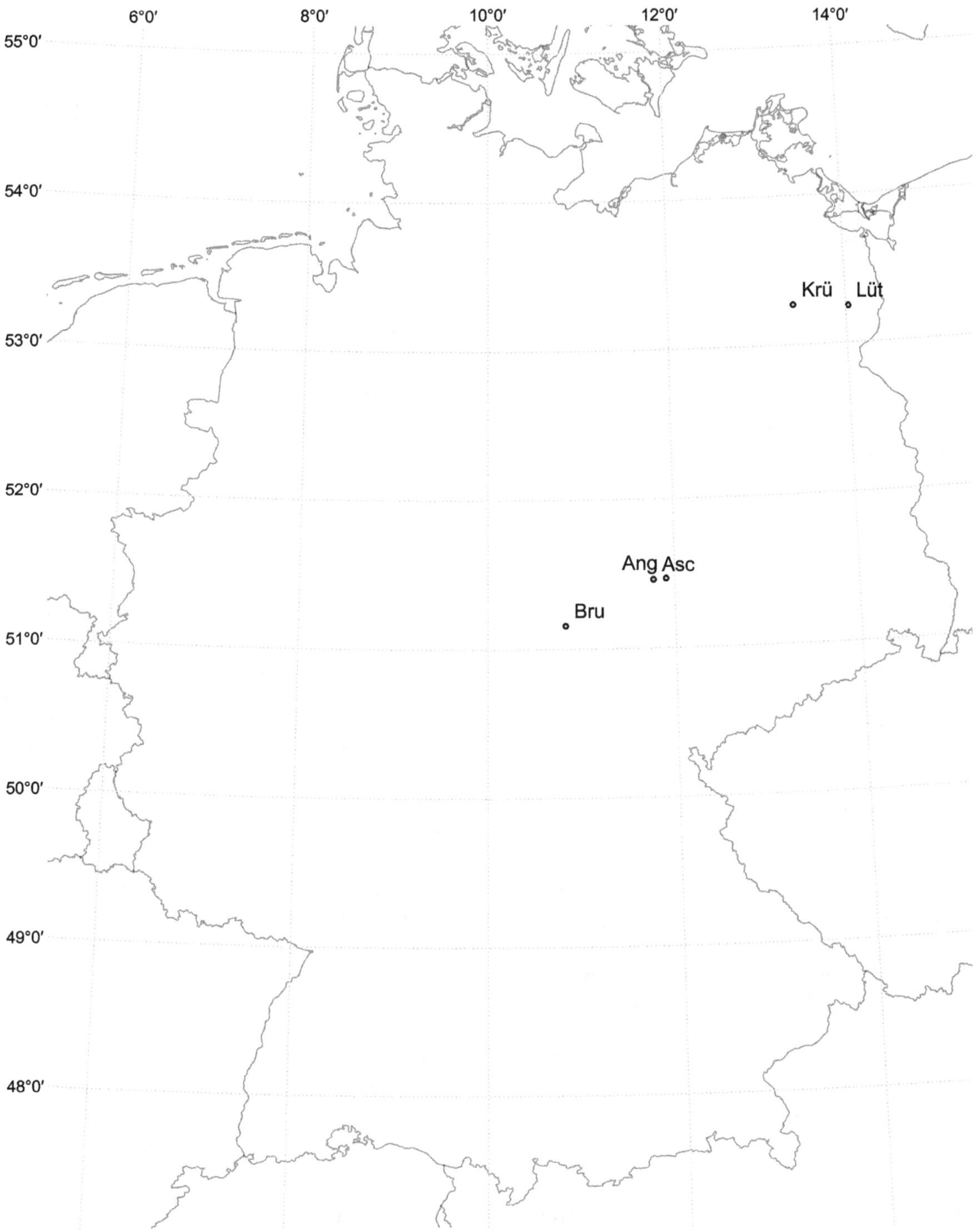


| site | abbreviation | latitude | longitude |
|---|---|---|---|
| Angersdorfer Teiche | Ang | 51°28'04.5" N | 11°54'24.6" E |
| Asche | Asc | 51°27'40.0" N | 11°46'18.4" E |
| Bruchwiesen | Bru | 51°09'25.3" N | 10°49'36.0" E |
| Krüselinsee | Krü | 53°16'02.3" N | 13°24'59.7" E |
| Lützlower See | Lüt | 53°14'44.7" N | 14°01'52.2" E |

**Fig. 1** Map of investigated sampling sites (QGIS version 3.2, GeoBasis-DE/BKG 2018). Sites are abbreviated as first three letters: Krüselinsee (Krü), Lützlower See (Lüt), Angersdorfer Teiche (Ang), Asche (Asc) and Bruchwiesen (Bru)

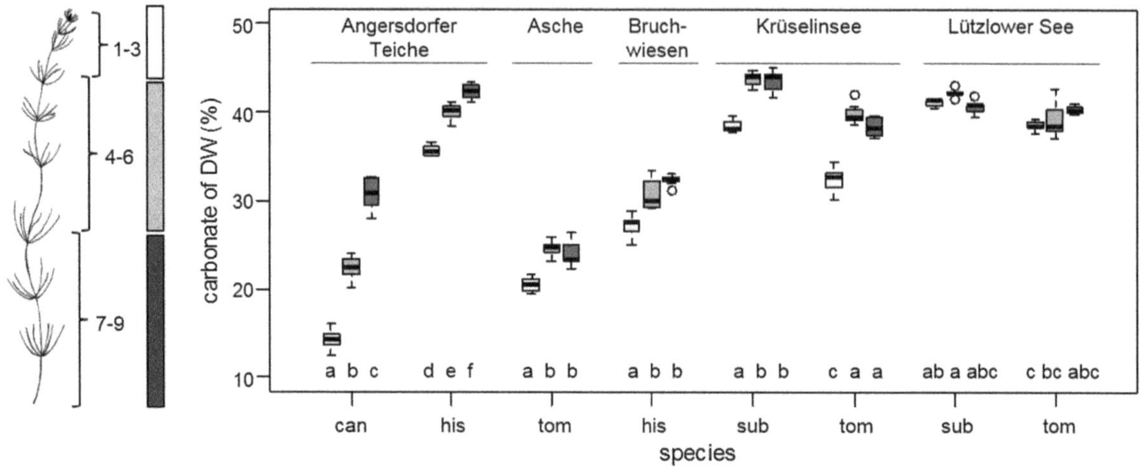


**Fig. 2** Comparison of carbonate content based on DW (%) of *C. canescens* (can), *C. hispida* (his), *C. subspinosa* (sub), and *C. tomentosa* (tom) from sampling sites (Angersdorfer Teiche, Asche, Bruchwiesen, Krüselinsee, and Lützlower See). Plants were analyzed in first to third (white), forth to sixth (light grey), seventh to ninth (dark grey boxplots) whorls and internodes. Box plots include whiskers (5–95% of variability) and outliers (points). Different letters indicate significant difference within the same sampling site (Tukey HSD post hoc test, $p < 0.05$)

compensation point ($I_c = R_d/\alpha$), and light saturation irradiance ($I_s = P_m/\alpha$). Photosynthetic parameters were based on chlorophyll *a* (chl *a*). Pigment content of first to third whorls and internodes were extracted with 3 ml of *N,N*-dimethylformamide. Samples were incubated overnight at 4 °C. Extinction at 470, 630, 647, 664, and 750 nm was measured using a Lambda 2 spectrometer (Perkin Elmer, Germany). Chl *a* content was calculated after Wellburn (1994).

Water samples were taken in 50 ml tubes above plant meadows, closely beneath the water surface. pH value and conductivity were measured directly at sampling sites with HACH HQ40d (Hach-Lange, Germany). The water samples were analyzed at Helmholtz Centre for Environmental Research (UFZ)—Central Laboratory for Water Analytics and Chemometrics. The concentrations of $Ca^{2+}$, $Cl^-$, $K^+$, $Mg^{2+}$, $Na^+$, and $SO_4^{2-}$ were measured with an ion chromatograph (Dionex ICS 3000, Thermo Fisher, USA). Total inorganic carbon (TIC) was determined with a Dimatoc 2000 (DIMATEC, Germany). Water samples were measured in accordance with EN ISO 14911:1999 and EN ISO 10304:2009-1.

### Statistical analyses

Data were analyzed for normality and homogeneity. In cases of normal and homogeneous distribution, data were analyzed with an analysis of variance (ANOVA) and post hoc Tukey HSD test. For no normal distributed data, a non-parametric test (Kruskal–Wallis test and Fisher Least Significant Difference (LSD) post hoc test) with Bonferroni correction for p-values adjustment was performed (de Mendiburu 2016). A principal component analysis (PCA) based on standardized data of element contents of Ca, Fe, K, Mg and TP of plant DW was conducted (Oksanen et al. 2017). Significance levels were set to a p-value $\leq 0.05$. All statistical analyses were performed using R (R Core Team 2017).

## Results

### Water chemistry

Water chemistry data of investigated sites are listed in Table 1. The sampling sites were distinct with respect to water chemistry, exhibiting enhanced ion concentrations in Angersdorfer Teiche, Asche, and Bruchwiesen.

**Table 1 Water chemistry data of sampling sites investigated and species sampled**

| Site | Species | pH | Cond. | $Ca^{2+}$ | $Cl^-$ | $K^+$ | $Mg^{2+}$ | $Na^+$ | $SO_4^{2-}$ | TIC |
|---|---|---|---|---|---|---|---|---|---|---|
| Angersdorfer Teiche | can, his | 9.1 | 10.15 | 660 | 2390 | 82.5 | 455 | 1560 | 2550 | 15.7 |
| Asche | tom | 7.9 | 5.03 | 431 | 472 | 33.7 | 462 | 626 | 2380 | 38.1 |
| Bruchwiesen | his | 7.4 | 1.39 | 276 | 41 | 3.0 | 51 | 23 | 577 | 64.1 |
| Krüselinsee | sub, tom | 9.3 | 0.26 | 29 | 18 | 2.4 | 6.2 | 9.3 | 29 | 22.2 |
| Lützlower See | sub, tom | 9.0 | 0.80 | 91 | 106 | 3.7 | 21 | 28 | 152 | 30.0 |

Sites and species are abbreviated as first three letters. Listed are pH, and conductivity ($mS\ cm^{-1}$) and concentrations of $Ca^{2+}$, $Cl^-$, $K^+$, $Mg^{2+}$, $Na^+$, $SO_4^{2-}$, and TIC ($mg\ L^{-1}$)

The highest ion concentrations of $Ca^{2+}$, $Cl^{-}$, $K^{+}$, $Na^{+}$ and $SO_4^{2-}$ were found in the gips pit Angersdorfer Teiche. $Mg^{2+}$ was an exception and the highest concentration was obtained in Asche. Bruchwiesen had high $Ca^{2+}$ and $SO_4^{2-}$ concentrations compared to Krüselinsee and Lützlower See. Both hard-water lakes had the lowest conductivity of investigated waters. Total inorganic carbon (TIC) concentration was highest in Bruchwiesen and lowest in Angersdorfer Teiche. pH was around 9 in Angersdorfer Teiche, Krüselinsee and Lützlower See, 7.9 in Asche, and lowest in Bruchwiesen with 7.4.

## Encrustation

Encrustation was analyzed in parts of plant thalli (Fig. 2). A pattern of lowest carbonate content in the first to third, middle in the forth to sixth, and highest in the seventh to ninth whorls and internodes was found for both species (*C. canescens* and *C. hispida*) from Angersdorfer Teiche (Fisher LSD post hoc test, $p < 0.01$). In Asche, Bruchwiesen, and Krüselinsee, the younger parts of *C. hispida*, *C. subspinosa*, and *C. tomentosa* were less encrusted than the middle and older parts of the plant thalli (Fisher LSD post hoc test, $p < 0.001$). In Lützlower See, *C. subspinosa* and *C. tomentosa* exhibited no significant differences in encrustation between their respective parts of the thalli.

Furthermore, encrustation was species-specific for species in Angersdorfer Teiche. *C. hispida* encrusted higher than *C. canescens*. Site-specific encrustation was found for *C. hispida* growing in Angersdorfer Teiche and Bruchwiesen with a lower carbonate proportion of *C. hispida* in Bruchwiesen (Fisher LSD post hoc test, $p < 0.01$). Also *C. tomentosa* encrustation was site-specific in Asche compared to individuals from Krüselinsee and Lützlower See. Encrustation of *C. tomentosa* was lower in Asche than in Krüselinsee and Lützlower See (Fisher LSD post hoc test, $p < 0.001$).

## Element composition

Element content based on plant dry weight was analyzed for species from habitats with high ion concentrations (Angersdorfer Teiche, Asche, and Bruchwiesen) to investigate the impact thereof (Fig. 3 and Table 2). In the PCA plot, *C. hispida* from Bruchwiesen was separated from the species of Angersdorfer Teiche and Asche. Habitat differences were pronounced by means of lowest contents of Mg and Na in *C. hispida* from Bruchwiesen (Table 2). Further, parts of the thallus were ordered along the axis 1 which was characterized most by Ca, P, and Na contents.

Regarding the element content, Ca was lowest in *C. canescens* and highest in *C. hispida*, both from Angersdorfer Teiche. K content was significantly higher in

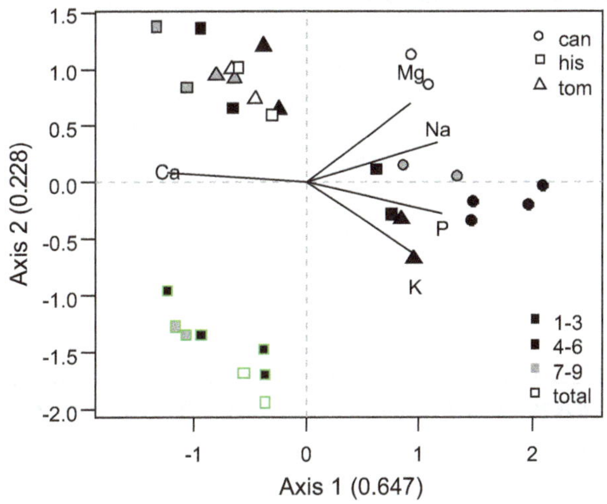


**Fig. 3** PCA plot of element contents (Ca, K, Mg, Na, and P) of charophytes from Angersdorfer Teiche (blue), Asche (red), and Bruchwiesen (green). Species were plotted with different symbols; background of symbols indicates parts of thallus, see legend

*C. canescens* compared to *C. hispida* (Angersdorfer Teiche) and *C. tomentosa* (Asche). Na content was different for all species investigated; highest in *C. canescens* (Angersdorfer Teiche) and lowest in *C. hispida* (Bruchwiesen). The ratio of K/Na was low (< 1) for both species from Angersdorfer Teiche and highest in *C. hispida* from Bruchwiesen. P content was significantly higher in *C. canescens* compared to *C. hispida* (Angersdorfer Teiche) and *C. tomentosa* (Asche).

## Photosynthesis

Photosynthesis was measured in plants from Angersdorfer Teiche, Asche and Bruchwiesen (Table 3). $P_m$ of *C. canescens* ($33.8 \pm 9.6$ 1 mmol $O_2$ $h^{-1}$g chl $a^{-1}$) and *C. hispida* ($42.1 \pm 6.1$ mmol $O_2$ $h^{-1}$ g chl $a^{-1}$) from Angerdorfer Teiche were not significantly different to *C. tomentosa* from Asche ($64.3 \pm 20.2$ mmol $O_2$ $h^{-1}$ g chl $a^{-1}$) and *C. hispida* from Bruchwiesen ($64.7 \pm 18.8$ mmol $O_2$ $h^{-1}$ g chl $a^{-1}$). α, $R_d$ and $I_s$ were also similar for species measured. Significant difference was found for the light compensation point; *C. canescens* from Angersdorfer Teiche had significantly lower $I_c$ than *C. hispida* from Bruchwiesen.

Photosynthesis of *C. tomentosa* from Asche was measured with the addition of $CaCl_2$, KCl, $MgCl_2$, NaCl and without any addition, as a control (Table 4). Differences in $O_2$ evolution was found with addition of $CaCl_2$, KCl and $MgCl_2$, NaCl at concentrations of 2.4, 3.6, 4.8 and 6.0 mmol $L^{-1}$. $O_2$ evolution of *C. tomentosa* was higher with the addition of KCl and $CaCl_2$ compared to $MgCl_2$ and NaCl. At 8.4 mmol $L^{-1}$ $O_2$ evolution of *C. tomentosa* was lowest with the addition of NaCl.

**Table 2 Comparison of element contents (mean ± SD) of charophytes from Angersdorfer Teiche, Asche, and Bruchwiesen**

| Site | Species | Ca | K | Mg | Na | P | K/Na |
|---|---|---|---|---|---|---|---|
| Angersdorfer Teiche | can | 94 ± 34.4 c | 16.4 ± 1.7 a | 12.6 ± 0.8 a | 26.0 ± 2.3 a | 1.6 ± 0.4 a | 0.6 ± 0.1 d |
| | his | 251.3 ± 41.8 a | 11.2 ± 3.5 b | 10.7 ± 0.7 b | 14.7 ± 4.2 b | 0.9 ± 0.4 b | 0.8 ± 0.1 c |
| Asche | tom | 158 ± 35.8 b | 11.7 ± 3.6 b | 11.7 ± 0.7 a | 8.4 ± 2.3 c | 0.9 ± 0.4 b | 1.4 ± 0.1 b |
| Bruchwiesen | his | 215.4 ± 17.3 a | 14.7 ± 2.2 ab | 6.3 ± 0.6 c | 2.9 ± 1.0 d | 1.0 ± 0.3 ab | 5.5 ± 1.4 a |

Ca, K, Na, Mg, and P contents are presented as g $kg^{-1}$ DW; K/Na mass ratios are calculated from weight. Species are abbreviated: *C. canescens* (can), *C. hispida* (his), and *C. tomentosa* (tom). Sample size was n = 8 for the element analysis. Different letters indicate significant differences in element content (Tukey HSD post hoc test, $p < 0.05$)

**Table 3 Comparison of photosynthetic parameters of *C. canescens* (can), *C. hispida* (his), and *C. tomentosa* (tom) from Angersdorfer Teiche (Ang), Asche (Asc) and Bruchwiesen (Bru)**

| Sample | $P_m$ | $a$ | $R_d$ | $I_c$ | $I_s$ |
|---|---|---|---|---|---|
| Ang_can | 33.8 ± 9.6 a | 0.3 ± 0.1 a | − 12.5 ± 4.5 a | 40.2 ± 7.9 a | 111.4 ± 23.6 a |
| Ang_his | 42.1 ± 6.1 a | 0.5 ± 0.1 a | − 21.9 ± 6.9 a | 47.6 ± 1.6 ab | 97.0 ± 28.8 a |
| Asc_tom | 64.3 ± 20.2 a | 0.3 ± 0.3 a | − 21.2 ± 6.9 a | 98.1 ± 47.5 ab | 347.0 ± 199.3 a |
| Bru_his | 64.7 ± 18.8 a | 0.2 ± 0.0 a | − 14.1 ± 1.0 a | 83.3 ± 11.7 b | 386.0 ± 142.6 a |

Maximum photosynthesis ($P_m$), dark respiration ($R_d$) in mmol $O_2$ $h^{-1}$ g chl $a^{-1}$ and light compensation point ($I_c$), light saturation irradiance ($I_s$) in µmol photons $m^{-2}$ $s^{-1}$ are presented. $a$ in mmol $O_2$ $h^{-1}$ g chl $a^{-1}$ (µmol photons $m^{-2}$ $s^{-1})^{-1}$ is the slope of the P/I curve at limiting irradiance. Sample size was n = 4 for the 40 min measurement of photosynthesis. Different letters indicate significant differences within photosynthetic parameters (Tukey HSD post hoc test, $p < 0.05$)

**Table 4 Difference in $O_2$ evolution of *C. tomentosa* with the addition of $CaCl_2$, KCl, $MgCl_2$ and NaCl referred to control (without addition)**

| Addition | 1.2 | 2.4 | 3.6 | 4.8 | 6.0 | 7.2 | 8.4 |
|---|---|---|---|---|---|---|---|
| $CaCl_2$ | 2.1 ± 2.8 a | 0.1 ± 3.6 a | − 1.2 ± 3.7 a | − 1.2 ± 3.6 a | − 1.3 ± 3.5 a | − 1.0 ± 3.9 a | − 0.2 ± 3.6 a |
| KCl | − 1.0 ± 3.3 b | 0.0 ± 4.0 a | 1.4 ± 4.9 b | 1.4 ± 5.2 a | 1.1 ± 5.5 a | 0.9 ± 5.7 ab | 1.2 ± 5.9 a |
| $MgCl_2$ | − 0.6 ± 2.3 b | − 3.8 ± 2.3 b | − 4.6 ± 2.9 c | − 4.0 ± 3.6 b | − 3.4 ± 4.2 b | − 2.7 ± 4.3 bc | − 1.7 ± 4.5 a |
| NaCl | − 0.9 ± 3.7 b | − 5.1 ± 5.5 b | − 6.8 ± 5.4 c | − 6.2 ± 5.8 b | − 5.9 ± 5.6 b | − 5.3 ± 5.9 c | − 4.1 ± 5.6 b |

1.2 mmol $L^{-1}$ of $CaCl_2$, KCl, $MgCl_2$ and NaCl was added stepwise. $O_2$ evolution (mmol $O_2$ $h^{-1}$ g chl $a^{-1}$) is presented as accumulated concentrations (1.2, 2.4, 3.6, 4.8, 6.0, 7.2 and 8.4 mmol $L^{-1}$). Sample size was n = 8 for the 20 min measurement of photosynthesis. Different letters indicate significant differences within a concentration level (Fischer LSD post hoc test, $p < 0.05$)

## Discussion

Encrustation was studied in brackish waters with strong ion anomalies. Increased ion concentrations were found in Angersdorfer Teiche, Asche and Bruchwiesen. For comparison, samples were also taken in two hard-waters lakes; Krüselinsee and Lützlower See. Carbonate precipitation was highest for charophytes in hard-water lakes and of *C. hispida* in Angersdorfer Teiche. This was not expected for *C. hispida* growing in Angersdorfer Teiche at a salinity of 6.3. Herbst et al. (2018) quantified encrustation of charophytes from freshwater and brackish water sites. Freshwater specimens had higher carbonate content of DW compared to brackish water specimens (Herbst et al. 2018). Thus, ion composition rather than ion concentration, is decisive for encrustation of charophytes. It could further explain the low carbonate precipitation of *C. tomentosa* in Asche, where $Mg^{2+}$ exceeded $Ca^{2+}$ concentrations, and consequently inhibits the encrustation as described by other authors (Siong and Asaeda 2009; Gomes and Asaeda 2010; Asaeda et al. 2014). Additionally, high $SO_4^{2-}$ concentrations in Asche and Bruchwiesen could reduce the encrustation of charophytes as Akin and Lagerwerff (1965) reported increased solubility of calcium carbonate at high $SO_4^{2-}$ concentrations.

The results emphasize the importance of water ion composition on carbonate precipitation of charophytes, which was strengthened regarding the site-specific pattern on encrustation along the age gradient of the plants. Here 'age' referred to the cell age along the plant thallus at a certain state. Site-specific patterns of encrustation were found: all parts of plant thalli (youngest, middle, and oldest) were different; the youngest differed from the other parts, and encrustation was similar for all parts of the thalli. Kawahata et al. (2013) confirmed age dependent encrustation along internodes of *C. globularis*. Younger internodes were less encrusted than older ones (Kawahata et al. 2013). Carbonate formations were also detected between the cortex and cell wall which indicated that encrustation was enclosed during forming of the cortex (Kawahata et al. 2013). Three levels of carbonate formation can therefore be distinguished; external precipitation, which can occur outside as well as intrathallose and internal calcification of oospores which are termed gyrogonites (Raven et al. 1986; Leitch 1991; Kawahata et al. 2013).

The age pattern on encrustation in this study showed a strong site-specificity which probably arise by combination of the plant growth rate and carbonate formation rate. In both hard-water lakes, Lützlower See and Krüselinsee, age pattern on encrustation differed. The youngest parts of *C. subspinosa* and *C. tomentosa* showed less encrustation than other parts in Krüselinsee compared to individuals from Lützlower See. In Lützlower See, encrustation was similar along plant thalli. Slow growth rate and fast encrustation rate, due to higher $Ca^{2+}$ concentration in Lützlower See, are likely to produce uniform encrustation of the whole plant.

However, in Bruchwiesen and Asche the youngest parts of plant thalli were also less encrusted than other parts, even under extreme high $Ca^{2+}$ concentrations. Both study sites were also influenced by high concentrations of $Mg^{2+}$ and $SO_4^{2-}$, which could adversely affect the carbonate formation rate. Furthermore, in the spring Bruchwiesen, temperatures did not exceed 12 °C in summer when sampling took place. Temperature, as another influencing factor, was shown to influence the precipitation of calcium carbonate (Ogata et al. 2010). Further investigations about habitat influence on encrustation of charophytes are needed to elucidate the interaction of various factors.

In Angersdorfer Teiche, the most pronounced difference in encrustation between species (*C. hispida* and *C. canescens*) and cell age (youngest, middle, and oldest part) was detected. The strong difference in species-specific encrustation could be explained by its low K/Na contents, which was 0.8 for *C. canescens* and 0.6 for *C. hispida* (Table 2). Winter and Kirst (1990) reported reduced vitality of charophytes at a ratio of $K^+/Na^+ < 1$, but found also ratios of 0.8–0.9 in *Tolypella glomerata* Leonh. 1863 and *T. nidifica* Braun 1856 (Winter et al. 1996) when exposed to a salinity of 12. Both *C. hispida*, as well as *C. canescens*, were probably growing close to their physiological limits in Angersdorfer Teiche. Photosynthesis, as a measure of the physiological state, confirmed lower $P_m$ in plants (*C. canescens* and *C. hispida*) from Angersdorfer Teiche compared to those from Asche (*C. tomentosa*) and Bruchwiesen (*C. hispida*).

It is known that $CO_2$ uptake for photosynthesis is connected to encrustation of charophytes (McConnaughey and Falk 1991; Raven and Beardall 2016). $H^+$ pumping acidified the plant surface and increases $CO_2$ concentration in the surrounding water which permeates into the cells (Beilby and Bisson 2012). The import of $CO_2$ into the chloroplast generates $OH^-$ which is excreted by the cell through $OH^-$ channels (Beilby and Bisson 2012; Absolonova et al. 2018). Alkaline zones are formed on the plant surface where carbonate is precipitated. Beside lower photosynthesis of plants from Angersdorfer Teiche, concentration of total inorganic carbon was also lowest of investigated sites. The results imply that carbonate uptake for photosynthesis is not the only mechanism responsible for encrustation of charophytes. $H^+$ is also generated by active ion transport which is linked to nutrient uptake and ion regulation (McConnaughey and Whelan 1997; Beilby and Casanova 2014). Consequently, ion regulation and osmoregulation capabilities will interfere with photosynthetic $CO_2$ recruitment and it can be argued, that the low encrustation values exhibited by *C. canescens*, the only brackish water species in this study, are caused by its different osmoregulation mechanism (Winter and Kirst 1991).

Photosynthesis was affected by the addition of $CaCl_2$, KCl, $MgCl_2$ and NaCl. Addition of $CaCl_2$, $MgCl_2$ and NaCl inhibited photosynthesis; NaCl had the most adverse effect. $Na^+$ toxicity is well-known; an exception is cyanobacteria which require $Na^+$ for the activity of nitrogenase (Thomas and Apte 1984). Furthermore, $Ca^{2+}$, $Mg^{2+}$, and $Na^+$ could not balance the effect of $Cl^-$, whereas $K^+$ compensated it. Addition of KCl enhanced photosynthesis when compared to the control measurements. It reflects the importance of ions concentration in habitats for charophyte growth.

## Conclusion

In summary, it was shown that encrustation of charophytes in water sites with strong ion anomalies could be as high as in hard-water lakes. It is assumed that ion composition, rather than ion concentration of $Na^+$, $Mg^{2+}$ and $SO_4^{2-}$, impact on the encrustation of charophytes. The age pattern on encrustation in this study showed a strong site-specificity, whereas encrustation of charophytes was species-specific. Encrustation, element composition and photosynthesis of charophytes were influenced by ion concentrations.

**Abbreviations**

ANOVA: analysis of variance; can: *C. canescens*; DW: dry weight; his: *C. hispida*; I: irradiance; $I_c$: light compensation point; $I_s$: light saturation irradiance; LOI: loss of ignition; P: photosynthesis; PCA: principal component analysis; $P_m$: maximum photosynthesis; $R_d$: dark respiration; SD: standard deviation; sub: *C. subspinosa*; TIC: total inorganic carbon; tom: *C. tomentosa*.

**Authors' contributions**

AH was involved in the design of this work, the data acquisition, the analysis and the interpretation of data and drafting the article. HS was responsible for the conception, the design and the interpretation of data. Both authors read and approved the final manuscript.

**Acknowledgements**

The authors want to thank P. Nowak for assistance in the field, W. von Tümpling (Helmholtz Centre for Environmental Research) for the water analysis, B. Balz and E. Heilmann (AUF, University of Rostock) for providing technical support for the biomass analysis and C. Porsche (MNF, University of Rostock) for advice on the statistical treatment of the data.

**Competing interests**

The authors declare that they have no competing interests.

**Funding**

One of the authors (AH) was supported by "Deutsche Bundesstiftung Umwelt" and "Professorinnenprogramm II of the University of Rostock", which is gratefully acknowledged.

**References**

Absolonova M, Beilby MJ, Sommer A, Hoepflinger MC, Foissner I (2018) Surface pH changes suggest a role for $H^+/OH^-$ channels in salinity response of *Chara australis*. Protoplasma 255:851–862

Akin GW, Lagerwerff JV (1965) Calcium carbonate equilibria in aqueous solutions open to the air. I. The solubility of calcite in relation to ionic strength. Geochim Cosmochim Acta 29:343–352

Asaeda T, Senavirathna MDHJ, Kaneko Y, Rashid MH (2014) Effect of calcium and magnesium on the growth and calcite encrustation of *Chara fibrosa*. Aquat Bot 113:100–106

Beilby MJ, Bisson MA (2012) pH banding in charophyte algae. In: Volkov A (ed) Plant electrophysiology, methods and cell electrophysiology. Springer, Berlin, pp 247–271

Beilby MJ, Casanova MT (2014) The physiology of characean cells. Springer, Berlin, pp 1–205

Blindow I (1992) Decline of charophytes during eutrophication: comparison with angiosperms. Freshw Biol 28:9–14

Blindow I, Hargeby A, Hilt S (2014) Facilitation of clear-water conditions in shallow lakes by macrophytes: differences between charophyte and angiosperm dominance. Hydrobiologia 737:99–110

de Mendiburu F (2016) Agricolae: statistical procedures for agricultural research. R package version 1.2-4

Doege A, van de Weyer K, Becker R, Schubert H (2016) Bioindikation mit Characeen. In: Characeen AG (ed) Armleuchteralgen—Die Characeen Deutschlands. Springer, Berlin, pp 79–138

Gomes PIA, Asaeda T (2010) Impact of calcium and magnesium on growth and morphological acclimations of *Nitella*: implications for calcification and nutrient dynamics. Chem Ecol 26:479–491

Heiri O, Lotter AF, Lemcke G (2001) Loss on ignition as a method for estimating organic and carbonate content in sediments: reproducibility and comparability of results. J Paleolimnol 25:101–110

Herbst A, Henningsen L, Schubert H, Blindow I (2018) Encrustations and element composition of charophytes from fresh or brackish water sites—habitat- or species-specific differences? Aquat Bot 148:29–34

Jeffrey C (1967) The origin and differentiation of the archegoniate land plants: a second contribution. Kew Bull 21:335–349

Kairesalo T, Gunnarsson K, St Jónsson G, Jónasson PM (1987) The occurrence and photosynthetic activity of epiphytes on the tips of *Nitella opaca* Ag. (Charophyceae). Aquat Bot 28:333–340

Kawahata C, Yamamuro M, Shiraiwa Y (2013) Changes in alkaline band formation and calcification of corticated charophyte *Chara globularis*. SpringerPlus 2:1–6

Kingsford RT, Porter JL (1994) Waterbirds on an adjacent freshwater lake and salt lake in arid Australia. Biol Conserv 69:219–228

Krause W (1981) Characeen als Bioindikatoren für den Gewässerzustand. Limnologica 13:399–418

Kuczyńska-Kippen N (2007) Habitat choice in rotifera communities of three shallow lakes: impact of macrophyte substratum and season. Hydrobiologia 593:27–37

Kufel L, Biardzka E, Strzałek M (2013) Calcium carbonate incrustation and phosphorus fractions in five charophyte species. Aquat Bot 109:54–57

Kufel L, Strzałek M, Biardzka E (2016) Site- and species-specific contribution of charophytes to calcium and phosphorus cycling in lakes. Hydrobiologia 767:185–195

Leitch AR (1991) Calcification of the charophyte oosporangium. In: Riding R (ed) Calcareous algae and stromatolites. Springer, Berlin, pp 204–216

McConnaughey TA, Falk RH (1991) Calcium-proton exchange during algal calcification. Biol Bull 180:185–195

McConnaughey TA, Whelan JF (1997) Calcification generates protons for nutrient and bicarbonate uptake. Earth Sci Rev 42:95–117

McCourt RM (1995) Green algal phylogeny. Trends Ecol Evol 10:159–163

Nishiyama T, Sakayama H, de Vries J et al (2018) The *Chara* genome: secondary complexity and implications for plant terrestrialization. Cell 174:448–464

Nõges P, Tuvikene L, Feldmann T, Tõnno I, Künnap H, Luup H, Salujõe J, Nõges T (2003) The role of charophytes in increasing water transparency: a case study of two shallow lakes in Estonia. Hydrobiologia 506–509:567–573

Ogata S, Kawasaki S, Hiroyoshi N, Tsunekawa M, Kaneko K (2010) Temperature dependence of calcium carbonate precipitation for biogrout. In: Vrkljan I (ed) Rock Engineering in difficult ground conditions. Taylor & Francis Group, London, pp 339–344

Oksanen J, Guillaume Blanchet F, Friendly M, Kindt R, Legendre P, McGlinn D, Minchin PR, O'Hara RB, Simpson GL, Solymos P, Stevens MHH, Szoecs E, Wagner H (2017) Vegan: community ecology package. R Pack Ver 2:4

Ozimek T, Kowalczewski A (1984) Long-term changes of the submersed macrophytes in eutrophic Lake Mikołajskie (North Poland). Aqua Bot 19:1–11

Pełechaty M, Pukacz A, Apolinarska K, Pełechata A, Siepak M (2013) The significance of *Chara* vegetation in the precipitation of lacustrine calcium carbonate. Sedimentology 60:1017–1035

Pieczyńska E, Ozimek T, Rybak JI (1988) Long-term changes in littoral habitats and communities in Lake Mikołajskie (Poland). Int Rev Ges Hydrobiol 73:361–378

Pukacz A, Pełechaty M, Frankowski M (2016a) Depth-dependence and monthly variability of charophyte biomass production: consequences for the precipitation of calcium carbonate in a shallow *Chara*-lake. Environ Sci Pollut Res 1:1. https://doi.org/10.1007/s11356-016-7420-8

Pukacz A, Pełechaty M, Frankowski M, Pronin E (2016b) Dry weight and calcium carbonate encrustation of two morphologically different *Chara* species: a comparative study from different lakes. Oceanol Hydrobiol Stud 45:377–387

Raven JA, Beardall J (2016) The ins and outs of $CO_2$. J Exp Bot 67:1–13

Raven JA, Smith FA, Walker NA (1986) Biomineralization in the Charophyceae sensu lato. In: Leadbeater BSC, Riding R (eds) Biomineralization in lower plants and animals. The systematics association. Clarendon Press, Oxford, pp 125–139

Sand-Jensen K, Jensen RS, Gomes M, Kristensen E, Martinsen KT, Kragh T, Baastrup-Spohr L, Borum J (2018) Photosynthesis and calcification of charophytes. Aquat Bot 149:46–51

Schubert H, Blindow I, Bueno NC, Casanova MT, Pełechaty M, Pukacz A (2018) Ecology of charophytes—permanent pioneers and ecosystem engineers. PiP 5:61–74

Siong K, Asaeda T (2009) Effect of magnesium on charophytes calcification: implications for phosphorus speciation stored in biomass and sediment in Myall Lake (Australia). Hydrobiologia 632:247–259

Stewart KD, Mattox KR (1975) Comparative cytology, evolution and classifica-

tion of the green algae with some consideration on the origin of other organisms with chlorophylls *a* and *b*. Bot Rev 41:104–135
R Core Team (2017) R: a language and environment for statistical computing. R Foundation for Statistical Computing, Vienna, Austria. https://www.r-project.org/
Thomas J, Apte SK (1984) Sodium requirement and metabolism in nitrogen-fixing cyanobacteria. J Biosci 6:771–794
Van den Berg MS, Coops H, Meijer M-L, Scheffer M, Simons J (1998a) Clearwater associated with a dense *Chara* vegetation in the shallow and turbid Lake Veluwemeer, the Netherlands. In: Jeppesen E, Søndergaard M, Søndergaard M, Kristoffersen K (eds) The structuring role of submerged macrophytes in lakes. Springer, New York, pp 339–352
Van den Berg MS, Scheffer M, Coops H, Simons J (1998b) The role of characean algae in the management of eutrophic shallow lakes. J Phycol 34:750–756
Wahlstedt LJ (1875) Monografi öfver Sveriges och Norges characeer. Boktryckeri-Aktie-Bolaget, Christianstad, Sweden, p 37
Walsby AE (1997) Numerical integration of phytoplankton photosynthesis through time and depth in a water column. New Phytol 136:189–209
Wellburn AR (1994) The spectral determination of chlorophylls *a* and *b*, as well as total carotenoids, using various solvents with spectrophotometers of different resolution. J Plant Physiol 144:307–313
Winter U, Kirst GO (1990) Salinity response of a freshwater charophyte, *Chara vulgaris*. Plant Cell Environ 13:123–134
Winter U, Kirst GO (1991) Partial turgor pressure regulation in *Chara canescens* and its implications for a generalized hypothesis of salinity response in charophytes. Bot Acta 104:37–46
Winter U, Soulié-Märsche I, Kirst GO (1996) Effects of salinity on turgor pressure and fertility in *Tolypella* (Characeae). Plant Cell Environ 19:869–879
Wolfstein K, Hartig P (1998) The photosynthetic light dispensation system: application to microphytobenthic primary production measurements. Mar Ecol Prog Ser 166:63–71

# 16

# Starch accumulation in hulless barley during grain filling

Xu-guang Zheng, Jun-cang Qi*, Hong-shan Hui, Li-hao Lin and Feng Wang

**Abstract**

**Background:** Starch consists of two types of molecules: amylose and amylopectin. The objective of this study was increase understanding about mechanisms related to starch accumulation in hulless barley (*Hordeum vulgare* L.) grain by measuring temporal changes in (i) grain amylose and amylopectin content, (ii) starch synthase activity, and (iii) the relative expressions of key starch-related genes.

**Results:** The amylopectin/amylose ratio gradually declined in both Beiqing 6 and Kunlun 12. In both cultivars, the activities of adenosine diphosphate glucose pyrophosphorylase, soluble starch synthase (SSS), granule bound starch synthase (GBSS), and starch branching enzyme (SBE) increased steadily during grain filling, reaching their maximums 20–25 days after anthesis. The activities of SSS and SBE were greater in Ganken 5 than in either Beiqing 6 or Kunlun 12. The expression of *GBSS I* was greater in Beiqing 6 and Kunlun 12 than in Ganken 5. In contrast, the expression of *SSS I*, *SSS II* and *SBE I* was greater in Ganken 5 than in Beiqing 6 and Kunlun 12. The peak in *GBSS I* expression was later than that of *SSS I*, *SSS II*, *SBE IIa* and *SBE IIb*. The *GBSS I* transcript in Kunlun 12 was expressed on average 90 times more than the *GBSS II* transcript.

**Conclusions:** The results suggest that SBE and SSS may control starch synthesis at the transcriptional level, whereas *GBSS I* may control starch synthesis at the post transcriptional level. *GBSS I* is mainly responsible for amylose synthesis whereas *SSS I* and *SBE II* are mainly responsible for amylopectin synthesis in amyloplasts.

**Keywords:** Hulless barley (*Hordeum vulgare* L.), Starch accumulation, The ratio of amylopectin and amylose, Enzyme activity, Gene expression

## Introduction

Starch is the main end product of carbon fixation during photosynthesis. Starch consists of two major components, amylose and amylopectin. Amylopectin is composed of short α-1,4-linked chains of glucose. About 5% of these chains are linked together by α-1,6 linkages (Manners 1989; Buléona et al. 1998; Preiss and Sivak 1998). Amylose is also composed of glucose chains; however less than 1% of these chains are linked by α-1,6 branches (Imberty et al. 1991). The structure and relative proportion of amylose and amylopectin are the primary determinants of the physical and chemical properties of starch.

Several enzymes are involved in starch biosynthesis. ADP-glucose pyrophosphorylase (AGPP) catalyzes the first reaction in starch synthesis, producing the activated glucosyl donor ADP-glucose (James et al. 2003). Granule-bound starch synthase (GBSS) is involved in amylose synthesis. Soluble starch synthase (SSS) catalyses the elongation of amylopectin chains. Starch branching enzyme (SBE) introduces branch points into the amylopectin chains. The precise mechanism that controls starch biosynthesis in grain is complex and not well understood.

Amylopectin is formed by multiple isoforms of SSS (*SSS I*, *SSS II* and *SSS III*) and SBE (*SBE I*, *SBE IIa* and *SBE IIb*). Several multiple isoform enzymes are plant species-specific. Among the species examined so far, each isoform of SSS and SBE plays a distinct role in amylopectin biosynthesis (Nakamura 2002).

Recent studies suggest that cereal endosperms have distinct cytosolic and plastidial forms of AGPP which

*Correspondence: shzqjc@qq.com
Department of Agronomy, College of Agriculture, Shihezi University, Shihezi, China

are encoded by separate large-subunit and small-subunit genes. The AGPP small subunit gene sequences from various eudicots and monocots differ primarily in exon (James et al. 2003). There are differing opinions about how *GBSS I* regulates starch synthesis. McCue et al. (2002) suggested that *GBSS I* controls starch synthesis in wheat at the transcriptional and post-transcriptional levels. Wang et al. (2014) studied starch biosynthesis in rice and proposed that *GBSS I* controls starch synthesis at the transcriptional level. There are few reports about the role of other enzyme genes on the synthesis and accumulation rate of amylose and amylopectin in grain. Information about the relative expression of genes encoding AGPP, GBSS, SSS and SBE at different development stages could provide insight about the mechanism controlling starch biosynthesis.

Hulless barley is mainly grown on the Tibetan Plateau. Hulless barley seeds consist primarily (63.2–65.3%) of starch. In fact, the process of grain filling in hulless barley is the process of starch accumulation. Hulless barley includes both waxy and non-waxy cultivars. The starch in waxy cultivars is composed of amylopectin (98.9%), whereas the starch in non-waxy cultivars is composed of both amylose (16.2–23.3%) and amylopectin (74.1–78.4%). The objectives of this experiment were (i) to compare the physiological and biochemical characteristics of starch biosynthesis during endosperm development in waxy and non-waxy cultivars of hulless barley and (ii) to study the differential expressions of AGPP, GBSS, SSS, and SBE genes at different grain filling stages.

## Materials and methods

### Plant materials and experiment design

The field experiment was conducted in 2015 at the Agriculture Experiment Station of Shihezi University, China. Three hulless barley cultivars were used in this study: Beiqing 6 (non-waxy), Kunlun 12 (non-waxy), and Ganken 5 (waxy).

The barley was sown in April. Each cultivar was grown in three replicate plots. The area of each plot was 5 $m^2$. The row spacing was 20 cm. There was 2 cm between plants within each row. The fertilizer and irrigation practices were the same as those used by farmers in the regions.

Plants were tagged as they flowered. Forty spikes with the same flowering date were collected from each plot 5, 10, 15, 20, 25, 30, and 35 days after anthesis (DAA) (Additional file 1: Figure S1). The fresh weight of the spikes was determined. Several kernels (5 g) were removed from each spike and then immediately frozen in liquid N. The samples were stored at −70 °C for determination of enzyme activity. The remaining kernels were placed in paper bags, dried at 105 °C for 1 h, and then dried at 80 °C until constant weight. These samples were used for measuring starch accumulation.

### Starch analysis and calculations

Starch, amylose, and amylopectin contents (mg $grain^{-1}$) were determined according to the "double-wave-length" method of He (1981). In this method, standard solutions were prepared of amylose (5, 10, 15, 20, 25, and 30 μg $mL^{-1}$) and amylopectin (40, 50, 60, 70, 80, 100, and 120 μg $mL^{-1}$). The absorbance of the amylose standards was measured at 554 nm ($A_{\lambda1}$) and 490 nm ($A_{\lambda2}$) (Fig. 1). A standard curve was then prepared by plotting the amylose concentration on the x-axis and the average absorbance difference (ΔA, calculated by subtracting $A_{\lambda2}$ from $A_{\lambda1}$) on the y-axis. The linear regression equation of the standard curve was $Y = 0.0079x - 0.0139$ ($R^2 = 0.99$). The standard curve for amylopectin was prepared in the same way only absorbance was measured at 542 nm ($A_{\lambda1}$) and 713 nm ($A_{\lambda2}$). The ΔA was calculated by subtracting $A_{\lambda2}$ from $A_{\lambda1}$. The linear regression equation of the standard curve was $Y = 0.0059x + 0.0191$ ($R_2 = 0.99$). Total starch contents were determined by summing the amylose content and amylopectin content.

Next, after crushing to pass a 60 mesh sieve, the grain samples (0.1 g) were put into 100 mL beakers. Small amounts of anhydrous ethanol were added to the beakers followed by 10 mL of 1 mol $L^{-1}$ NaOH. The beakers were then put into a water bath at 65 °C. After the starch had dissolved, the beakers were removed from the water bath and distilled water was added to bring the sample volume to 50 mL. The samples were then mixed and allowed to stand at room temperature.

A 3-mL aliquot of each solution was mixed with 20–30 mL distilled water. The pH of the solutions was

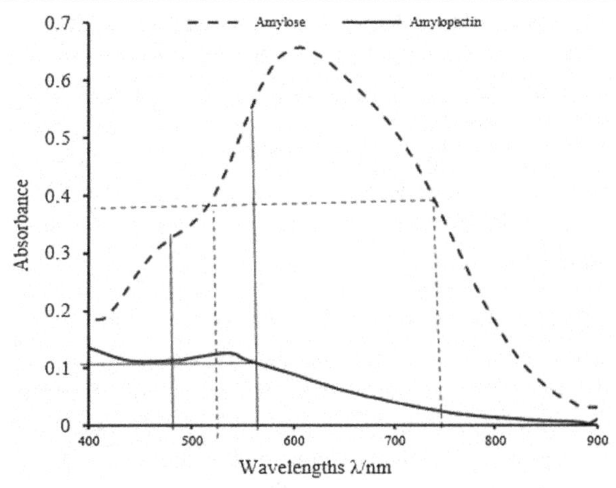


**Fig. 1** The main wavelength and the particular wavelength for amylose were 554 and 490 nm and the main wavelength and the particular wavelength for amylopectin were 542 and 713 nm

adjusted to 3.5 with 0.1 mol $L^{-1}$ HCl. A 0.5 mL aliquot of iodine reagent (0.1% $I_2$ + 1% KI) was added to the samples and then distilled water was added to bring the final volume to 50 mL. The samples were mixed and allowed to stand at room temperature for 20 min. Standard solutions were prepared of amylose (0–30 μg $mL^{-1}$) and amylopectin (40–120 μg $mL^{-1}$).

The absorbance the main and particular wavelengths for amylose were 554 and 490 nm, respectively. The main and particular wavelengths for amylopectin were 542 and 713 nm, respectively (Fig. 1). Total starch content (%) was calculated as the sum of the amylose content (%) and the amylopectin content (%). Starch accumulation (mg $grain^{-1}$) was calculated by multiplying the starch content (%) by the seed weight (mg $grain^{-1}$).

The absorbance of the sample solution at λ1, λ2, λ3 and λ4 was determined by spectrophotometer. The straight and amylopectin content of the samples was calculated according to the double wavelength standard curve of straight and amylopectin. The sum of the two samples was determined by the total starch content.

$$\mathrm{w1} = \frac{(A554\,\mathrm{nm} - A490\,\mathrm{nm}) + 0.0139}{0.079 * M * 10};$$

$$\mathrm{w2} = \frac{(A542\,\mathrm{nm} - A713\,\mathrm{nm}) - 0.0191}{0.059 * M * 10};\ w = w1 + w2$$

Note: W is the total starch content/%; W1 is the amylose content/%; W2 is the amylopectin content/%; M is the sample mass/g; 10 is the unit conversion factor.

The grain starch accumulation process can be described by the logistic equation $W = K/(1 + ae^{-bt})$, where $W$ is starch accumulation, $t$ is days after anthesis, $K$ is the maximum accumulation amount, and $a$ and $b$ are constants. The equation can be rewritten as: $y = k/(1 + e^A + BT)$ where $A = \ln(a)$ and $B = -b$. The following characteristics were calculated for starch, amylopectin and amylose accumulation using CurveExpert 1.4 software: (i) time when the starch accumulation rate was at its maximum (Tmax) = −A/B; (ii) starch accumulation at 90% of its final amount (D) = [ln (1/9) − A]/B; (iii) the maximum starch accumulation rate ($V_{max}$) = −KB/4; (iv) the mean starch accumulation rate ($V_{mean}$) = K/D; and (v) the initial growth power ($C_0$) = $K/(1 + e^A)$ (Wang et al. 2014).

#### Enzyme isolation and assays

Enzyme extraction was done using the method of Cheng and Jiang (2001). Twenty fresh kernels were placed into 5 mL buffer solution (50 mmol $L^{-1}$ Hepes–NaOH, pH 7.5; 8 mmol $L^{-1}$ $MgCl_2$, 2 mmol $L^{-1}$ $EDTA_2$, 10 g $L^{-1}$ PVP-30 and 1 mmol $L^{-1}$ DTT). The samples were homogenized by grinding in an ice bath and then centrifuged at 10,000 rpm for 30 min at 4 °C. The supernatant, referred to as enzyme preparation 1, was used to determine the activities of AGPP, SSS, and SBE activity. The sediment was resuspended with 5 mL of buffer solution and then centrifuged at 1000 rpm for 3 min at 4 °C. The supernatant, referred to as enzyme preparation two, was used to determine GBSS activity.

The AGPP activity was measured by mixing enzyme preparation 1 (20 μL) into a reaction mixture(110 μL) consisting of 100 mmol $L^{-1}$ Hepes–NaOH (pH 7.4), 1.2 mmol $L^{-1}$ ADPG, 3 mmol $L^{-1}$ PPi, 5 mmol $L^{-1}$ $MgCl_2$ and 4 mmol $L^{-1}$ DTT. The total reaction volume was 130 μL. The solutions were kept at 30 °C for 20 min and then the reactions were terminated by placing the samples in boiling water for 1 min. The solutions were centrifuged at 10,000 rpm for 10 min. A 100-μL aliquot of the supernatant was collected and mixed with 5.2 μL of colorizing agent consisting of 5.76 mmol/L NADP, 0.08 unit P-gucomutase, and 0.07 unit G6P-dehydrogenase. The absorption of the samples was measured at 340 nm. One unit of AGPP activity was defined as the amount of enzyme causing a one unit per min increase in absorbance.

The GBSS activity was measured by mixing enzyme preparation 1 (20 μL) into a reaction mixture (36 μL) consisting of 50 mmol $L^{-1}$ Hepes–NaOH (pH 7.4), 1.6 mmol $L^{-1}$ ADPG, 0.7 mg amylopectin, and 15 mmol $L^{-1}$ DTT. The total reaction volume was 56 μL. The solutions were kept at 30 °C for 20 min. The reactions were terminated by placing the reaction mixtures in boiling water for 1 min followed by cooling in a water bath. Next, the samples were mixed with 20 μL of a reaction solution consisting of 50 mmol $L^{-1}$ Hepes–NaOH (pH 7.4), 4 mmol $L^{-1}$ PEP, 200 mmol $L^{-1}$ KCl, 10 mmol $L^{-1}$ $MgCl_2$, and 1.2 units of pyruvate kinase. The solutions were kept at 30 °C for 20 min. The reactions were terminated by placing the mixture in boiling water for 1 min. The samples was centrifuged at 10,000 rpm for 10 min. A 60-μL aliquot of the supernatant was collected and then mixed with 43 μL of color liquid including 50 mmol $L^{-1}$ Hepes–NaOH (pH 7.4), 10 mmol $L^{-1}$ glucose, 20 mM $MgCl_2$, 2 mmol/L NADP, 1.4 units hexokinase, and 0.35 units G-6-P-dehydrogenase. The solutions were kept at room temperature for 10 min. The enzyme activity was assayed spectrophotometrically at 340 nm. One unit of GBSS activity was defined as the amount of enzyme causing a one unit per min increase in absorbance (Nakamura et al. 1989).

The SBE activity was measured by mixing 50 μL of enzyme preparation 1 into a reaction mixture containing 50 mM Hepes–NaOH (pH 7.4), 5 mM G1P, 1.25 Mm AMP, and phosphorylase a (54 unit). The total reaction volume was 1.45 mL. The solutions were kept at 30 °C for 20 min and then the reactions were terminated by placing the samples in boiling water for 1 min. The reaction was terminated by adding 50 μL of 1 mol $L^{-1}$ HCl. The solutions were then mixed with 500 μL dimethylsulfoxide and 700 μL of iodine reagent (0.1% $I_2$ + 1% KI). The enzyme activity was assayed spectrophotometrically at 540 nm. One unit of SBE activity was defined as the amount of enzyme causing a one unit per min increase in absorbance.

The procedure for measuring SSS activity used enzyme preparation 2 rather than enzyme preparation 1. In all other aspects, the procedure to determine SSS activity was the same as that used to determine GBSS activity. One unit of GBSS activity was defined as the amount of enzyme causing a one unit per min increase in absorbance at 340 nm.

### Gene expression and calculation of the expression rate

Gene expression was performed using a reverse transcriptase polymerase chain reaction (RT-PCR). Total RNA from endosperm was isolated using Fruit-mate™ for RNA purification (TaKaRa) and RNAiso Plus (TaKaRa Code: 9108). The RNA was reverse transcribed with the Script cDNA synthesis kit (TaKaRa). The synthesized cDNA was subjected to PCR for 45 cycles. The conserved regions of the gene sequences of *AGPP* (Tan et al. 2010), *GBSS I, GBSS II, SSS I, SSS II, SSIII, SBE I, SBE IIa, SBE IIb* (Wang et al. 2014) were obtained from wheat and used to design primers for detecting gene expression in wheat endosperm. The wheat actin gene (NCBI Accession No. DN551593) was used as a reference control. The fragment size of the sequences, annealing temperature (Tm), and real-time PCR efficiency are listed in Table 1. Aliquots of the individual PCR products were resolved by agarose gel electrophoresis and the bands were visualized with a PhotoDoc-It imaging system (UVP, LLC, Upland, CA, USA).

The RT-PCR efficiencies (E) were calculated from the slopes in Light Cycler software using the equation $E = 10[-1/\text{slope}]$. Calculations for the test precision and test variability are based on the crossing-point (CP) variation from the CP mean value. A mathematical model was used to determine the relative expression ratio (R) of an unknown sample to the control, and expressed in comparison to a reference gene (Wang et al. 2014).

$$\text{Ratio} = \left(E_{\text{target}}\right)^{\Delta CP(\text{control}-\text{sample})} / (E_{\text{ref}})^{\Delta CP(\text{control}-\text{sample})}.$$

Actin gene was used as a reference control because it has highly conserved proteins that are involved in cell motility, structure and integrity.

### Statistical analysis

The data were analyzed with SPSS 19.0 and Excel 2013 software. Samples were analyzed in triplicate and all data are presented as means. Logistic equations were fit with CurveExpert v. 1.4 software to describe the relationship between starch accumulation quantity (y) and days after flowering (x).

## Results

### Starch accumulation

Starch, amylose, and amylopectin contents increased across time in all three cultivars (Fig. 2a, c, e). Starch accumulation at flowering was less in Ganken 5 than in either Beiqing 6 or Kunlun 12. The starch content was less in Beiqing 6 than in Kunlun 12 at 10, 15, 20, and 25 DAA; however the difference between the two cultivars became less with time. The starch content of Beiqing 6 at 30 and 35 DAA was 1.62 and 3.91% greater, respectively, compared with Kunlun 12. The temporal changes in amylopectin content were similar to those of starch. The amylopectin contents of Beiqing 6 at 30 and 35 DAA

**Table 1 Primer sequence for starch synthesis related gene**

| Gene | NCBI accession | Forward primers (5′–3′) | Reverse primers (3′–5′) | Annealing temperature (°C) |
|---|---|---|---|---|
| AGPP | AK359692 | TTCAGTTTCATGACCGTTC | AGTATCCATCTGTCTCCCTC | 60 |
| *GBSS I* | AY050174 | GACACTATCGTGGAAGGCAAG | TTGACCATCTCATGGTACGC | 60 |
| *GBSS II* | AF109395 | CACAGAATGCCAGAGGCATAG | GAACAGATGGGAATCACTCCA | 60 |
| *SSS I* | AJ292521 | GCAAAAGGAGAGGAGGGTACA | ACGTATGGTCTTTCGTCATGC | 60 |
| *SSS II* | AB201445 | GCTACACCAACTTCTCCCTG | GATGATCTCCACGCCCTTCT | 60 |
| *SSS III* | AF258608 | GACATGTGGTTTTGCTTGGTT | AGCCAGCGTATATCAGGTGAG | 60 |
| *SBE I* | AF286317 | ATGTTTGGTGGACATGGAAGA | ATGTTTGGTGGACATGGAAGA | 60 |
| *SBE IIa* | Y11282 | ATGTTTGGTGGACATGGAAGA | ATGTTTGGTGGACATGGAAGA | 60 |
| *SBE IIb* | AY740401 | ATGTTTGGTGGACATGGAAGA | ATGTTTGGTGGACATGGAAGA | 60 |
| Actin | DN551593 | GGAAAAGTGCAGAGAGACACG | TACAGTGTCTGGATCGGTGGT | 60 |

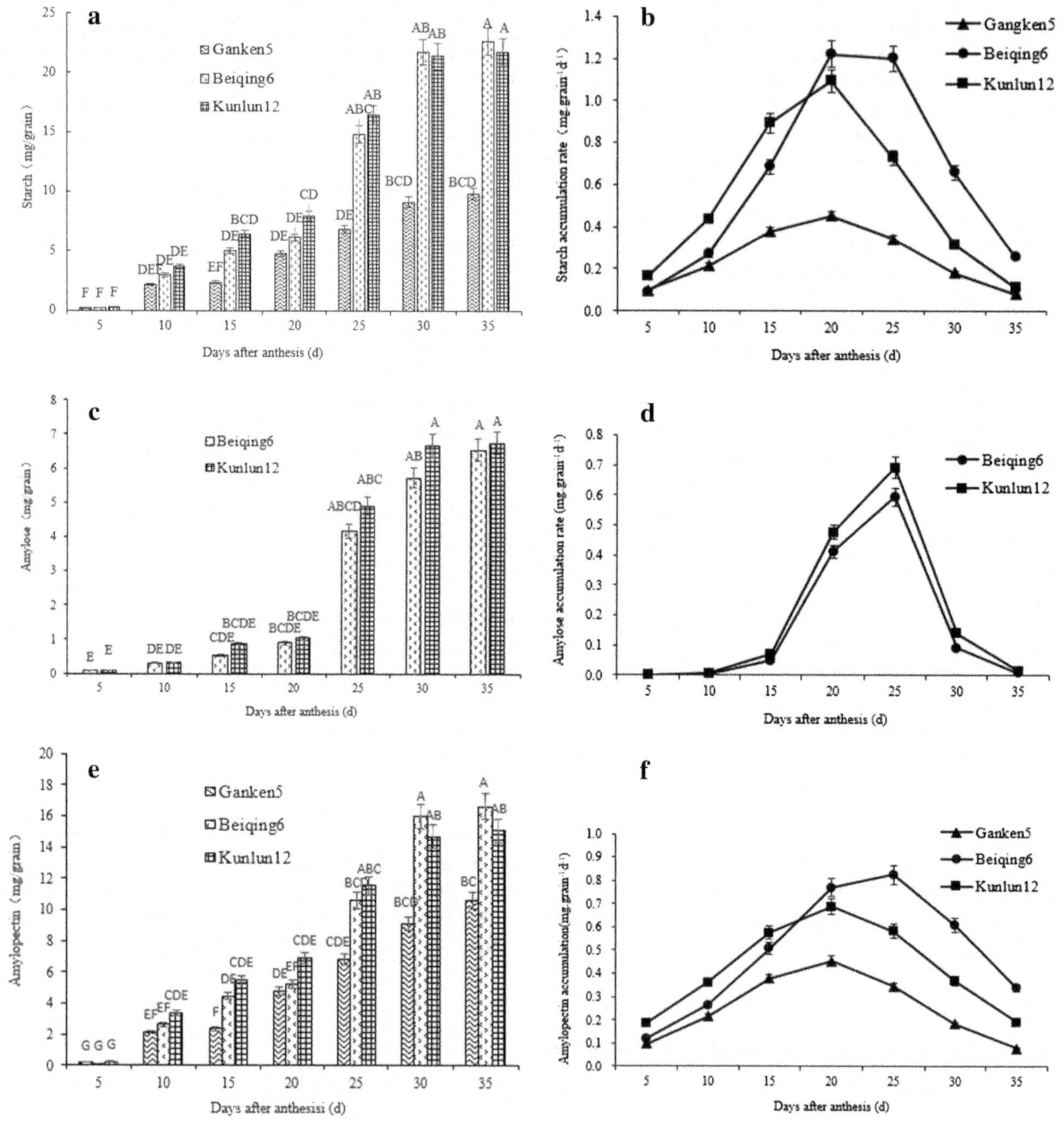


**Fig. 2** Temporal changes in the content and accumulation rate in three hulless barley cultivars. **a** Starch content; **b** Starch accumulation rate; **c** Amylose content; **d** Amylose accumulation rate; **e** Amylopectin content; **f** Amylopectin accumulation rate. *Bars* with *different letters* are significantly different at the 0.01 level of significance. *Error bars* indicate standard deviation

were 8.12 and 9.43% greater, respectively, than of Kunlun 12. The amylose content was greater in Kunlun 12 than in Beiqing 6 from 5DAA to 35DAA during grain filling. The amylose content of Ganken 5 was too low to be detected.

Grain starch contents are affected by the starch accumulation rate during grain filling. There were temporal differences in the accumulation rates of starch, amylose, and amylopectin among the three hulless barley cultivars (Fig. 2b, d, f). These differences became significant 15–25 DAA. The accumulation rates of starch and amylopectin decreased in the order Kunlun 12 > Beiqing 6 > Ganken 5 until 18 DAA. During the remainder of grain filling, the accumulation rates decreased in the order Beiqing 6 > Kunlun 12 > Ganken 5. The temporal changes in starch and amylopectin accumulation rate were consistent with the changes in starch and amylopectin content. The amylose accumulation rate was greater in Kunlun 12 than in Beiqing 6 during grain filling and, as a result, Kunlun 12 had the highest final amylose content (7.21 mg $grain^{-1}$).

The data in Fig. 2a, c, e were fit with logistic equations to describe the temporal changes in grain starch content. The equations and their parameters are shown in Table 2. For starch, the maximum accumulation rate, the average accumulation rate, and the initial growth power were all less in Ganken 5 than in Beiqing 6 and Kunlun 12. The initial growth power of Ganken 5 was less than that of the other cultivars because (i) the starch in Ganken 5 is mainly composed of amylopectin and (ii) amylopectin synthesis is more complex than amylose synthesis. The maximum starch accumulation rate, the time to the maximum accumulation rate, the accumulation duration, the average accumulation rate, and the initial growth power were greater in Kunlun 12 than in Beiqing 6. Thus, the final starch content of non-waxy hulless barley depends on both the amylose accumulation rate and the initial growth potential. For amylopectin, the maximum accumulation rate, the time to reach the maximum accumulation rate, the accumulation duration, the average accumulation rate, and the initial growth power were greater in Ganken 5 than in Beiqing 6 and Kunlun 12. This indicated that the final amylopectin content depends on the maximum amylopectin accumulation rate and on the average accumulation rate. The onset and termination times of amylopectin accumulation did not affect the final amylopectin content.

The amylopectin/amylose ratio increased to a maximum and then decreased in both Beiqing 6 and Kunlun 12 (Fig. 3). This shows that the amylopectin synthesis rate was greater than the amylose synthesis rate between 5 and 10 DAA, whereas the opposite was true between 10 and 25 DAA. The amylopectin/amylose ratio at 10 DAA was 10.72 for Kunlun 12 and 8.64 for Beiqing 6. The ratio slowly declined after 15 DAA. There was no significant difference in the amylopectin/amylose ratio of Kunlun 12 and Beiqing 6 between 15 and 30 DAA.

### Starch synthesis enzymes

#### *Adenosine diphosphate glucose pyrophosphorylase (AGPP) activity*

The AGPP activity increased to a maximum and then decreased in all three cultivars (Fig. 4a). There were differences among cultivars both in the maximum activity and in the time of maximum activity. Beiqing 6 had the highest maximum AGPP activity, followed by Kunlun 12 and then Ganken 5. The enzyme activity of AGPP of Beiqing 6 reached 26.28 nmol $grain^{-1}$ $d^{-1}$ at 25 DAA, whereas Kunlun 12 reached 21.86 nmol $grain^{-1}$ $d^{-1}$ and Ganken 5 reached 21.10 nmol $grain^{-1}$ $d^{-1}$ at 20 DAA.

**Table 2 Logistic equations describing grain starch accumulation in three hulless barley cultivars**

| Cultivars | Poly-saccharide | Logistic equation | Correlation coefficient | $T_{max}$/d | D/d | $V_{mean}$/ (mg $grain^{-1}$ $d^{-1}$) | $V_{max}$/ (mg $grain^{-1}$ $d^{-1}$) | $C_0$ |
|---|---|---|---|---|---|---|---|---|
| Ganken 5 | Starch | $y = 9.358/(1 + e^{3.780 - 0.194X})$ | 0.982** | 19.46 | 30.78 | 0.304 | 0.454 | 0.209 |
| | Amylose | – | – | – | – | – | – | – |
| | Amylopectin | $y = 9.358/(1 + e^{3.779 - 0.194X})$ | 0.982** | 19.46 | 30.79 | 0.304 | 0.454 | 0.209 |
| Beiqing 6 | Starch | $y = 23.839/(1 + e^{5.510 - 0.222X})$ | 0.982** | 22.84 | 32.76 | 0.728 | 1.32 | 0.15 |
| | Amylose | $y = 5.834/(1 + e^{11.522 - 0.497X})$ | 0.990** | 23.17 | 27.59 | 0.212 | 0.725 | 0.0001 |
| | Amylopectin | $y = 19.008/(1 + e^{4.139 - 0.177X})$ | 0.981** | 23.39 | 35.80 | 0.531 | 0.841 | 0.298 |
| Kunlun 12 | Starch | $y = 19.432/(1 + e^{4.352 - 0.227X})$ | 0.978** | 19.13 | 28.79 | 0.675 | 1.105 | 0.247 |
| | Amylose | $y = 7.046/(1 + e^{10.573 - 0.454X})$ | 0.982** | 23.31 | 28.16 | 0.25 | 0.799 | 0.0002 |
| | Amylopectin | $y = 16.389/(1 + e^{3.380 - 0.168X})$ | 0.977** | 20.10 | 33.18 | 0.494 | 0.689 | 0.54 |

y, grain starch accumulation; x, days after anthesis

$T_{max}$ time to reach the maximum accumulation rate, *D* accumulation duration, $V_{mean}$ mean accumulation rate, $V_{max}$ maximum accumulation rate, $C_0$ initial growth potential

** Significant at the 0.01 probability level

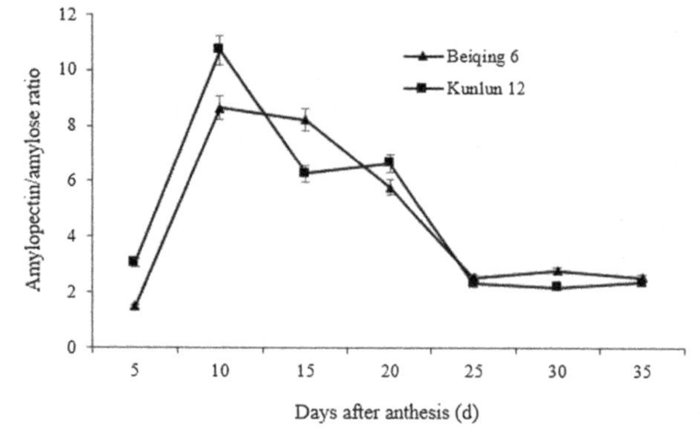


**Fig. 3** The ratio of amylopectin and amylose in Beiqing 6 and Ganken 5. *Error bars* indicate standard deviation

### *Soluble starch synthase (SSS) activity*

The SSS activity in all three cultivars increased to a maximum about 20 DAA and then decreased (Fig. 4b). There were significant differences in SSS activity among the cultivars. During grain filling, SSS activity changed less in Beiqing 6 than in either Ganken 5 or Kunlun 12. The SSS is mainly involved in amylopectin synthesis. At 20 DAA, Ganken 5 had the highest SSS activity (24.21 nmol grain$^{-1}$ d$^{-1}$) followed by Kunlun 12 (20.39 nmol grain$^{-1}$ d$^{-1}$) and then Beiqing 6 (18.00 nmol grain$^{-1}$ d$^{-1}$). This is because amylopectin is the major component of starch in Ganken 5. Overall, SSS activity was greater in Ganken 5 than in either Kunlun 12 or Beiqing 6 during the entire grain filling process.

### *Granule-bound starch synthase (GBSS) activity*

No GBSS activity was detected in Ganken 5 during grain filling (Fig. 4c). There was no significant difference in GBSS activity between Kunlun 12 and Beiqing 6 during early grain filling. However, during late grain filling, GBSS activity was significantly greater in Kunlun 12 than in Beiqing 6. The maximum GBSS activities were observed 20 DAA in Beiqing 6 (17.95 nmol grain$^{-1}$ d$^{-1}$) and 25 DAA in Kunlun 12 (19.87 nmol grain$^{-1}$ d$^{-1}$). The amylose content of Kunlun 12 was also greater than that of Beiqing 6. Overall, these results indicated that GBSS regulates amylose synthesis in barley and determines the amylose content of the grain. The role of GBSS is more important during late grain filling than during early grain filling.

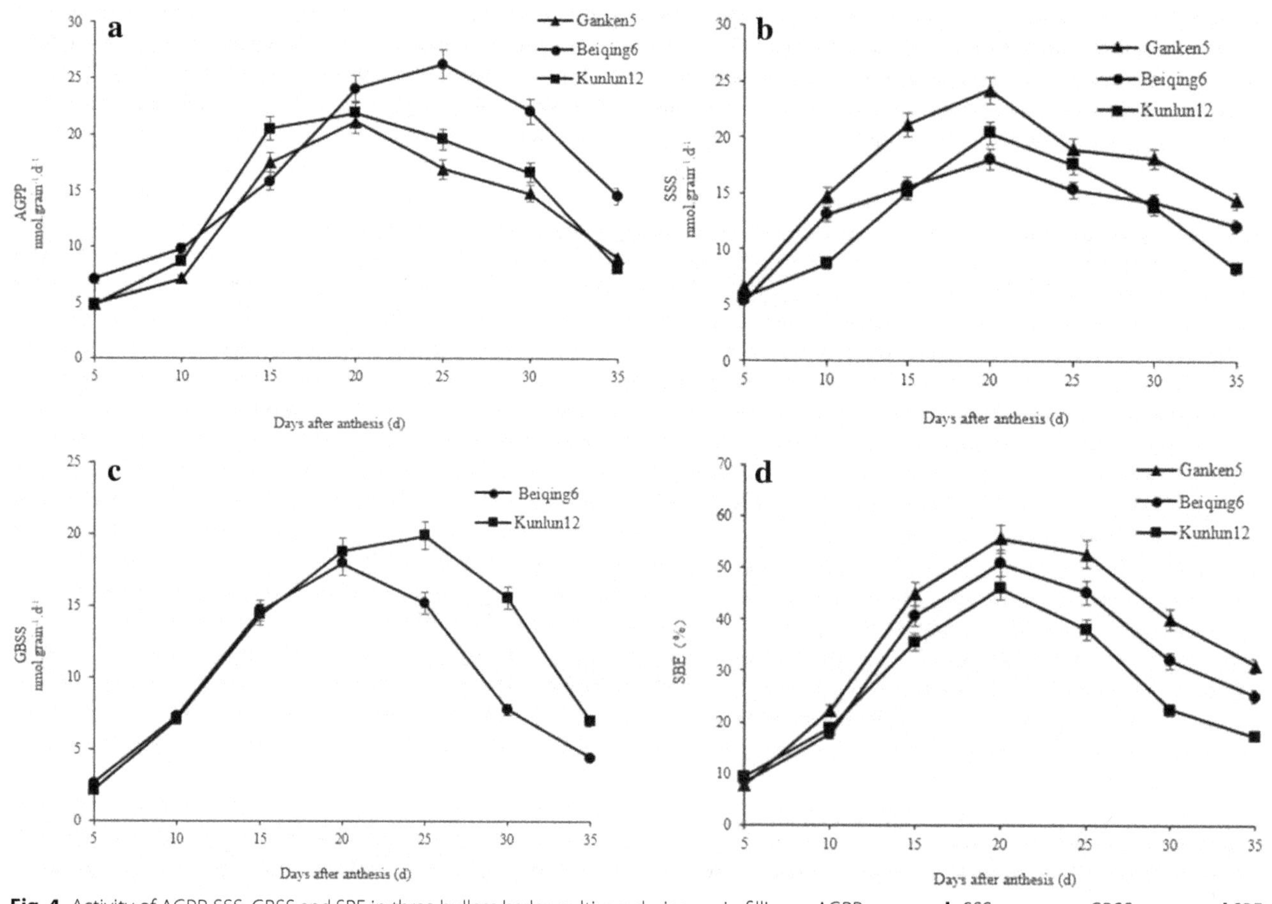


**Fig. 4** Activity of AGPP, SSS, GBSS and SBE in three hulless barley cultivars during grain filling. **a** AGPP enzyme; **b** SSS enzyme; **c** GBSS enzyme; **d** SBE enzyme. *Error bars* indicate standard deviation

#### *Starch branching enzyme activity*

The SBE directly affects grain amylopectin content. The SBE can cut α-1, 6-glycosidic bonds, thereby forming branching sugar chains. The SBE activity in all three cultivars increased to a maximum about 20 DAA and then declined (Fig. 4d). Ganken 5 had the highest maximum SBE activity, followed by Beiqing 6 and then Kunlun 12. These results indicated that the waxy cultivar, Ganken 5, had more ability to synthesize amylopectin than the non-waxy cultivars, Beiqing 6 and Kunlun 12.

#### *Correlation between starch synthase activity and starch accumulation rate*

Previous studies indicated that starch accumulation rates were more consistently related to the activity of AGPP than to the activities of SSS, GBSS, and SBE (Liu et al. 2004). However, in the present study, the activities of AGPP, SSS, GBSS, and SBE were all significantly or highly significantly correlated with starch accumulation. This indicates that these enzymes play an important regulatory role in starch synthesis in hulless barley. The activities of AGPP, GBSS, and SBE were significantly positively correlated with amylose accumulation rate. The correlation between SSS activity and amylose accumulation rate was not significant, probably because SSS is mainly involved in amylopectin synthesis (Table 3).

#### *Gene expression*

AGPP catalyzes the first reaction in starch synthesis, producing the activated glucosyl donor ADP-glucose (ADPG). AGPP is comprised of two large subunits and two small subunits, each of which is encoded by distinct genes. Our study indicated no significant differences in *AGPP* expression between the waxy hulless barely, Ganken 5, and the non-waxy hulless barely, Beiqing 6 and Ganken 12 (Fig. 5a).

GBSS consists of two forms: *GBSS I* and *GBSS II*. The expression of *GBSS I* appears to be mostly confined to storage tissues. *GBSS II* is encoded by a separate gene and is generally expressed more in leaves and other non-storage tissues that accumulate transient starch. The results of the present study indicated that the relative expression of *GBSS I* increased to a maximum at 20 DAA and then declined in the non-waxy cultivars Beiqing 6 and Kunlun 12 (Fig. 5b). In contrast, the relative expression of the waxy cultivar Ganken 5 was much less and did not change with time. The relative expression of *GBSS II* increased with time in all three cultivars, reaching maximums between 20 and 30 DAA (Fig. 5c). The relative expression of *GBSS I* was on average 90 times greater than that of *GBSS II* in Beiqing 6 and Kunlun 12 during grain filling. Furthermore, the relative expression of *GBSS II* was much less than that of the other starch synthesis genes. Overall, low *GBSS I* expression in Ganken 5 combined with the cultivar's low amylose concentrations suggests that *GBSS I* is mainly responsible for amylose synthesis in cereal endosperm. Although its relative expression is low in endosperm, *GBSS II* may be essential to starch synthesis in other non-storage tissues. Low *GBSS I* expression in Ganken 5 indicates why amylose concentrations are non-detectable in waxy Ganken 5.

The SSS genes are exclusively involved in amylopectin biosynthesis. The distribution of the SSS genes between the stroma and starch granules within plastids varies

**Table 3 Correlation analysis between enzyme activity and starch accumulation rate in three hulless barley cultivars at grain filling**

| Enzyme | Cultivar | Amylose accumulation rate | Amylopectin accumulation rate | Starch accumulation rate |
|---|---|---|---|---|
| AGPP | Ganken 5 | – | 0.878** | 0.878** |
| | Beiqing 6 | 0.829* | 0.983** | 0.925** |
| | Kunlun 12 | 0.668 | 0.934** | 0.871* |
| SSS | Ganken 5 | – | 0.857* | 0.857* |
| | Beiqing 6 | 0.556 | 0.848* | 0.838* |
| | Kunlun 12 | 0.57 | 0.886** | 0.841* |
| GBSS | Ganken 5 | – | 0.952** | 0.592** |
| | Beiqing 6 | 0.748* | 0.854* | 0.924** |
| | Kunlun 12 | 0.751* | 0.993** | 0.970** |
| SBE | Ganken 5 | – | – | – |
| | Beiqing 6 | 0.765* | 0.878** | 0.935** |
| | Kunlun 12 | 0.760* | 0.966** | 0.944** |

* and ** indicate significance at the 0.05 and 0.01 probability levels, respectively

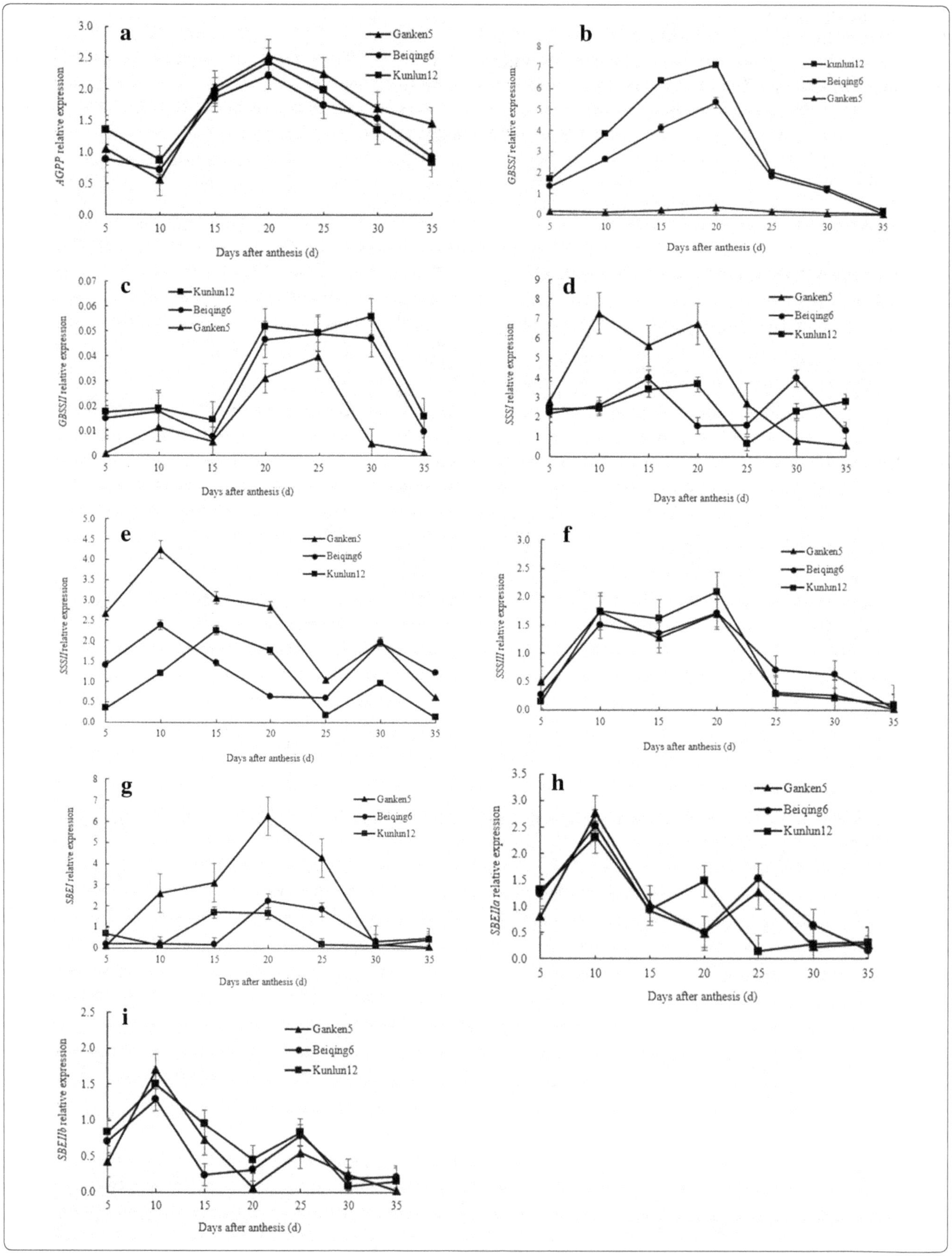
a
Ganken5
Beiqing6
Kunlun12
AGPP relative expression
Days after anthesis (d)
b
kunlun12
Beiqing6
Ganken5
GBSSI relative expressiom
Days after anthesis (d)
c
Kunlun12
Beiqing6
Ganken5
GBSSII relative expression
Days after anthesis (d)
d
Ganken5
Beiqing6
Kunlun12
SSI relative expression
Days after anthesis (d)
e
Ganken5
Beiqing6
Kunlun12
SSII relative expression
Days after anthesis (d)
f
Ganken5
Beiqing6
Kunlun12
SSIII relative expression
Days after anthesis (d)
g
Ganken5
Beiqing6
Kunlun12
SBEI relative expression
Days after anthesis (d)
h
Ganken5
Beiqing6
Kunlun12
SBEIIa relative expression
Days after anthesis (d)
i
Ganken5
Beiqing6
Kunlun12
SBEIIb relative expression
Days after anthesis (d)
5 10 15 20 25 30 35

(See figure on previous page.)
**Fig. 5** Changes in enzyme gene expression during grain filling. **a** *AGPP* relative expression; **b** *GBSS I* relative expression; **c** *GBSS II* relative expression; **d** *SSS I* relative expression; **e** *SSS II* relative expression; **f** *SSS III* relative expression; **g** *SBE I* relative expression; **h** *SBE IIa* relative expression; **i** *SBE IIb* relative expression. *Error bars* indicate standard deviation

among plant species, tissue types, and plant development stages. In our study, the relative expression of *SSS I* and *SSS II* was significantly greater in Ganken 5 than in Beiqing 6 and Kunlun 12 between 5 and 25 DAA (Fig. 5d, e). The relative expression of *SSS I, SSS II,* and *SSS III* exhibited similar patterns across time in Beiqing 6 and Kunlun 12 (Fig. 5d–f). The relative expression of *SSS I* in Ganken 5 was already high at the earliest phase of seed formation (5 DAA). The *SSS I* expression increased to peak at 10 DAA, but then declined. The *SSS I* expression of Ganken 5 was significantly less than that of Beiqing 6 and Kunlun 12 between 25 and 35 DAA. The SSS genes are exclusively involved in amylopectin biosynthesis and *SSS I* is the major SSS form in cereal endosperm. This may be one reason for the decline in the amylopectin/amylose ratio in Beiqing 6 and Kunlun 12.

The SBE gene can combine α-1,6 linkages by cleaving internal α-1,4 bonds and then transferring the released reducing ends to C6 hydroxyls. Two SBE gene classes (i.e., *SBE I* and *SBE II*) differ in the length of glucan chains transferred in vitro and in their substrate specificities. In cereals, there are two closely related forms of *SBE II* (i.e., *SBE IIa* and *SBE IIb*). These two forms differ in chain-length specificity in vitro, with *SBE IIb* transferring shorter chains than *SBE IIa* during extended incubation. In the present study, temporal changes in *SBE I, SBE IIa,* and *SBE IIb* were similar in all three cultivars (Fig. 5h, i, j). However, the relative expression of *SBE I* was significantly greater in waxy Ganken 5 than in Beiqing 6 and Ganken 5. This suggested that *SBE I* has a major influence on SBE activity and on amylopectin accumulation during endosperm development.

## Discussion

Starch biosynthesis in grain has been studied extensively in wheat and rice but not in hulless barely (Li and Sun 1980; Nakamura and Yuki 1992). We observed that starch and its components increased with time during grain filling. The patterns were similar to those observed in wheat and rice (Li and Sun 1980; Jian et al. 2002). The starch and amylose accumulation rates were always greater in the non-waxy cultivars (i.e., Beiqing 6 and Kunlun 12) than in the waxy cultivar (Ganken 5). The amylopectin content of non-waxy Beiqing 6 was significantly greater than that of waxy Ganken 5 but not greater than that of non-waxy Kunlun 12.

The roles of AGPP, SSS, GBSS, and SBE in starch biosynthesis have been investigated in previous studies (Li and Sun 1980; Doehlert et al. 1988; Nakamura and Yuki 1992). However, the contribution of each enzyme to starch synthesis is still disputed. Our results showed that the maximum activities of AGPP in Beiqing 6 was 26.28 nmol grain$^{-1}$ d$^{-1}$ and in Kunlun 12 was 21.86 nmol grain$^{-1}$ d$^{-1}$ were greater in Ganken 5 was 21.10 nmol grain$^{-1}$ d$^{-1}$. Furthermore, the activities of AGPP, SSS, GBSS, and SBE were significantly positively correlated with the starch accumulation rate during grain filling. This indicated that AGPP, SSS, and GBSS play important roles in starch biosynthesis in hulless barley. Preiss et al. (1988) reported that AGPP was the rate-limiting enzyme in starch biosynthesis and that AGPP activity was related to the synthesis rate and final starch amount. Okita (1992) observed that SSS activity was consistent with starch accumulation rate. We observed peaks in the activities of AGPP, SSS, and GBSS between 20 and 25 DAA in all three cultivars. This was also the time when starch accumulation rates were highest.

AGP is composed of two large subunits and two small subunits, each of which is encoded by distinct genes (*AGPP-L* and *AGPP-S*). The enzyme is known to be largely extraplastidial (i.e., 85–95% cytosolic) in cereal endosperm, but plastidial in other cereal tissues and in all tissues of non-cereal plants (Fig. 5a) (Beckles et al. 2001). AGPP catalyzes the first reaction in starch synthesis, producing the activated glucosyl donor ADP-glucose. Cao (2012) observed that waxy and non-waxy wheat differed significantly both in *AGPP-L* expression and in *AGPP-S* expression. However, in this study, AGPP activity and the relative expression of AGPP exhibited similar temporal changes in all three cultivars. This may be because AGPP expression was different in wheat than in hulless barley. Additional research needs to be done to confirm this.

Studies of waxy mutations in wheat and other cereals have shown that null mutations in genes encoding granule-bound starch synthase I (*GBSS I*) result in amylose-free starch in endosperm and pollen grains. The waxy locus in cereals is encoded by *GBSS I*, which catalyzes the elongation of amylose. In addition to its role in amylose biosynthesis, *GBSS I* is also responsible for the extension of long glucans within the amylopectin fraction (van de Wal et al. 1998). In the present study, GBSS enzyme activity peaked later than SBE and SSS (Fig. 5b, d, i, j).

This suggested that SBE and SSS may control starch synthesis at the transcriptional level, and that *GBSS I* may control starch synthesis at the post transcriptional level.

Biochemical evidence suggests that *SSS I* is primarily responsible for the synthesis of the shortest glucan chains with a polymerization degree of 10 glucosyl units or less (Li et al. 2000). Further extension of the chains is achieved by the activities of the *SS-II* and *SS-III* isoforms (Li et al. 2000). In the present study, SSS genes were exclusively involved in amylopectin biosynthesis in hulless barley endosperm. Furthermore, *SSS I* was the major SSS form. The relative expression of *SSS I* decreased steadily across time (Fig. 5d), and there was a corresponding decrease in the amylopectin synthesis rate (Fig. 2f). This may be one reason for the decline in the amylopectin/amylose ratio across time (Fig. 3).

Two SBE gene classes (*SBE I* and *SBE II*) differ in the length of glucan chains transferred in vitro and in their substrate specificities. *SBE II*, compared with *SBE I*, transfers shorter chains, exhibits greater affinity towards amylopectin, and exhibits greater rates of branching with amylose (Guan and Preiss 1993). The results of the present study indicated similar expression profiles for *SBE IIa* and *SBE IIb* in endosperm. Both *SBE II* genes were expressed at high levels from 5 DAA to 10 DAA, but were then down-regulated after 15 DAA (Fig. 5i, j). Declines in the expression of the *SBE II* genes may be another reason for the decline in the amylopectin/amylose ratio across time (Fig. 3). Overall, the results suggested that *SBE II* plays an important role during early stages of endosperm development in hulless barley.

## Conclusion

Overall, the results indicated that the activities of AGPP, SSS, GBSS and SBE had some correlation with the rates of starch synthesis during grain filling in both the waxy and non-waxy cultivars. GBSS has an important effect on amylose synthesis, especially during late grain filling. SSS and SBE are associated with amylopectin biosynthesis. It remains unclear how AGPP, SSS, GBSS and SBE activities are coordinated to control starch biosynthesis rates. Further research will aim to identify direct interactions between starch biosynthetic enzymes, as well as the factors that regulate the enzyme activities.

Transcriptome analysis has the advantage of being able to quantify changes in gene transcript levels at different developmental stages in wild-type species, their mutants, and cultivars with different genetic backgrounds. Therefore, expression profiling could (i) lay a foundation for identifying genes involved in the regulation of starch metabolism and (ii) provide valuable insight into the mechanism of metabolic regulation of starch biosynthesis under various physiological conditions.

The amylopectin/amylose ratio gradually declined in both Beiqing 6 and Kunlun 12 (Fig. 4). Furthermore, the peak in *GBSS I* expression was later than that of *SSS I, SSS II, SBE IIa,* and *SBE IIb*. This may explain the decline in the ratio of amylopectin to amylose during grain filling in Beiqing 6 and Kunlun 12 (Additional file 1: Figure S1). The grain of Ganken 5 has high amylopectin content but almost no amylose. *GBSS I* expression was less in Ganken 5 than in Beiqing 6 and Kunlun 12. Overall, we conclude that *GBSS I* is mainly responsible for amylose synthesis whereas *SSS I* and *SBE II* are mainly responsible for amylopectin synthesis in amyloplasts.

#### Abbreviations

AGPP: adenosine diphosphate glucose pyrophosphorylase; SSS: soluble starch synthase; GBSS: granule bound starch synthase; SBE: starch branching enzyme; DAA: days after anthesis.

#### Authors' contributions

XZ carried out the molecular studies and drafted the manuscript. JQ participated in the design of the study and modified the manuscript. HH participated in catechin analysis. LL participated in the design of the study and performed the statistical analysis. FW conceived of the study and participated in its design. All authors read and approved the final manuscript.

#### Acknowledgements

We would like to thank Dr. Gale from Department of Agronomy at Shihezi University for reviewing the paper and providing valuable comments regarding to this research.

#### Competing interests

The authors declare that they have no competing interests.

#### Funding

This work was supported by The Earmarked Fund for Modern Agro-industry Technology Research System (CARS-05) and Animal and Plant Breeding Program of Research and Development Project of Science and Technology in Shihezi University (gxjs2015-yz02).

#### References

Beckles DM, Craig J, Smith AM (2001) ADP-glucose pyrophosphorylase is located in the plastid in developing tomato fruit. Plant Physiol 126(1):261–266. doi:10.1104/pp.126.1.261

Buléona A, Colonnaa P, Planchot V, Ball S (1998) Starch granules: structure and biosynthesis. Int J Biol Macromol 23(2):85–112. doi:10.1016/S0141-8130(98)00040-3

Cheng F, Jiang D (2001) The dynamic change of starch synthesis enzymes during the grain filling stage and effects of temperature upon it. Acta Agron Sin 27(2):201–206

Doehlert DC, Kuo TM, Felker FC (1988) Enzymes of sucrose and hexose metabolism in developing kernels of two inbreds of maize. Plant Physiol 86(4):1013–1019. doi:10.1104/pp.86.4.1013

Guan HP, Preiss J (1993) Differentiation of the properties of the branching isozymes from maize (*Zea mays*). Plant Physiol 102(4):1269–1273. doi:10.1104/pp.102.4.1269

He Z-f (1981) Determination of amylose, amylopectin and total starch in grains by dual wavelength. Prog Biochem Biophys 1:70–72

Imberty A, Buléon A, Tran V et al (1991) Recent advances in knowledge of starch structure. Starch Stärke 43(10):375–384. doi:10.1002/star.19910431002

James MG, Denyer K, Myers AM (2003) Starch synthesis in the cereal endosperm. Curr Opin Plant Biol 6(3):215–222. doi:10.1016/S1369-5266(03)00042-6

Jian S, Guo W, Hu H et al (2002) Relationship between activities of key enzymes involving in starch synthesis and starch accumulation in wheat. J Yangzhou Univ 24(4):49–53

Li L, Sun B (1980) Study on determination of micro ATP with luciferin–luciferase reagent. Prog Biochem Biophys 6:60–62

Li Z, Mouille G, Kosar-Hashemi B et al (2000) The structure and expression of the wheat starch synthase III gene. Motifs in the expressed gene define the lineage of the starch synthase III gene family. Plant Physiol 123(2):613–624. doi:10.1104/pp.123.2.613

Liu X et al (2004) Activities of the enzymes involved in starch synthesis and starch accumulation in the grains of wheat cultivars, GC8901 and SN1391. Agric Sci China Spons Chin Acad Agric Sci 4(5):348–354

Manners DJ (1989) Recent developments in our understanding of amylopectin structure. Carbohyd Polym 11(2):87–112. doi:10.1016/0144-8617(89)90018-0

McCue KF, Hurkman WJ, Tanka CK et al (2002) Starch-branching enzymesSbe1 andSbe2 from wheat (Triticum aestivum cv. Cheyenne): molecular characterization, development expression, and homoeologue assignment by differential PCR. Plant Mol Biol Report 20(2):191–192. doi:10.1007/BF02799436

Nakamura Y (2002) Towards a better understanding of the metabolic system for amylopectin biosynthesis in plants: rice endosperm as a model tissue. Plant Cell Physiol 43(7):718–725. doi:10.1093/pcp/pcf091

Nakamura Y, Yuki K (1992) Changes in enzyme activities associated with carbohydrate metabolism during the development of rice endosperm. Plant Sci 82(1):15–20. doi:10.1016/0168-9452(92)90003-5

Nakamura Y, Yuki K, Park S-Y et al (1989) Carbohydrate metabolism in the developing endosperm of rice grains. Plant Cell Physiol 30(6):833–839. doi:10.1093/oxfordjournals.pcp.a077813

Okita TW (1992) Is there an alternative pathway for starch synthesis? Plant Physiol 100(2):560–564. doi:10.1104/pp.100.2.560

Preiss J (1988) Biosynthesis of starch and its regulation. Biochem Plants 14:181–254

Preiss J, Sivak M (1998) Biochemistry, molecular biology and regulation of starch synthesis. Genetic engineering. Springer, Berlin, pp 177–223

Tan C-x, Feng C-n, Guo W-s et al (2010) Expression of starch synthase genes and starch synthesis in wheat grains. J Jinling Inst Technol 1:47–53

van de Wal M, D'Hulst C, Vincken J-P et al (1998) Amylose is synthesized in vitro by extension of and cleavage from amylopectin. J Biol Chem 273(35):22232–22240. doi:10.1074/jbc.273.35.22232

Wang Z-b, Li W-h, Qi J-c et al (2014) Starch accumulation, activities of key enzyme and gene expression in starch synthesis of wheat endosperm with different starch contents. J Food Sci Technol 51(3):419–429. doi:10.1007/s13197-011-0520-z

Ying-ni Cao (2012) Effect of the activities of key enzymes involved in starch biosynthesis and expression profiles of genes which encoded starch synthase in developing wheat grain [D]: Northwest A&F University, Xianyang Shi

# Assimilates mobilization, stable canopy temperature and expression of expansin stabilizes grain weight in wheat cultivar LOK-1 under different soil moisture conditions

Mahesh Kumar[1*], Susheel Kumar Raina[1,2], Venkadasamy Govindasamy[1,3], Ajay Kumar Singh[1], Ram Lal Choudhary[4], Jagadish Rane[1] and Paramjit Singh Minhas[5]

**Abstract**

**Background:** Grain yield of wheat is primarily determined by both grain number and grain weight, which often influence each other in response to environmental stimuli. Some of the genotypes are capable of maintaining high single grain weight (SGW) across the environments. Understanding mechanisms and factors associated with the superiority of such genotypes over others is necessary to enhance productivity of wheat.

**Results:** Experiments were conducted to elucidate the physiological basis of high SGW of LOK-1, a wheat cultivar grown in dry and hot environments in the central and peninsular zones of India. SGW of LOK-1 was least affected by removal of spikelets indicating little competition between the grains within the spike for assimilates. Reduction in SGW due to defoliation was less and the contribution of stem reserves to the grain development was high in LOK-1 relative to other cultivars. It seems that high level of expression of genes such as expansin *(TaExpA6)* contributes to the high SGW of LOK-1.

**Conclusions:** Source was not a limiting factor for grain growth of LOK-1 in contrast to other cultivars, whereas sink appeared to be a limiting factor in recently released/identified cultivars. Differences in the amounts of water soluble stem carbohydrate reserves translocated to grain could be one of the factors contributing to higher grain weight in LOK-1. High level expression of *TaExpA6*, one of the genes contributing to the elongation of endosperm, seems to be crucial for grain growth in wheat.

**Keywords:** Wheat, LOK-1, Source-sink relations, Grain growth, Grain weight, Canopy temperature

## Background

To meet the food demand of the world population projected to reach over 9 billion by 2050, production of wheat (*Triticum aestivum* L.) needs to be increased by 70–100% (Godfray et al. 2010). This has to be accomplished with available natural resources including the land vulnerable to abiotic stresses caused by supra-optimal ambient temperature and soil moisture deficit. Since the predicted climate change can amplify the frequency and magnitude of these stresses, enhanced efforts are needed for genetic improvement of stress tolerance in temperate crops like wheat. The key components of wheat grain yield viz., grain number and grain weight are highly vulnerable to high temperatures and soil moisture deficit (Dreccer et al. 2009; Vignjevic et al. 2015). However, some genotypes of wheat can maintain high SGW across the varying environments (Mohammad et al. 2012). Several attempts have been made to elucidate the basis of variation in yield components, particularly the size of the grain (Lopes et al. 2012; Bustos et al. 2013; Rebetzke et al. 2013; Aisawi et al. 2015). Often, the seed

*Correspondence: mahesagrawal@gmail.com
[1] School of Drought Stress Management, ICAR-National Institute of Abiotic Stress Management (NIASM), Malegaon, Baramati, Pune 413 115, India
Full list of author information is available at the end of the article

weight has been associated with both source and sinks limitations (Reynolds et al. 2007; Álvaro et al. 2008). This can also be influenced by differences in assimilates supply from current photosynthesis and pre-anthesis assimilates stored in the stem (Blum 1998; Rane et al. 2003; Fischer 2011; Vignjevic et al. 2015). Other traits such as cooler canopy also facilitates better grain development and hence the productivity of wheat (Mason and Singh 2014).

Recently, genes contributing to SGW have been demonstrated in crops like rice, barley and wheat (Ashikari et al. 2005; Weng et al. 2008; Zalewski et al. 2010; Zhang et al. 2012). Starch is the major storage reserve of wheat grains. Grain weight is primarily determined by the photo synthetic productivity, assimilation and translocation of starch in the developing grain during the crop growth (Jenner et al. 1991). In higher plants, chlorophyll is the major pigment contributing to photosynthesis which is pre-requisite for starch biosynthesis. The stability and net content of chlorophyll can be modulated by cytokinin, a phytohormone (Chang et al. 2015). *TaCKX6-D1*, a rice orthologue of *OsCKX2* (*O. sativa Cytokinin oxidase 2*), has been associated with grain weight in wheat (Zhang et al. 2012). Specific suppression of *OsCKX2* enhanced tiller number and grain weight in transgenic rice (Yeh et al. 2015). Amylopectin, the more abundant polymer of starch is synthesized by the coordinated actions of AGPase, soluble starch synthase (SS), starch branching enzyme (BE) and starch debranching enzyme (DBE) (Kang et al. 2013). Other genes coding for expansins have also been implicated in wheat grain development. Lizana et al. (2010) reported abundance of *TaExpA6* transcripts in the pericarp and endosperm of wheat at 187 thermal time (°Cd) (10 days after anthesis) close to the peak expression detected by RT-PCR. All these reports prompted us to investigate the role of genes like *TaCKX6-D1*, expansin and starch synthase in contributing to higher grain growth in LOK-1 cultivar of wheat. However, many of the hypotheses related to grain weight cannot be generalized due to genotype by environment interaction and location specific trait evaluation.

LOK-1, one of the popular cultivars of wheat, in India for the last 20 years owing to its superior quality grains. This cultivar yields stable and large grains across the environment though the yield potential is relatively less than the recently released cultivars for the region (AICW & BIP Report 2012–13).

We conducted the experiments to elucidate the physiological and molecular basis of differences between SGW of LOK-1 and that of other recently released cultivars. The main objective was to identify factors responsible for high SGW in the former genotype by assessing contribution of source, sink, current photosynthesis assimilate, pre-anthesis stem reserves, canopy temperature and genes associated with the grain development in wheat.

## Methods

### Growing conditions and plant materials

Field experiments were conducted at ICAR-NIASM experimental farm, Baramati, Maharashtra located in western part of India (18°9′N, 74°28′E) during three growing seasons, 2012–2013, 2013–2014, and 2014–2015. The experimental site was featured by 60–70 cm deep black soil with about 30% silt and 40% clay layered over native basaltic soil. Each experiment consisted of a split plot design with four replications and plots of 2 by 1.5 m (4 or 6 rows 0.23 m apart). Locally adapted high yielding cultivars namely NIAW-34, NIAW-301, HD-2189 and LOK-1 were chosen for the studies (Table 1; Additional file 1: Table S1). Experiments were executed between 10–20th of November and 30th of March in all cases. Planting was carried out manually with a seed rate of 100 kg $ha^{-1}$. Mean temperature during grain filling was around 25.5 °C during the 3 years, with average maximum temperature of 29 °C. Seasonal precipitations ranged from 8 to 76 mm. In addition, flood-irrigation with 250 mm of water was provided during each crop season to ensure

**Table 1 Genetic, physiological and yield characteristics of four wheat cultivars used in the study**

| Cultivar name | Pedigree | Year of release | Days to anthesis[a] | Days to physiological maturity[a] | Number of tillers/$m^2$ | | Yield (g/$m^2$) | |
|---|---|---|---|---|---|---|---|---|
| | | | | | Well water | Water stress | Well water | Water stress |
| LOK-1 | S-308 × S-331 (Sonalika × Chhotilarma) | 1979 | 57–65 | 81–86 | 335 | 289 | 426 | 396 |
| HD-2189 | HD 1963/HD 1931 | 1980 | 68–72 | 86–102 | 348 | 261 | 454 | 405 |
| NIAW-34 | CIANO-79/(SIB) Parula | 1997 | 62–70 | 88–108 | 309 | 273 | 394 | 368 |
| NIAW-301 | Seri-82/Hork | 2002 | 65–70 | 90–105 | 325 | 286 | 404 | 370 |
| CD (p = 0.05) | | Between stress levels at same variety | | | 21.9 | | 24.95 | |
| | | Between varieties at same stress level | | | 24.15 | | 27.59 | |

[a] Based on experimental data

sufficient soil moisture. There were no rains during the grain development phase and the post anthesis water stress was imposed by withholding irrigation a week before anthesis. Experimental plots were fertilized before planting with 60 kg Nitrogen, 60 kg $P_2O_5$ and 40 kg $K_2O$ $ha^{-1}$. Additional nitrogen of 30 kg $ha^{-1}$ was applied at crown root initiation and at maximum tillering stage represented by Zadoks scale 1.3 and 2.8 respectively (Zadoks et al. 1974). Weeds were controlled manually throughout the experimental period.

## Measurements and analyses

### *Soil moisture measurements*

Soil samples from each experimental unit were collected in aluminum boxes with secure lids every week at two depths (15 and 30 cm) by using augers. The samples were weighed immediately and then oven dried at 105 °C for 72 h for determining soil moisture content by gravimetric method (Fig. 1).

### *Grain growth*

At anthesis, 65–70 identical main shoot spikes of each cultivar from every experimental units were ear marked. Five spikes were harvested to measure grain dry weight at every 3rd or 4th day beginning from day of anthesis till 30 days after anthesis (physiological maturity). The harvested spikes were dried in an oven at 70 °C for 72 h. Grains were separated from the middle six spikelets of each of the five spikes for measuring SGW (Laxman et al. 2014). Grain dry weight (mg) of each cultivar plotted against growing degree days after anthesis (GDDAA).

### *Sink limitation*

At 50% anthesis, four pairs of identical main-stem spikes (a total of 8) were selected and tagged in each plot (Ma et al. 1996). Since the spikes in each pair had the same number of spikelets, alternate spikelets from one of the spikes were removed by hand to reduce the sink size by 50% while other spike was kept as a control. It was assumed that the removal of half of the spikelets will double the assimilate supply to the remaining spikelets on the spikes if source is a limiting factor. At maturity, the ears were harvested separately and grains were separated from the spike manually for recording the number of spikelets, grains per spike and grain weight. SGW was measured by dividing the grain weight per spike at approximately 12% moisture level by the number of grains per spike.

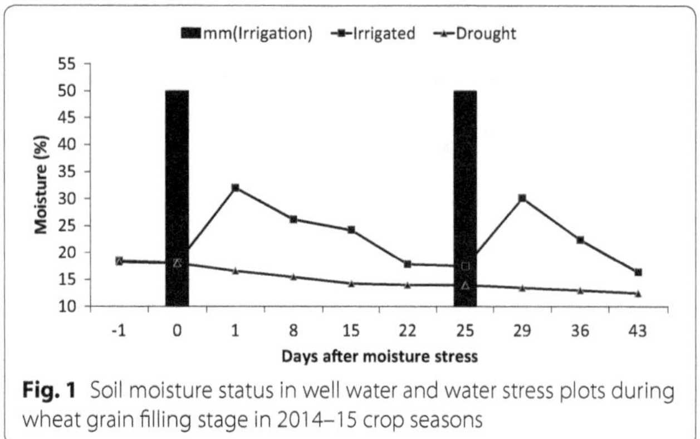


**Fig. 1** Soil moisture status in well water and water stress plots during wheat grain filling stage in 2014–15 crop seasons

### *Source limitation*

Seven days after anthesis, four pairs of main stem spikes (a total of 8 ears) were selected and tagged in each plot (Ma et al. 1996). In each plant only main stem was retained and rests of the tillers were removed to ensure the uniformity in sink and source during this experiment. Leaves of one of the main stem in each pair were completely removed to prevent the supply of current photosynthesis assimilates for the developing grains, while identical main stem of neighboring plant in each plot was kept as control. At harvest, all the pairs of main stems were collected and grains were separated by hand for measuring grain number and grain weight.

### *Net $CO_2$ assimilation rate*

Net $CO_2$ assimilation rate was recorded through photosynthesis system (GFS 3000; Walz, Germany) during 10.00 AM to 2.00 PM in the fully expanded flag leaf (Biswas et al. 2013). Data were recorded three times in the flag leaf at 55, 71 and 88 DAS. Reference $CO_2$ levels were maintained at 400 μmol $s^{-1}$, and chamber flow was set at 750 μmol $s^{-1}$. Ambient photosynthetically active radiation (PAR) during the measurements ranged from 1000 to 1200 μmol $m^{-2}$ $s^{-1}$, while ambient humidity ranged between 30 and 40%.

### *Canopy temperature*

Canopy temperature was recorded by thermal imager (VarioCam®hr inspect 575, Jenoptic, Germany) that operates in the wave bands of 8–14 μm with a thermal resolution of 0.01 °C. Thermal images with spatial resolution of 768 × 576 pixels were captured seven times covering crop growth stages just before and after anthesis. A tripod perpendicular to the area being imaged was used to ensure constant distance and angle between the camera and crop canopy for during all the measurements. Dry and wet references were used to mimic leaves with fully closed and fully opened stomata, respectively (Jones et al. 2002) and to avoid extreme conditions while capturing images. Emissivity for all the measurements were set at 0.96 (Jones 2004). IRBIS® software (Jenoptic, Germany) was used to analyse thermal images. A total of six areas of interest in each thermal image were outlined manually for analysis. Images in the visible wavelength were referred to exclude noise from ground area. Minimum canopy temperature in each of the sections in each thermal image was considered to get a value that represents cooling capacity of cultivar. Since the variation among the cultivars for this trait at any given point of time was marginal, we opted to compute area under curve of canopy temperature vs days with reference to anthesis for assessing genetic variation in plant's capacity to keep the canopy cooler.

#### Estimation of stem water soluble carbohydrates (WSC)

Five plants at anthesis and at harvest were randomly selected and separated into leaf blades, and stems along with leaf sheath. All samples were dried at 70 °C for 48 h and weighed. The WSC in the stems was measured at both anthesis and maturity. WSC were extracted from 0.1 g of ground stem material by extracting once with 8 mL of 80% ethanol at 80 °C in water bath followed by two extractions with 8 mL distilled water at 70 °C. Each extraction was carried out for at least 1 h. The extract was centrifuged at room temperature for 10 min at 4000*g* and the extracts were combined. Total soluble sugars in the samples were analyzed by the anthrone method (Yemm and Willis 1954). Mobilization efficiency of assimilates was estimated by the proportion (%) of WSC at physiological maturity to post anthesis for maximum WSC (Ehdaie et al. 2006).

#### qRT-PCR analysis of genes associated with grain development

Developing grains of all the four cultivars were collected 8 days after anthesis to assess the expression of *TaCKX6-D1*, *TaExpA6* and *TaSSIIA* genes. For each variety, three biological samples, each represented by the mixed samples at the same phonological growth stage from at least three individual spikes, were used for qRT-PCR. Three independent amplifications were performed on each biological sample. Total RNA was extracted with RNA extraction kit (Qiagen, Germany). DNA was removed by digestion with DNAse I (Thermo Scientific, USA) before reverse transcription. The RNA samples were quantified using a UV–Visible spectrophotometer (SMA-3000) and 2 μg of total RNA were used for first-strand cDNA synthesis using Superscript cDNA synthesis kit (Invitrogen, USA). Allele-specific primer pairs for *TaCKX6-D1* (Zhang et al. 2012) and *TaExpA6* (Lizana et al. 2010) were used to detect transcripts of *TaCKX6-D1* and *TaExpA6* although primers for starch synthase IIA (*TaSSIIA*) were designed with the help of Primer3 plus software using the sequence available in NCBI. Amplification of glyceraldehyde-3-phosphate dehydrogenase (*GAPDH*) was used as an internal control for relative quantification. qRT-PCR was performed on a Biorad make CFX96 machine using the iQ SYBR Green master mix (Biorad, USA). The procedure used was initial polymerase activation for 30 s at 95 °C followed by 40 cycles of 95 °C for 10 s and 60 °C for 30 s. Data from the quantitative analysis of expression were transformed by the $2^{-\Delta\Delta Ct}$ method (Livak and Schmittgen 2001). The details of primers used in the study are given in Table 2.

#### Experimental design and statistical analysis

The field experiment was conducted in split plot design in four replications. However, individual plants were considered as replications except for canopy temperatures and net $CO_2$ assimilation rate. Data were analysed by two way ANOVA technique using the SPSS statistical software (version 16) to determine the effect of each treatment. If the F-ratio was significant, a multiple mean comparison was performed using Duncan Multiple range Test (DMRT) at α-0.05. Simple statistics with analysis of variance were also computed to differentiate the treatment effects by using statistical analysis software package SAS® 9.3.

## Results

#### Weather during crop season

During the crop season, average minimum temperatures ranged from 14.3 to 17.0 °C while the average maximum temperature ranged from 30.5 to 32.3 °C

**Table 2 Details of the primers used in the study**

| Primer name | Primer sequence | Product size (bp) | References |
|---|---|---|---|
| C19EF4 | 5′-CGACGAGATCTTACGGTTCT-3′ | ~230 | Zhang et al. (2012) |
| C19ER4 | 5′-GACCGATGGATCAGCCA-3′ | | |
| GAPDHF | 5′-TTAGACTTGCGAAGCCAGCA-3′ | ~81 | This study |
| GAPDHR | 5′-AAATGCCCTTGAGGTTTCCC-3′ | | |
| TaExp6F | 5′-CAATCCTCCCCGCGA AC-3′ | ~217 | Lizana et al. (2010) |
| TaExp6-R | 5′-GGTCCCCTTCACCGACAT-3′ | | |
| TaSSIIAF | 5′- ACGTCATGAACGTGGTCGTC-3′ | ~198 | This study |
| TaSSIIAR | 5′-CCTGTCCAGCAGCCTTGTA-3′ | | |

(Table 3; Fig. 2). Crop season 2012–13 was the hottest and 2014–15 was the coolest among all three crop season throughout the crop growth and more so at grain development.

### Changes in unit grain weight across grain growth period

Results clearly revealed that grains of LOK-1 accumulated more photosynthesis/starch assimilates than other cultivars under both well watered and water stress conditions in the field during grain growth (Fig. 3). The difference between SGW of LOK-1 and other variety was conspicuous during all three crop seasons. Rate of grain growth as represented by slope of trend line plotted for grain weight vs growing degree days after anthesis further revealed that LOK-1 accumulates assimilates faster than other cultivars.

### Effect of spikelet removal on SGW

Reduction in average number of grains per spike as a result of removal of spikelets ranged from 20.5 to 31.4 across the genotypes and soil moisture levels. Consequently, increase in SGW in response to spikelet removal ranged from 4.8 to 24.8% under soil moisture stress and 2.1–8.0% under ample soil moisture conditions (Fig. 4A). However, there was no significant impact of removal of spikelets on SGW of LOK-1 under both well water and water stress condition. In contrast, there was significant increase in the SGW of HD-2189 and NIAW- 34 particularly under soil moisture deficit (Fig. 4B).

SGW of LOK-1 was not sensitive to variation in grain number caused by removal of grains in contrast to other tested cultivars (Additional file 2). This suggested that more grain weight in LOK-1 was not necessarily due to less number of grain, but as a result of intrinsic sink capacity to accommodate the assimilates relatively in larger quantity as compared to other genotypes.

### Effect of defoliation on SGW

Defoliation caused a reduction in SGW to an extent of 12.5–17.6% in all the cultivars under well water conditions, while the reduction was 17.4–23.0% under limited moisture conditions (Fig. 5A, B). LOK-1 and NIAW-301 exhibited marginal reduction in grain weight compared to others indicating that these cultivars utilize reserve carbohydrates for grain development and leaves as source was not a limiting factor.

Impact of defoliation was also noticed on number of grains per spike, with the largest reduction in HD-2189 as compared to others (Fig. 5C, D). However, defoliation mediated reduction in grain number under both moisture conditions in other cultivars was not significant.

### Effect of soil moisture on Net $CO_2$ assimilation rate ($P_n$)

Mean net $CO_2$ assimilation rate ranged from 24.3 to 28.0 and 14.2–16.2 µmol $m^{-2}$ $s^{-1}$ under well water and

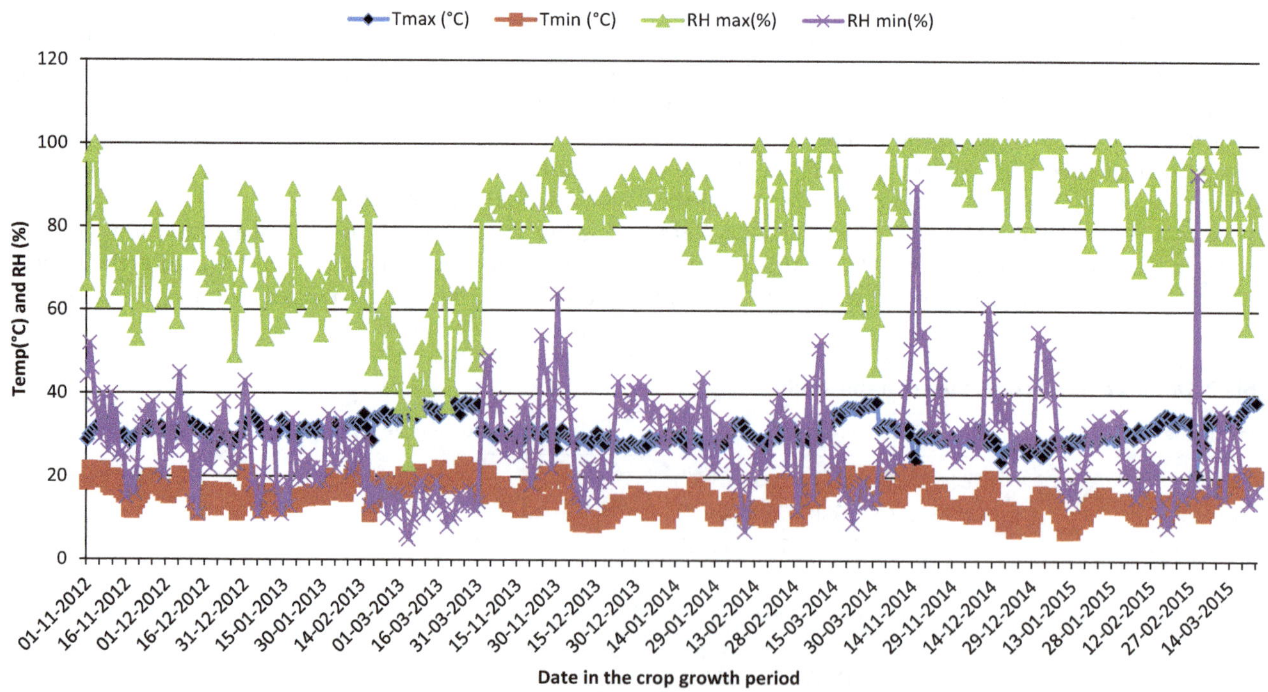


**Fig. 2** Temperature (max and min) Relative humidity (max and min) of the day during three crop seasons for the period from 2012–15

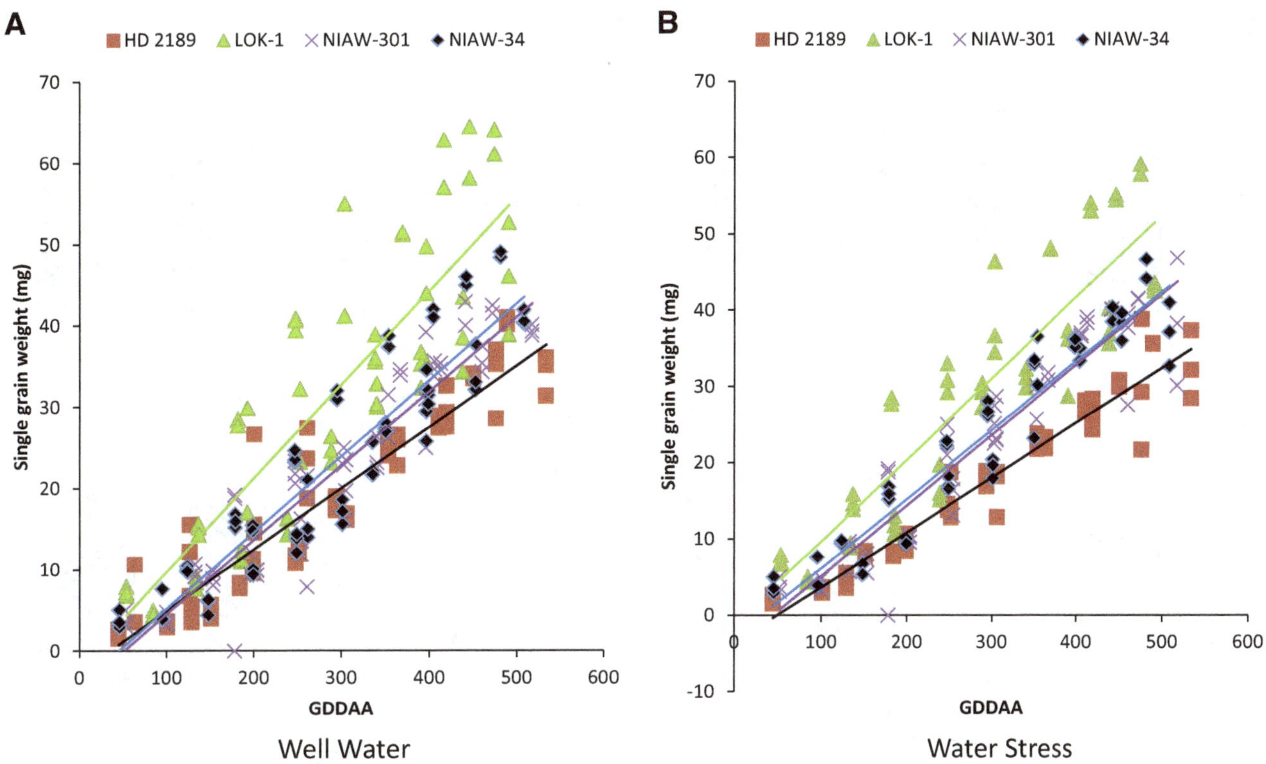


**Fig. 3** Changes in grain dry weight of 4 wheat cultivars with GDDAA during three crop seasons (2012–2015) under well watered (**A**) and water stress condition (**B**). Observation was recorded at 3–4 days interval after anthesis (each point represent average of 6–8 grain from each 10–12 spikes)

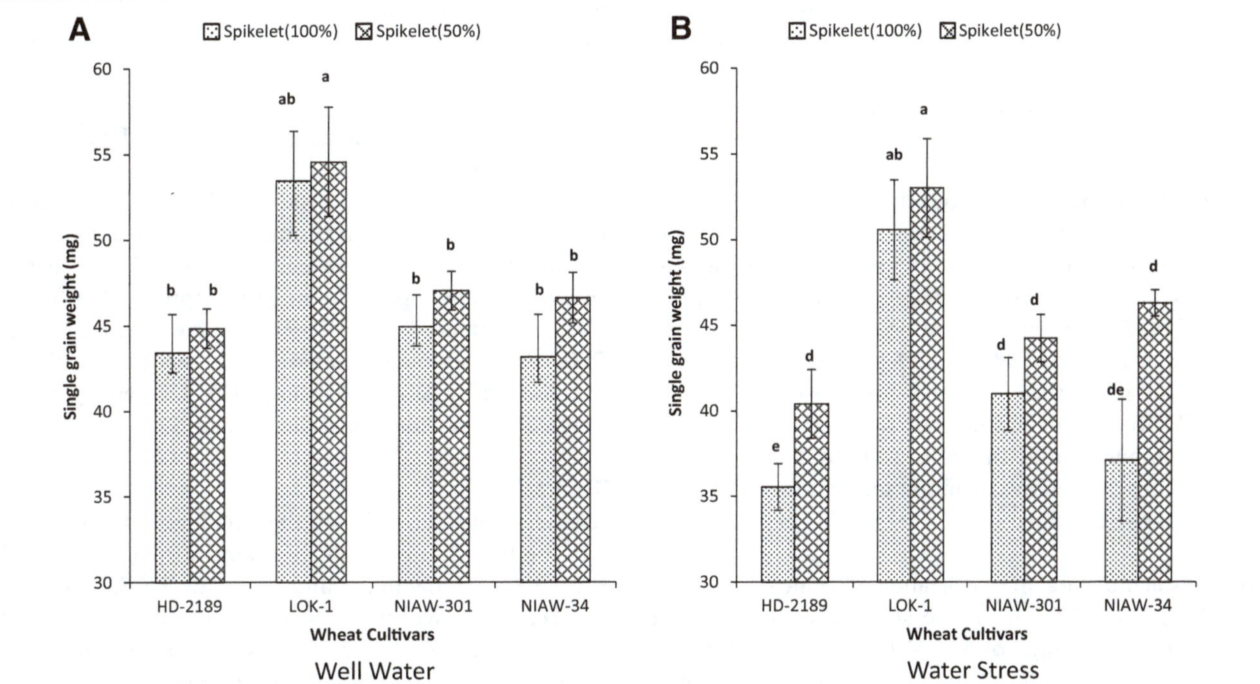


**Fig. 4** Genetic variation in grain weight at harvest with 100 or 50% grains per spike on main stem during grain growth under well watered (**A**) and post anthesis moisture stress (**B**) in four wheat cultivars. (n = 8). Means with *same letter* are not significantly different at $p < 0.05$

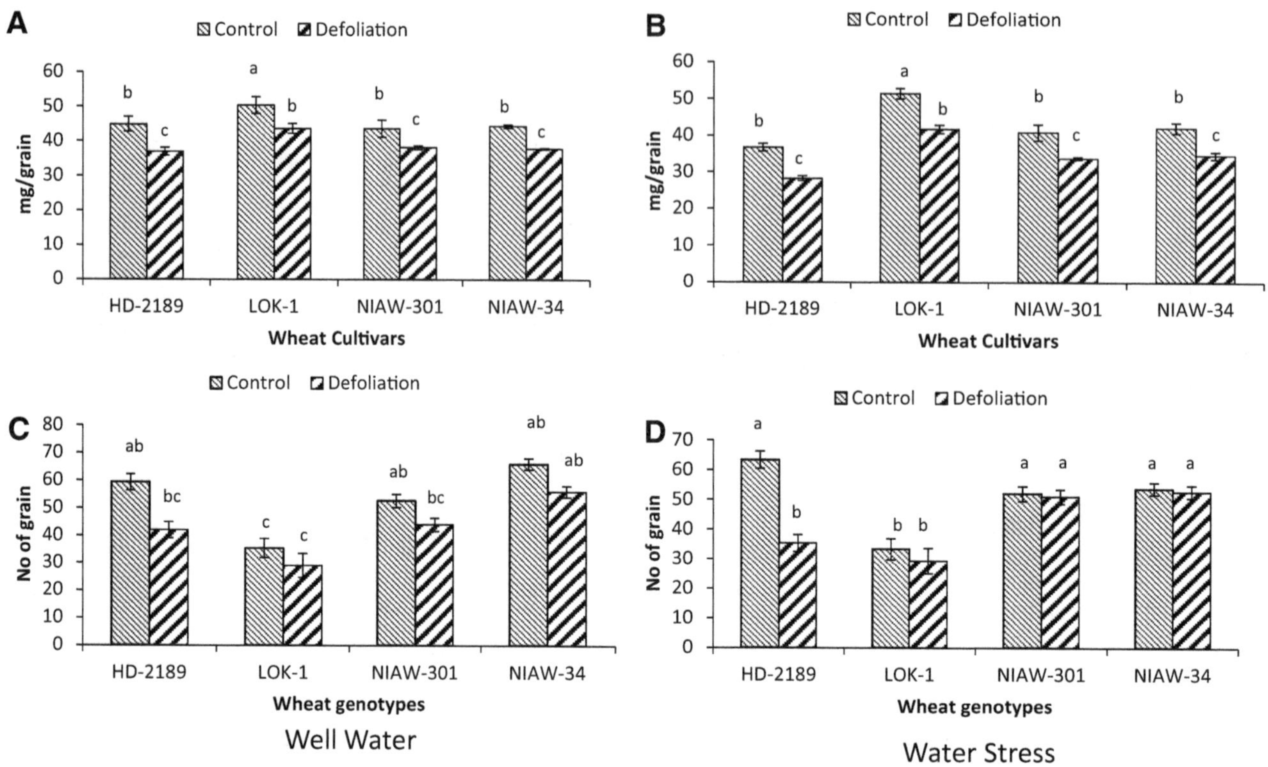


**Fig. 5** Genetic variation in single grain weight (**A, B**) and grain number (**C, D**) in presence and absence of leaves on main stem during grain growth under well water (**A, C**) and post anthesis water stress (**B, D**) in four wheat cultivars. Defoliation was done 7 days after anthesis. Means with *same letter* are not significantly different at $p < 0.05$

deficit soil moisture conditions respectively (Fig. 6A, B). In general under water stress net $CO_2$ assimilation rate in decreasing trend. Under well water conditions LOK-1 had higher Pn after anthesis (Fig. 6A) as compared to other cultivars.

### Effect of soil moisture on canopy temperature

Canopy temperature for all the four cultivars was recorded throughout the growing period in well watered as well as water stress conditions. Area under curve plotted with minimum temperature observed in the canopy vs days after anthesis (DAA) indicated that canopy temperature did not vary across the genotypes (Fig. 7). However, soil moisture deficit did not significantly affect the canopy temperature of the cultivar LOK-1 and NIAW-301. In other two cultivar canopy temperature is higher under water stress plots.

### Effect of soil moisture on mobilization efficiency

Mobilization of WSC from the stem to economic part was found in the range of 50.2–72.6% in the well water conditions and 68.5–82.7% under soil moisture limited conditions after anthesis. WSC mobilization in LOK-1 was up to 44.6 and 20.7%, in NIAW-34 was up to 38.6 and 12.15% higher than the same in other cultivars in the presence and absence of post anthesis soil moisture stress respectively (Fig. 8).

### Expression analysis of genes associated with the grain development

The wheat cultivar, LOK-1 exhibited maximum accumulation of *TaCKX6-D1* transcripts with an approx. 1.5-fold increase compared to HD-2189, NIAW-301 (Fig. 9). *TaExpA6* transcripts accumulated approx. 2.5-fold in LOK-1 compared to HD-2189. However, maximum *TaSSIIA* transcript accumulation was recorded in NIAW-34 followed by NIAW-301 and LOK-1 revealing almost identical levels (Additional file 1: Table S2).

## Discussion

Several studies have shown that the higher the temperature the lighter the weight of grains at maturity (Sofield et al. 1977; Chowdhury and Wardlaw 1978; Stone and Nicolas 1994; Wardlaw 1994; Savin and Nicolas 1996; Rane et al. 2007). In addition, soil moisture stress also severely affects the development of grain weight in wheat under different environmental conditions (Konopka et al. 2007). However, despite adverse environmental

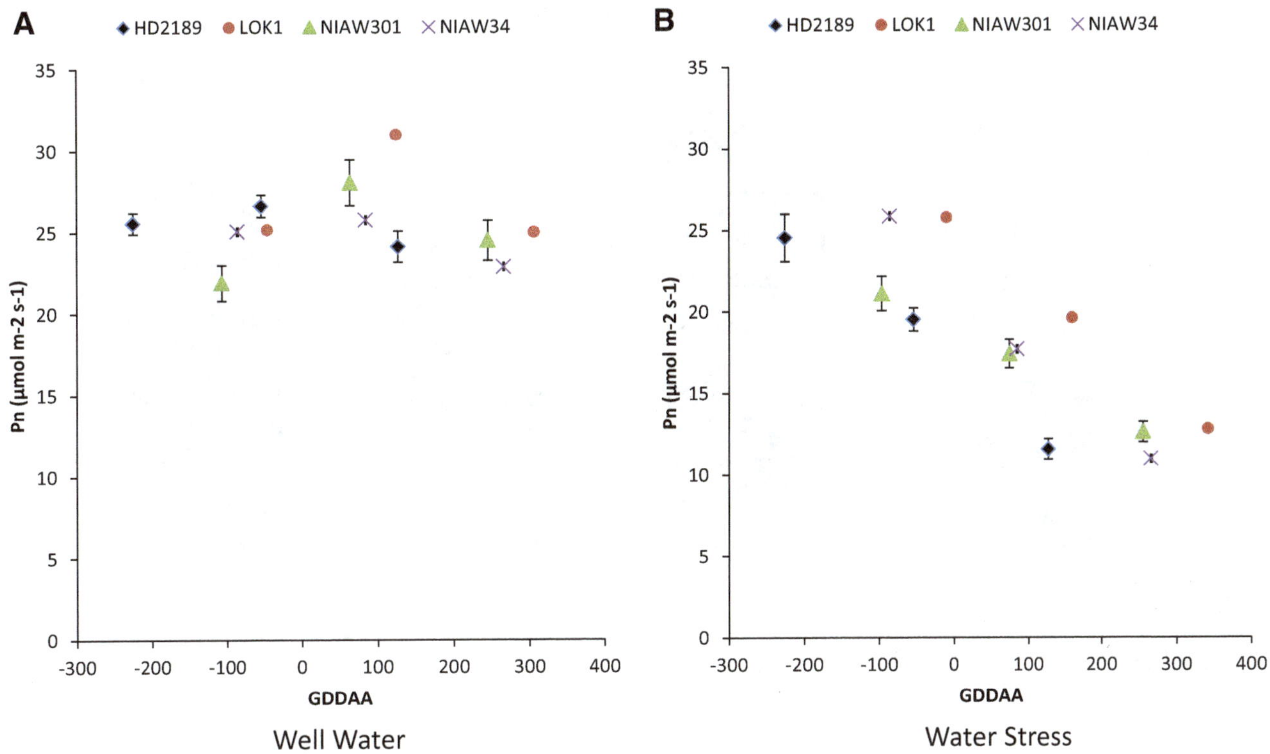


**Fig. 6** Relationship between net $CO_2$ assimilation rate and thermal time under well water (**A**) and water stress (**B**) conditions in four cultivars. Each point is mean of 6-8 data point

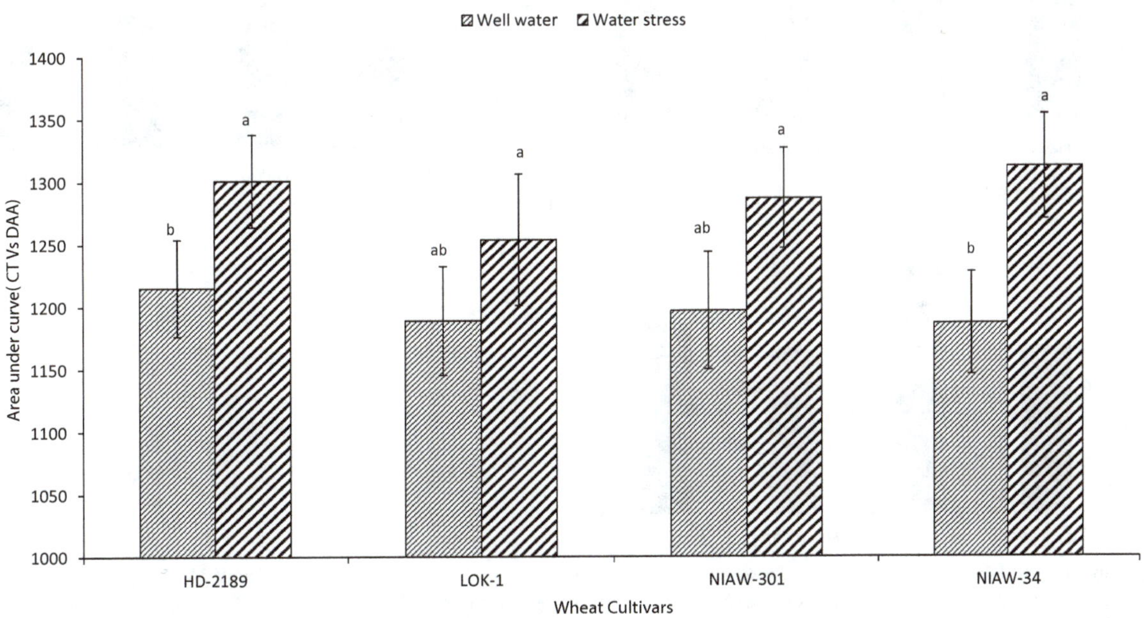


**Fig. 7** Genetic variation in canopy temperature (min) under well watered and post anthesis water stress in four wheat cultivars. Area under curve was estimated from the graph of canopy temperature vs days After Anthesis. Means with *same letter* are not significantly different at $p < 0.05$

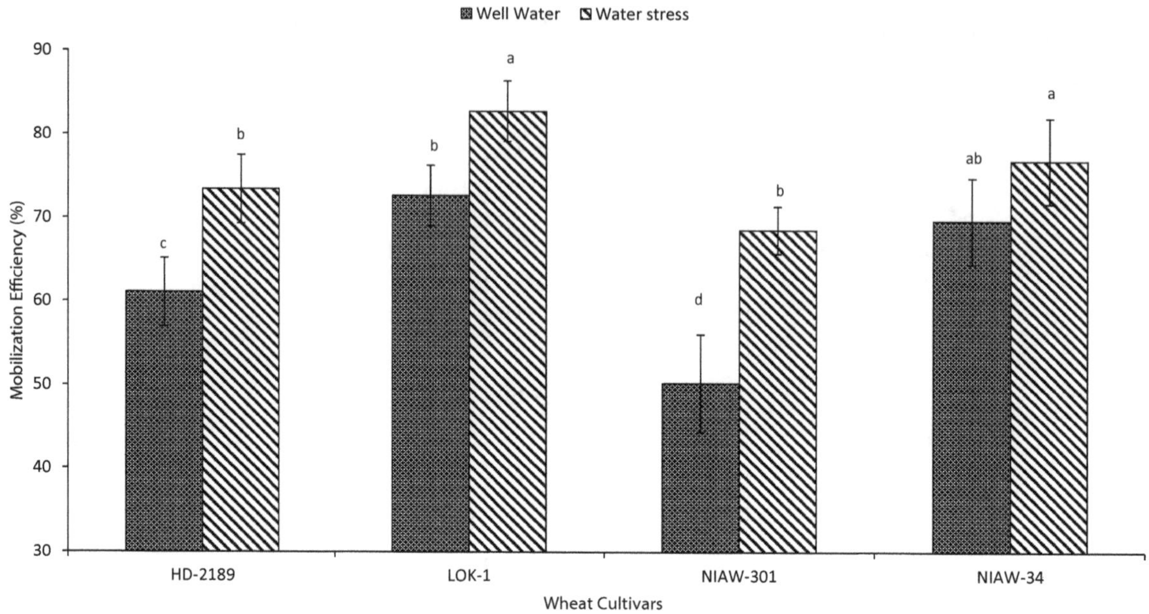


**Fig. 8** Mobilization efficiency (%) of stem WSC in four wheat cultivars under well watered and soil moisture stress conditions. Means with same letter are not significantly different at $p < 0.05$

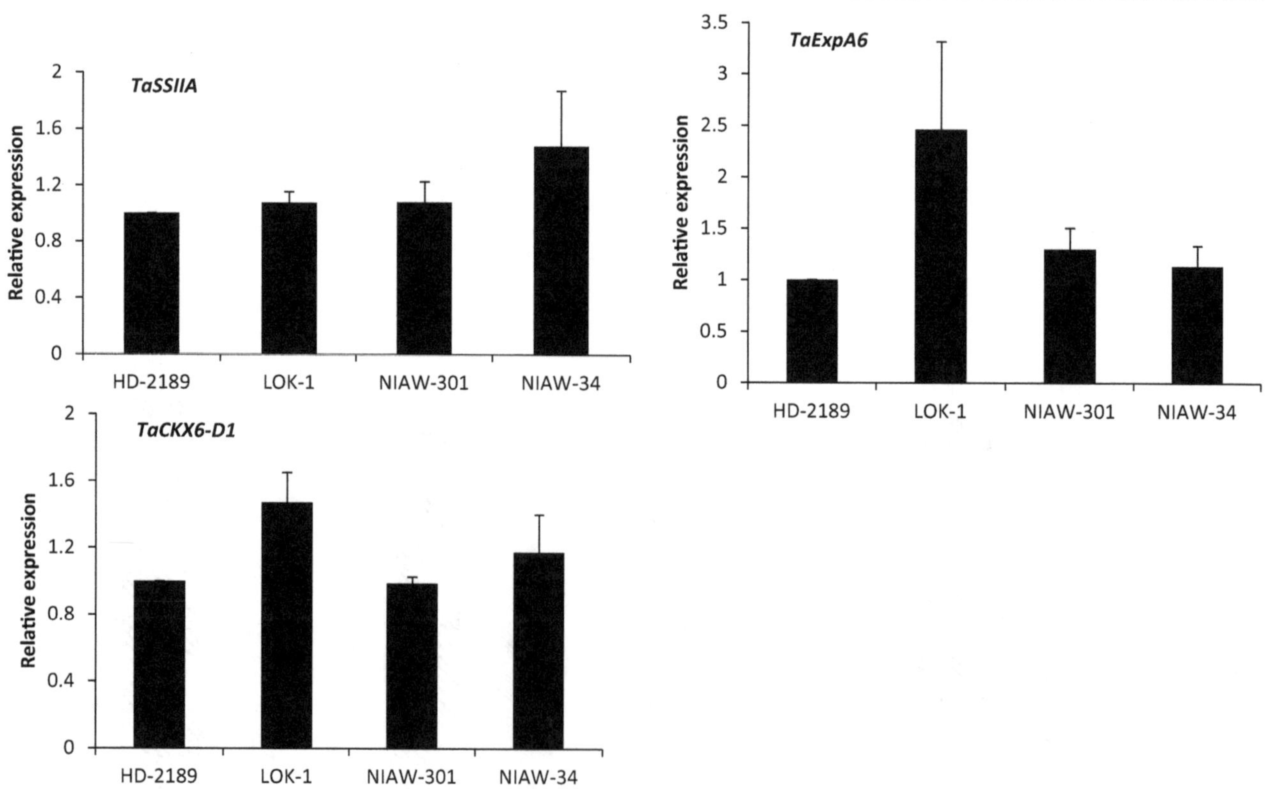


**Fig. 9** Quantitative RT-PCR analysis of *TaCKX-D1*, *TaEXpA6* and *TaSSIIA* genes in HD-2189, LOK-1, NIAW-301 and NIAW-34 wheat genotypes. Expression levels were normalized against expression of *T. aestivum* GAPDH gene as an internal control and are shown relative to HD-2189 as control genotype. The relative level of each gene in control genotype was standardized as 1. Values are presented as the mean and the *errors bars* indicate standard errors of triplicate samples

conditions some wheat genotypes maintain their grain weight. For our investigations, we choose bold grained cultivar LOK-1, one of the mega cultivars of wheat being grown in hot and dry conditions prevailing in Central and Peninsular part of India for more than two decades although it has now turned susceptible to diseases. Though its grain yield potential is less, grain weight of LOK-1 is relatively higher than those of other recently developed cultivars even under hot and dry environment. Since the grain weight and its stability are critical determinants of final grain yield we probed possible reasons for higher and stable grain weight of LOK-1 relative to three other cultivars, which are popular in these regions.

### Weather conditions during the crop seasons on SGW and total grain yield

Ambient temperatures during 2012–13 crop seasons were higher by about 2 °C than the consecutive 2 years. In addition, the humidity was substantially low and there were negligible rains during the first year (Table 3). In contrast, there were no much differences between 2013–14 and 2014–15 with respect to weather parameters. Consequently, the grain weight in general was lower during the first year relative to the same recorded during the next consecutive years. The warmer climate is affecting growth and development of plant and grain filling (Porter and Moot 1998). Irrespective of weather, the LOK-1 had higher single grain weight than other cultivars both under water stressed and well watered condition. Rate of grain filling was faster in LOK-1 in the first year as compared to other years under both water stressed and well watered conditions (Fig. 3a, b). Differences among varieties in SGW in the response of moisture stress are thought to occur because moisture stress tolerant varieties maintain faster rates of grain growth temperature-sensitive varieties, at high temperatures (Wheeler et al. 1996).

### Higher SGW was not due to fewer grains per spike

The two major yield components viz., grain number and SGW are determined at different times of the growing season and therefore are subjected to different conditions and stresses. While success of floret set at pre-anthesis largely determines grain number (Miralles et al. 1998; Gonzalez et al. 2003), SGW is dependent on the extent to which post-anthesis conditions favor grain-fill. However, grain number and weight are interrelated and these results in a capacity to compensate to a certain extent for loss in grain number by increase in SGW and many combinations of these two components are possible (Peltonen-Sainio et al. 2007). With increase in grain number SGW decreases concomitantly in wheat, but independently of any competitive relationship between growing grains (Acreche and Slafer 2006). Under well water conditions LOK-1 maintained higher SGW than other cultivars during all the three crop seasons. Even post anthesis water stress did not affect the grain filling of LOK-1 when compared with other cultivars. This signifies the recalcitrance of SGW trait of LOK-1 to environmental stress.

**Table 3 Different weather parameters during three wheat crop growing seasons for the period of 2012–2015**

| Weather parameters | 2012–2013 | 2013–2014 | 2014–2015 |
|---|---|---|---|
| Maximum temperature (°C) | 32.3 | 30.4 | 30.5 |
| Minimum temperature (°C) | 17.0 | 15.0 | 14.3 |
| RH maximum (%) | 65.2 | 83.9 | 91.6 |
| RH minimum (%) | 23.3 | 29.6 | 29.8 |
| Rain (mm) | 8.7 | 76.2 | 74.1 |

Increase in SGW in response to degraining can range from 0.5 to 13.2% (Aisawi et al. 2015). Degraining increased SGW by 3.3–8% in recently released cultivars such as HD-2189 and NIAW-34 but it was ineffective in LOK-1. Our attempt to manipulate sink size by partial grain removal revealed that LOK-1 maintained a higher SGW than that of other cultivars in all the range of grain numbers per spike indicating absence of compensation mechanisms involving these two critical traits.

### Current photosynthesis and pre-anthesis assimilates supply for grain development

Grain filling in small-grain cereals occur by acquisition of assimilates from both the current photosynthesis and through translocation of stored non-structural assimilates accumulated before anthesis. In environments featured by gradual increase in temperature and decrease in soil moisture, current photosynthesis during grain filling declines and the remobilization of stored non-structural carbohydrates becomes particularly important for grain filling (Blum 1998). Previous studies on durum wheat have shown that total dry matter translocated from the main stem to the developing grains reach values of up to 25% of the main stem biomass translocated (Álvaro et al. 2008), which are lower than those found in modern bread wheat cultivars (Ehdaie et al. 2006). Stem reserves can contribute 20–40% of the final grain weight in non-stressed environments (Van Herwaarden et al. 1998; Dreccer et al. 2009; Vignjevic et al. 2015) and this can be up to 70% under stressed conditions during grain filling (Tahir and Nakata 2005; Ehdaie et al. 2006; Rebetzke et al. 2008). Hence, we measured current photosynthesis and grain weight in the absence or presence of active leaves during grain growth to explain difference in SGW among the cultivars. There was difference in net rate of $CO_2$ assimilation in flag leaves between LOK-1 and other cultivars during grain development even under post-anthesis water stress. This was apparent particularly during the peak phase of assimilates accumulation in grains indicating that this might be one of the reasons for higher single grain weight. In addition, important contribution of pre-anthesis reserves was also evident in LOK-1 from both defoliation and WSC experiments. Further, in contrast to other cultivars, LOK-1 had constant SGW even after reduction in sink size resulted from removal of spikelets at anthesis, which suggested that there was no competition for assimilates among the grains within the spike. It is largely accepted that stem reserve mobilization in grains is affected by sink size, environment and variety. Mobilization efficiency and sink strength (number of grains per spike and grain weight) are two component traits involved in the extent of contribution of stored reserves to grain yield in wheat (Ehdaie et al. 2006). NIAW-34 with higher grain number per spike and LOK-1 with higher SGW explains similar mobilization efficiency in both the cultivar.

#### Impact of canopy temperature on SGW

Capacity to keep the canopy cool has been often associated with grain yield in wheat (Mason and Singh 2014). Canopy temperature has been shown to correlate with physiological traits contributing to grain yield under field conditions (Reynolds et al. 2007; Lopes and Reynolds 2010; Rebetzke et al. 2013). We hypothesized that the higher SGW in LOK-1 could be due to its capacity to keep its canopy cooler for longer duration during grain development. Instead of routine method of associating grain yield with canopy temperature measured at any given point of time during crop growth, we considered duration of cooler canopy as represented by area under curve of canopy temperature vs days after anthesis. Here we found that genotype tested in present experiment have similar canopy temperature dynamics during their growth period. However in LOK-1 behave almost similar in both moisture stress as well as normal soil moisture conditions.

#### Genes associated with the grain growth and Development

In order to get an insight into molecular mechanism governing grain growth in wheat, we studied the transcript abundance of *TaCKX6-D1*, *TaExpA6* and *TaSSIIA* genes in the four genotypes used in present studies. Since minimum grain growth rate was recorded in HD-2189, it was considered as control genotype for comparing the relative transcript abundance of different genes. Cytokinin oxidases/dehydrogenases (CKX) catalyze the irreversible degradation of the CKs iso-pentenyl adenine, zeatin and their ribosides in a single enzymatic step by oxidative side-chain cleavage (Schmülling et al. Schmu¨lling et al. 2003). Gene coding for this enzyme has been found to play a role in determining the grain development in rice (Ashikari et al. 2005), barley (Zalewski et al. 2010) and wheat (Hess et al. 2002; Zhang et al. 2012). Hence, an attempt was made in this study to elucidate the role if any of these genes in determining SGW of wheat. Previously, maximum accumulation of cytokinin oxidase/dehydrogenase (CKX) transcripts was reported 8 days after pollination in wheat. When we measured transcripts abundance for *TaCKX 6-D1* gene 8 days after anthesis in four wheat cultivars, maximum expression was observed in LOK-1 indicating possibility of diminishing CKs activity in this cultivar due to increased inhibition of cytokinin catabolism. This is in consistency with earlier report that CKs gradually decrease after the soft-dough stage (Hess et al. 2002). However, further investigations are needed to probe deletions, if any in the sequences of the gene or factors that bring changes at post translational levels which might be contributing to enhanced grain size in wheat.

Cell growth in plant is primarily constrained by cell wall which must be loosened to allow for cell expansion and expansin gene play a role in the cell wall loosening (Cosgrove 2005). Since expansins have been associated with the grain size in wheat (Lizana et al. 2010), we quantified the expression of *TaExpA6*, one of the genes contributing to the elongation of endosperm, which is crucial for grain growth. The possible role of this gene in determining the size of the grain was evident from approx. 2.5 fold enhanced expression in LOK-1 compared to HD-2189.

Starch synthase genes are known to regulate biosynthesis of amylose as well as elongation of glucan/glucose side chains in amylopectin during starch synthesis in the grains of wheat. Wheat grains exposed to elevated temperatures exhibited reduced grain growth and decreased activity of soluble starch synthase in grains (Prakash et al. 2009). In the present study, maximum expression of *TaSSIIA* was recorded in NIAW-34, although there was no appreciable change in its levels across the other genotypes tested- NIAW-301, LOK-1 and HD-2189, indication of a probably minor role for this gene in the development of grain size.

## Conclusions

- Our experiment confirmed that LOK-1, a wheat cultivar known for bold seeded grain, accumulates more biomass as compared to recently released cultivars for the hot and dry regions of India, both under well water and soil moisture stress conditions.
- Source was not a limiting factor for grain growth of LOK-1 in contrast to other cultivars.

- Sink appeared to be a limiting factor in recently released high yielding wheat cultivars like NIAW-34 but not in LOK-1.
- The difference in single grain weights between LOK-1 and any of the other varieties is not due to differences in canopy temperature; however, this genotype has greater capacity to maintain its canopy temperature cooler even under soil moisture deficit.
- Differences in the amounts of water soluble stem carbohydrate reserves could be one of the factors contributing to higher single grain weight in LOK-1.
- *TaExpA6*, one of the expansin genes contributing to the elongation of endosperm, seems crucial for grain growth.

### Abbreviations

SGW: single grain weight; GDDAA: growing degree days after anthesis; WSC: water soluble carbohydrates; DAS: days after sowing.

### Authors' contributions

MK designed experimental plan, conducted experiment and written the manuscript, VG helped in field experimental plan, data analysis and manuscript writing; SKR and AKS performed molecular analysis, RLC helped in field experiment management and data collection, JR guided in designing experimental plan, statistical analysis and manuscript writing. PSM guided experiment and edited the manuscript. All authors read and approved the final manuscript.

### Author details

[1] School of Drought Stress Management, ICAR-National Institute of Abiotic Stress Management (NIASM), Malegaon, Baramati, Pune 413 115, India. [2] ICAR-Central Institute of Temperate Horticulture, Old Air Field, PO Rangreth, Srinagar, J & K 190007, India. [3] Division of Microbiology, ICAR-Indian Agricultural Research Institute, New Delhi 110012, India. [4] School of Edaphic Stress Management, ICAR-National Institute of Abiotic Stress Management, Malegaon, Baramati 413 115, India. [5] ICAR-National Institute of Abiotic Stress Management, Malegaon, Baramati 413 115, India.

### Acknowledgements

The authors gratefully acknowledge financial support received from Indian Council of Agricultural Research, New Delhi, India in the form of an institutional project (Project Code No. IXX09675). Authors also acknowledge SASM for providing meteorological data used in this manuscript.

### Competing interests

The authors declare that they have no competing interests.

### References

Acreche MM, Slafer GA (2006) Grain weight response to increases in number of grains in wheat in Mediterranean area. Field Crops Res 98:52–59

Aisawi KAB, Reynolds MP, Singh RP, Foulkes MJ (2015) The physiological basis of the genetic progress in yield potential of CIMMYT spring wheat cultivars from 1966 to 2009. Crop Sci 55:1749–1764

Álvaro F, Isidro J, Villegas D, Garcíadel MLF, Royo C (2008) Breeding effects on grain filling, biomass partitioning, and remobilization in Mediterranean durum wheat. Agron J 100:361–370

Ashikari M, Sakakibara H, Lin S, Yamamoto T, Takashi T, Nishimura A et al (2005) Cytokinin oxidase regulates rice grain production. Science 309:741–745

Biswas DK, Xu H, Li YG, Ma BL, Jiang GM (2013) Modification of photosynthesis and growth responses to elevated $CO_2$ by ozone in two cultivars of winter wheat with different years of release. J Exp Bot. doi:10.1093/jxb/ert005

Blum A (1998) Improving wheat grain filling under stress by stem reserve mobilisation. Euphytica 100:77–83

Bustos DV, Hasan AK, Reynolds MP, Calderini DF (2013) Combining high grain number and weight through a DH-population to improve grain yield potential of wheat in high-yielding environments. Field Crops Res 145:106–115

Chang C, Lu J, Zhang HP, Ma CX, Sun G (2015) Copy number variation of cytokinin oxidase gene Tackx4 associated with grain weight and chlorophyll content of flag leaf in common wheat. PLoS ONE 10(12):e0145970. doi:10.1371/journal.pone.0145970

Chowdhury SI, Wardlaw IF (1978) The effect of temperature on kernel development in cereals. Aust J Agric Res 29:205–223

Cosgrove DJ (2005) Growth of the plant cell wall. Nat Rev Mol Cell Biol 6:850–861

Dreccer MF, Herwaarden AFV, Chapman SC (2009) Grain number and grain weight in wheat lines contrasting for stem water soluble carbohydrate concentration. Field Crops Res 112(1):43–54

Ehdaie B, Alloush GA, Madore MA, Waines JG (2006) Genotypic variation for stem reserves and mobilization in wheat: I. Post anthesis changes in internode dry matter. Crop Sci 46:735–746

Fischer RA (2011) Wheat physiology: a review of recent developments. Crop Pasture Sci 62(2):95–114

Godfray HC, Beddington JR, Crute IR, Haddad L, Lawrence D, Muir JF, Pretty J, Robinson S, Thomas SM, Toulmin C (2010) Food security: the challenge of feeding 9 billion people. Science 327:812–818

Gonzalez FG, Slafer GA, Miralles DJ (2003) Grain and floret number in response to photoperiod during stem elongation in fully and slightly vernalized wheats. Field Crops Res 81:17–27

Hess JR, Carman JG, Banowetz GM (2002) Hormones in wheat kernels during embryony. J Plant Physiol 159:379–386

Jenner CF, Ugalde TD, Aspinall D (1991) The physiology of starch and protein deposition in the endosperm of wheat. Aust J Plant Physiol 18(3):211–226

Jones HG (2004) Irrigation scheduling: advantages and pitfalls of plant-based methods. J Exp Bot 55(407):2427–2436

Jones HG, Stoll S, Santos T, de Sousa C, Manuela M, Chaves M, Grant OM (2002) Use of infrared thermography for monitoring stomatal closure in the field: application to grapevine. J Exp Bot 53(378):2249–2260

Kang G, Liu G, Peng X, Wei L, Wang C, Zhu Y, Ma Y, Jiang Y, Guo T (2013) Increasing the starch content and grain weight of common wheat by overexpression of the cytosolic AGPase large subunit gene. Plant Physiol Biochem 73:93–98. doi:10.1016/j.plaphy.2013.09.003

Konopka I, Ta´nska M, Pszcz´ołkowska A, Fordo´nski G, Kozirok W, Olszewski J (2007) The effect of water stress on wheat kernel size, color and protein composition. Pol J Nat Sci 22:157–171

Laxman SV, Solanki YPS, Redhu AS (2014) Phenological development, grain growth rate and yield relationships in wheat cultivars under late sown condition. Indian J Plant Physiol. 19:222

Livak KJ, Schmittgen TD (2001) Analysis of relative gene expression data using real-time quantitative PCR and the $2^{-\Delta\Delta CT}$ method. Methods 25(4):402–408

Lizana XC, Riegel R, Gomez LD, Herrera J, Isla A, McQueen-Mason SJ, Calderini DF (2010) Expansins expression is associated with grain size dynamics in wheat (*Triticum aestivum* L.). J Exp Bot 61(4):1147–1157

Lopes MS, Reynolds MP (2010) Partitioning of assimilates to deeper roots is associated with cooler canopies and increased yield under drought in wheat. Funct Plant Biol 37:147–156

Lopes MS, Reynolds MP, Manes Y, Singh RP, Crossa J, Braun HJ (2012) Genetic yield gains and changes in associated traits of CIMMYT spring bread wheat in a "Historic" set representing 30 years of breeding. Crop Sci 52:1123–1131

Ma YZ, Mackown CT, Sanford DAV (1996) Kernel growth of in vitro cultured wheat spikes of cultivars with divergent source-sink limitations. Plant Sci 119:135–148

Mason RE, Singh RP (2014) Considerations when deploying canopy temperature to select high yielding wheat breeding lines under drought and heat stress. Agron 4:191–201

Miralles DJ, Katz SD, Colloca A, Slafer G (1998) Floret development in near isogenic wheat lines differing in plant height. Field Crops Res 59:21–30

Mohammad E, Ali A, John EL, Alireza A, Kazem P, Mehdi J (2012) Variation in grain weight among Iranian wheat cultivars: the importance of stem carbohydrate reserves in determining final grain weight under source limited conditions. Aust J Crop Sci 6(11):1508–1515

Peltonen-Sainio P, Kangas A, SaloY Jauhiainen L (2007) Grain number dominates grain weight in temperate cereal yield determination: evidence based on 30 years of multi-location trials. Field Crops Res 100:179–188

Porter JR, Moot DJ (1998) Research beyond the means: climatic variability and plant growth. In: Dalezios NR (ed) International symposium on applied agrometeorology and agroclimatology. Office for Official Publication of the European Commission, Luxembourg, pp 13–23

Prakash P, Kumari A, Singh DV, Pandey R, Sharma-Natu P, Ghildiyal MC (2009) Starch synthase activity and grain growth in wheat cultivars under elevated temperature: a comparison of responses. Indian J Plant Physiol 14:364–369

Rane J, Chauhan H, Shoran J (2003) Post anthesis stem reserve mobilization in wheat genotypes tolerant and susceptible to high temperature. Indian J Plant Physiol 8(Special Issue):383–385

Rane J, Pannu RK, Sohu VS, Saini RS, Mishra B, Shoran J, Crossa J, Vargas M, Joshi AK (2007) Performance of yield and stability of advanced wheat genotypes under heat stress environments of the Indo-Gangetic plains. Crop Sci 47:1561–1573

Rebetzke GJ, van Herwaarden AF, Jenkins C, Weiss M, Lewis D, Ruuska S, Tabe L, Fettell NA, Richards RA, Rebetzke GJ, van Herwaarden AF, Jenkins C, Weiss M, Lewis D, Ruuska S, Tabe L, Fettell NA, Richards RA (2008) Quantitative trait loci for water-soluble carbohydrates and associations with agronomic traits in wheat. Aust J Agric Res 59(10):891

Rebetzke GJ, Rattey AR, Farquhar GD, Richards RA, Condon AG (2013) Genomic regions for canopy temperature and their genetic association with stomatal conductance and grain yield in wheat. Funct Plant Biol 40:14–33

Reynolds MP, Pierre CS, Saad ASI, Vargas M, Condon AG (2007) Evaluating potential genetic gains in wheat associated with stress-adaptive trait expression in elite genetic resources under drought and heat stress. Crop Sci 47:S172–S189

Savin R, Nicolas M (1996) Effects of short periods of drought and high temperature on grain growth and starch accumulation of two malting barley cultivars. Aust J Plant Physiol 23:201–210

Schmu¨lling T, Werner T, Riefler M, Krupkova E, Bartrina Y, Manns I (2003) Structure and function of cytokinin oxidase/dehydrogenase genes of maize, rice, *Arabidopsis* and other species. J Plant Res 116:241–252

Sofield I, Evans LT, Cook MG, Wardlaw IF (1977) Factors influencing the rate and duration of grain filling in wheat. Aust J Plant Physiol 4:785–797

Stone PJ, Nicolas MM (1994) Wheat cultivars vary widely in their responses of grain yield and quality to short periods of post-anthesis heat stress. Aust J Plant Physiol 21:887–900

Tahir ISA, Nakata N (2005) Remobilization of nitrogen and carbohydrate from stems of bread wheat in response to heat stress during grain filling. J Agron Crop Sci 191(2):106–115

Van Herwaarden AF, Angus JF, Richards RA, Farquhar GD (1998) "Haying-off", the negative grain yield response of dry land wheat to nitrogen fertilizer. II. Carbohydrate and protein dynamics. Aust J Agric Res 49:1083–1093.

Vignjevic M, Wang X, Olesen JE, Wollenweber B (2015) Traits in spring wheat cultivars associated with yield loss caused by a heat stress episode after anthesis. J Agro Crop Sci 201:32–48

Wardlaw IF (1994) The effect of high temperature on kernel development in wheat: variability related to pre-heading and post-anthesis conditions. Aust J Plant Physiol 21:731–739

Weng J, Gu S, Wan X, Gao H, Guo T, Su N, Lei C, Zhang X, Cheng Z, Guo X (2008) Isolation and initial characterization of GW5, a major QTL associated with rice grain width and weight. Cell Res 18:1199–1209

Wheeler TR, Batts GR, Ellis RH, Hadley P, Morison JIL (1996) Growth and yield of winter wheat (*Triticum aestivum* L.) crops in response to $CO_2$ and temperature. J Agric Sci 127:37–48

Yeh SY, Chen HW, Ng CY, Lin CY, Tseng TH, Li WH, Ku MS (2015) Down-regulation of cytokinin oxidase 2 expression increases tiller number and improves rice yield. Rice 8(1):36. doi:10.1186/s12284-015-0070-5

Yemm EW, Willis AJ (1954) The estimation of carbohydrates in plant extracts by anthrone. Biochem J 57:508–514

Zadoks JC, Chang TT, Konzak CF (1974) A decimal code for the growth stages of cereals. Weed Res 14(6):415–421

Zalewski W, Galuszka P, Gasparis S, Orczyk W, Nadolska-Orczyk A (2010) Silencing of the HvCKX1 gene decreases the cytokinin oxidase/dehydrogenase level in barley and leads to higher plant productivity. J Exp Bot 61:1839–1851

Zhang L, Zhao YL, Gao LF, Zhao GY, Zhou RH, Zhang BS, Jia JZ (2012) TaCKX6-D1, the ortholog of rice OsCKX2, is associated with grain weight in hexaploid wheat. New Phytol 195(3):574–584

# 18

# Meso- or xeromorphic? Foliar characters of Asteraceae in a xeric scrub of Mexico

Patricia Rivera[1,2*], José Luis Villaseñor[1] and Teresa Terrazas[1]

**Abstract**

**Background:** The anatomical traits associated with water deficit are also observed in plants growing in poor soils. The species may resist water deficit through three main strategies: escape, avoid or tolerate. The Pedregal de San Ángel Ecological Reserve (REPSA), Mexico, is an environment with low nutrient soil and low water availability. It is set on the basalt formation derived from the Xitle volcano eruption. The main vegetation type is characterized as xerophytic shrub. Thus we expect that species growing in this community will show leaf xeromorphic traits and may have any of the three response strategies. We analyzed the foliar anatomy of 52 species of the Asteraceae family at the REPSA because it is the most abundant angiosperm family in the site, showing a wide variety of growth forms and anatomical variation.

**Results:** The foliar anatomies of the studied Asteraceae were highly variable as well as their quantitative traits as revealed by principal component analysis. This agrees with previous studies that found great anatomical variation within the family. Leaves have multiple layered palisade parenchyma and parenchyma bundle sheaths and could not be categorized as xeromorphic because they possess mesomorphic leaf features as simple lamina, single-layered epidermis, and soft large-size glabrous leaves with high specific leaf area.

**Conclusions:** The combination of mesomorphic and few xeromorphic foliar traits with other characters at the genus and tribal level probably has been essential in Asteraceae to colonize various environments, including those with low water and nutrient availability.

**Keywords:** Astereae, Eupatorieae, Heliantheae, Foliar anatomy, Poor soils, Water stress

## Background

Plants growing under xeric conditions can have a set of anatomical, physiological or phenological adaptations allowing them to escape, avoid or tolerate water stress (Santos and Ochoa 1990; Fahn and Cutler 1992; Dickison 2000; Valladares et al. 2004). Drought-escaping species generally are annual or biannual herbs that develop their life-cycle in short time periods when conditions are favorable (Kramer 1983; Santos and Ochoa 1990; Fahn and Cutler 1992; Valladares et al. 2004) or they usually possess perennation structures, such as bulbs, rhizomes or runners that remain latent until conditions are favorable again. Plants showing the avoidance-of-stress strategy possess some features enabling them to minimize or compensate for water loss or to increase water uptake and avoid desiccation. Indicator traits of the drought avoidance mechanism include deep roots, small or strongly lobulated leaves, mesophyll cells smaller in size, thick cell-walls and cuticles, multiseriate epidermis, strongly developed palisade parenchyma, stomata sunken or in crypts and abundant trichomes (Esau 1976; Kramer 1983; Santos and Ochoa 1990; Gibson 1996; Dickison 2000; Valladares et al. 2004; Fang and Xiong 2015). Drought tolerating plants can sustain a certain level of physiological activities under stress conditions. Tolerating plants show tissues resistant to dehydration and characters such as stress-induced leaf abscission, succulent leaves, lignified cell-walls, mucilage accumulation and the occurrence of buliform cells (Kramer 1983; Santos

*Correspondence: rivera.perezpatricia@gmail.com
[1] Departamento de Botánica, Instituto de Biología, Universidad Nacional Autónoma de México, Apartado Postal 70-367, 04510 Mexico City, Mexico
Full list of author information is available at the end of the article

and Ochoa 1990; Dickison 2000; Valladares et al. 2004; Fang and Xiong 2015).

Traits of stress-avoiding and stress-tolerating plants have also been related to soil nutrient deficit (Dickison 2000; Fonseca et al. 2000; Wright et al. 2002). In fact, soil nutrient availability explains better the distribution of foliar characters in plants around the world than mean annual temperature, mean annual precipitation and irradiance together (Ordoñez et al. 2009). It has been hypothesized that xeromorphism has evolved from the adaptation of mesomorphic plants to low nutrient soil (Fahn and Cutler 1992) or to both, the lack of water and poor soils. These two conditions are found at the Pedregal de San Ángel Ecological Reserve in the Basin of Mexico.

The Pedregal de San Ángel Ecological Reserve (REPSA) is a protected area for the vegetation at the southern edge of the Mexico basin (Fig. 1a). The REPSA is located in the southwest region of the Mexico City basin (19°18′21″ to 19°20′11″N and 99°10′15″ to 99°12′4″W), at an elevation interval of 2292–2365 m, and it encompasses an area of 237.3 ha (Fig. 1a). The vegetation type has been characterized as xerophilous scrub (Rzedowski 2006). This ecosystem is on a set of basaltic formations produced by the solidification of lava flow during the eruption of the Xitle volcano about 1670 years ago (Siebe 2000). Although mean annual precipitation is 883 mm, the bedrock does not allow water retention and the soil is shallow. So aridity in this site is more edaphic than climatic. It is a unique system in the world because it is within the biogeographic province of the Transmexican Volcanic Belt (Morrone 2006) and the flora composition has Neotropical and Nearctic affinities (Rzedowski 1998; Rzedowski and Rzedowski 2005). The vegetation surrounding the REPSA includes various types of forests, including oak forest, fir forest, and cloud forest (Rzedowski and Rzedowski 2005; Lot and Cano-Santana 2009), which is consistent with the vegetation type expected given the elevation (over 2240 masl), the mean annual rainfall (770–1200 mm), and the mean temperature (11–15 °C) of the site (Fig. 1b). Therefore, it is assumed that species hat colonized after the eruption came mainly from the surrounding forests; however, the current vegetation type is xerophilous scrub (Castillo-Argüero et al. 2007; Lot and Cano-Santana 2009). At the REPSA, the Asteraceae family shows the highest number of species compared to the 73 other vascular plant families found there (Castillo-Argüero

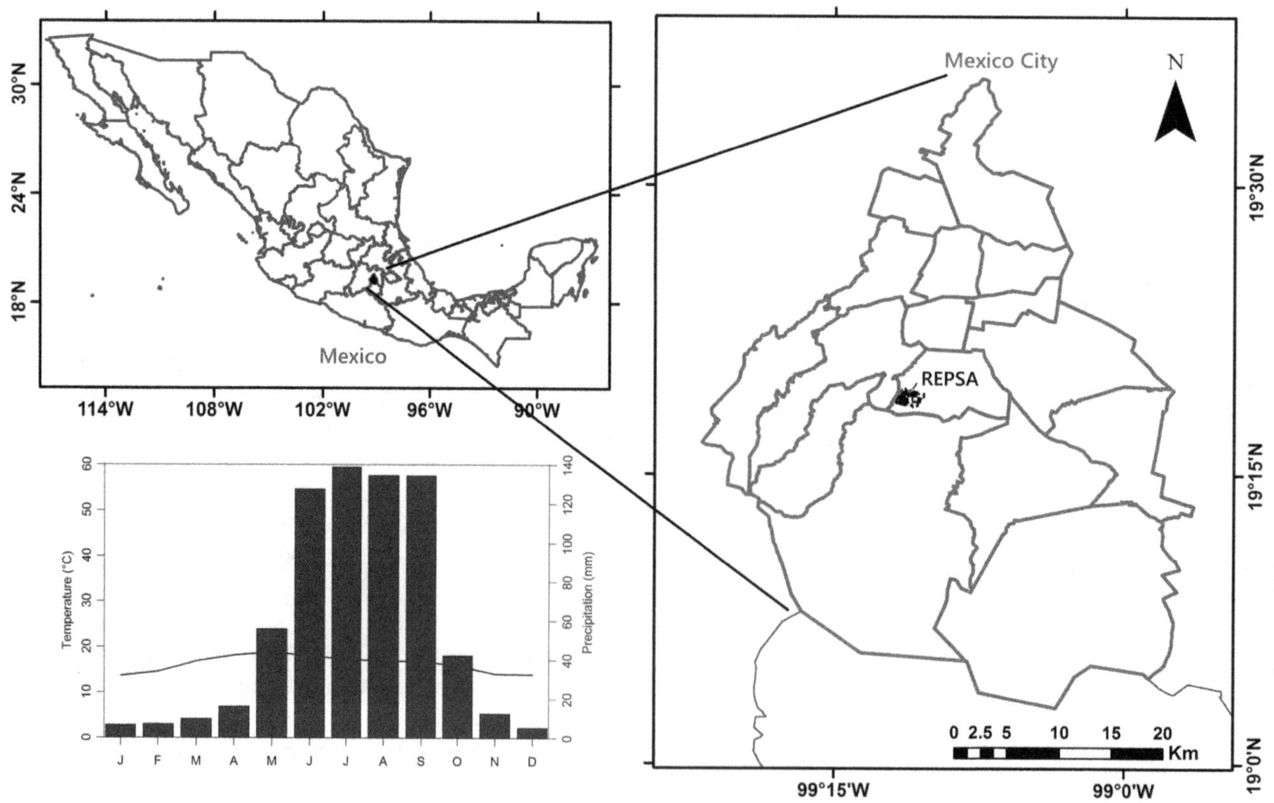


**Fig. 1** Location and some characteristics of the study site, the Pedregal de San Ángel Ecological Reserve-REPSA. **a** Geographic location of the REPSA in Mexico. The REPSA is marked as the *black shaded area* in Mexico City. **b** Climograph of the REPSA

et al. 2004, 2007; Lot and Cano-Santana 2009). Céspedes (2010) reported 93 species belonging to 51 genera classified in 13 tribes of Asteraceae in this site. Moreover, the morphological diversity of the family is well represented, as there is a wide variety of growth forms, polyploid species and species with high chromosomic numbers (Soto-Trejo et al. 2011).

The foliar anatomy of Asteraceae has been studied mainly in species with economical value (Ragonese 1988; Ferreira et al. 2002; Freire et al. 2005, 2007; Milán et al. 2006). The leaves are usually dorsiventral, hypo- or amphistomatic and have anomocytic stomata. However, it is difficult to assert generalizations regarding the foliar anatomy of Asteraceae because there is so much variation in characters such as stomata distribution, trichome density and type, hypodermis development, presence of secretory structures and parenchymatic vascular bundles sheaths (Anderson and Creech 1975; Metcalfe and Chalk 1979; Freire et al. 2005). The usefulness of foliar characters for taxonomic classification at the genus or species level (Ferreira and De Oliveira 1989; Castro et al. 1997; Luque et al. 1999; Delbón et al. 2007; Adedeji and Jewoola 2008) or for phylogenetic and ecological studies (Boeger and Wisniewski 2003; Horn et al. 2009) has been overlooked because most studies have focused exclusively on examining the epidermal surface (Ferreira et al. 2002; Freire et al. 2005, 2007; Adedeji and Jewoola 2008; Gil et al. 2012; Redonda-Martínez et al. 2016) and ecological or leaf evolution traits are missing for Asteraceae as they have been generated for other plant families (Luckow 2002; Schmerler et al. 2012; Brodribb et al. 2013).

We hypothesize that Asteraceae species at this ecological protected area (REPSA) will show anatomical traits for avoiding or tolerating stress, the aim of this study was to characterize the foliar anatomy of a sample of 52 species belonging to 41 genera and 13 tribes of Asteraceae living in the xeric environment of this reserve.

## Methods

The present study was conducted in the REPSA, within the central campus of the National Autonomous University of Mexico (UNAM, Fig. 1a). Samples were collected during two rainy seasons from August 2008 to December 2009. Three to six fully developed leaves showing no evidence of injury were removed from three individuals of each selected species. All the leaves were photographed with a digital camera. Leaf area was determined using an image analyzer according to the procedure described by Garnier et al. (2001). After being photographed, two to three the leaf laminas per species were dried in oven at 60 °C for at least 48 h to constant weight and then weighed on an analytical balance to determine the dry mass of the leaf. Specific leaf area (SLA) was calculated as the ratio between the leaf area and leaf dry mass [SLA = leaf area ($cm^2$)/dry leaf mass (g)]. The remaining leaves (1–4 per species) were fixed with a formaldehyde-glacial acetic acid-ethyl alcohol solution (Ruzin 1999). They were then rinsed with tap water to remove fixative residues and stored in a glycerin-ethyl-alcohol-water solution (1:1:1) until sectioning. Voucher specimens information is given in Additional file 1.

Portions of the middle region of the leaf, including the intercostal area from the midvein to the margin were cut, rinsed and dehydrated in increasing concentrations of tert-butanol (10–100%) with a Leica automatic tissue processor (TP1020) remaining for 24 h in each concentration. The tissues were embedded in paraffin (melting point: 56 °C) and transverse and paradermal sections 10–12 μm in thickness were cut with a Leica rotatory microtome (RM2125). The resulting sections were stained with safranin-fast green (Johansen 1940) and mounted on synthetic resin. In the paradermal sections, the cuticle, guard cell length and epidermal cell shapes were examined. In the transverse sections, the thickness of the following features was measured: the total leaf, mesophyll, palisade parenchyma, spongy parenchyma, cuticle, and height of the abaxial and adaxial epidermis cells. The measurements were obtained through an Olympus microscope BX-51 attached to an image analyzer (Image Pro-plus version 6.1, Media Cybernetics 2006). For each variable, 25 measures were recorded. Other features, such as the number of palisade parenchyma layers or the spongy parenchyma type (open or compact), were also determined. Laminar size classification followed Webb (1959), detecting 6 categories: nanophyll (0.25–2.25 $cm^2$), microphyll (2.25–20.25 $cm^2$), notophyll (20.25–45 $cm^2$), mesophyll (45–182.25 $cm^2$), macrophyll (182.25–1640.25 $cm^2$) and megaphylly (>1640.25 $cm^2$). Sclerophylly index (SI) was calculated following Boeger and Wisniewski (2003) where SI = leaf dry mass (g)/2 leaf area ($cm^2$), considering SI > 0.6 as sclerophylly and SI < 0.6 as mesophylly). Stomatal pore area index (SPI) was calculated following Tian et al. (2016) where SPI = stomatal density*stomatal length*$10^{-4}$. Leaf lamina and midvein descriptions followed Dickison (2000) and tribe classification followed Funk et al. (2009).

Characters were square root and log10-transformed prior to statistical analyses. Pearson and Sperman correlations were calculated between pairs of leaf traits. In addition we produce a principal component analysis (PCA) to evaluate which leaf traits explained the variation in this site. Analyses were performed with R (R Core Team 2015) and SAS (Statistical Analysis System software version 9, SAS Institute 2002).

## Results

### Leaf morphology

The leaves were simple in all the species. However, 29% of species, including most of the studied species of the Tageteae tribe, were deeply lobed. Based on the laminar size, most species were classified in four categories: microphyll (29% of the species), mesophyll (29%), notophyll (23%) and macrophyll (13%). Only two species (*Montanoa grandiflora* and *Tithonia tubiformis*) were classified as megaphyll and one species (*Tagetes micrantha*) as nanophyll leaves. Laminar shape was variable, from ovate, obovate to elliptic. Toothed or untoothed leaf margins were observed. Serrate margins were common and crenate margins were present in some species. Leaf morphological characters by species are given in Additional file 2.

### Epidermis

On the surface, the abaxial and adaxial epidermises were glabrous in 95% of the examined species. The remaining 5% were glandular, tector or pluricellular uniseriate or possessed multiseriate trichomes. Trichomes could be found on one or both leaf surfaces, but they were more frequent on the abaxial surface. The epidermal cell shape of both surfaces was elongated polygonal or isodiametrical polygonal, with straight (Fig. 2a–c) or wavy anticlinal walls (Fig. 2d–f). Both amphistomatic and hypostomatic leaves were observed. When amphistomatic, the stomata were more abundant on the abaxial surface. The stomata were anomocytic and rarely anisocytic. The average stomatal density was 362 stomata/mm$^2$, and the average per tribe is given in Table 1.

On the transverse section, the examined species had cuticle thicknesses varying between 0.19 and 1.15 µm. In 25% of the species, the cuticle appeared smooth. Cuticular striae (Fig. 3a–c) were found in greater abundance in cells of the veins and around trichomes. The epidermises were simple in all taxa. Abaxial (7.48–35.79 µm in height) and adaxial (6.22–40.69 µm in height) epidermises were variable in size. The epidermal cells were generally rectangular in shape and rarely square. In some species, the epidermal cells appeared ovoidal in the periclinal walls due to the elevation of the outer periclinal wall. The outer periclinal wall was often thicker than the internal one,

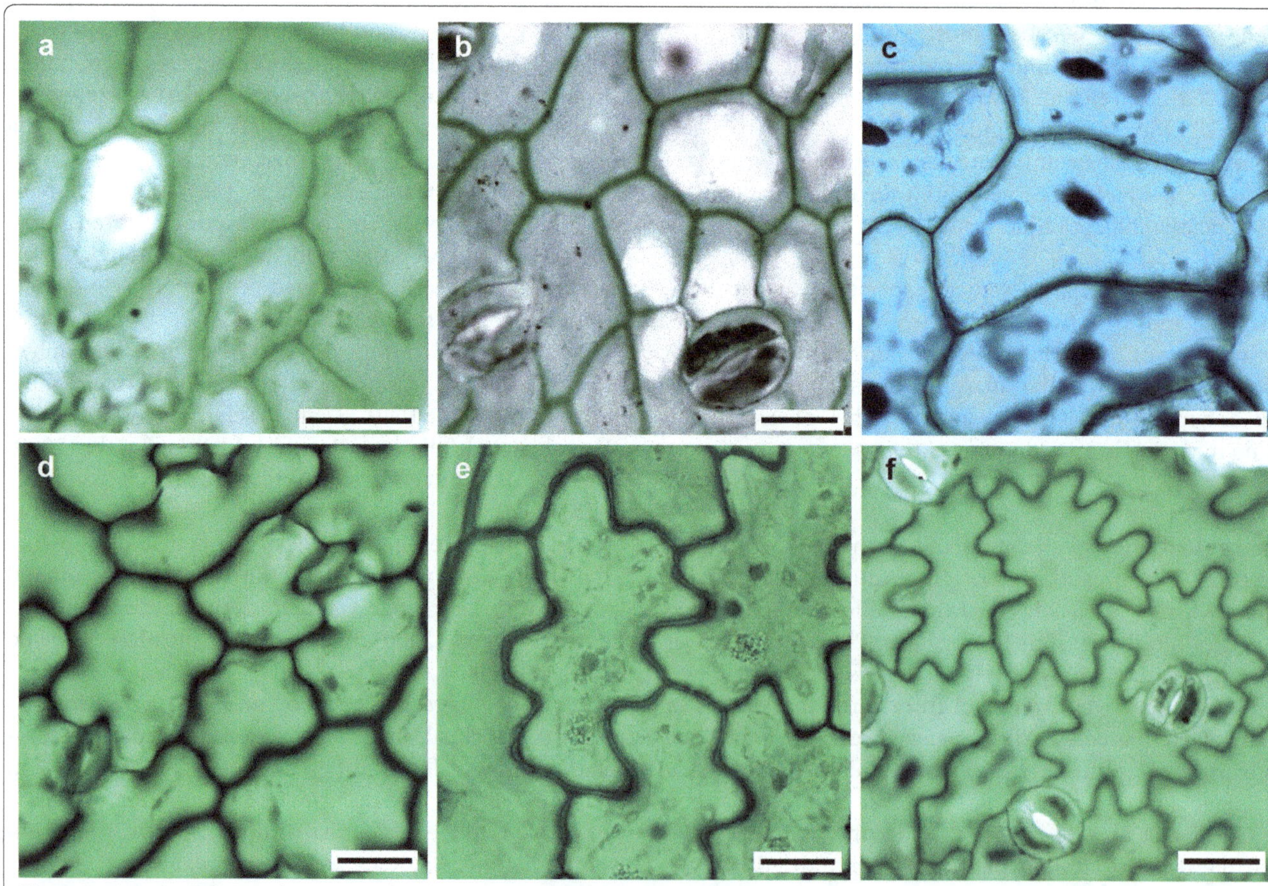


**Fig. 2** Epidermal cells forms of Asteraceae. **a** *Conyza canadensis* (Tribe Astereae). **b** *Barckleyanthus salicifolius* (Senecioneae). **c** *Pittocaulon praecox* (Senecioneae). **d** *Pectis prostrata* (Tageteae). **e** *Ageratina adenophora* (Eupatorieae). **f** *Lactuca serriola* (Cichorieae). *Bar* is 20 µm

**Table 1 Leaf characters by tribe in Asteraceae of the xeric scrub at the REPSA, Mexico**

| Tribe | Anthemidae | Astereae | Bahieae | Carduae | Cichorieae | Coreopsidae | Eupatorieae | Gnaphalieae | Heliantheae | Millerieae | Nassauvieae | Senecioneae | Tageteae |
|---|---|---|---|---|---|---|---|---|---|---|---|---|---|
| No. species/character | 2 | 5 | 2 | 1 | 3 | 5 | 12 | 2 | 10 | 2 | 1 | 3 | 4 |
| Leaf size classification | Micro* | Noto (3), micro (2) | Micro | Macro | Macro (2), meso | Meso (2), micro (2), noto (1) | Noto (5), meso (4), micro (2), macro (1) | Micro | Meso (4), noto (2), mega (2), macro (1), micro (1) | Meso, noto | Meso | Macro (2), meso | micro (2), meso (1), nano (1) |
| Lobation | Lobed | Unlobed (4), lobed | Lobed | Lobed | Lobed (2), unlobed | Lobed | Unlobed | Unlobed | Unlobed (7), lobed | Unlobed | Unlobed | Unlobed (2), lobed | Unlobed (2), lobed |
| Margin type | Untoothed | Untoothed (3), toothed | Untoothed | Tooted | Toothed (2), untoothed | toothed (3), untoothed | Toothed (11spp), untoothed | Untoothed | Toothed (8) or untoothed | Toothed | Toothed | Toothed (2spp) or untoothed | Toothed (2spp) or untoothed |
| Tooth type | NA | Serrated (2 spp) | NA | Serrated | Eentate | Serrated (3) | Serrated (11) | NA | Serrated (6), crenated (2) | Serrated | Serrated | Serrated (1) or dentate (1) | Serrated (2) |
| Leaf type by the position of stomata | Amphi* | Amphi | Amphi | Amphi | Amphi (2), hypos | Amphi (3), hypos | Hypos (8), amphi | Amphi (1), hypos | Amphi (7), hypos | Amphi | Amphi | Hypos (2), amphi | Amphi |
| | Bifacial or equifacial | Bifacial | Bifacial or equifacial | Bifacial | Bifacial | Bifacial or equifacial | Bifacial | Bifacial | Bifacial | Bifacial | Bifacial | Bifacial | Bifacial or equifacial |
| Total lamina thickness, mean µm | 102.8 | 217.3 | 326.7 | 413.1 | 182.8 | 246.4 | 148.5 | 126.8 | 200.5 | 197.07 | 231.7 | 238.2 | 274.9 |
| Cuticle surface | Striated (1), smooth | Striated | Striated | Striated | Striated (2), smooth | Striated | Striated (11), smooth | Smooth | Striated (7), smooth | Smooth | Striated | Striated | Striated (3), smooth |
| Cuticle thickness, range µm | 0.31–0.43 | 0.27–0.56 | 0.32–0.37 | 0.32 | 0.30–0.45 | 0.22–0.41 | 0.20–0.60 | 0.33–0.42 | 0.21–1.15 | 0.23–0.24 | <0.1 | 0.36–0.46 | 0.19–0.44 |
| Epidermal cell walls (surface view) | Straight to wavy | Straight to wavy | Wavy | Straight | Straight to wavy | Straight to wavy | Wavy | Wavy | Straight to wavy | Wavy | Wavy | Straight to wavy | Straight to wavy |
| Position of stomata regarding ordinary epidermal cells | Same level | Same level | Same level | Same level | Same level | Same level | Same level (10), higher | Same level | Same level (10), higher | Same level | Same level | Same level | Same level |

**Table 1 continued**

| Tribe | Anthemidae | Astereae | Bahieae | Carduae | Cichorieae | Coreopsidae | Eupatorieae | Gnaphalieae | Heliantheae | Millerieae | Nassauvieae | Senecioneae | Tageteae |
|---|---|---|---|---|---|---|---|---|---|---|---|---|---|
| Stomatal density (abaxial epidermis, mean stomata/$mm^2$) | 244 | 369 | 366 | 349 | 322 | 383 | 421 | 349 | 394 | 279 | 174 | 256 | 349 |
| Guard cell length, mean µm | 21.36 | 21.04 | 23.8 | 24.38 | 18.7 | 21.61 | 19.64 | 17.89 | 20.31 | 20.18 | 25.09 | 23.09 | 19.39 |
| Palisade parenchyma thickness, mean µm | 22.48 | 69.99 | 117.65 | 195.58 | 46.2 | 98.03 | 47.29 | 44.43 | 76.27 | 54.34 | 52.40 | 45.47 | 72.68 |
| Spongy parenchyma thickness, mean µm | 53.38 | 48.20 | 58.08 | 165.68 | 90.94 | 92.89 | 71.48 | 52.97 | 81.54 | 113.32 | 140.46 | 157.6 | 93.12 |
| Stomatal pore area index (SPI) % | 10.37 | 15.53 | 15.70 | 20.75 | 15.06 | 17.27 | 14.56 | 11.13 | 15.09 | 11.39 | 10.98 | 13.57 | 13.80 |

In parenthesis number of species for each tribe

* See text for complete names

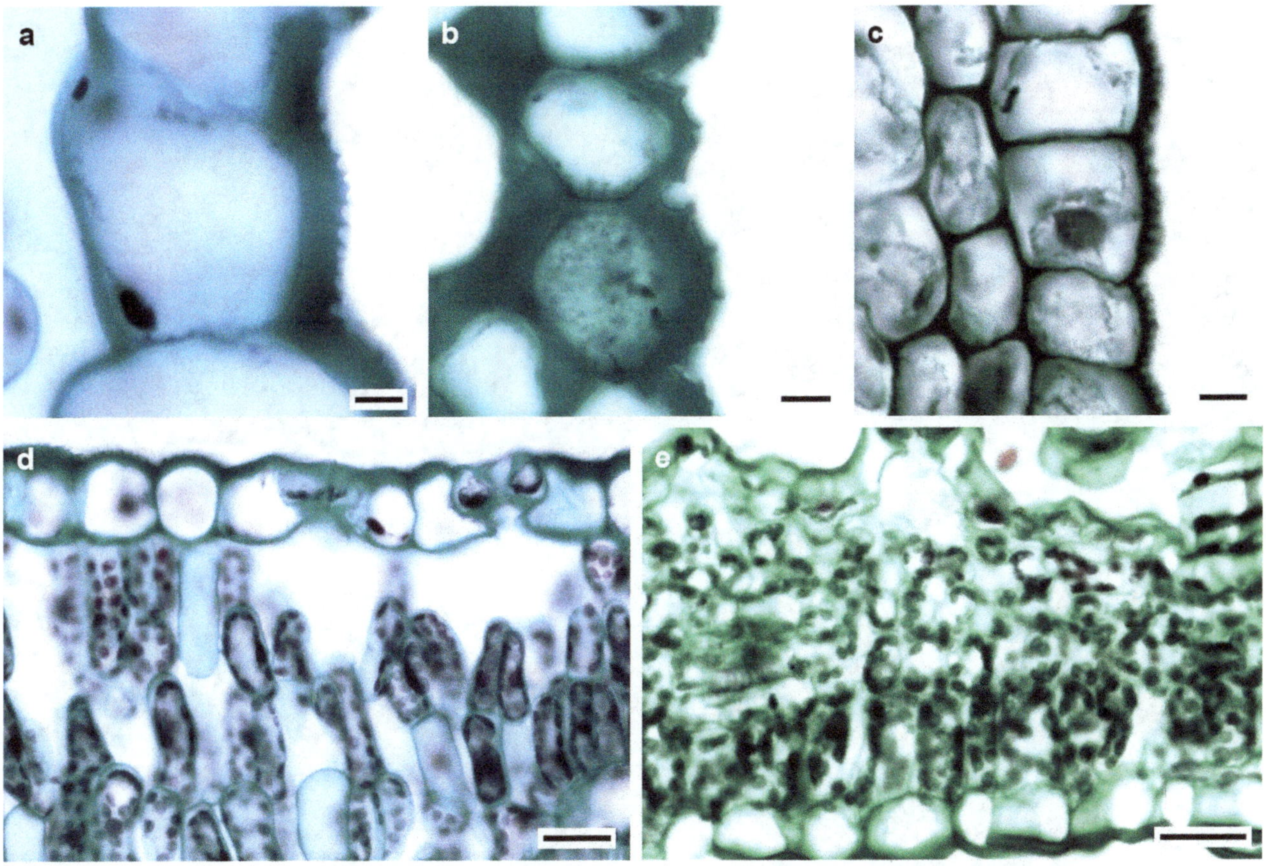


**Fig. 3** Cuticular striae and transverse view of stomata of some Asteraceae species. **a** *Baccharis salicifolia* (Tribe Astereae). **b** *Ambrosia psilostachya* (Heliantheae). **c** *Barckleyanthus salicifolius* (Senecioneae). **d** *Baccharis salicifolia* (Astereae). **e** *Brickellia secundiflora* (Eupatorieae). *Bar* is 5 μm in **a–c**, 20 μm in **d**, **e**

as observed in *Dyssodia papposa*, *Laennecia sophiifolia* and *Schkuhria pinnata*. In some species belonging to the Coreopsideae, Heliantheae and Tageteae tribes, some contents were stained a dark color in the cell lumen of the epidermis. The stomata were located at the same level as other epidermal cells (Fig. 3d) except for some species of the Eupatorieae and Heliantheae tribes, for which they were localized at a higher level than other epidermal ordinary cells (Fig. 3e).

### Mesophyll

A hypodermis was only observed in *Pectis prostrata* from the Tageteae tribe, where the hypodermis was one cell thick. The mesophyll was dorsiventral in most species and only isofacial in five species (Fig. 4a–d). The palisade parenchyma was formed by one to five cell strata, and, in most cases, it occupied more than half of the total mesophyll thickness. In some species of Astereae and Heliantheae, the palisade parenchyma occupied the most of the mesophyll (Fig. 4). The spongy parenchyma was usually compact in 65% of the species and loose in the rest. In some species from Anthemidae, Eupatorieae, Senecioneae and Tageteae tribes, it constituted the whole mesophyll (Fig. 5). There was an aerenchyma in *Jaegeria hirta* (Millerieae). No mineral contents in the mesophyll cells were observed.

### Vascular bundles

Closed collateral vascular bundles were observed (Figs. 4, 5). In most species, they were surrounded by a sheath of one to three parenchyma layers, which were more conspicuous in species from the Astereae and Tageteae tribes. In the Eupatorieae, Heliantheae, Cichorieae and Senecioneae tribes, sheath extensions towards both surfaces were found (Fig. 5c, d). Secretory canals in some species of Astereae, Eupatorieae and Senecioneae were observed within the sheath (Fig. 4b).

### Midvein

The midvein occupies the central position of the leaf. The epidermis of the midvein in all species was the same as in the lamina, except for *Acourtia cordata* (Nassauvieae),

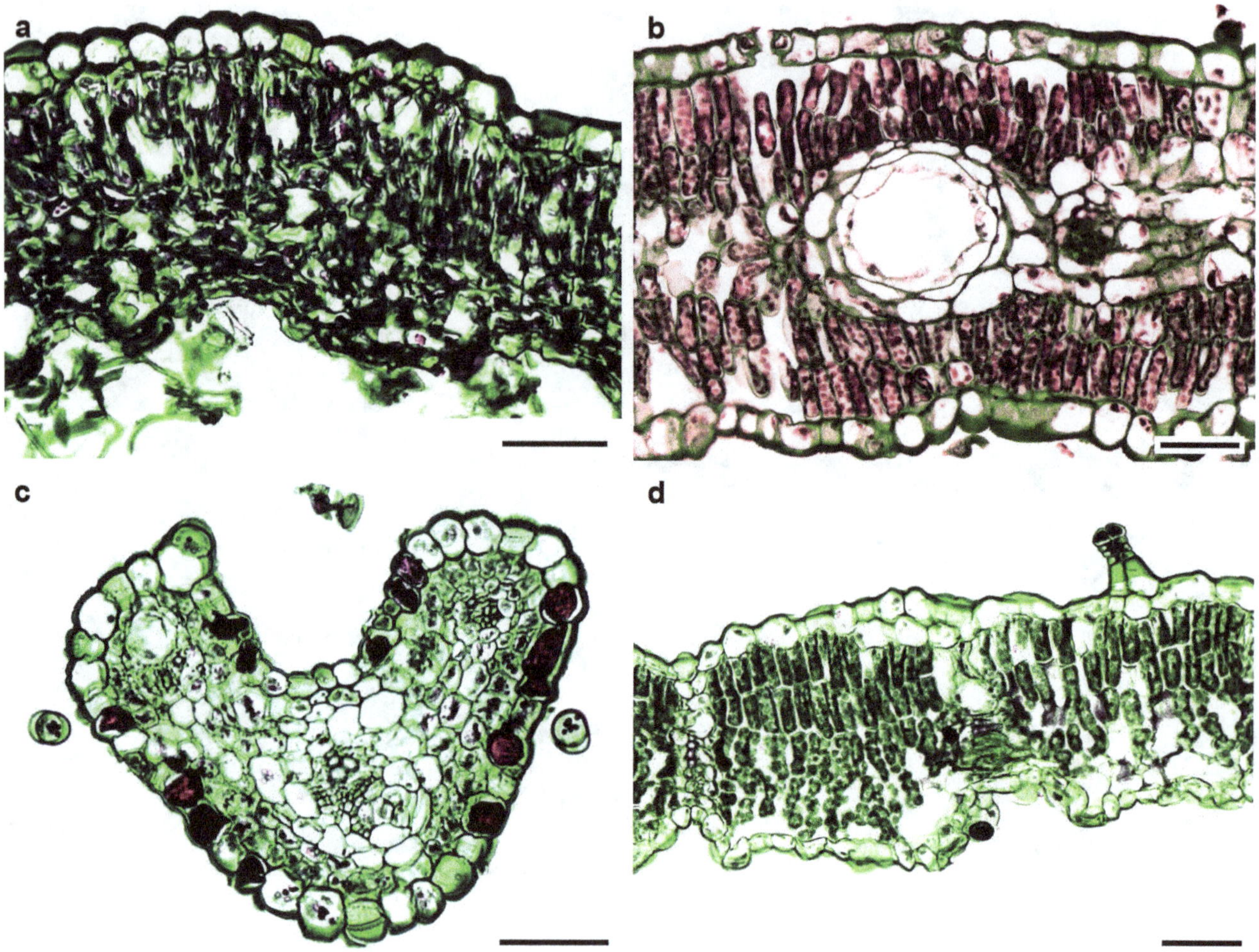


**Fig. 4** Variation of mesophyll in Asteraceae. **a** *Artemisia ludoviciana* (Tribe Anthemideae). **b** *Baccharis salicifolia* (Astereae). **c** *Cosmos bippinatus* (Coreopsideae). **d** *Brickellia veronicifolia* (Eupatorieae). *Bar* is 20 μm

which had a thicker outer epidermal wall. In this area, the mesophyll and the palisade parenchyma were interrupted by angular collenchyma with 9–16 cell layers. The number of bundles composing the midvein was variable (Fig. 6). The vascular tissue generally formed an arch, with parenchyma at both ends of the vascular bundles, except for *A. cordata*, where fibers were observed (Fig. 6a, c–e). No foliar characters were unique to any tribe, and most character states were shared by most tribe members as seen in Table 1.

### Statistical analyses

We found significant correlations between some of the studied characters. For example, there is a negative and significant relationship between the guard cell length and stomatal density (Fig. 7). We also found positive significant correlations between leaf thickness and the thickness of palisade and spongy parenchyma ($r = 0.66$, $P < 0.001$) and between guard cell length and the thickness of the leaves ($r = 0.43$, $P = 0.003$), other correlations are shown in Fig. 7. PCA explained in four axes 69% of the total variance (Table 2). The first axis explains the leaf thickness whereas the second axis explains the positive values for stomatal density and palisade thickness and negative values for guard cell length (Fig. 8).

## Discussion

The foliar anatomy of Asteraceae was variable. In the epidermis, differences were found in the shape of epidermal cells, seen in surface view, and in stomata location, as well as epidermis and cuticle thickness. The cell types constituting the mesophyll were variable as well as their abundance. These observations were consistent with the high foliar variation reported for the family (Anderson and Creech 1975; Metcalfe and Chalk 1979; Ferreira et al. 2002; Freire et al. 2005, 2007; Milán et al. 2006; Adedeji and Jewoola 2008). The striae in the leaf cuticle have been previously observed by other authors (Adedeji and

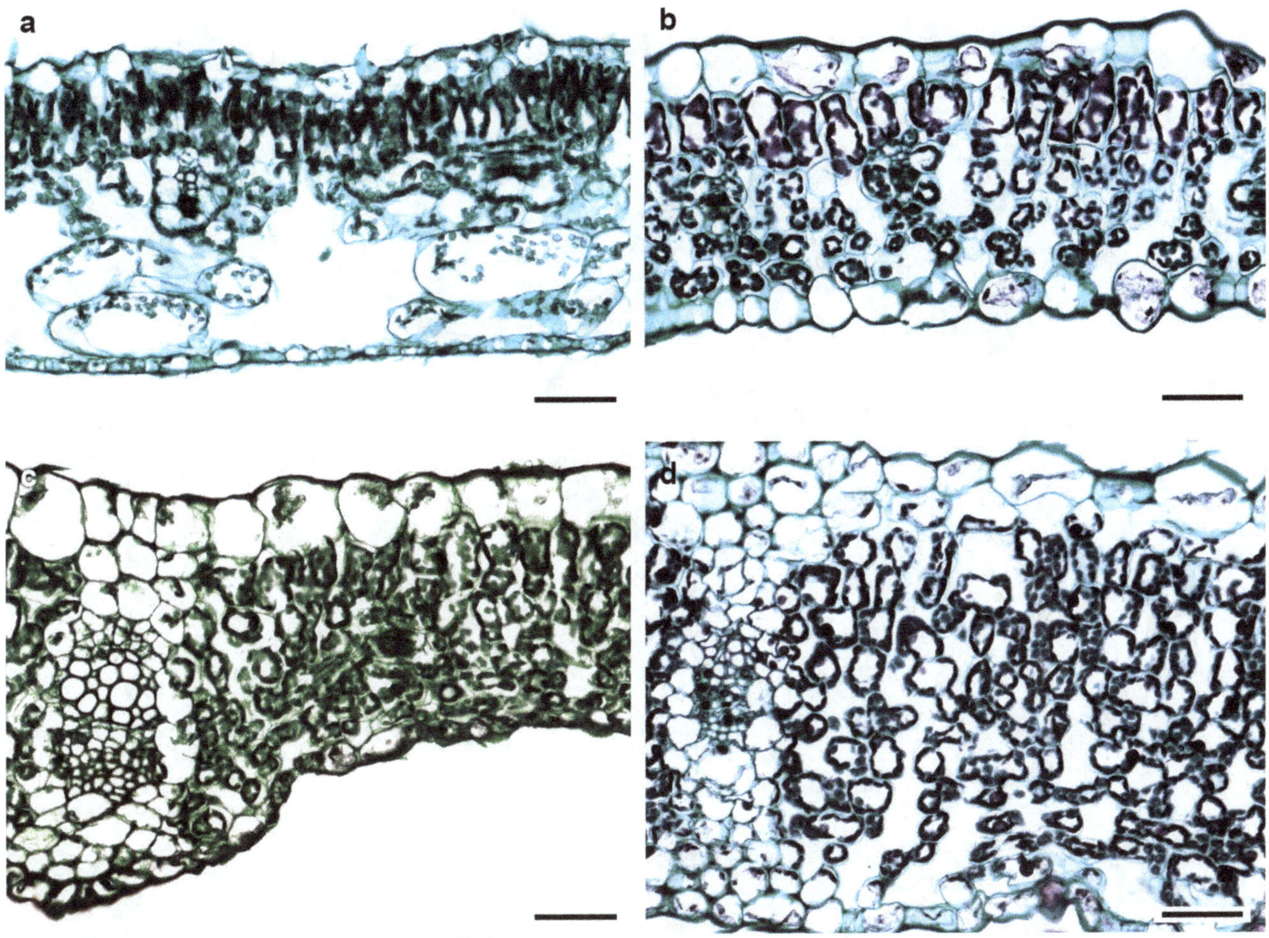


**Fig. 5** Variation of mesophyll in Asteraceae. **a** *Jaegeria hirta* (Tribe Millerieae). **b** *Montanoa grandiflora* (Heliantheae). **c** *Sonchus oleraceus* (Cichorieae). d. *Pittocaulon praecox* (Senecioneae). Bar is 20 µm

Jewoola 2008), who have suggested that the striae could be used to separate species. The observed stomatal types and the stomata being preferably distributed on the abaxial surface agrees with previous reports for the family in other works (Metcalfe and Chalk 1979; Adedeji and Jewoola 2008).

It was difficult to find a defined anatomical pattern for the family. However, we found that the combination of anatomical features is usually more homogenous within the tribes than the family level. For example, most examined species show bifacial leaves, except for some members of the Bahieae, Coreopsidae and Tageteae tribes, which had equifacial leaves. It should also be stressed that all the members of Astereae had amphistomatic bifacial leaves with striated cuticles, while striae were absent in all members of Gnaphalieae. These observations should be confirmed by examining more Astereae and Gnaphalieae members.

### Relationship between the leaves and the habitat

Given that the Asteraceae examined here were found in a habitat with low water availability and poor soils, it was expected that the foliar anatomy would display xerophytic features (Shields 1951; Kramer 1983; Esau 1976; Santos and Ochoa 1990; Fahn and Cutler 1992; Dickison 2000; Cutler et al. 2007). This assertion was not supported by the anatomical features found. The leaves of the Asteraceae studied are better classified as mesomorphic and do not correspond with the vegetation community or the water deficit or poor soils of the volcanic outcrop they leave.

#### *Leaf size*

It was suggested that a small leaf size is a xeromorphic trait because it is correlated with a reduction in transpiration rate (Fahn and Cutler 1992), so we expected to observe leptophyll, nanophyll or microphyll leaves.

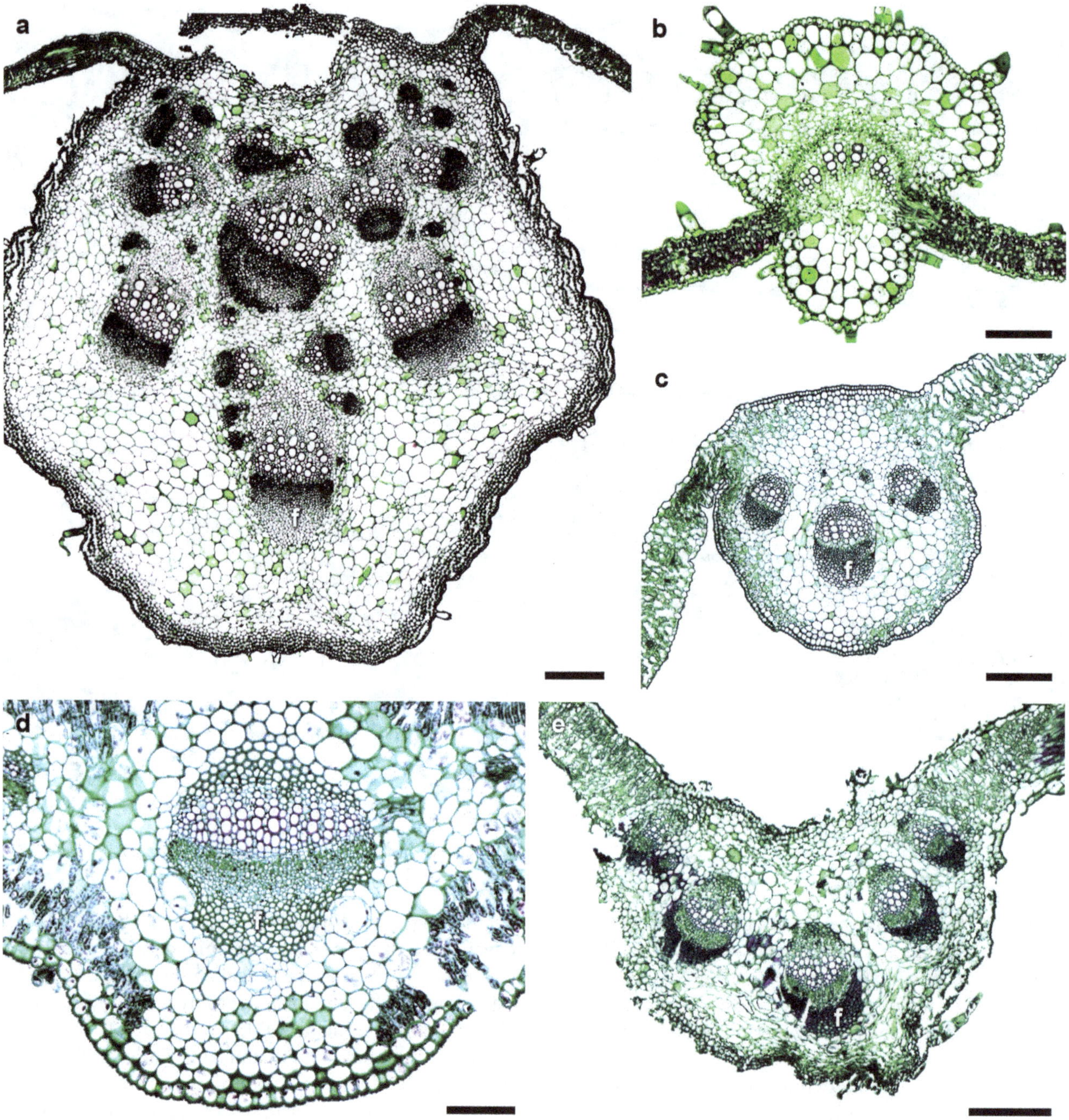


**Fig. 6** Midvein of some Asteraceae species. a. *Montanoa grandiflora* (Tribe Heliantheae). **b** *Ageratina pichinchensis* (Eupatorieae). **c** *Picris echioides* (Cichorieae). **d** *Baccharis salicifolia* (Astereae). **e** *Acourtia cordata* (Nassauvieae). *Bar* is 300 µm in **a**–**c**, 100 µm in **d**

However, most of the studied species (46%) have large-sized laminae (mesophyll, macrophyll and rarely megaphyll). Twenty-three percent of the studied species had medium-sized laminae (notophyll), and only 31% of the studied species had small leaves. From the species with larger leaves, 29% had deeply lobed margins, such as *Dahlia coccinea* and *Tagetes tenuifolia*. Half of the species with the larger leaves were ephemeral herbs that only grow during the rainy season, such as *Tithonia tubiformis* (megaphyll) and *Cirsium vulgare* (macrophyll). The other 21% of the species with larger leaves were deciduous that usually respond rapidly to water stress by producing

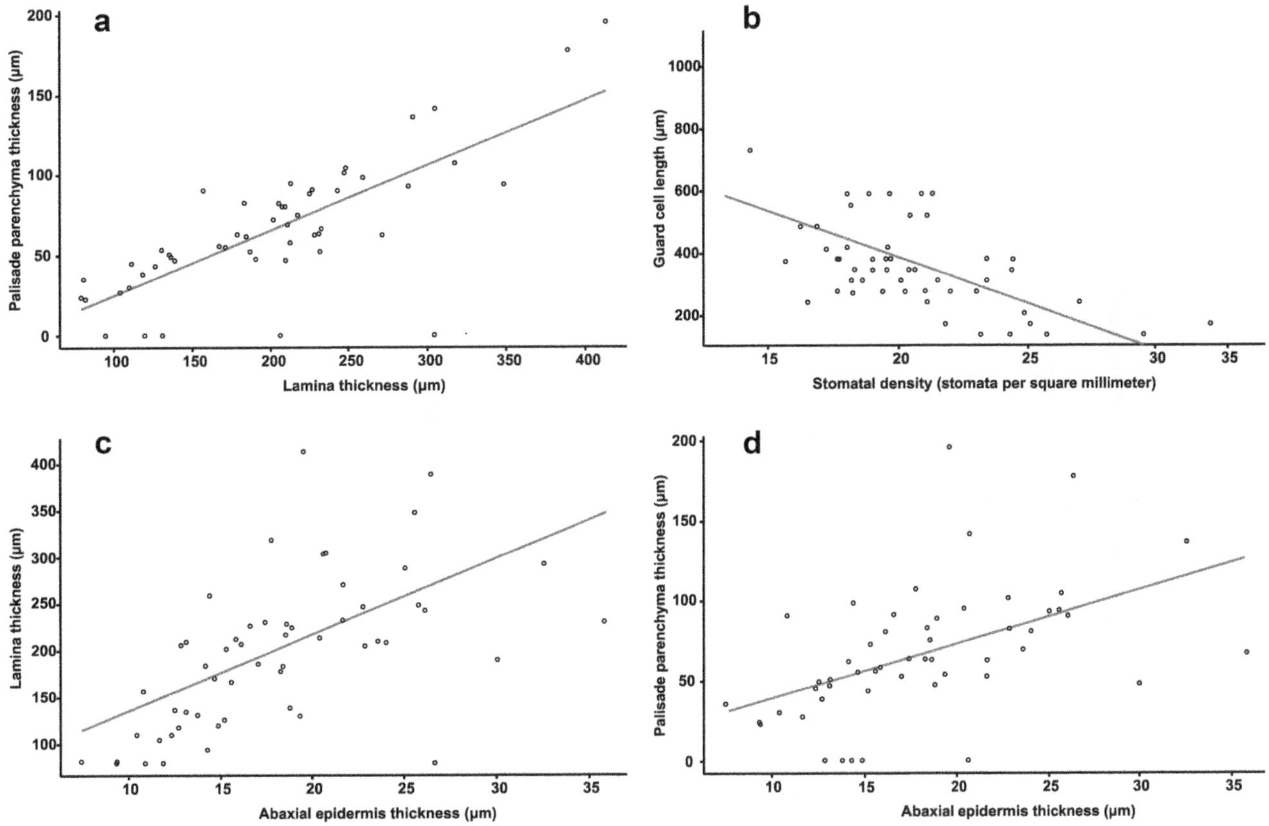


**Fig. 7** Correlations among pair of anatomical characters. **a** Lamina vs palisade parenchyma thickness. **b** Stomatal density vs guard cell length. **c** Abaxial epidermis vs lamina thickness. **d** Abaxial epidermis vs palisade parenchyma thickness

leaves exclusively during the rainy season and losing them when the rain stops, such as *Pittocaulon praecox* (macrophyll) and *Roldana lobata* (macrophyll) that is a perennial

**Table 2 Eigenvector for principal component analysis for the 11 variables in the first four components analyzed for the leaf traits of the 52 species of Asteraceae at the REPSA, Mexico**

| Characters | Prin 1 | Prin 2 | Prin 3 | Prin 4 |
|---|---|---|---|---|
| Variation explained (%) | 28.48 | 17.11 | 13.15 | 10.28 |
| Eigenvalue | 3.42 | 2.05 | 1.48 | 1.23 |
| Lamina size | −0.127 | 0.332 | 0.455 | −0.664 |
| Specific leaf area | −0.216 | −0.176 | 0.029 | 0.489 |
| Sclerophyll index | −0.197 | 0.332 | 0.455 | −0.066 |
| Lamina thickness | −0.144 | −0.134 | 0.575 | −0.024 |
| Cuticle thickness | −0.016 | 0.294 | −0.058 | −0.430 |
| Abaxial epidermis height | 0.426 | 0.059 | 0.278 | 0.127 |
| Stomatal density | −0.159 | 0.518 | −0.137 | 0.214 |
| Guard cell length | 0.281 | −0.487 | 0.133 | −0.195 |
| Mesophyll thickness | 0.510 | 0.075 | 0.054 | 0.142 |
| Palisade thickness | 0.216 | 0.455 | 0.067 | 0.204 |
| Spongy thickness | 0.138 | −0.105 | −0.524 | 0.269 |

herb. Laminar area was expressed as types and this trait was not revealed by our PCA as a variable that contributes to explain the variance in the first four components. The projections of the margin of the lamina (i.e., the lobes and teeth) are interpreted as being related to heat loss morphology (Freire et al. 2005, 2007; Fahn and Cutler 1992; Adedeji and Jewoola 2008; Gil et al. 2012; Schmerler et al. 2012; Redonda-Martínez et al. 2016). Eighty-three percent of the studied species had toothed or lobed margins, which is a common combination in Asteraceae (Bailey and Sinnott 1916; Rojas et al. in review).

#### *Palisade parenchyma*

Gibson (1996) mentioned that non-succulent xerophytes have thicker leaves than mesophytes because they have well developed multi-layered palisade parenchyma. This agrees with our findings because the palisade parenchyma was well developed in most tribes (Table 1), and in some species as *Baccharis salicifolia* (Astereae) and *Lagascea rigida* (Heliantheae) the mesophyll is composed exclusively of palisade parenchyma. The palisade thickness was within the range (94–355 μm) found in other plant families growing in dry environments (Nevo et al.

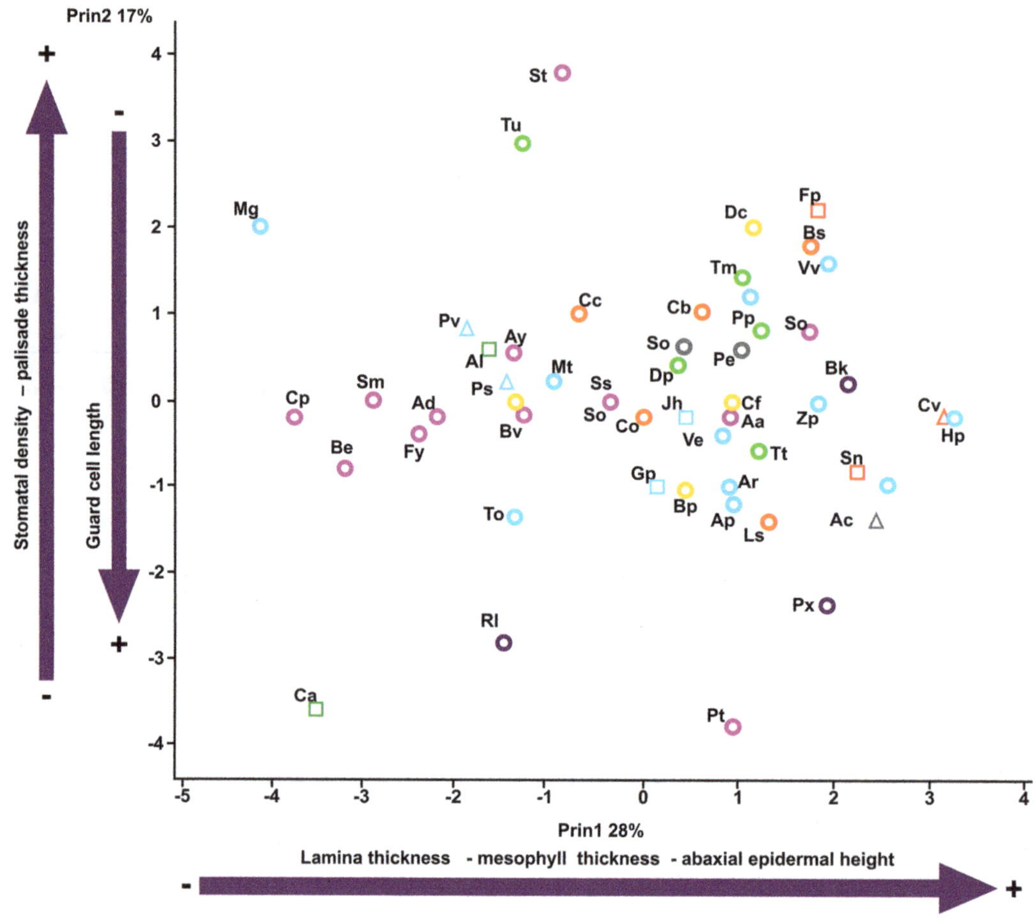


**Fig. 8** Biplot of the two principal component axis based on leaf characters plotted for species (letters, see information in Additional file 2) and tribes indicated by *colors* and *symbols*

2000; Rotondi et al. 2003; Bacelar et al. 2004; Gratani and Varone 2004). Palisade thickness was one of the variables with high loading to explain the variation found in the species studied, this agrees with the results of Tian et al. (2016) who found that leaf thickness is closely related to palisade and spongy thickness. This close relationship appears to be a strategy together with phenology to be efficient during the rainy season that is short.

### *SLA, SI and SPI*

Specific leaf area is positive correlated with moisture and nutrient availability (Ackerly et al. 2002) and is mainly determined by leaf density and thickness (Ackerly et al. 2002; Meziane and Shipley 2001). SLA was expected to be low (less than 10 $mm^2/mg$, Ackerly 2004) because it enhances the photosynthetic rates and the resistance to cell wall collapse under water-stress (Turner 1994; Ackerly 2004). Most of the studied species (78%) have high values of SLA compared with other species of semi-arid environments and are more similar to values found by other authors in temperate and cold forests (15.4–36.3 $mm^2/mg$, Chen et al. 2011; Tian et al. 2016). SLA have been negatively correlated with leaf lifespan (Shipley et al. 2006; Tian et al. 2016) and this relationship have been explained as the result of increased photosynthetic capacity per unit of leaf dry mass in plants with shorter leaf lifespan. Since the most of the studied species by us are herbs, the high SLA found can be explained by this correlation. Also, compared to the tree species studied by Tian et al. (2016) SLA of the studied species was significantly higher. This can be interpreted as evidence of the correlation between growth form and SLA.

The SI was expected to be higher than 0.6 which is indicative of sclerophyllous leaves. Sclerophylls are typically associated to low nutrient concentration (Turner 1994). However in all the studied species the SI values was lower than expected (Additional file 2). Together these two variables indicate that the studied leaves can be considered mallacophylls (Turner 1994) and mesomorphic. The SPI found for the plants in this study was higher than that found for Tian et al. (2016) for some tree species within different ecosystems across China. They discuss that a higher SPI may be an adaptive strategy of leaf stomatal traits that results from higher stomatal conductance and therefore increased photosynthetic capacity which maximize carbon gain during the short growing season (Tian et al. 2016).

#### *Cuticle and epidermis*

We expected to see thick cuticles and a multiseriate epidermis on the leaf surface because those adaptations minimize transpiration rates. However, cuticle thickness was thin compared to other xerophytic plants, which are 2–22 μm thick (Rotondi et al. 2003; Bacelar et al. 2004; Gratani and Varone 2004). Moreover, the thin cuticles found in the REPSA are similar to those reported for species of *Ambrosia*, *Artemisia* and *Encelia* growing in a desert of North America (Gibson 1996). We also expected to see the leaf surface covered with abundant trichomes, but most of the observed species (97%) were glabrous or had trichomes that were scarcely distributed, except for *Pseudognaphalium*, *Artemisia* and *Conyza* which are herbs growing in open spaces. Most authors have based their taxonomic discussions on trichome morphology (Freire et al. 2005, 2007; Adedeji and Jewoola 2008; Gil et al. 2012; Redonda-Martínez et al. 2016). However in the studied species they were not abundant.

#### *Stomata*

Fahn and Cutler (1992) have hypothesized that amphistomatic leaves may have evolved in response to increasing aridity during the Tertiary period because they increase leaf conductance to $CO_2$. Therefore, amphistomatic leaves favored high maximum leaf conductances and are present in species growing in arid environments (Camargo and Marenco 2011), especially in herbs and shrubs from different environments such as successional forest, deserts or swamps (Mott et al. 1982; Mott and Michaelson 1991). In this study, 60% of the observed species had amphistomatic leaves and agrees with Mott and Michaelson (1991) findings in *Ambrosia cordifolia* because most of these species live in full sun during the short rainy season. The stomata in the studied species were at the same level as other epidermal cells. This pattern is different from the one observed in xeromorphic plants, where stomata are usually sunken or protected in crypts. The mean stomatal density found in this study (362 stomata/mm$^2$) was within the range found for other plants growing in arid environments (133–537 stomata/mm$^2$. Fahn and Cutler 1992; Bacelar et al. 2004; Yiotis et al. 2006; Gil et al. 2012). However, the stomatal densities observed in this environment (Table 1) are lower but not significantly different than those found in mesic environments (462–846 stomata/mm$^2$, Popma et al. 1992; Camargo and Marenco 2011). Wood (1934) suggested that although stomatal density increases with aridity, variations in stomatal density in sclerophyllous forests of South Australia were more related to intra-family characteristics rather than environmental conditions. It is possible that the observed variation in stomatal density has a phylogenetic signal, thus further analysis with a larger Asteraceae sampling growing in different communities is needed to support this assertion. The observed stomatal lengths are similar to the values reported by other authors for Asteraceae growing in arid and temperate environments (16–24 μm, Bacelar et al. 2004; Yiotis et al. 2006; Gil et al. 2012). Both stomatal density and length showed a significant negative scaling as found for other species of different families (Hetherington and Woodward 2003; Pearce et al. 2006; Camargo and Marenco 2011) and were important traits that explained part of the variance according to PCA suggesting that they adjust with palisade thickness and spongy parenchyma to maintain efficient photosynthesis.

#### *Dark staining deposits and oils*

It has been suggested that the presence of specialized cell types, such as oil containing cells or tannin cells, may be advantageous in dry environments because they protect the mesophyll cells against excess radiation or against herbivores and help to reduce the evaporation rate by interfering with water movement through the leaves (Fahn and Cutler 1992; Jordaan and Theunissen 1992; Turner 1994). Although we did not perform histochemical tests for tannins, the dark staining deposits that we described may be these contents. Thirty percent of the studied species had oil glands associated to vascular bundles, including *B. salicifolia* and *Barkleyanthus salicifolius*. Dark staining deposits were found only in two species (*Cosmos parviflorus* and *D. papposa*).

#### *Allometric relationships within the leaf*

The correlation between abaxial epidermis thickness and palisade thickness can be viewed from a functional perspective since it can be interpreted as a relationship between the gas exchange mechanism and the photosynthetic tissue resulting in greater efficiency of the leaf (Shields 1951; Rotondi et al. 2003). Our results of the PCA confirms the close allometric relationships between lamina thickness, palisade and spongy parenchyma and abaxial epidermis found by John et al. (2013) in a sample including various angiosperm families. This relationship has been interpreted as evidence of coordinated changes of the tissues composing the leaf (Brodribb et al. 2013).

Particularly the correlation between guard cell length and leaf thickness may be showing that the changes in cell size are key to the coordinated variation of leaf size.

### Leaves classification

Some of the studied species showed indicator traits of each of three types of response strategies to water stress and absence of soil. In species of the Eupatorieae (2 spp: *Ageratina adenophora* and *Stevia tomentosa*), Tageteae (3 spp: *D. papposa, T. micrantha, Pectis prostrata*), and Heliantheae (*L. rigida*) tribes, some foliar features corresponding to the avoidance strategy—water loss minimization—were observed. Those characteristics were hypostomatic or bifacial leaves with well developed, many-layers-thick palisade parenchyma, generally striated cuticles, thick epidermises, and vascular bundles surrounded by a well differentiated parenchyma sheath. In accordance with the foliar anatomy some other plant features indicate that these species prefer the avoidance strategy: Eupatorieae species that are generally shrubs or perennial herbs, there are rhizomes allowing them to store water and all leaves from the examined species of Tageteae were lobulate. The tolerance strategy was observed in all the species from the Senecioneae tribe. They showed traits such as leaf abscission during the higher stress season (November–May), mucilage accumulation (in the shoot), and perennation structures that allow regrowth. Some species from the Astereae tribe (4 spp, including *Conyza canadensis* and *L. sophiifolia*), and some members of Coreopsidae (e.g. *C. parviflorus*) and Heliantheae (e.g. *Tithonia tubiformis*) tribes, showed an escape strategy because they usually developed their life cycle during the rainy season (June–October) and produced seeds before water stress conditions arrived.

Most of the species showed a combination of characters that made their classification difficult. For example, species from the Bahieae and Carduae tribes are annual herbs that grow mainly from June to October; thus, they could be considered escapists. However, these species showed some avoidance features, such as lobulate leaves with striated cuticles and a well-developed palisade parenchyma. This agrees with the asseveration from some authors (Jones 1992; Chaves et al. 2003) about the three strategies not being mutually exclusive but instead we can find a combination of indicator traits of all three strategies within the same plant. The different character combinations for each species could be explained as a result of a fast evolutionary process as proposed by Stebbins (1952) for some members of Cichorieae (Asteraceae).

## Conclusions

The foliar anatomy of Asteraceae was variable, even among members of the same tribe growing in the same locality. If the observed variations were the result of adaptation to a xeric and poor-soil environment, then the observed features would correspond to non-succulent xeromorphic leaves but it was not the case. The majority of the studied species possess mesomorphic leaf features as simple lamina, single-layered epidermis, and soft (malacophyllous) large-size leaves with high SLA. Although some characters of drought resistance can be observed as for well-developed palisade and parenchyma bundle sheaths. The aforementioned character combination made difficult to classify the species studied within one of the three main response strategies to water stress. We suggest that the occurrence of specific character combinations in each species is due to a very fast evolutionary process involving the growth form, the adaptation to the environment and the phylogeny of the family. A character evolution analysis in an Asteraceae phylogeny including genera endemic to Mexico is needed in order to support this hypothesis (Terrazas, on going research).

The combined study of morphological and anatomical traits and its correlations is important to understand the constraints imposed over the diversity generated by phylogeny and adaptation. For example, although leaf anatomy is not xeromorphic in the studied species, it is advantageous for the species living in the REPSA to have the combination of traits mentioned. Those traits allow them to survive in full sun during the short rainy season having efficient photosynthetic capacity. Assuming that species from the forests around the Xitle colonized the REPSA, it is not surprising that they retained some of their mesic characters and also adapted to the low water environment. There were also species with Neotropical and Nearctic affinities that probably had to modify their leaves to colonize new environments. These species probably retained the capacity to adapt to dry environments, and that is one of the reasons for their success. It is known that in ecological succession, pioneer species need to have low nutritional requirements and efficient metabolisms to survive. The invasiveness of Asteraceae species is probably related to their capacity to grow in this poor-soil environment.

### Abbreviations

REPSA: Pedregal de San Ángel Ecological Reserve; UNAM: National Autonomous University of Mexico; SLA: specific leaf area; SI: sclerophylly index; SPI: stomatal pore area index; PCA: principal component analysis.

### Authors' contributions

PR and TT designed the study, PR carried out the laboratory work, analyzed the data and drafted the manuscript. TT analyzed the data, revised and corrected the manuscript. JLV approved the manuscript. All authors read and approved the final manuscript.

### Author details

[1] Departamento de Botánica, Instituto de Biología, Universidad Nacional Autónoma de México, Apartado Postal 70-367, 04510 Mexico City, Mexico. [2] Coordinación del Posgrado en Ciencias Biológicas, Circuito de Posgrados Ciudad Universitaria, Coyoacán, 04510 Mexico City, Mexico.

### Acknowledgements

This study is part of the requirements for PR to obtain the Doctor of Science degree from the Posgrado en Ciencias Biológicas UNAM (PCB). We thank Oscar Hinojosa for his help in the collection and identification of the plants, Dalia Grego for her support in many aspects of the laboratory work, Enrique Ortiz for doing the climograph and the map and Itzi Fragoso for her help with map edition. We appreciate the comments of the reviewers.

### Competing interests

The authors declare that they have no competing interests.

### Funding

PR thanks CONACYT for a scholarship (288322). JLV thanks the UNAM-DGAPA-PAPIIT (IN213916) program for financial support to acquire material and collect the plants.

### References

Ackerly DD (2004) Adaptation, niche conservatism and convergence: comparative studies of leaf evolution in the Californian chaparral. Amer Nat 163:654–671

Ackerly DD, Knight CA, Weiss SB, Barton K, Starmer KP (2002) Leaf size, specific leaf area and microhabitat distribution of chaparral woody plants: contrasting patterns in species level and community level analyses. Oecologia 130:449–457

Adedeji O, Jewoola OA (2008) Importance of leaf epidermal characters in the Asteraceae family. Not Bot Horti Agrobot Cluj Napoca 36:7–16

Anderson LC, Creech JB (1975) Comparative leaf anatomy of *Solidago* and related Asteraceae. Am J Bot 62:486–493

Bacelar EA, Correia CM, Moutinho-Pereira JM, Goncalves BC, Lopes JI, Torres-Pereria JMG (2004) Sclerophylly and leaf anatomical traits of five field-grown olive cultivars growing under drought conditions. Tree Physiol 24:233–239

Bailey IW, Sinnott EW (1916) The climatic distribution of certain types of angiosperms leaves. Am J Bot 3:24–39

Boeger MRT, Wisniewski C (2003) Comparacao da morfologia foliar de espécies arbóreas de tres estádios sucessionais distintos de floresta ombrófila densa (Floresta Atlántica) no Sul do Brasil. Rev Brasil Bot 26:61–72

Brodribb TJ, Jordan GJ, Carpenter RJ (2013) Unified changes in cell size permit coordinated leaf evolution. New Phytol 199:559–570

Camargo MAB, Marenco RA (2011) Density, size and distribution of stomata in 35 rainforest tree species in Central Amazonia. Acta Amazonica 41:205–212

Castillo-Argüero S, Montes-Cartas G, Romero-Romero MA, Martínez-Orea Y, Guadarrama-Chávez P, Sánchez-Gallén I, Núñez-Castillo O (2004) Dinámica y conservación de la flora del matorral xerófilo de la Reserva Ecológica del Pedregal de San Angel (D.F.) México. Bol Soc Bot México 74:51–75

Castillo-Argüero S, Martínez-Orea Y, Romero-Romero MA, Guadarrama-Chávez P, Núñez-Castillo O, Sánchez-Gallén I, Meave JA (2007) La reserva ecológica del Pedregal de San Ángel: Aspectos florísticos y ecológicos. México: Secretaria Ejecutiva de la Reserva Ecológica del Pedregal de San Ángel de Ciudad Universitaria, Universidad Nacional Autónoma de México

Castro MM, Leitão-Filho HF, Monteiro WR (1997) Utilização de estruturas secretoras na identificação dos gêneros de Asteraceae de uma vegetação de cerrado. Rev Brasil Bot 20:163–174

Céspedes CLA (2010) Riqueza florística de Asteraceae en los fragmentos de vegetación de la Reserva Ecológica del Pedregal de San Ángel, México D.F. M. Sc. Thesis. México: Facultad de Ciencias, Universidad Nacional Autónoma de México

Chaves MM, Maroco JP, Pereira JS (2003) Understanding plant responses to drought-from genes to the whole plant. Funct Plant Biol 30:239–264

Chen FS, Niklas KJ, Zeng DH (2011) Important foliar traits depend on species-grouping: analysis of a remnant temperate forest at the Keerqin Sandy Lands, China. Plant Soil 340:337–345

Cutler DF, Botha T, Stevenson DW (2007) Plant anatomy: an applied approach. Blackwell Publishing, Massachusetts

Delbón NM, Cosa T, Dottori N, Stiefkens L (2007) Análisis comparativo de los caracteres epidérmicos en *Flourensia campestris* y *F. oolepis* (Asteraceae). Bol Soc Argent Bot 42:245–250

Dickison WC (2000) Integrative plant anatomy. Academic Press, San Diego

Esau K (1976) Anatomía vegetal. Ediciones Omega, España. English edition: Esau K (1976) Plant anatomy (trans: Pons J). John Willey & Sons Inc., New York

Fahn A, Cutler DF (1992) Xerophytes. Gebrüder Borntraege, Berlin

Fang Y, Xiong L (2015) General mechanisms of drought response and their application in drought resistance improvement in plants. Cell Mol Life Sci 72(673):689

Ferreira JL, De Oliveira FV (1989) Identificaçao de *Sonchus oleraceus* L. (serralha): Principais características estruturais e químicas. Rev Bras Farmacogn 2/4:67–77

Ferreira EA, Procópio SO, Silva EAM, Silva AA, Rufino RJN (2002) Estudos anatômicos de folhas de espécies de plantas daninhas II. Planta Daninha 20:327–335

Fonseca CR, Overton JM, Collins B, Westoby M (2000) Shifts in trait-combinations along rainfall and phosphorus gradients. J Ecol 88:954–977

Freire S, Arambarri AM, Bayón N, Sancho G, Urtubey E, Monti C, Novoa MC, Colares MN (2005) Epidermal characteristics of toxic plants for cattle from the Salado river basin (Buenos Aires, Argentina). Bol Soc Argent Bot 40:241–281

Freire S, Urtubey E, Guiliano DA (2007) Epidermal characters of *Baccharis* (Asteraceae) species uses in traditional medicine. Caldasia 29:23–38

Funk VA, Susanna A, Stuessy TF, Robinson H (2009) Classification of Compositae. In: Funk VA, Susanna A, Stuessy TF, Bayer RJ (eds) Systematics, evolution, and biogeography of Compositae. Sheridan Books Inc, Ann Arbor

Garnier E, Shipley B, Roumet C, Laurent G (2001) A standarized protocol for the determination of specific leaf area and leaf dry matter content. Func Ecol 15:688–695

Gibson AC (1996) Structure—function relations of warm desert plants. Springer Verlag, Berlin

Gil SP, Seisdedos L, Reyna ME, Cerana MM (2012) Epidermis foliar de tres especies de asteráceas nativas de Argentina con potencial ornamental. Phyton (B. Aires) 81:205–210

Gratani L, Varone L (2004) Leaf key traits of *Erica arborea* L., *Erica multiflora* L. and *Rosmarinus officinalis* L. co-occurring in the Mediterranean maquis. Flora 199:58–69

Hetherington AM, Woodward FI (2003) The role of stomata in sensing and driving environmental change. Nature 424:901–908

Horn JW, Fisher JB, Tomlinson PB, Lewis CE, Laubengayer K (2009) Evolution of lamina anatomy in the palm family (Arecaceae). Am J Bot 96:1462–1486

Johansen DA (1940) Plant microtechnique. MacGraw Hill, New York

John GP, Scoffoni C, Sack L (2013) Allometry of cells and tissues within leaves. Am J Bot 100:1936–1948

Jones HG (1992) Plants and microclimate. Cambridge University Press, Cambridge

Jordaan A, Theunissen JD (1992) Phenolic deposits and tannin in the leaves of five xerophytic species from southern Africa. Bot Bull Academia Sinica 33:55–61

Kramer PJ (1983) Water relations of plants. Academic Press, New York

Lot A, Cano-Santana Z (2009) Biodiversidad del ecosistema del pedregal de San Ángel. Universidad Nacional Autónoma de México, México

Luckow M (2002) Anatomical features of the leaves in the *Dichrostachys* group

(Leguminosae: Mimosoideae) and their utility for Phylogenetic studies. Syst Bot 27:29–40
Luque R, De Menezes NL, Semir J (1999) Anatomía foliar de *Lychnophora* Mart. (Vernonieae: Asteraceae). Plantula 2:141–152
Metcalfe C, Chalk L (1979) Anatomy of the dicotyledons, 2nd edn. Oxford University Press, New York
Meziane D, Shipley B (2001) Direct and indirect relationships between specific leaf area, leaf nitrogen and leaf gas exchange. Effects of irradiance and nutrient supply. Ann Bot 88:915–927
Milán P, Hayashi AH, Appezzato-da-Glóra B (2006) Comparative leaf morphology and anatomy of three Asteraceae species. Braz Arch Biol Technol 49:135–144
Morrone JJ (2006) Biogeographic areas and transition zones of Latin America and the Caribbean islands based on panbiogeographic and cladistic analyses of the entomofauna. Annu Rev Entomol 51:467–494
Mott KA, Michaelson O (1991) Amphistomy as an adaptation to high light-intensity in *Ambrosia cordifolia* (Compositae). Am J Bot 78:76–79
Mott KA, Gibson AC, I'Leary JW (1982) The adaptive significance of amphistomatic leaves. Plant Cell Environ 5:455–460
Nevo E, PavlíekT, Beharav A, Bolshakova MA, Martyn GI, MusatenkoLI Sytnik KM (2000) Drought and light anatomical adaptive leaf strategies in three woody species caused by microclimatic selection at "Evolution Canyon", Israel. Israel J Plant Sci 48:33–46
Ordoñez JC, Van Bodegom PM, Witte JPM, Wright IJ, Reich PB, Aerts R (2009) A global study of relationships between leaf traits, climate and soil measures of nutrient fertility. Global Ecol Biogeogr 18:137–149
Pearce DW, Millard S, Bray DF, Rood SB (2006) Stomatal characteristics of riparian poplar species in a semi-arid environment. Tree Physiol 26:211–218
Popma J, Bongers F, Werger MJA (1992) Gap-dependence and leaf characteristics of trees in a tropical lowland rain forest in Mexico. Oikos 63:207–214
R Core Team (2015) R: A language and environment for statistical computing. R Foundation for Statistical Computing, Vienna, Austria. https://www.R-project.org/
Ragonese AM (1988) Canales secretores en los órganos vegetativos de *Eupatorium inulaefolium* H.B.K. (Compositae). Acta Farm Bonaer 7:161–168
Redonda-Martínez R, Villaseñor JL, Terrazas T (2016) Trichome diversity in the subtribe Leiboldiinae (Vernonieae, Asteraceae). J Torrey Bot Soc 143:298–310
Rojas A, Terrazas T, Villaseñor JL (in review). Foliar architecture in members of Senecioneae (Asteraceae). Rev Mex Biodivers
Rotondi A, Rossi F, Asunis C, Cesaraccio C (2003) Leaf xeromorphic adaptations of some plants of a coastal Mediterranean macchia ecosystem. J Mediterr Ecol 4:25–35
Ruzin ES (1999) Plant microtechnique and microscopy. Oxford University Press, New York
Rzedowski J (1998) Diversidad y orígenes de la flora fanerogámica de México. In: Ramamoorthy TP, Bye R, Lot A, Fa J (eds) Diversidad biológica de México: Orígenes y distribución. Instituto de Biología, Universidad Nacional Autónoma de México, México, D.F
Rzedowski J (2006) Vegetación de México. 1st digital edition. Comisión Nacional para el Conocimiento y Uso de la Biodiversidad, México
Rzedowski GC, Rzedowski J (2005) Flora fanerogámica del Valle de México. 2nd ed., 1st reimp., Instituto de Ecología, A.C. y Comisión Nacional para el Conocimiento y Uso de la Biodiversidad, Pátzcuaro
Santos MS, Ochoa N (1990) Adaptación de las plantas al déficit hídrico. Ciencia 41:333–344
Schmerler SB, Clement WL, Beaulieu JM, Chatelet DS, Sack L, Donoghue MJ, Edwards EJ (2012) Evolution of leaf form correlates with tropical-temperate transitions in *Viburnum* (Adoxaceae). Proc R Soc Lond B Bio 279:3905–3913
Shields LM (1951) Leaf xeromorphy in dicotyledon species from a gypsum sand deposit. Am J Bot 38:175–190
Shipley B, Lechowicz MJ, Wright I, Reich PB (2006) Fundamental trade-offs generating the worldwide leaf economics spectrum. Ecology 87:535–541
Siebe C (2000) Age and archaeological implications of Xitle volcano, southwestern Basin of Mexico-City. J Volcanol Geotherm Res 104:45–64
Soto-Trejo F, Palomino G, Villaseñor JL (2011) Números cromosómicos de Asteraceae de la Reserva Ecológica del Pedregal de San Ángel (REPSA), México, Distrito Federal. Rev Mex Biodivers 82:383–393
Stebbins GL (1952) Aridity as a stimulus to plant evolution. Amer Nat 86:33–44
Tian M, Yu G, He N, Hou J (2016) Leaf morphological and anatomical traits from tropical to temperate coniferous forests: mechanisms and influencing factors. Sci Rep 6:19703
Turner IM (1994) Sclerophylly: primary protective? Funct Ecol 8:669–675
Valladares F, Vilagrosa A, Peñuelas J, Ogaya R, Camarero JJ, Corcuera L, Sisó S, Gil-Pelegrín E (2004) Estrés hídrico: ecofisiología y escalas de la sequía. In: Valladares F (ed) Ecología del bosque mediterráneo en un mundo cambiante. Ministerio de Medio Ambiente, EGRAF, S. A, Madrid
Webb LJ (1959) A physiognomic classification of Australian rain forests. J Ecol 47:551–570
Wood JG (1934) The physiology of xerophytism in Australian plants. The stomatal frequencies, transpiration and osmotic pressures of sclerophyll and tomentose-succulent leaved plants. J Ecol 22:69–87
Wright IJ, Westoby M, Reich PB (2002) Convergence towards higher leaf mass per area in dry and nutrient poor habitats has different consequences for leaf life span. J Ecol 90:534–543
Yiotis C, Manetas Y, Psaras GK (2006) Leaf and green stem anatomy of the drought deciduous Mediterranean shrub *Calicome villosa* (Poiret) Link. (Leguminosae). Flora 201:102–107

# 19

# Allelopathy and resource competition: the effects of *Phragmites australis* invasion in plant communities

Md Nazim Uddin[1,2*] and Randall William Robinson[1,2]

**Abstract**

**Background:** *Phragmites australis*, a ubiquitous wetland plant, has been considered one of the most invasive species in the world. Allelopathy appears to be one of the invasion mechanisms, however, the effects could be masked by resource competition among target plants. The difficulty of distinguishing allelopathy from resource competition among plants has hindered investigations of the role of phytotoxic allelochemicals in plant communities. This has been addressed via experiments conducted in both the greenhouse and laboratory by growing associated plants, *Melaleuca ericifolia*, *Rumex conglomeratus*, and model plant, *Lactuca sativa* at varying densities with the allelopathic plant, *P. australis*, its litter and leachate of *P. australis* litter. This study investigated the potential interacting influences of allelopathy and resource competition on plant growth–density relationships.

**Results:** In greenhouse, the root exudates mediated effects showed the strongest growth inhibition of *M. ericifolia* at high density whereas litter mediated results revealed increased growth at medium density treatments compared to low and high density. Again, laboratory experiments related to seed germination and seedling growth of *L. sativa* and *R. conglomeratus* exhibited phytotoxicity decreased showing positive growth as plant density increased and vice versa. Overall, the differential effects were observed among experiments but maximum individual plant biomass and some other positive effects on plant traits such as root and shoot length, chlorophyll content occurred at an intermediate density. This was attributed to the sharing of the available phytotoxin among plants at high densities which is compatible to density-dependent phytotoxicity model.

**Conclusions:** The results demonstrated that plant–plant interference is the combined effect of allelopathy and resource competition with many other factors but this experimental design, target-neighbor mixed-culture in combination of plant grown at varying densities with varying level of phytotoxins, mono-culture, can successfully separate allelopathic effects from competition.

**Keywords:** Allelopathy, Density-dependent phytotoxicity, Ecosystems, Invasion, *Phragmites australis*, Resource competition

## Background

Allelopathic interference by invasive plant species has potential to impact seed germination, seedling growth, development and establishment of neighbouring plant species, as well as of the same species, in both natural and agricultural systems (Bich and Kato-Noguchi 2014; Dorning and Cipollini 2006; Lara-Núñez et al. 2006). Allelopathy has been considered an important attribute to the success of an invasive species in natural ecosystems (Callaway and Ridenour 2004; Kimura et al. 2015; Lorenzo et al. 2010). The sources of allelochemicals released into the rhizosphere include leaching from leaves and other aerial parts, volatilization, root exudation and litter decomposition (Hussain and Reigosa 2012; Uddin et al. 2012; Weir et al. 2004).

*Correspondence: Nazim.Uddin@vu.edu.au; naz_es@yahoo.com
[2] Institute for Sustainability & Innovation, Victoria University, Melbourne, VIC 8001, Australia
Full list of author information is available at the end of the article

*Phragmites australis*, a ubiquitous wetland plant, is considered one of the most invasive species in the world (Uddin et al. 2012) however, the origin of the species is still unclear (Plut et al. 2011). A perennial graminaceous plant, to 3 m tall, it reproduces mainly through rhizomes and, at low frequency, through seeds. *P. australis* grows in all temperate zones of the world, especially North America, most countries in Europe, some parts of Canada and Australia (Hocking et al. 1983; Kulmatiski et al. 2011), being especially common in south-eastern Australia (Kettenring et al. 2011; Morris et al. 2008). The worldwide, regional and local distribution and abundance of *P. australis* has expanded over the last 150 years and in most areas it forms dense monocultures (Saltonstall and Miller 2005). Due to the impacts of *P. australis* invasions, habitats have been diminished or altered significantly for other flora and fauna causing loss of biodiversity and ecosystem functions (Mack et al. 2000). Several studies have identified chemicals within *P. australis* organs which have antialgal, antifungal or antibacterial effects (Li and Hu 2005). Previous allelopathic studies have shown that water extracts, decomposed materials, root exudates and specific identified chemicals of *P. australis* organs have strong phytotoxic effects on germination, growth, and establishment of other plant species (Kettenring et al. 2011; Rudrappa et al. 2007, 2009; Uddin et al. 2012, 2014a, b, c) and thus, it is assumed that *P. australis* achieves its competitive advantages over invasion process into wetlands through allelopathy (Bains et al. 2009; Rudrappa et al. 2007).

While *P. australis* has clearly shown phytotoxic potential, the effects should be considered in more ecologically realistic ways by differentiating allelopathic interactions from resource competition. The allelopathic effects might be masked by resource competition among target plants (Barto and Cipollini 2009; Weidenhamer et al. 1989). A better understanding of dose–response relationships of allelochemicals would help to clarify this issue. Toxin dilution is thought to occur because plants share and compete not only for resources but also for toxin (Hansi et al. 2014; Hoffman and Lavy 1978; Suman et al. 2002). Population density modifies phytotoxic effects through dilution of available toxins among plants (i.e. phytotoxicity decreases as plant density increases) (Thijs et al. 1994). Thus, the dose of a phytotoxin received by a plant is inversely related to plant density. This toxin dilution study can be performed using density-dependent experiments, and may be a potential tool for exploring the effects of phyto-toxins on plant growth as well as for differentiating the resource competition from allelopathy. As the study of allelopathic interactions may be hindered by the lack of proper experimental methods, it may be more productive to first demonstrate explicit interference, by allelochemicals rather than rely solely on explanations that involve resource competition or other mechanisms (Barto and Cipollini 2009; Thijs et al. 1994).

Although the use of activated carbon (AC) as a soil amendment, has use in determining distinct differences between allelopathy and resource competition (Inderjit and Callaway 2003), AC may change the availability of soil nutrients (Weißhuhn and Prati 2009). Weidenhamer (2006) proposed allelopathic effects might be differentiated experimentally using the density-dependent nature of phytotoxic effects, in turn, causing deviations from predicted growth–density relationships. The effects by allelochemicals depends on density of neighbour-target plant species and might be masked by resource competition at high density (Weidenhamer 1996). Density-dependent models suggest that yield decreases with increasing density due to resource competition acting as a dominating factor. Alternatively, results due to allelopathy show a slow decrease of yield or even increase in yield as density increases, until density reaches a point where resource competition among neighbouring target plant becomes the dominating factor. The yields of exposed plants to pure chemicals (Andersen 1981), ground tissue of allelopathic plants (Tseng et al. 2003) and soil mediated allelopathic plants (Weidenhamer et al. 1989) are consistent with the assumptions of a density-dependent phytotoxicity model. Andersen (1981) found that reduced plant seedlings of soybeans may partially reduce the negative effects of herbicides as well as Weidenhamer et al. (1989) and Tseng et al. (2003) stated that phytotoxicity decreased as plant density increased.

Therefore, this study has been designed to determine the occurrence and magnitude of potential allelopathic effects mediated by *P. australis* root exudates, its litter and extracts of litter with a wide range of doses through a density-dependent approach. This method might be effective in distinguishing the allelopathic interactions of *P. australis* with neighbouring plant species from resource competition. We hypothesized that phytotoxic effects of allelochemicals depend on the neighbouring plant density, due to phytotoxins dilution among individual plants.

## Methods

### Study site, plant litter and soil sample collection

Fallen leaves of *P. australis* were collected in June 2011 from natural stands adjacent to Cherry lake (37°51′ 30″S, 144°50′ 5″E), a coastal wetland in Altona, Melbourne, Australia. All samples were placed into sealable plastic bags for transportation to the laboratory. Plant samples were sorted from other plant residue and debris, then kept at room temperature to air dry until constant dry weight. After desiccation, sorted samples were cut into small pieces (<2 cm) and preserved in plastic ziplock bags until use. Soil samples were collected from the top layer of *P. australis* free areas of the same study site, separated from other organic materials, dried at room temperature and kept in ziplock bags after passing through a 2 mm sieve.

### Choice of target species

Seeds of several species were used to determine any differential response in native, introduced and model species to allelopathy. Seed capsules of the native species, *Melaleuca ericifolia* and introduced species, *Rumex conglomeratus* were collected in May 2011 from Cherry Lake and stored in paper bags at room temperature for 1 week. Seeds were shaken from the capsules and sieved to remove empty capsules and other detritus. Associated species were used because of ecological relevance. Seeds of a model species, *Lactuca sativa* were purchased from a commercial source (DT Brown Seeds, South Windsor, NSW, Australia). Lettuce was selected as it is widely used in phyto-toxicity bioassays. Model species is easily grown, minimizing the risk of observed growth differences, due to factors other than treatments applied in the experiments.

### Greenhouse experiments

#### *Root exudates mediated effects on* M. ericifolia *seedlings*

Spring buds of *P. australis* with rhizome attached were collected on 8 September 2011 from Cherry Lake. Each live rhizome was cut to contain exactly one active node, weighed and planted within 6 h of collection in 7 L plastic pots lined with watertight plastic bags filled with 4 L substrate [a 1:7 mixture of unsterilized river sand and potting mix soil respectively (Earth-wise Growing Essential, Australian Prime Fibre Pty Ltd, Queensland-4660)] and 20 g inoculum collected from *P. australis* field. Potting mix contained organic materials (pine bark), living organisms (bacteria, fungi and protozoa), minerals, fertilizers additives and more in details in our previous studies (Uddin et al. 2014b). The control pots containing substrates only were kept with other treated pots in the same greenhouse conditions until *M. ericifolia* juvenile plants were replanted. To all of the pots, 1 g $L^{-1}$ of mixed pelletised fertilizer (Pivot fertilizer-900; N-P-K: 16-8-9) was incorporated into the tilled topsoil bimonthly. Pots were kept in a naturally-lit greenhouse at 23 ± 3 and 12 ± 2 °C day/night temperature and watered regularly with an auto irrigation system equipped by micro sprinklers at the soil surface to keep soil moist at a level of 55 ± 5%, similar to field soil. Soil moisture was monitored weekly by measuring water content in soil randomly collected from the pots. Pots were randomly shuffled every week to minimise the spatial effects and unwanted germinant (weeds) were removed. Tube stock of *M. ericifolia* (6 months old), grown in potting mix, were purchased from 'Go Native Landscapes Pty Ltd' (Inverloch, Victoria, 3996, Australia) grown from seeds collected from *M. ericifolia* stands at Dowd Morass wetland in Inverloch, Victoria, Australia. On 10 November 2011, the purchased *M. ericifolia* juvenile plants were planted into the prepared pots with and without *P. australis* by three neighbour densities (one, two, and four plants per pot) with three replicates of each treatment. After 6 months of *M. ericifolia* growth, all plants (*P. australis* and *M. ericifolia*) were harvested and data collected on above-ground biomass (AGB), below-ground biomass (BGB), root–shoot length, plant height, stem diameter, and number of growth points (terminal branches).

#### *Litter mediated effects on* M. ericifolia *seedlings*

*Phragmites australis* free soil (200 g) was set in 1.5 L pot, moistened to approximately field saturation level and kept for 2 days in the greenhouse. Equal size 2-month old *M. ericifolia* seedlings grown in naturally-lit greenhouse were then replanted in each pot with a density of one, two, and four plants at eight replicates. After 3 weeks of plant acclimation, equal size litter (≤2 cm) was put on soil surface in the pot with a concentration of 4 g/100 g of soil, but no litter for control and kept as per the above conditions. The amount of litter present on the soil surface was identified on previous field observations to be the amount of litter fall in a 1-$m^2$ quadrate. After 4 months, the survived plants with mixed replicates (low density = 8, medium density = 3 and high density = 4) were harvested and measured for the various phenotypic characteristics.

### Laboratory experiments

#### *Litter leachate mediated effects on* R. conglomeratus *seedlings*

Litter leachate was made by soaking litter in distilled water with a concentration of 10% (100 g litter in 1 L water). Left to soak for 24 h, the leachate was filtered with cheese-cloth, centrifuged at 3000 rpm and filtrate then used for the experiment with pH adjustment at 6.5 with 1 N NaOH and 1 N HCl to avoid non-relevant effects of leachate due to pH. Fifty grams of *P. australis* free soil was moistened with 30 mL litter leachate at four different concentrations (0, 2.5, 5.0, and 10.0%) in 250 mL container, and incubated for 24 h. In general, the most of the bioassays are conducted in the allelopathy study at concentrations of 1–5% (the weight or mass of the plant matter per volume of solvent) (Reigosa et al. 2013). Therefore, we used a wide range of doses to ensure encompassing lowest dose for an observable effect, as well as the highest dose for maximal effect (Belz et al. 2007). Equal numbers

(15 seeds with three replicates) of *R. conglomeratus* seeds were sown in each pot to ensure emergence of adequate numbers of even-age seedlings. All pots were then placed in a growth chamber (Westinghouse, Electrolux home products, Australia) set to 25/15 °C day/night temperature and a 12 h photoperiod with illumination of 84 μmol $s^{-1}$ $m^{-2}$. One week after emergence, seedlings were thinned to a density of one, two, four, and eight in each pot. Water loss by evaporation was measured by weight and compensated with leachate twice in a week. Plants were harvested and measured the phenotypic and physiological characteristics after 6 weeks of treatments.

#### *Unburnt versus burnt litter extracts mediated effects on* L. sativa *seedlings*

*Phragmites australis* litter was placed in a furnace at 300 °C for 1 h to produce burnt litter and a phyto-toxicity test was conducted with extracts of both type of litter (unburnt and burnt). Extracts were made by mixing unburnt and burnt litter powder passed through 0.5 mm sieve with distilled water to a concentration of 10%. After a 24 h period, it was filtered with cheese cloth and centrifuged at 3000 rpm. The resultant filtrate then used as the extract for the experiment with pH adjustment at 6.5 with 1 N NaOH and 1 N HCl, to avoid non-relevant effects of extracts due to pH. Extract (5 mL) at four different concentrations (0, 2.5, 5.0, and 10.0%) was placed into a sterile 9 cm Petri dish containing two sterile sheets of filter paper (Whatman No. 1). At least, three replicates were used for each treatment with density of one, two, four and eight pre-germinated *L. sativa* seedlings. Petri dishes were sealed with parafilm (Pechiney, Plastic Packaging Company, Menasha, WI 54952) then placed in polyethylene bags to prevent water loss by evaporation and to avoid contamination by fungi and bacteria. The prepared dishes were arranged in a completely randomized design (CRD) and placed in a growth chamber according to above mentioned conditions. The Petri dishes were randomized each day to minimize the spatial effect. After 7 days of experiment, phenotypic characteristics of the grown of the plants were measured.

#### *Litter mediated effects on* L. sativa *seed germination*

Seed germination bioassay was conducted using the 'sandwich' method adopted from Fujii et al. (2004). Four different concentrations of air dried litter by weight (0, 2.5, 5.0, and 10%) were placed in between two layers of 0.5% agarose (total 10 mL) in a container of 4.5 by 5.5 cm. Agarose was autoclaved at 121 °C for 15 min and subsequently cooled at room temperature. Agar media was used due to higher bio-availability and greater exposer of the allelochemicals to the receiver plants through diffusion (Duke 2015). In addition, microbial community in substrate convert phytotoxins to more active or less active compounds but autoclaving may eliminate the possibility (Duke et al. 2009). Using autoclaved agarose with litter, the seeds of *L. sativa* at a density of four, eight, and 16 were placed on each container and incubated for 7 days as above condition with three replicates. The germination and biometric parameters were measured at the end of the experiment.

#### Phenolics determination

Total phenolics (TP) and water soluble phenolic (WSP) content were measured in unburnt and burnt litter with three replicates, due to a causal relationship between phenolic content and phytotoxicity in plant (An et al. 2001; Huang et al. 2003). Approximately 100 mg portion of powder (burnt and unburnt litter) was weighed out and transferred to Eppendorf tube. After addition of 5 mL of 70% acetone for TP and distilled water for WSP, these were incubated at 4 °C for 1 h to extract phenolics followed by centrifuging at 15,000 rpm for 10 min at 4 °C. The 0.5 mL of the supernatant was taken and made up to 1 mL with distilled water. Then 5 mL of 2% $Na_2CO_3$ in 0.1 N NaOH was added and mixed using vortex mixer (Vortex Mixer, VOU1, Ratek Instruments Pty Ltd., Australia). To the mixture obtained by the above process 0.5 mL Folin-Ciocalteu reagent was added and mixed. After 2 h, absorbance was read at 760 nm. Based on the standard curve, phenolic was determined as gallic acid equivalents of the sample (mg TP and WSP per g sample).

#### Chlorophyll measurement

Photosynthesis is the basic physiological process of plant growth and it has been inhibited by allelochemicals through influencing the chlorophyll content of the exposed plants (Qian et al. 2009; Zhou and Yu 2006). To gain an understanding of the physiological mechanism influenced by allelochemicals, chlorophyll *a* (Chl *a*), chlorophyll *b* (Chl *b*) and total chlorophyll were determined. Approximately 15 mg of fresh leaf of affected *R. conglomeratus* was placed in 7 mL of *N–N* dimethylformamide (DMF) for 24 h in darkness at 4 °C for extraction (Moran and Porath 1980). The absorbance of the extracts was measured spectrophotometrically at 664 and 647 nm and chlorophyll was determined following the equation proposed by Inskeep and Bloom (1985).

#### Statistical analyses

All the experiments were conducted in a completely randomized design with at least three replicates. All data were referred as percentage of the values of control treatments at each density and analysed using IBM SPSS statistics 21.0 and Microsoft Excel 2010. Two-way ANOVA was used to show the effects of *P. australis* litter leachate on seedling densities of *R. conglomeratus*. Three-way ANOVA was designed to perform the effects of different materials (unburnt and burnt) with different concentrations of *P. australis* litter extract on growth of *L. sativa* seedling densities. Again, using two-way ANOVA we measured the effects of different concentrations of *P. australis* litter mediated agarose on germination and growth of *L. sativa* with different seed densities. In all cases, variance homogeneity was tested using Levenes's test and transformed if necessary (square root). Significance tests were performed using univariate analysis of variance with 2-sided Duncan's tests at the 0.05 probability level. Furthermore, the competitive effect of a neighbour species on a target species was quantified through linear regression procedures. The slope of the regression and the correlation coefficient that evaluates the dependence of each measured variables of exposed plants such as biomass (above-ground, below-ground and total biomass) and root length on the concentration of the residues and residue extracts as well as plant and seed density, were calculated. Again, an effort has been made to determine whether the difference in slopes between treatment and control was statistically significant by Real Statistics Resource Pack Software plug-in Microsoft Excel 2010 with log-transformed data (Zaiontz 2014). Slopes were compared via the slopes test [(=SlopesTest(x1, y1, x2, y2)] and the T-distribution [(=TDIST(x, df, tails)] following the method for comparing slopes of two independent samples developed by Zaiontz (2014).

## Results

### Greenhouse experiments

Root exudates of *P. australis* affected the biomass and other morphological characters of *M. ericifolia* at various densities (Table 1). Most of the growth parameters decreased with increasing plant density indicating that plant density was the most significant factor for influencing the growth of *M. ericifolia* such as AGB ($F_{2,6} = 26.96$, $P < 0.01$), BGB ($F_{2,6} = 52.91$, $P < 0.001$), total biomass ($F_{2,6} = 29.04$, $P < 0.001$), root length ($F_{2,6} = 39.95$, $P < 0.001$), plant height ($F_{2,6} = 14.65$, $P < 0.005$), and growth points ($F_{2,6} = 111.83$, $P < 0.001$). The slopes of all the growth parameters like AGB ($t = 2.24$), BGB ($t = 3.12$), total biomass ($t = 2.90$), root length ($t = 2.82$), plant height ($t = 2.92$), and growth points ($t = 2.90$) were significant compared to control with degree of freedom *df* 14 and significance level $\alpha$ 0.05.

Treated soil mixed with or without the leaf litter of *P. australis* reduced growth of *M. ericifolia* by inhibiting biomass, root length, plant height, stem diameter and number of growth points at various densities (Table 1). A differential in reduction was observed in the whole experiment but medium density treatments had increased growth compared to low and high density treatments. Total dry biomass per plant was reduced by 94, 77 and 90% at low, medium and high density respectively. The reduction of root length (65%), plant height (44%), stem diameter (61%) and number of growth points (89%) was more severe at low density than high density, but comparatively less inhibited at medium density. Growth parameters like AGB ($F_{2,12} = 20.90$, $P < 0.001$), BGB ($F_{2,12} = 16.39$, $P < 0.001$), total biomass ($F_{2,12} = 20.71$, $P < 0.001$), root length ($F_{2,12} = 18.22$, $P < 0.001$), plant height ($F_{2,12} = 6.65$, $P < 0.01$), stem diameter ($F_{2,12} = 28.83$, $P < 0.001$), and number of growth points ($F_{2,12} = 6.41$, $P < 0.01$) varied significantly across the densities. Regression analyses among densities and growth parameters of *M. ericifolia* showed that the difference in slopes in the pots to *P. australis* treatments compared to the control were not significant such as AGB ($t = 1.96$), BGB ($t = 1.70$), total biomass ($t = 1.87$), plant height ($t = 1.97$), and number of growth points ($t = 0.26$) except root length ($t = 4.29$) with *df* 24 and $\alpha$ 0.05.

### Laboratory experiments

Plants grown in *P. australis* litter leachate treated soil showed that the growth, biomass and other morphological and physiological parameters of *R. conglomeratus*

**Table 1 Changes of biomass and morphological characters at various density of *Melaleuca ericifolia* in treatments of *Phragmites australis* root exudates and litter as % of control**

| Plant density/ pot | Source of materials | Aboveground mass | Belowground mass | Total biomass | Root length | Plant height | Growth point |
|---|---|---|---|---|---|---|---|
| Low | Root exudates | 94.47 ± 3.09 a | 88.03 ± 6.16 a | 93.87 ± 2.54 a | 100.00 ± 7.22 a | 90.56 ± 3.09 a | 98.27 ± 5.38 a |
| | Plant litter | 6.18 ± 0.84 a | 7.06 ± 1.13 a | 6.35 ± 0.86 a | 34.02 ± 2.71 a | 55.93 ± 4.08 a | 11.38 ± 1.47 a |
| Medium | Root exudates | 62.98 ± 4.26 a | 67.90 ± 10.55 a | 63.42 ± 4.77 a | 85.86 ± 7.89 a | 93.74 ± 8.28 a | 71.72 ± 8.27 b |
| | Plant litter | 22.85 ± 4.40 b | 25.15 ± 4.80 b | 23.27 ± 4.26 b | 61.75 ± 3.49 b | 82.74 ± 1.49 b | 17.94 ± 2.18 b |
| High | Root exudates | 165.45 ± 16.70 b | 273.33 ± 24.04 b | 171.65 ± 17.10 b | 195.56 ± 12.37 b | 126.92 ± 2.21 b | 194.42 ± 3.80 c |
| | Plant litter | 10.81 ± 1.24 a | 10.02 ± 2.15 a | 10.63 ± 1.35 c | 66.50 ± 8.80 b | 66.51 ± 6.42 ab | 8.22 ± 0.71 a |

Average (±standard error) values in the column followed by same letter are not significantly different at the 0.05 level by Duncan's test were affected by both leachate concentration and density of test plant (Fig. 1; Table 2; Additional file 1: Figure S1). The degree of inhibition and stimulation by leachate was density-dependent. Generally, all growth parameters (AGB, BGB, total biomass, root length, plant height, number of leaves and chlorophyll content) increased with increasing plant density at lower and medium leachate concentration, but there were occurrences at a higher level. The highest biomass reduction was observed at the low plant density (one seedling/pot) treatment by 51 at 2.5% extract amended soil compared to control soil whereas 49 and 35% were in medium (two seedlings/pot) and high (eight seedlings/pot) density respectively but stimulated at intermediate (four seedlings/pot) density by 49%. All results showed a reduced leachate effect in relation to plant density and there was a clear trend to stimulation of growth as density increases. The variation in the magnitude of stimulation with density demonstrates dose-dependency of the stimulatory response, whereas no stimulatory effects were observed in the control. Deviations in the relationships between plant growth parameters and density were pronounced, however, comparisons of the regression lines showed that the differences in observed slopes were not significant with

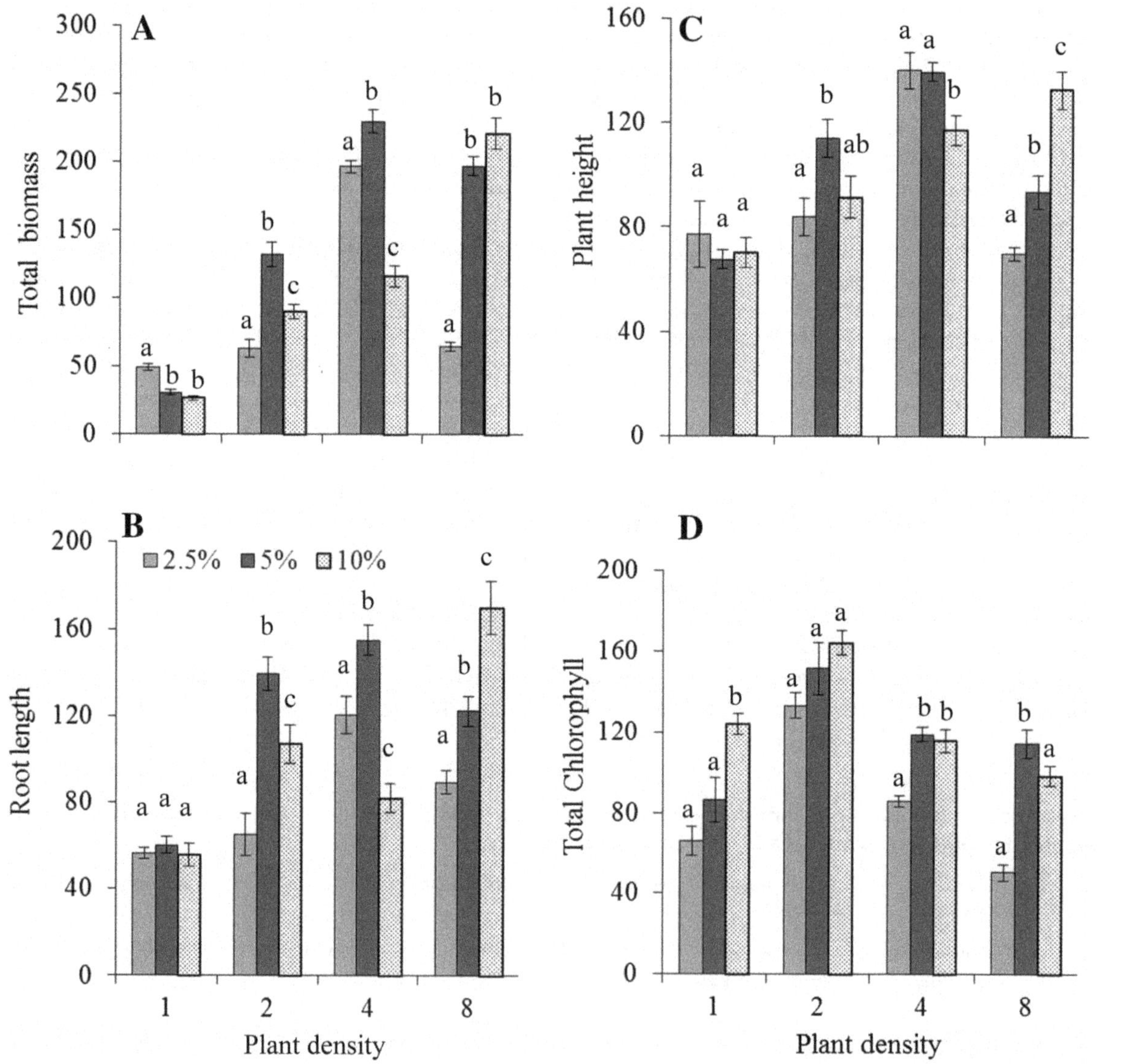


**Fig. 1** Effects of *Phragmites australis* litter leachate on **A** total biomass, **B** root length, **C** plant height and **D** total chlorophyll of *Rumex conglomeratus* seedlings at different densities. Values (weight per plant) are means as % of control treatments at each density ± standard errors ($n = 3$). *Letters* indicate homogenous subgroups ($P \leq 0.05$) at each density in Duncan's test

**Table 2 Results of two-way ANOVA (*F*-ratios) showing the effects of *Phragmites australis* litter leachate (C) on seedling densities (D) of *Rumex conglomeratus***

| Factor | *df* | *F* ratio and probability | | | | | | |
|---|---|---|---|---|---|---|---|---|
| | | Above-ground biomass | Below-ground biomass | Total biomass | Root length | Plant height | Number of leaf/ plant | Total chlorophyll |
| Concentration (C) | 2 | 68.15*** | 40.02*** | 71.26*** | 23.49*** | $3.11^{ns}$ | 9.00*** | 40.39*** |
| Density (D) | 3 | 316.71*** | 267.80*** | 313.54*** | 51.72*** | 39.01*** | 5.50** | 49.22*** |
| C × D | 6 | 74.66*** | 28.92*** | 70.88*** | 17.87*** | 9.12*** | 5.72*** | 3.99** |
| Error | 24 | | | | | | | |
| Total | 36 | | | | | | | |

Data used here as % of control. Level of significance: ***, **, * and $^{ns}$ indicate significant difference at $P \leq 0.001$, $P \leq 0.01$, $P \leq 0.05$ and non-significant respectively

exception of root length ($t = 2.92$) at 2.5% leachate, AGB ($t = 6.54$), BGB ($t = 2.38$), total biomass ($t = 4.19$) and root length ($t = 2.97$) at 5.0% leachate whereas at 10.0% leachate most of the slopes were significant except number of leaf ($t = 1.47$) and total chlorophyll ($t = 1.54$) with $df$ 20 and $\alpha$ 0.05.

The effects of unburnt and burnt litter extracts of *P. australis* on the growth of *L. sativa* seedlings indicated significant reduction in terms of AGB, BGB, total biomass, root and shoot length (Table 3; Fig. 2; Additional file 1: Figure S2). Greater inhibition was observed with increasing concentration of the aqueous extracts. Materials (unburnt versus burnt), concentrations and densities all showed differential effects individually and interactively on biomass and root length of *L. sativa* (Table 3). BGB was inhibited more than AGB, but AGB–BGB ratio showed significant individual and interactive effects (Table 3). Inhibition was higher in unburnt residue extract, than burnt extracts (Fig. 2; Additional file 1: Figure S2) but the degree and magnitude of inhibition on AGB, BGB, total biomass and root length was not strongly correlated with the TP and WSP of the extracts. TP in unburnt versus burnt was 17.77 ± 0.64 versus 4.15 ± 0.24 mg $g^{-1}$ whereas WSP was 4.59 ± 0.1 versus 2.3 ± 0.3 mg $g^{-1}$ (Fig. 3). Although phenolics were significantly lower in burnt residue extracts, the inhibition shown by burnt extracts was not significantly lower than unburnt residue (Fig. 2; Additional file 1: Figure S2, available as additional material). The observed deviations (inhibitory and stimulatory) from the growth–density relationship were measured by the comparison of the slopes among treatments. The slope difference was significant at 10% unburnt extract in AGB ($t = 3.90$), BGB ($t = 3.85$) and total biomass ($t = 4.51$), but BGB was highly affected even at 2.5% ($t = 3.12$) and 5.0% extract ($t = 2.67$) with $df$ 20 and $\alpha$ 0.05. Whereas, only significant slope difference was observed by burnt extract on BGB in each extract concentration such as 2.5% ($t = 3.44$), 5.0% ($t = 3.61$) and 10.0% ($t = 4.57$), with $df$ 20 and $\alpha$ 0.05. Medium density had larger biomass and root length compared to low and higher density (Fig. 2; Additional file 1: Figure S2, available as additional material). Total biomass of *L. sativa* was significantly higher in burnt extracts than unburnt (Fig. 2). The total biomass increased with

**Table 3 Results of three-way ANOVA (*F*-ratios) showing the effects of different materials (unburnt and burnt) (M) with different concentrations (C) of *Phragmites australis* litter extract on growth of *Lactuca sativa* seedling densities (D)**

| Factor | *df* | *F* ratio and probability | | | | |
|---|---|---|---|---|---|---|
| | | Above-ground biomass (AGB) | Below-ground biomass (BGB) | Total biomass | AGB–BGB ratio | Root length |
| Materials (M) | 1 | 64.76*** | $1.92^{ns}$ | 25.55*** | 64.27*** | 246.88*** |
| Concentration (C) | 2 | 10.54*** | 50.81*** | 9.71*** | 93.17*** | 180.34*** |
| Density (D) | 3 | 37.02*** | 23.85*** | 40.55*** | 5.20** | 154.91*** |
| M × C | 2 | 11.03*** | 4.44* | 4.55* | 27.20*** | 19.55*** |
| M × D | 3 | 7.41*** | $1.57^{ns}$ | 3.41* | 11.98*** | 16.25*** |
| C × D | 6 | 6.91*** | 3.78** | 6.10*** | $2.06^{ns}$ | 6.19*** |
| M × C × D | 6 | $1.00^{ns}$ | $1.34^{ns}$ | $1.01^{ns}$ | 2.96* | 2.29* |
| Error | 48 | | | | | |
| Total | 72 | | | | | |

Data used as % of control. Level of significance: ***, **, * and $^{ns}$ indicate significant difference at $P \leq 0.001$, $P \leq 0.01$, $P \leq 0.05$ and non-significant respectively

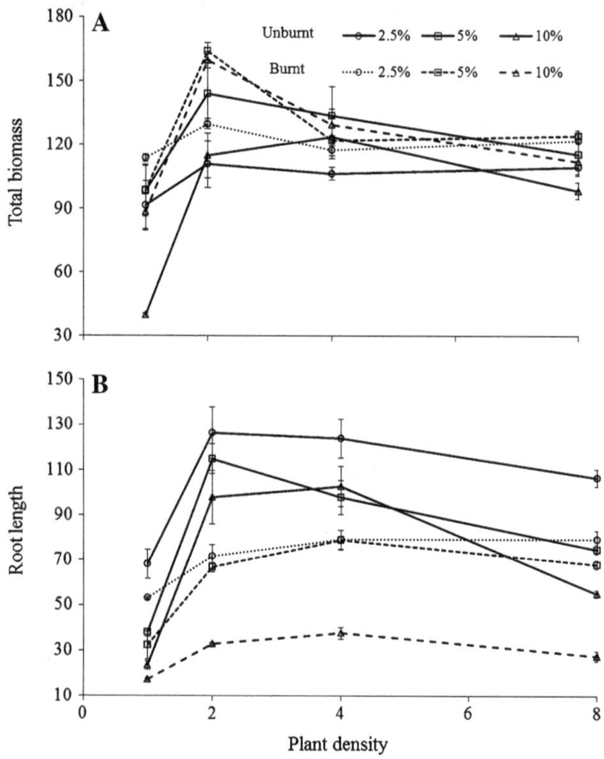


**Fig. 2** Relationship of **A** total biomass and **B** root length with seedling densities of *Lactuca sativa* grown at different concentrations of *Phragmites australis* litter (unburnt versus burnt) extracts. Values (weight per plant and length per plant) are means as % of control treatments at each density ± standard errors (*n* = 3)

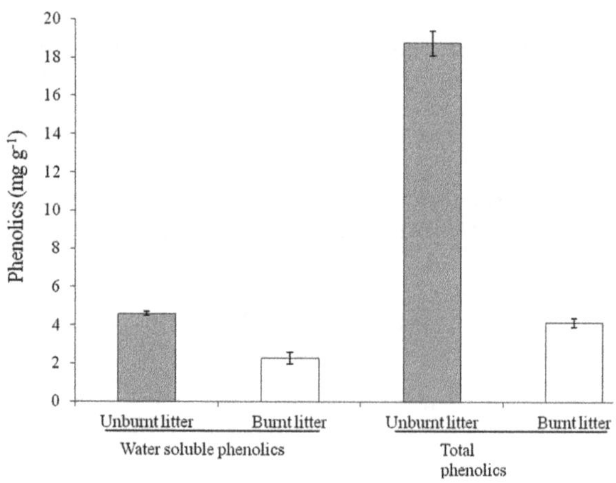


**Fig. 3** Total phenolics (TP) and water soluble phenolics (WSP) *Phragmites australis* litter (unburnt versus burnt). Values are means ± standard errors (*n* = 3)

increasing density as compared to the control, except for density one in both unburnt and burnt litter extract.

Germination bioassay showed differential effects on germination and growth parameters (Table 4; Fig. 4; Additional file 1: Figure S3). Germination percentage and AGB were significantly affected by both interactive and individual effect of residue concentration and density. All other measured parameters (BGB, total biomass, root and shoot length) varied significantly across the residue concentrations and density individually but not interactively (Table 4). BGB was the lowest at lower density and decreased (13, 31 and 68%) with increasing concentration (1.25, 2.5 and 5.0% respectively) compared to control while highest at the intermediate density. Root and shoot length was always higher along low concentrations at each density but they were significantly higher at medium densities (Fig. 4C, D). Again, these lowered with increasing and decreasing densities (Fig. 4). The difference in slopes was significant in AGB ($t = 3.55$), BGB ($t = 3.56$), total biomass ($t = 4.14$) and root length ($t = 3.17$) only at 5.0% concentration with $df$ 14 and $\alpha$ 0.05.

## Discussion

Analysis of growth–density relationships is useful tool for understanding the resource competition and allelopathic interference between plants of the suspected invasive species (Weidenhamer et al. 1989). The issue 'separating allelopathy from resource competition' is controversial in natural ecosystems (Inderjit and del Moral 1997; Weidenhamer 2006) but it is important in plant–plant interactions to evaluate the relative contribution and identify the mechanisms involved in their biological invasion processes. The replacement series design (Dekker et al. 1983) and using activated carbon (Ridenour and Callaway 2001) have been proposed that could yield insights into the nature of plant–plant interactions, and provide evidence for allelopathy. However, both methods have some complications in allelopathy studies (Goldberg and Werner 1983; Goldberg and Fleetwood 1987; Lau et al. 2008) whereas density-dependent phytotoxicity test is able to demonstrate in differentiating the relative contributions of those mechanisms in plant–plant interactions (Weidenhamer et al. 1989).

In general, the yield (in terms of growth and development) of plants decreases with increasing density; in contrast, density-dependent phytotoxicity studies reveal that plant growth may be positively influenced up to the point where resource competition acts as the dominant factor. Density-dependent phytotoxicity studies imply

**Table 4 Results of two-way ANOVA (*F*-ratios) showing the effects of different concentrations (C) of *Phragmites australis* litter mediated agarose on germination and growth of *Lactuca sativa* with different seed densities (D)**

| Factor | *df* | *F* ratio and probability | | | | | |
|---|---|---|---|---|---|---|---|
| | | Germination percentage | Above ground biomass | Below ground biomass | Total biomass | Root length | Root length |
| Concentration (C) | 2 | 24.17*** | 71.35*** | 30.54*** | 53.59*** | 91.53*** | 29.67*** |
| Density (D) | 2 | 7.61*** | 7.95** | 7.19** | 6.19** | 14.99*** | 20.82*** |
| C × D | 4 | 2.76* | 4.51* | $0.43^{ns}$ | $2.05^{ns}$ | $2.40^{ns}$ | $1.95^{ns}$ |
| Error | 18 | | | | | | |
| Total | 27 | | | | | | |

Data used as % of control. Level of significance: ***, **, * and $^{ns}$ indicate significant difference at $P \leq 0.001$, $P \leq 0.01$, $P \leq 0.05$ and non-significant respectively

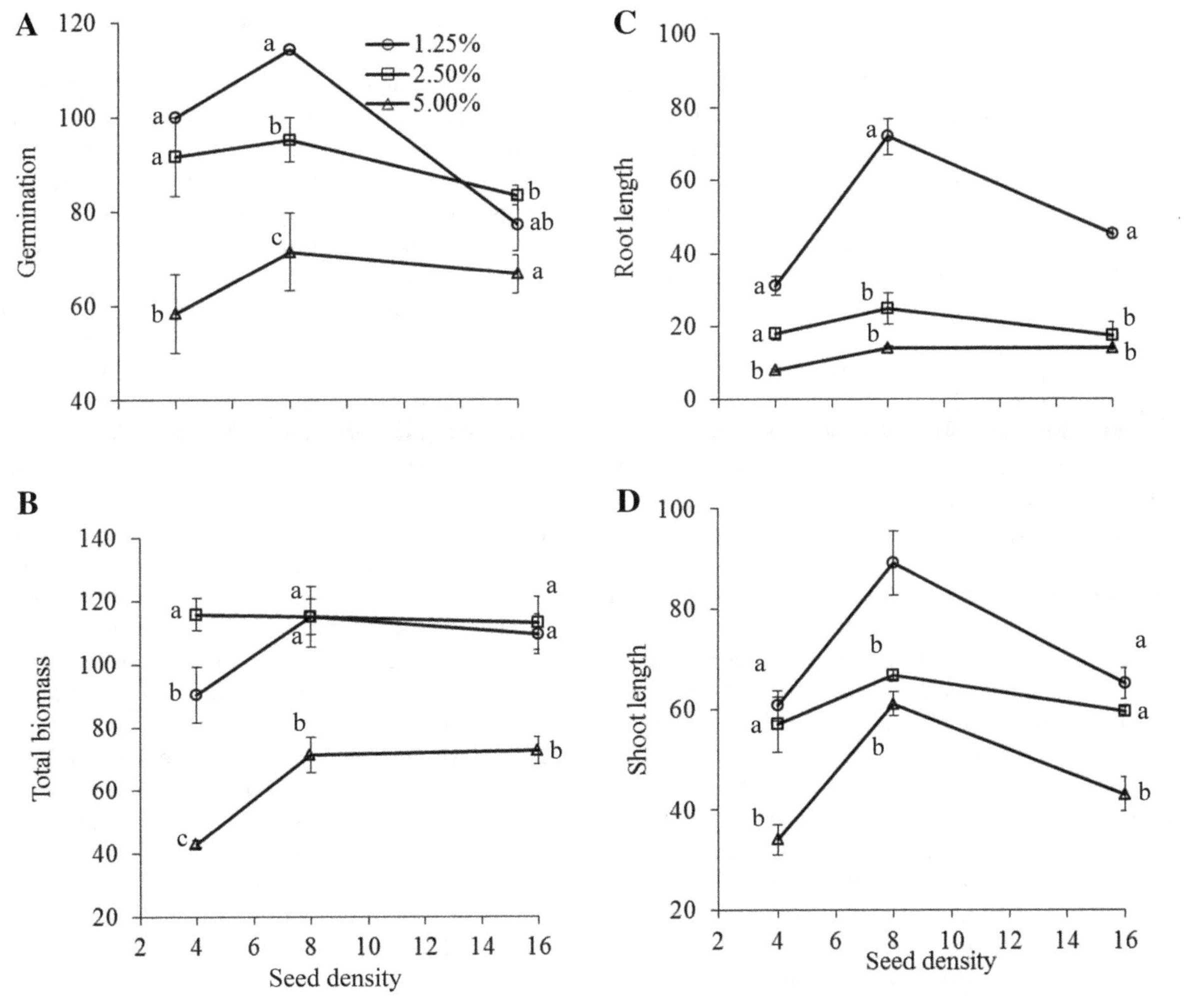


**Fig. 4** **A** Germination percentage, **B** total mass, **C** root length, and **D** shoot length in different concentrations of *Phragmites australis* litter mediated agarose at different seed densities of *Lactuca sativa*. Values (per plant) are means as % of control treatments at each density ± standard errors ($n = 3$). *Letters* indicate homogenous subgroups ($P \leq 0.05$) at each density in Duncan's test

a positive feedback between population density and phytotoxins present in a system, as the toxin is shared among increased plant biomass with each plant receiving a proportionately smaller amount of toxin. Density-dependent phytotoxicity stands in contrast to resource competition as increased growth of plants at low density is dependent on large part to the amount of resources available. Despite the allelopathic potential of *P. australis* on associated and model plant species, as shown by the growth of *M. ericifolia*, *R. conglomeratus* and *L. sativa* observed in this study was masked by the resource competition but the allelopathic effects of *P. australis* are well supported (Rudrappa et al. 2007; Uddin et al. 2012, 2014a, b, c). These studies showed that water extracts of different organs, residue decomposition and root secreted phytotoxins had negative effect on germination, growth, and development of other plant species.

In greenhouse experiments of this study, the strongest growth inhibition was observed in high density treatments when compared to low and medium density. This demonstrates resource competition is the dominating factor, consistent with other studies (Inderjit and del Moral 1997; Uddin et al. 2014b). Our previous studies demonstrated that allelopathy through root exudates of *P. australis* had relatively low contribution in suppression of *M. ericifolia* in comparison to other competitive effects. Again the *P. australis* litter mediated soil experiment showed the highest root suppression of *M. ericifolia* potentially due to allelochemicals leached from litter mediated soil as the intermediate density of *M. ericifolia* showed increased growth compared to low and high density. The findings are well supported by the total assumptions of density-dependent phytotoxicity concept proposed by Weidenhamer (2008). This study states that growth is reduced at low but diminished at high density compared to control; and plant growth is highest at intermediate density, due to a reversal in slope of the predicted growth–density line.

In addition, the laboratory experiments showed a clear density-dependent phytotoxic effect, a result well aligned with other studies (Hansi et al. 2014; Lambertini et al. 2012) where allelochemicals, herbicides and inorganic compounds such as copper showing phytotoxicity is density-dependent. Our results suggest the relationship between growths, in terms of biomass, root length, plant height and plant density of *R. conglomeratus* exhibits a reversal in slope indicating the presence of phytotoxins in the litter leachate mediated soil used in the bioassay. The significant variation in phenolic content of unburnt versus burnt residues did not reflect the associate effects on plant growth suggesting heat induced transformation of phenolic compounds might be effective in suppression of plant growth, even though these residues may contain small amount of phenolic compounds. This result is well aligned with the study of Zhang et al. (2012) who found that there was no significant difference between unburnt versus burnt residues of *Flaveria bidentis* (L.) Kuntze on the growth of wheat (*Triticum aestivum* L.) seedlings. However, our studies found higher inhibitory effects in unburnt than burnt residue extract but it was not a true reflection of causal relationships between total phenolics and growth variables. Seed germination study showed that germination percentage and root–shoot length of *L. sativa* are a function of both concentration and the amount of phytotoxin available per seed. The inhibition increased as the concentration increased at lower density but stimulation was observed with intermediate density in most of the cases. This suggests that lower seed density increases the availability of phytotoxin per seed. Weidenhamer et al. (1987) found that even lower phytotoxin concentration may cause similar or greater inhibitory effects than higher concentrations, when the amount of phytotoxin per seed is greater.

In general, allelopathy research is more concerned with using concentrations, the introduction of soil microbes, and an autotoxicity test, involving a wide range of associated plant species in the bioassays, by questioning whether those involved are ecologically relevant. Despite this, it may be difficult to determine with some precision as to those occurring naturally in the field, but this has importance in determining causal relationships and minimizing effects due to unnaturally occurring situations. Therefore, allelopathy studies should consider more complex bioassays involving soil microbial communities, field concentrations of allelochemicals, multiple test species and using native leachate as a control that might represent the ecological phenomena in the field. For example, soil microorganisms might play an important role in influencing the bioavailability of allelochemicals in soil (Bauer et al. 2012; Ehlers 2011; Inderjit 2005), which could be achieved by the addition of microbial inoculum in experimental soil substrate collected from test species grown in the field. So, further research related to incorporation of soil microorganisms might be imperative to advance the allelopathy as one of its invasion mechanisms.

An effort has been made to overcome the concern through measuring the concentration of allelochemicals in the *P. australis* rhizosphere soil (Uddin et al. 2012), considering the osmotic potential of higher concentrations (Uddin et al. 2014a), adjusting pH, and measuring the quantity of litter biomass produced per unit of soil or covered area. These criteria in our previous studies have been taken into consideration in this current study. Moreover, separation of allelopathic effects from resource competition is a vital point in allelopathy research which has been addressed in this study, indicating phytotoxins secreted by different means from *P. australis* are responsible for invasion process except root exudations. Despite the results indicating less inhibition of root exudates by *P. australis* on *M. ericifolia* transplanted plants (Uddin et al. 2014b) but toxin may arise from other sources such as residue decomposition into soil, inhibiting germination processes and other growth parameters (Uddin et al. 2014c). These results are well aligned with other allelopathy studies of *Agropyron repens* in which Welbank (1960) found that decaying roots and rhizomes of *Agropyron* markedly inhibit the root and shoot growth of rape seedlings but no significant inhibition by root secretion. On the other hand, plant–plant allelopathic interactions may be explained by species-specific (Hierro and Callaway 2003; Prati and Bossdorf 2004) and contextual relationships (Bauer et al. 2012) that may prove the consistency of whole results of our studies. Finally, it can be said that the possibility of allelopathy as a probable cause of plant growth inhibition in some natural systems is not denied, but it has not yet been proven to be the sole factor of interference in any study. There are other possible explanations of the effect, e.g. volatilisation, mechanical root interaction etc.

## Conclusions

The overall observation of growth reductions in test plant species at low densities was inconsistent with the standard resource competition hypothesis and provides support for the hypothesis of chemical interference by *P. australis*. Although, the growth response of test species did not follow the consistency in all experiments, in most cases, the results demonstrate the density-dependent phytotoxicity concept. Therefore, these studies may provide an understanding of plant–plant allelopathy interactions and may distinguish the mechanisms involved in plant interference i.e. resource competition and allelopathy. Our findings may be useful to evaluate the response of agricultural plants such as *L. sativa* to weed residues, and may also provide insight evidence of allelopathic potential in *P. australis* invaded wetlands. In addition, the density-dependent phytotoxicity phenomenon may bring important ecological implications as a methodological approach in allelopathy (Weidenhamer and Romeo 1989).

#### Authors' contributions

MNU performed the most part of the experiments, analyzed the data and wrote the manuscript. RWR edited the manuscript, gave support and conceptual advices. Both authors read and approved the final manuscript.

#### Author details

[1] Department of Ecology & Environmental Management, College of Engineering & Science, Victoria University, Melbourne, VIC 8001, Australia. [2] Institute for Sustainability & Innovation, Victoria University, Melbourne, VIC 8001, Australia.

#### Acknowledgements

The authors thank to all laboratory technicians in the ecology group in Victoria University, Melbourne for their kind help in sample collection from the field and for technical support associated with all experiment settings and parameter assessments. All field works were carried out under the regulations of Melbourne Water, Victoria, Australia.

#### Competing interests

The authors declare that they have no competing interests.

#### Data collection

I confirm we have collected all plant and soil materials from field under appropriate permissions and/or licences for the study from Melbourne Water, Australia.

#### Funding

The authors thank to International Postgraduate Research Scholarship (IPRS), Australia for financial support to the first author.

#### References

An M, Pratley JE, Haig T (2001) Phytotoxicity of Vulpia residues: IV. Dynamics of allelochemicals during decomposition of Vulpia residues and their corresponding phytotoxicity. J Chem Ecol 27:395–409

Andersen RN (1981) Increasing herbicide tolerance of soybeans (*Glycine max*) by increasing seeding rates. Weed Sci 29:336–338

Bains G, Kumar AS, Rudrappa T, Alff E, Hanson TE, Bais HP (2009) Native plant and microbial contributions to a negative plant–plant interaction. Plant Physiol 151:2145–2151

Barto E, Cipollini D (2009) Density-dependent phytotoxicity of impatiens pallida plants exposed to extracts of *Alliaria petiolata*. J Chem Ecol 35:495–504

Bauer J, Shannon S, Stoops R, Reynolds H (2012) Context dependency of the allelopathic effects of *Lonicera maackii* on seed germination. Plant Ecol 213:1907–1916

Belz RG, Velini ED, Duke SO (2007) Dose/response relationships in allelopathy research. Allelopathy: new concepts and methodologies. Science Publishers, Enfield, pp 3–29

Bich TTN, Kato-Noguchi H (2014) Isolation and identification of a phytotoxic substance from the emergent macrophyte *Centrostachys aquatica*. Bot Stud 55:1

Callaway RM, Ridenour WM (2004) Novel weapons: invasive success and the evolution of increased competitive ability. Front Ecol Environ 2:436–443

Dekker J, Meggitt W, Putnam A (1983) Experimental methodologies to evaluate allelopathic plant interactions. J Chem Ecol 9:945–981

Dorning M, Cipollini D (2006) Leaf and root extracts of the invasive shrub, *Lonicera maackii*, inhibit seed germination of three herbs with no autotoxic effects. Plant Ecol 184:287–296

Duke SO (2015) Proving allelopathy in crop-weed interactions. Weed Sci 63:121–132

Duke SO, Blair AC, Dayan FE, Johnson RD, Meepagala KM, Cook D, Bajsa J (2009) Is (−)-catechin a novel weapon of spotted knapweed (*Centaurea stoebe*)? J Chem Ecol 35:141–153

Ehlers BK (2011) Soil microorganisms alleviate the allelochemical effects of a thyme monoterpene on the performance of an associated grass species. PLoS ONE 6:e26321

Fujii Y, Shibuya T, Nakatani K, Itani T, Hiradate S, Parvez MM (2004) Assessment method for allelopathic effect from leaf litter leachates. Weed Biol Manag 4:19–23

Goldberg DE, Fleetwood L (1987) Competitive effect and response in four annual plants. J Ecol 75:1131–1143

Goldberg DE, Werner PA (1983) Equivalence of competitors in plant communities: A null hypothesis and a field experimental approach. Am J Bot 70:1098–1104

Hansi M, Weidenhamer JD, Sinkkonen A (2014) Plant growth responses to inorganic environmental contaminants are density-dependent: experiments with copper sulfate, barley and lettuce. Environ Pollut 184:443–448

Hierro JL, Callaway RM (2003) Allelopathy and exotic plant invasion. Plant Soil 256:29–39

Hocking PJ, Finlayson CM, Chick AJ (1983) The biology of Australian weeds. 12. *Phragmites australis* (Cav.) Trin. ex Steud. Aust Inst Agric Sci J 49:123–132

Hoffman DW, Lavy TL (1978) Plant competition for atrazine. Weed Sci 26:94–99

Huang Z, Haig T, Wu H, An M, Pratley J (2003) Correlation between phytotoxicity on annual ryegrass (*Lolium rigidum*) and production dynamics of allelochemicals within root exudates of an allelopathic wheat. J Chem Ecol 29:2263–2279

Hussain MI, Reigosa MJ (2012) Seedling growth, leaf water status and

signature of stable carbon isotopes in C3 perennials exposed to natural phytochemicals. Aust J Bot 60:676–684
Inderjit (2005) Soil microorganisms: an important determinant of allelopathic activity. Plant Soil 274:227–236
Inderjit, Callaway RM (2003) Experimental designs for the study of allelopathy. Plant Soil 256:1–11
Inderjit, del Moral R (1997) Is separating resource competition from allelopathy realistic? Bot Rev 63:221–230
Inskeep WP, Bloom PR (1985) Extinction coefficients of chlorophyll a and b in *N,N*-dimethylformamide and 80% acetone. Plant Physiol 77:483–485
Kettenring KM, McCormick MK, Baron HM, Whigham DF (2011) Mechanisms of *Phragmites australis* invasion: feedbacks among genetic diversity, nutrients, and sexual reproduction. J Appl Ecol 48:1305–1313
Kimura F, Sato M, Kato-Noguchi H (2015) Allelopathy of pine litter: delivery of allelopathic substances into forest floor. J Plant Biol 58:61–67
Kulmatiski A, Beard KH, Meyerson LA, Gibson JR, Mock KE (2011) Nonnative *Phragmites australis* invasion into Utah wetlands. West N Am Nat 70:541–552
Lambertini C, Sorrell BK, Riis T, Olesen B, Brix H (2012) Exploring the borders of European Phragmites within a cosmopolitan genus. AoB Plants 2012:pls020. doi:10.1093/aobpla/pls020
Lara-Núñez A, Romero-Romero T, Ventura JL, Blancas V, Anaya AL, Cruz-Ortega R (2006) Allelochemical stress causes inhibition of growth and oxidative damage in *Lycopersicon esculentum* Mill. Plant Cell Environ 29:2009–2016
Lau JA, Puliafico KP, Kopshever JA, Steltzer H, Jarvis EP, Schwarzländer M, Strauss SY, Hufbauer RA (2008) Inference of allelopathy is complicated by effects of activated carbon on plant growth. New Phytol 178:412–423
Li F-M, Hu H-Y (2005) Isolation and characterization of a novel antialgal allelochemical from *Phragmites communis*. Appl Environ Microbiol 71:6545–6553
Lorenzo P, Pazos-Malvido E, Reigosa MJ, González L (2010) Differential responses to allelopathic compounds released by the invasive *Acacia dealbata* Link (Mimosaceae) indicate stimulation of its own seed. Aust J Bot 58:546–553
Mack RN, Simberloff D, Mark Lonsdale W, Evans H, Clout M, Bazzaz FA (2000) Biotic invasions: causes, epidemiology, global consequences, and control. Ecol Appl 10:689–710
Moran R, Porath D (1980) Chlorophyll determination in intact tissues using *N,N*-dimethylformamide. Plant Physiol 65:478–479
Morris K, Boon PI, Raulings EJ, White SD (2008) Floristic shifts in wetlands: the effects of environmental variables on the interaction between *Phragmites australis* (Common Reed) and *Melaleuca ericifolia* (Swamp Paperbark). Mar Freshw Res 59:187–204
Plut K, Paul J, Ciotir C, Major M, Freeland JR (2011) Origin of non-native *Phragmites australis* in North America, a common wetland invader. Fundam Appl Limnol/Archiv für Hydrobiologie 179:121–129
Prati D, Bossdorf O (2004) Allelopathic inhibition of germination by *Alliaria petiolata* (Brassicaceae). Am J Bot 91:285–288
Qian H, Xu X, Chen W, Jiang H, Jin Y, Liu W, Fu Z (2009) Allelochemical stress causes oxidative damage and inhibition of photosynthesis in *Chlorella vulgaris*. Chemosphere 75:368–375
Reigosa M, Gomes AS, Ferreira AG, Borghetti F (2013) Allelopathic research in Brazil. Acta Bot Bras 27:629–646
Ridenour WM, Callaway RM (2001) The relative importance of allelopathy in interference: the effects of an invasive weed on a native bunchgrass. Oecologia 126:444–450
Rudrappa T, Bonsall J, Gallagher JL, Seliskar DM, Bais HP (2007) Root-secreted allelochemical in the noxious weed *Phragmites australis* deploys a reactive oxygen species response and microtubule assembly disruption to execute rhizotoxicity. J Chem Ecol 33:1898–1918
Rudrappa T, Choi YS, Levia DF, Legates DR, Lee KH, Bais HP (2009) *Phragmites australis* root secreted phytotoxin undergoes photo-degradation to execute severe phytotoxicity. Plant Signal Behav 4(6):506–513
Saltonstall K, Burdick D, Miller S, Smith B (2005) Native and introduced Phragmites: challenges in identification, research, and management of the common reed. National Estuarine Research Reserve Technical Report Series
Suman A, Shahi HN, Singh P, Gaur A (2002) Allelopathic influence of *Vigna mungo* (black gram) seeds on germination and radical growth of some crop plants. Plant Growth Regul 38:69–74
Thijs H, Shann JR, Weidenhamer JD (1994) The effect of phytotoxins on competitive outcome in a model system. Ecology 75:1959–1964
Tseng M-H, Kuo Y-H, Chen Y-M, Chou C-H (2003) Allelopathic potential of *Macaranga tanarius* (L.) Muell.–Arg. J Chem Ecol 29:1269–1286
Uddin MN, Caridi D, Robinson RW (2012) Phytotoxic evaluation of *Phragmites australis*: an investigation of aqueous extracts of different organs. Mar Freshw Res 63:777–787
Uddin MN, Robinson RW, Caridi D (2014a) Phytotoxicity induced by *Phragmites australis*: an assessment of phenotypic and physiological parameters involved in germination process and growth of receptor plant. J Plant Interact 9:338–353
Uddin MN, Robinson RW, Caridi D, Al Harun MAY (2014b) Suppression of native *Melaleuca ericifolia* by the invasive *Phragmites australis* through allelopathic root exudates. Am J Bot 101:479–487
Uddin MN, Robinson RW, Caridi D, Harun MA (2014c) Is phytotoxicity of *Phragmites australis* residue influenced by decomposition condition, time and density? Mar Freshw Res 65:505–516
Weidenhamer JD (1996) Distinguishing resource competition and chemical interference: overcoming the methodological impasse. Agron J 88:866–875
Weidenhamer JO (2006) Distinguishing allelopathy from resource competition: the role of density allelopathy. In: Reigosa MJ, Pedrol N, González L (eds) Allelopathy: A physiological process with ecological implications, vol 137. Springer, Netherlands, p 85
Weidenhamer JD (2008) Allelopathic mechanisms and experimental methodology. In: Zeng RS, Mallik AU, Luo SM (eds) Allelopathy in sustainable agriculture and forestry. Springer, New York, pp 119–135
Weidenhamer JD, Romeo JT (1989) Allelopathic properties of *Polygonella myriophylla* field evidence and bioassays. J Chem Ecol 15:1957–1970
Weidenhamer JD, Morton TC, Romeo JT (1987) Solution volume and seed number: often overlooked factors in allelopathic bioassays. J Chem Ecol 13:1481–1491
Weidenhamer JD, Hartnett DC, Romeo JT (1989) Density-dependent phytotoxicity: distinguishing resource competition and allelopathic interference in plants. J Appl Ecol 26:613–624
Weir TL, Park S-W, Vivanco JM (2004) Biochemical and physiological mechanisms mediated by allelochemicals. Curr Opin Plant Biol 7:472–479
Weißhuhn K, Prati D (2009) Activated carbon may have undesired side effects for testing allelopathy in invasive plants. Basic Appl Ecol 10:500–507
Welbank P (1960) Toxin production from *Agropyron repens* biology of weeds. In: Symp. Brit. ecol. Soc. pp 158–164
Zaiontz C (2014) Real statistics using Excel. http://www.real-statistics.com/. Accessed 10 Oct 2016
Zhang F-J, Guo J-Y, Chen F-X, Guo A-Y, Wan F-H (2012) Assessment of allelopathic effects of residues of *Flaveria bidentis* (L.) Kuntze on wheat seedlings. Arch Agron Soil Sci 58:257–265
Zhou Y, Yu J (2006) Allelochemicals and photosynthesis allelopathy. Springer, Berlin, pp 127–139

# Anti-hepatitis, antioxidant activities and bioactive compounds of *Dracocephalum heterophyllum* extracts

Qiang-Qiang Shi[1,2], Jun Dang[1], Huai-Xiu Wen[1], Xiang Yuan[1,2], Yan-Duo Tao[1] and Qi-Lan Wang[1*]

**Abstract**

**Background:** *Dracocephalum heterophyllum* was a traditional Tibetan medicine possesses various pharmacological effects involved in anti-inflammatory, antibacterial activities. However, its anti-hepatitis, antioxidant activity and bioactive compounds have not been reported, the objective of this research work was to investigate the pharmacological activity and bioactive compounds of *D. heterophyllum* extracts.

**Results:** In the present study, the anti-hepatics and antioxidant activities of four *D. heterophyllum* extracts (i.e. petroleum ether extracts, ethyl acetate extracts, *n*-BuOH extracts, and water extracts) were conducted. The main chemical constituent of petroleum ether and ethyl acetate extracts were also isolated using chromatographic techniques and identified by NMR spectroscopic methods. The anti-hepatitis assay showed that the petroleum ether and ethyl acetate extracts of *D. heterophyllum* significantly prolonged the mean survival times and reduced the mortality of mouse hepatitis model induced by concanavalin A (ConA). The levels of alanine transaminase, aspartate transaminase in blood serum could be decreased obviously by ethyl acetate extracts compared with ConA group ($P < 0.01$). The histological analysis demonstrated that the ethyl acetate extracts could inhibit apoptosis and necrosis caused by ConA. In addition, the antioxidant activities of the four extracts of *D. heterophyllum* were measured by DPPH assay, ABTS assay, anti-lipidperoxidation assay, ferric reducing antioxidant power assay, ferrous metal ions chelating assay and determination of total phenolic contents. The results showed that the ethyl acetate extract had the highest antioxidant activities, followed by petroleum ether extract. Finally, nine mainly compounds were isolated from the Petroleum ether and ethyl acetate extracts, including four triterpenes: oleanolic acid (**1**), ursolic acid (**2**), pomolic acid (**3**), 2α- hydroxyl ursolic acid (**4**), three flavonoids: apigenin-7-*O*-rutinoside (**5**), luteolin (**8**), diosmetin (**9**) and two phenolic acids: rosmarinic acid (**6**), methyl rosmarinate (**7**).

**Conclusion:** The Ethyl acetate extract of *D. heterophyllum* had the highest anti-hepatitis and antioxidants activities, followed by petroleum ether extract. The bioactive substances may be triterpenes, flavonoids and phenolic acids, the ethyl acetate extracts of *D. heterophyllum* may be possible candidates in developing anti-hepatitis medicine.

**Keywords:** Anti-hepatitis activity, Antioxidant activity, Chemical constituents, *Dracocephalum heterophyllum*

## Background

Hepatitis is an inflammation of the liver. The condition can progress to fibrosis (scarring), cirrhosis or liver cancer. Hepatitis viruses are the most common cause of hepatitis in the world. There are five main hepatitis viruses, referred to as types A, B, C, D and E. In particular, types B and C lead to chronic disease in hundreds of millions of people and, together, are the most common cause of liver cirrhosis and cancer. More than 20 million people worldwide are infected with hepatitis virus. And infection from these viruses results in approximately 1.45 million deaths each year. Effective prophylactic vaccines have been available since the 1980s (Hollinger and Liang 2001). Nonetheless, for many developing countries,

*Correspondence: wql@nwipb.cas.cn
[1] Key Laboratory of Tibetan Medicine Research, Northwest Institute of Plateau Biology, CAS, Xining, China
Full list of author information is available at the end of the article

large-scale vaccination programs were hardly affordable, and an enormous number of chronic hepatitis virus carriers will be in need of better medication for decades to come. Current therapies are based on the systemic administration of high doses of interferon-α (IFN-α) or on nucleoside analogs. However, both therapies have a sustained response rate of only about 30 %, combinations exert no clear synergism, and lamivudine therapy leads to the rapid emergence of resistant virus variants (Pumpens et al. 2002; Zoulim 2001). Hepatitis B virus (HBV) is a hepadnavirus DNA virus with species specificity and tissue specificity, normally infect only humans and chimpanzees. There is no other feasible small experimental animal infection model. It is also very difficult to infect the cells cultured in vitro. For now, though there are many hepatitis animal model, such as duck hepatitis B model (Schultz et al. 2004), woodchuck model (Wang et al. 2004), chimpanzees model (Wieland et al. 2004), HBV transgenic mice model (Chisari 1995), there are still some different levels of flaws.

*Dracocephalum heterophyllum* is a traditional Tibetan medicine growing on Qinghai-Tibet Platean with special living environment of high elevation and strong sunlight irradiation. The plant is distributed widely in Sitsang, Qinghai, Sinkiang and Gansu province of China. In traditional Tibetan Medicine, *D. heterophyllum* is known as Ao-Ga or Ji-Mei-Qing-Bao, which has been used as an ethnomedicine to treat various ailments such as jaundice, hepatopathy, cough, lymphangitis, mouth ulcers and tooth diseases. There had been reports said that the herb had antiviral activity (Zhang et al. 2009), antianoxic effect (Peng 1984), antiasthmatic, anticoughing and disinfectant action (Mahmood et al. 2005), and the essential oil of it also had antimicrobial and antioxidant activities (Zhang et al. 2008).

The aim of this paper was to evaluate the anti-hepatitis activities of *D. heterophyllum* in the mouse fulminant hepatitis model induced by concanavalin A (ConA), and measured the antioxidant activities of this herbs in a series of in vitro assay such as free radical scavenging experiments (DPPH and ABTS assay), anti-lipidperoxidation experiments (FTC assay), ferric reducing antioxidant power assay (FRAP), metal chelating assay and determination of total phenolic contents (TPC). Finally, the bioactive substances were also separated and purified using chromatographic techniques.

## Methods

### Chemical reagent

Female Bal B/C mice were bought from Beijing Vitalriver Experimental Animals Ltd. (Beijing, China), The animals were performed according to guidelines laid down by the Animal Care and Use Committee of Shenzhen Institutes of Advanced Technology, Chinese Academy of Sciences (Approval IDs: SCXY2012-0119) that follows internationally acceptable standards on animal care and use in laboratory experimentation. Concanavalin A (ConA) was obtained from Sigma-Aldrich Co. (Shanghai, China), ALT kit, AST kit, TUNEL kit, DAPI kit were purchased from Nanjing Jiancheng Bioengineering Institute (Nanjing, China), 1,1-Diphenyl-2-picryl-hydrazyl (DPPH), 2-azino-bis(3-ethylbenzthiazoline-6-sulfonic acid) (ABTS), 3-(2-pyridyl)-5,6-bis (4-phenyl-sulfonic acid)-1,2,4-triazine (ferrozine), 2,4,6-Tris(2-pyridyl)-s-triazine (TPTZ), linoleic acid, α-Tocopherol, butylated hydroxytoluene (BHT), butylated hydroxyanisole (BHA), potassium persulfate ($K_2S_2O_8$), ammonium thiocyanate were ($NH_4SCN$), ferrous chloride tetrahydrate ($FeCl_2{\cdot}4H_2O$), Ferric chloride tetrahydrate ($FeCl_3{\cdot}6H_2O$), sodium tungstate dihydrate ($Na_2WO_4{\cdot}2H_2O$), sodium molybdate dehydrate ($Na_2MoO_4{\cdot}2H_2O$), sodium carbonate anhydrous ($Na_2CO_3$) were all purchased from Aladdin Industrial Corporation (Shanghai, China). Gallic acid was obtained from National Institute for Food and Drug Control (Beijing, China). All other chemical reagent and buffer used were analytical grade and obtained from Beijing Chemical Co. (Beijing, China).

### Sample preparation

The whole grass of *D. heterophyllum* was harvested in August 2014, from North Mountain in Huzhu, Qinghai province, China. The sample was identified by MEI Lijuan (Northwest Institute of Plateau Biology, Qinghai, China). The fresh samples were air-dried in hade, then ground into a homogeneous powder in a mill.

8.86 kg of air-dried and powered *D. heterophyllum* was extracted by 95 % ethanol at 70 °C through heating reflux. The samples were filtered with filter paper while the residue was further extracted under the same conditions 3 times. The filtrates were collected, and then ethanol was removed by a rotary evaporator (EYELA, Japan) at 50 °C to get the crude extract of *D. heterophyllum*.

The crude ethanol extract (1026 g) of *D. heterophyllum* was suspended into 500 mL water. The suspension was successively extracted 3 times by the same volume of petroleum ether, ethyl acetate and n-butanol at room temperature to get four fractions. Then the four fractions were dried by a rotary evaporator (EYELA, Japan), respectively. The four extracts were stored at 4 °C until used.

### Anti-hepatitis activity assay

#### *The survival experiment of mice with lethal doses of ConA*

Concanavalin A (ConA) is a plants lectin, and is known for its ability to stimulate mouse T cell subsets giving rise to four functionally distinct T-cell populations, including precursors to suppressor T-cell; one subset of human suppressor T-cells as well is sensitive to ConA (Dwyer and Johnson 1981). Female BALB/c mice would develop severe liver injury as assessed by transaminase release within 8 h when an intravenous dose concanavalin A (Con A) was given. Histopathologically, only the liver was affected. Con A-induced liver injury depends on the activation of T lymphocytes by macrophages in the presence of ConA (Tiegs et al. 1992). The model might allow the study of the pathophysiology of autoimmune hepatitis and viral hepatitis.

Female Balb/C mice 9–10 weeks old weighing about 25–29 g were housed for 1 day to acclimatize. The mice were randomly divided into nine groups each comprised of eight mice. The first group, ConA group, was only injected with 1 mg/mL ConA in caudal vein at the lethal dose of 20 mg/Kg. The next four groups, drug group, were only injected with 10 mg/mL four extracts of *D. heterophyllum* in abdominal cavity at the dose of 50 mg/Kg, respectively. The last four groups, ConA+ drug group, were injected with 10 mg/mL four extracts of *D. heterophyllum* in abdominal cavity at the dose of 50 mg/Kg, respectively, and 2 h later, were injected with 1 mg/mL ConA in caudal vein at the dose of 20 mg/Kg to induced hepatitis. The mice were fed and observed for 24 h to determine their mortality.

#### *The level of transaminase in mice serum*

According to the (Table 1), 32 female Balb/C mice were randomly divided into four groups each comprised of eight mice. The first group, control group, was only injected with DMSO and PBS in abdominal cavity. The second group, ConA group, was only injected with 1 mg/mL ConA in caudal vein at the dose of 12.5 mg/Kg. The third group, drug group, was only injected with 10 mg/mL Ethyl acetate extracts of *D. heterophyllum* in abdominal cavity at the dose of 50 mg/Kg. The last group, ConA+ drug group, were injected with 10 mg/mL ethyl acetate extracts of *D. heterophyllum* in abdominal cavity at the dose of 50 mg/Kg, and 2 h later, was injected with 1 mg/mL ConA in caudal vein at the dose of 12.5 mg/Kg. The blood was taken using orbital venous plexus blood collection in the time of 8, 16, 24 h after injected ConA. The blood volume was collected 200–300 μL, and centrifuged 5 min at 3000 rpm. The upper serum was moved to the new Eppendorf tube, and saved at −20 °C, temporarily. The ALT and AST kit were used for quantitation of transaminase in mice serum of different groups.

**Table 1 The treatment on the hepatitis model induced by conA**

| Groups | The number of mice | Method of administration | Testing index |
|---|---|---|---|
| Control group | 8 | Abdominal cavity | ALT/AST |
| ConA group | 8 | Caudal vein | ALT/AST |
| Drug group | 8 | Abdominal cavity | ALT/AST |
| ConA+ drug group | 8 | Caudal vein + abdominal cavity | ALT/AST |

*ConA* concanavalin A; *ALT* alanine transferase; *AST* aspartate transaminase; *Drug* ethyl acetate extracts of *D. heterophyllum*

#### Morphologic, histological and hepatocytes apoptosis analysis

Pretreating the mice base on the (Table 1). Test mice breaking the neck to death after induced by ConA 24 h later, removing the liver, observing the hepatic pathological changes of liver using morphologic method. Then the livers were fixed in 10 % formalin, embedded in paraffin and cut into slices. HE staining were performed and the result of the staining was analyzed with microscopic examination. The method of TdT mediated dUTP nick end labeling (TUNEL) and 4′,6-diamidino-2-phenylindole (DAPI) were used to detect apoptosis of mouse liver.

#### Antioxidant activity

#### *DPPH free radical scavenging activity*

The free radical scavenging activity of four *D. heterophyllum* extracts were estimated using in vitro 1,1-diphenyl-2-picryl-hydrazil (DPPH) free radical assay following the methodology described by Blois (1958). DPPH free radical reagent solution was prepared by dissolving DPPH in ethanol to obtain 0.1 mM concentration. Four extracts test solution were prepared by dissolving different concentrations of four extracts (10, 50, 100, 150 200, 300, 500 μg/mL) in ethanol. To 2 mL of test solution 2 mL of freshly prepared DPPH reagent was added. The reaction mixture was incubated in dark at room temperature. Thirty minutes later, the absorbance was measured at 517 nm using 752 N UV–Vis spectrophotometer (Shanghai, China). Aliquots of α-tocopherol, BHT and BHA was served as standards for the assay. A mixture of DPPH and ethanol was served as control and a mixture of *D. heterophyllum* extracts solution and ethanol was served as a blank for the sample. All experiments were repeated three times and take the mean to eradicate any discrepancies. % inhibition of DPPH was calculated by the following formula:

$$\text{Inhibition of DPPH (\%)} = \left[\left(A_{Control} - A_{Sample}\right)/A_{Control}\right] \times 100$$

where $A_{Control}$ is the absorbance of the control reaction, while $A_{Sample}$ is the absorbance at 517 nm with *D. heterophyllum* extracts.

***ABTS free radical scavenging activity***

The ABTS·$^+$ free radical scavenging activity was measured as the methodology described by Li et al. (2012) with some modifications. The ABTS·$^+$ was produced by mixing 0.2 mL ABTS diammonium salt (7.4 mM) with 0.2 mL potassium persulfate (2.6 mM). The mixture was kept in the dark at room temperature for 12 h to allow completion of radical generation, then diluted with 95 % ethanol (about 40–50 times) so that its absorbance at 734 nm was 0.70 ± 0.02. To determine the scavenging activity, 4 mL aliquot of ABTS·$^+$ reagent was mixed with 1 mL of sample alcoholic solutions (10–500 μg/mL). After incubation for 6 min, the absorbance at 734 nm was read on 752 N UV–Vis spectrophotometer (Shanghai, China). Aliquots of α-tocopherol, BHT and BHA was served as standards for the assay. A mixture of ABTS·$^+$ reagent and ethanol was served as control and a mixture of *D. heterophyllum* extracts solution and ethanol was served as a blank for the sample. All experiments were repeated three times and take the mean to eradicate any discrepancies. The percentage inhibition of the samples was calculated as:

$$\text{Inhibition of ABTS}\cdot^{+}(\%) = \left[(A_{Control} - A_{Sample})/A_{Control}\right] \times 100$$

where $A_{Control}$ is the absorbance of the control reaction, while $A_{Sample}$ is the absorbance at 734 nm with *D. heterophyllum* extracts.

**Anti-lipidperoxidation activity by FTC**

The anti-lipidperoxidation activity of four *D. heterophyllum* extracts and standards was measured by ferric thiocyanate method (Kikuzaki and Nakatani 1993), which was slightly modified. A mixture of 1 mL of sample alcoholic solutions (25 and 50 μg/mL), 1 mL of 2.5 % linoleic acid solution in ethanol, 2 mL of a 0.05 M phosphate buffer (pH 7.0) and 1 mL of distilled water was placed in a centrifuge tube (15 mL) with a screw cap. The mixed solution (5 mL) was incubated in an oven at 40 °C in the dark. On the other hand, the 5 mL control was composed of 1 mL of ethanol, 1 mL of 2.5 % linoleic acid solution in ethanol, 2 mL of a 0.05 M phosphate buffer (pH 7.0) and 1 mL of distilled water. To 0.1 mL of this solution was added 9.5 mL of 75 % ethanol and 0.1 mL of 30 % ammonium thiocyanate. Exactly 3 min after adding 0.1 mL of 0.02 M $FeCl_2$ in 3.5 % HCl, the absorbance of red color was measured at 500 nm. This step was repeated every 24 h until 1 day after absorbance of the control reached maximum. The percentage inhibition of lipid peroxidation in linoleic acid emulsion was calculated by following equation.

$$\text{Inhibition of lipid peroxidation}(\%) = \left[(A_{Control} - A_{Sample})/A_{Control}\right] \times 100$$

where $A_{Control}$ is the absorbance of the control reaction and $A_{Sample}$ is the absorbance at 500 nm with *D. heterophyllum* extracts or standards compounds.

**Ferric reducing antioxidant power assay (FRAP)**

The ability to reduce ferric ions was measured according to a modified method developed by Benzie and Strain (1996). To prepare the FRAP reagent, 0.1 mM acetate buffer (pH 3.6), 10 mM 2,4,6-tripyridyl-s-triazine (TPTZ) in 40 mM HCl, and 20 mM $FeCl_3{\cdot}6H_2O$ were mixed together in a ratio of 10:1:1 (v/v/v). A 100 μL aliquot of sample solutions (100–500 μg/mL) were added to 3 mL freshly prepared FRAP reagent. The absorbance of the reaction mixture was measured at 593 nm after 10 min incubation at 37 °C. Experiments were performed in triplicate. Aliquots of α-tocopherol, BHT and BHA was served as standards for the assay. The $FeSO_4{\cdot}7H_2O$ solutions (0.2–2 mM/L) was used to performed the calibration curves. The ferric reducing antioxidants power were calculated from the linear calibration curve and expressed by FRAP value: 1FRAP value = 1 mmol/L $FeSO_4$.

***Ferrous metal ions chelating activity***

The ferrous metal chelating activity of four *D. heterophyllum* extracts and standards was estimated by the method of Dinis et al. (1994). Where in the ferrous ion chelating activity was measured by the absorbance of the ferrous iron-ferrozine complex at 562 nm. Briefly, four *D. heterophyllum* extracts (100–500 μg/mL) in 0.4 mL were added to 2 mM $FeCl_2$ solution (0.05 mL), then 5 mM ferrozine (0.2 mL) was added in. The mixed solution was adjusted to 4 mL with ethanol. The reaction mixture was incubated in dark at room temperature for 10 min. Absorbance of the mixture was then measured at 562 nm. All tests and analyses were run in triplicate and averaged. The percentage of inhibition of the ferrozine–$Fe^{2+}$ complex formation was calculated by the following formula (Gulcin 2006):

$$\text{Ferrous ions chelating effect}(\%) = \left[(A_{Control} - A_{Sample})/A_{Control}\right] \times 100$$

where $A_{Control}$ is the absorbance of the control reaction, while $A_{Sample}$ is the absorbance at 562 nm with *D. heterophyllum* extracts. The control contains $FeCl_2$ and ferrozine, complex formation molecules.

***Determination of total phenolic contents (TPC)***

The total phenolic contents of four *D. heterophyllum* extracts were determined using the Folin-Ciocalteu method (Ragazzi and Veronese 1973) with a little modification. To prepare the Folin-Ciocalteu reagent, 25 g sodium tungstate, 6.25 g sodium molybdate, 175 mL distilled water, 8.5 % phosphoric acid solution, and 25 mL concentrated hydrochloric acid were mixed together. The above mixture was refluxed slowly in a water bath for 10 h. After cooling, 37.5 g lithium sulfate, 12.5 mL distilled water, and 50 mL hydrogen peroxide were added in, then continued to heat in boiling water for 15 min without cap. Finally dilute with water to 250 mL, and stored at 4 °C until used. 100 µL sample solution (1 mg/mL) was mixed with 500 µL Folin-Ciocalteu reagent and diluted with 1000 µL distilled water, then 1.5 mL of $Na_2CO_3$ solution (20 %, w/v) was added. Absorbance of the mixture was measured at 765 nm after 2 h incubation in dark at home temperature. The determinations were performed in triplicate, and the calculations were based on a calibration curve obtained with gallic acid (100–800 µg/mL).

#### *Separation and purification of the bioactive compounds*

The petroleum ether extract (500 g) was subjected to a silica gel column (petroleum ether-acetone, 95:5–5:95) to afford fractions (I–VI). Fraction III (150 g), the mixture of two compounds, was purified on a preparative $C_{18}$ HPLC column with a isocratic of MeOH–$H_2O$ (87:13) to yield compound **1** (3 mg) and compound **2** (33 mg). The two compounds were difficult to separate. Fraction IV was purified on a preparative $C_{18}$ HPLC column with a gradient of $C_2H_3N$–$H_2O$ (65:35–95:5) to yield compound **3** (130 mg) and compound **4** (75 mg). Fraction VI was purified on a preparative $C_{18}$ HPLC column with a gradient of $C_2H_3N$–$H_2O$ (20:80–30:70) to yield compound **5** (200 mg).

The ethyl acetate extract (19 g) was subjected to medium pressure liquid chromatography with a gradient of MeOH–$H_2O$ (3:7–8:2) to afford fractions (I–VII). Fraction III (3 g) was purified on a preparative $C_{18}$ HPLC column with a gradient of $C_2H_3N$–$H_2O$ (10:90–30:70) to yield compound **6** (1.5 g). Fraction IV (2.3 g) was purified on a preparative $C_{18}$ HPLC column with a isocratic of $C_2H_3N$–$H_2O$ (25:75) to yield compound **7** (853 mg). Fraction V (1.1 g) was purified on a preparative $C_{18}$ HPLC column with a isocratic of $C_2H_3N$–$H_2O$ (28:72) to yield compound **8** (58 mg) Fraction IV (1.1 g) was purified on a preparative $C_{18}$ HPLC column with a isocratic of $C_2H_3N$–$H_2O$ (27:73) to yield compound **9** (87.9 mg).

## Results and discussion

### Extraction and fractionation of *Dracocephalum heterophyllum*

From 8.86 kg of *D. heterophyllum*, 1026 g of 95 % ethanol extract was obtained. The yield was 11.6 %. The crude extract (1026 g) was then suspended in water and extracted with petroleum ether, ethyl acetate and *n*-butanol sequentially to get 500 g petroleum ether fraction (48.7.6 %), 19 g ethyl acetate fraction (2 %), 41.8 g *n*-butanol fraction (4.1 %) and 115.3 g water fraction (11.2 %).

### Anti-hepatitis activity

#### *The survival experiment of mice with lethal doses of ConA*

As show in Fig. 1, the survival rate of mice in the first group, ConA group, is lower than 50 % in 8 h, and the mice were all die in 16 h. The mice in the second group, drug group, were all still living in 24 h. Its means that the four extracts of *D. heterophyllum* have no toxic effect on mice. The survival time of mice in the third group, ConA+ drug group, is all longer than ConA group on different levels, respectively. In particular, the survival rate of mice which were injected with petroleum ether extract is nearer to 50 % in 24 h. And the survival rate of mice which were injected with ethyl acetate extract has beyond 65 % in 24 h. The experimental results suggest, the petroleum ether and ethyl acetate extracts of *D. heterophyllum* significantly prolonged the mean survival times and reduced the mortality of ConA induced mice.

#### *The level of transaminase in mice serum*

In biochemistry, a transaminase or an aminotransferase is an essential enzyme in the process of normal cellular metabolism, and mainly remains in hepatocytes. In medicine, the presence of elevated transaminases in serum can be an indicator of liver damage. Two important transaminase enzymes are alanine transaminase (ALT), also called alanine aminotransferase (ALAT) or serum glutamate-pyruvate transaminase (SGPT); and aspartate transaminase (AST), also known as serum glutamic oxaloacetic transaminase (SGOT). As show in Fig. 2, the level of ALT and AST in ConA group was significantly higher than control group ($P < 0.01$). Its means that ConA can induce live damage, increase the contents of ALT and AST in serum. But, the level of ALT and AST in ConA+ drug group was obviously lower than ConA group ($P < 0.01$). Its suggest that ethyl acetate extract of *D. heterophyllum* can improve live injury induced by ConA.

### Morphologic, histological and hepatocytes apoptosis analysis

As show in Fig. 3, in contrast with control group, the liver of ConA group had serious pathological changes, the liver issue become harden, the color become deepen, external surface of liver studded with a lot of regenerated nodules. Meanwhile the pathological histology analysis results showed that inflammatory cellular infiltration and necrosis were observed in ConA group. However, compared with ConA group the liver lesions of the ConA+ drug group were improved dramatically, and the inflammatory cells and necrosis of hepatic cells were significantly reduced.

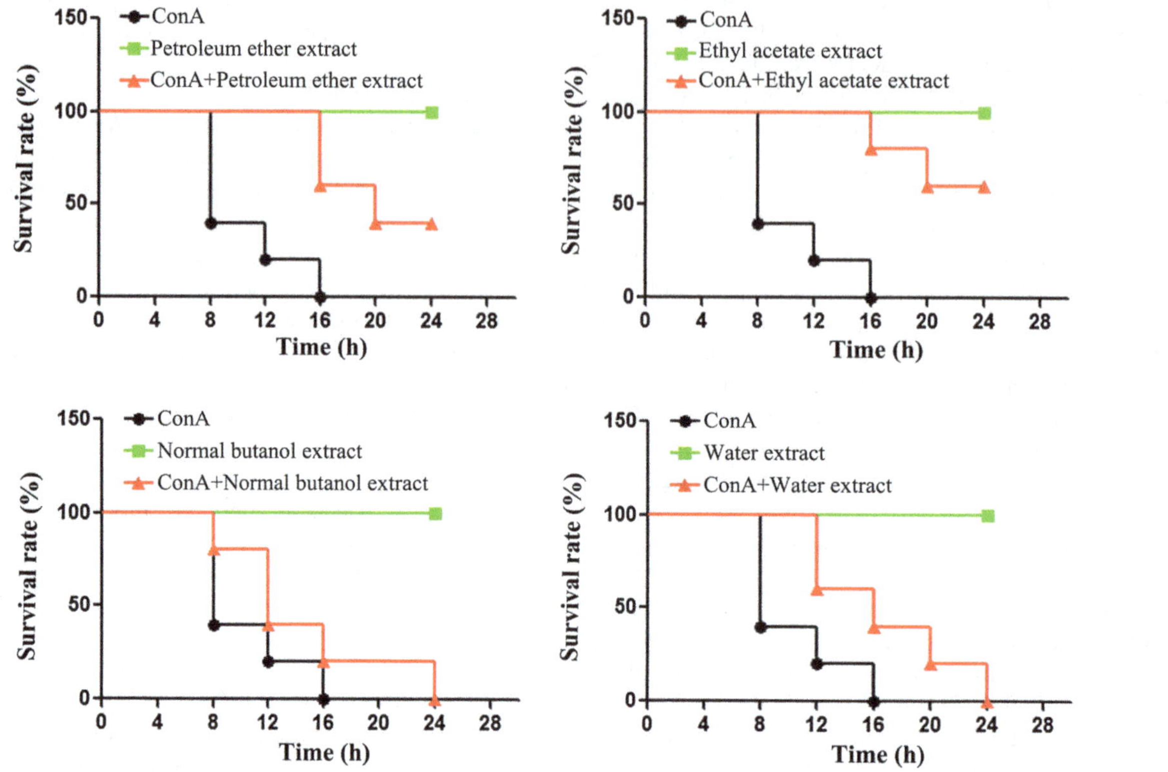


**Fig. 1** The experiments of survival rate. Female Balb/C mice were injected with four extracts of *D. heterophyllum* (50 mg/kg) at 2 h before the injection of a lethal dose of Con A (20 mg/kg). The survival rate was monitored at different times after Con A administration

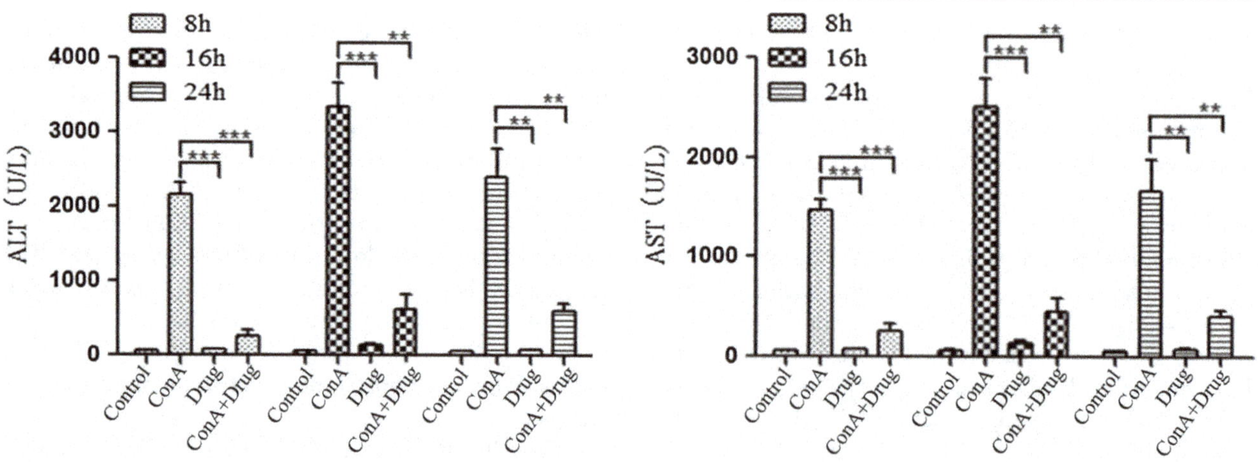


**Fig. 2** The level of ALT and AST in mice serum. Female Balb/C mice were injected with ethyl acetate extract of *D. heterophyllum* (50 mg/kg) at 2 h before the injection of a dose of Con A (12.5 mg/kg). Serum transaminase ALT and AST levels were determined 8, 16, 14 h after the Con A injected. Data expressed as mean ± SD (n = 8; **P < 0.01 and ***P < 0.001)

The result of TUNEL assay indicated that there were many apoptotic cells in the ConA group compared with control group. And the DAPI staining showed nuclear condensation and fragmentation in ConA group. But, in ConA+ drug group, there were only A few apoptotic cells were detected. The results suggested that ConA can induce a large amount of apoptotic cells, and the ethyl acetate extracts of *D. heterophyllum* can improve apoptotic (Fig. 4).

Liver damage caused by hepatitis is associated with excessive activation of the immune responses. It has

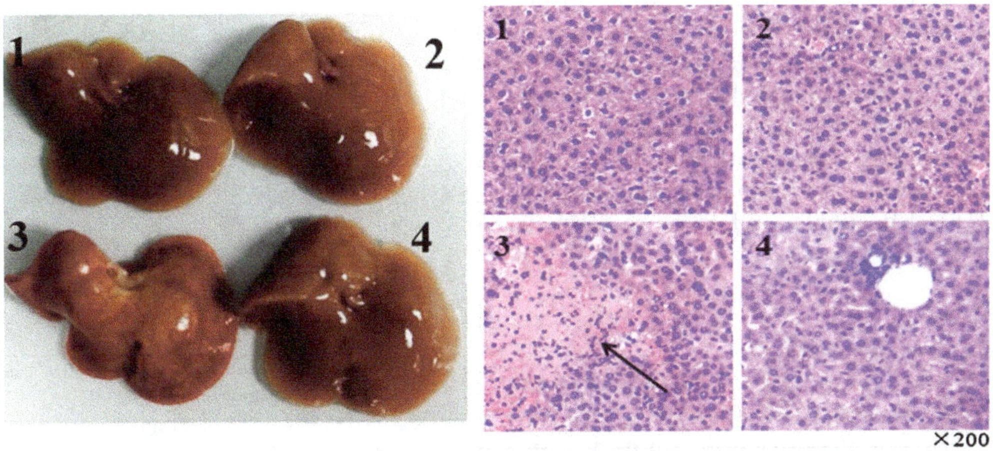


**Fig. 3** Morphological change and histological pathologic qualitative evaluation for the livers of experimental group and controlled group. Mice were sacrificed at 24 h after the Con A injection. The liver were harvested from control (*1*), drug (*2*), Con A (*3*) and Con A + drug (*4*) mice respectively. And the liver tissues of the four groups were fixed and strain with hematoxylin and eosin (H and E). The arrow indicates massive cell death in the liver section. Original magnification ×200

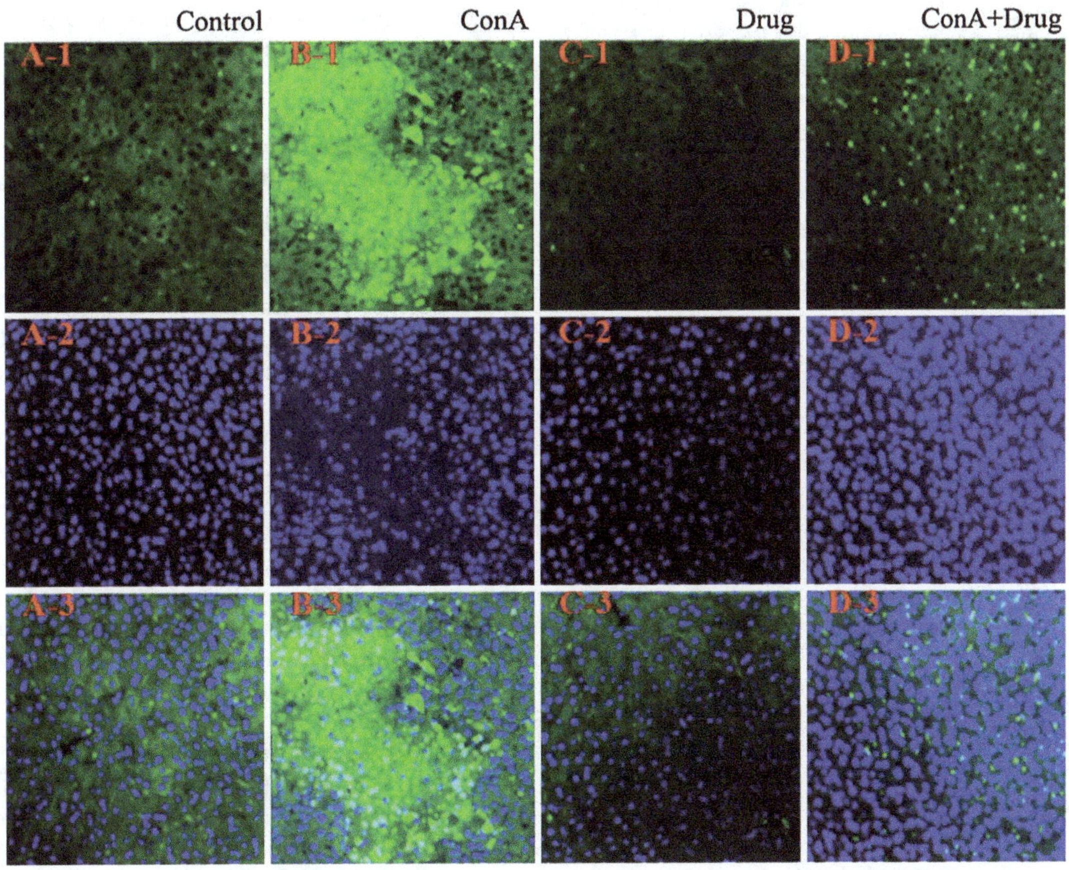


**Fig. 4** Analysis of liver cell apoptosis. Mice were sacrificed at 24 h after Con A injection. The livers were harvested from control (**A**), Con A (**B**), drug (**C**) and ConA+ drug (**D**) groups respectively. Liver tissues from the four groups were fixed and stained with TUNEL (1), DAPI (2) and overlap (3). Original magnification ×200

been reported that TNF-α participate in various forms of liver damage, such as viral, toxic and autoimmune hepatitis, and play an important role in ConA-induced hepatitis. It has also been reported that Kupffer cells secrete a large amount of TNF-α to aggravate the liver damage. Kupffer cells play an important role in T cell activation-induced liver injury. However, the mechanism may not be that simple, and more research is needed.

### Antioxidant activity

#### *DPPH free radical scavenging activity*

In this study, the antioxidant activities of four *D. heterophyllum* extracts and standards such as α-tocopherol, BHT and BHA were determined using the DPPH coloring method in different concentrations (10–500 μg/mL). Figure 5 illustrates the radical-scavenging activity of the different fractions of *D. heterophyllum* and three standards antioxidants. The scavenging effect of four *D. heterophyllum* extracts and standards on the DPPH radical decreased in the order: EtOAc > α-tocopherol > BHT > BHA > *n*-BuOH > petroleum ether > water, which was 88.8, 81.6, 80.4, 57.3, 33.7, 12, 9.1 % at a concentration of 50 μg/mL, respectively. Figure 5 also illustrated that the free radical scavenging activity of these samples increased with increasing concentration obviously.

#### *ABTS·⁺ free radical scavenging activity*

In this experiment, BHT, BHA and α-tocopherol were served as standard antioxidants. As seen in (Fig. 6), the extracts of *D. heterophyllum* had effective ABTS·⁺ free radical scavenging activity, and the scavenging activity of four fractions of *D. heterophyllum* and standards

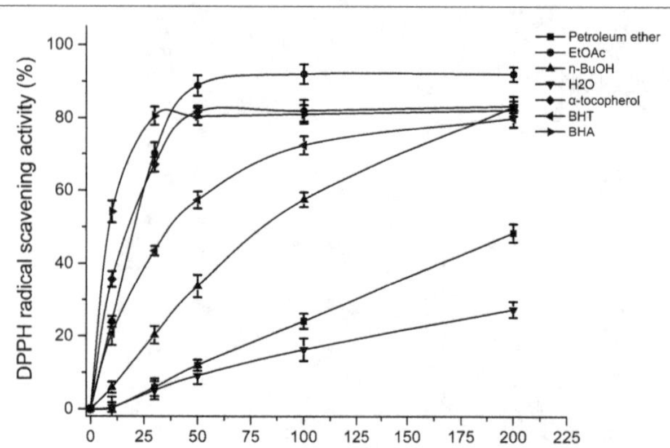


**Fig. 5** Free radical scavenging activity of various fractions of *D. heterophyllum*. 0, 10, 30, 50, 100, 200 μg/mL of four *D. heterophyllum* extracts were measured by the DPPH method at 517 nm. BHT, BHA and α-tocopherol were used as standard antioxidants

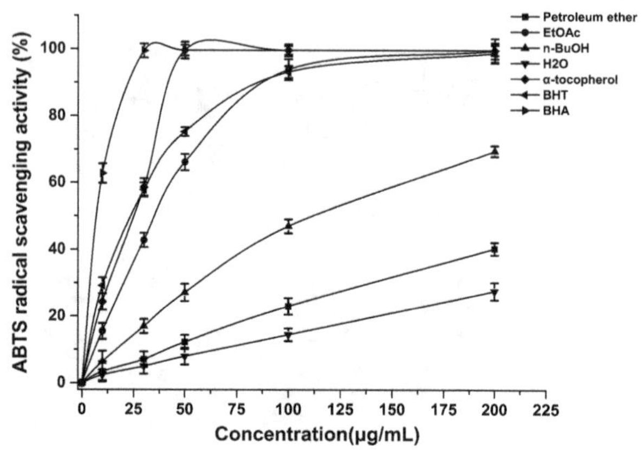


**Fig. 6** Free radical scavenging activity of various fractions of *D. heterophyllum*. 0, 10, 30, 50, 100, 200 μg/mL of four *D. heterophyllum* extracts were measured by the ABTS method at 734 nm. BHT, BHA and α-tocopherol were used as standard antioxidants

antioxidants on the ABTS·⁺ decreased in that order: α-tocopherol = BHA > BHT > EtOAc > n-BuOH > petroleum ether > water, which were 99.56, 99.56, 75.2, 66.2, 27.1, 12.2 and 8.0 %, at the concentration of 50 μg/mL, respectively.

### Anti-lipidperoxidation activity

The petroleum ether fraction, EtOAc fraction, *n*-BuOH fraction, water fraction of *D. heterophyllum* and standards compounds exhibited effective Anti-lipidperoxidation activity. At the concentration of 50 μg/mL, the effects of four fractions of *D. heterophyllum* and α-tocopherol, BHT and BHA on lipid peroxidation of the linoleic acid emulsion are shown in (Fig. 7). The percentage inhibition of peroxidation in the linoleic acid system by 50 μg/mL concentrations of petroleum ether fraction, EtOAc fraction, *n*-BuOH fraction and water fraction of *D. heterophyllum* was found to be 78.2, 98.7, 83.0 and 67.5 %, respectively. On the other hand, the percentage inhibition of 50 μg/mL concentrations of standards α-tocopherol, BHT and BHA was found to be 91.5, 97.5 and 93.75 %, respectively. The result showed that the EtOAc fraction of *D. heterophyllum* had the highest anti-lipidperoxidation activity.

### Ferric reducing ability of plasma (FRAP)

The $FeSO_4 \cdot 7H_2O$ solutions (0.2–2 mM/L) was used to performed the calibration curves, regression equations of it was $Y = 0.68006X + 0.00916$, $R^2 = 0.999$. The ferric reducing antioxidant power of four *D. heterophyllum* extracts and standards antioxidants were expressed as FRAP value: 1 FRAP value = 1 mmol/L $FeSO_4$.

α-tocopherol, BHA and BHT had the reducing antioxidant power of 2.1 ± 0.01, 1.8 ± 0.01 and 1.3 ± 0.02 FRAP

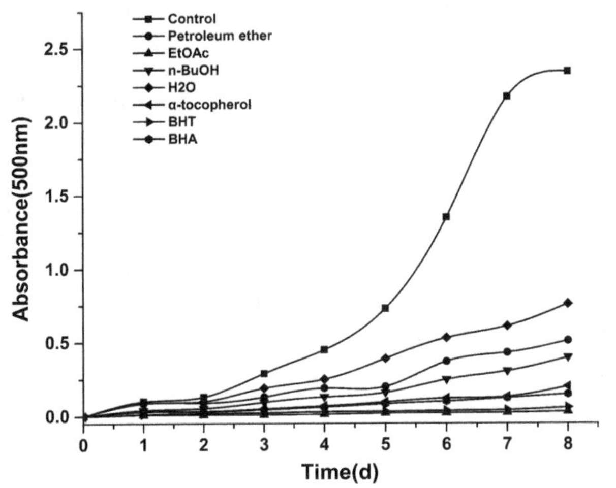


**Fig. 7** Anti-lipidperoxidation activity of various fractions of *D. heterophyllum* (50 μg/mL) was measured by the FTC method at 500 nm. BHT, BHA and α-tocopherol were used as standard antioxidants

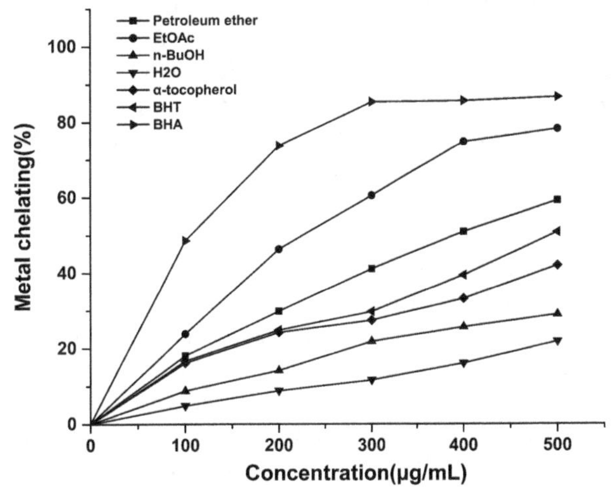


**Fig. 9** Ferrous ions chelating effect of different concentrations of Petroleum ether fraction, EtOAc fraction, *n*-BuOH fraction, water fraction α-tocopherol, BHT and BHA

value. The EtOAc fraction had highest reducing antioxidant power with 1.6 ± 0.01 FRAP value, and followed by petroleum ether, *n*-BuOH and water fraction. The (Fig. 8) illustrates the dose response of each individual antioxidant tested was linear, showing that reducing antioxidant activity is not concentration-dependent, at least over the concentration ranges tested in this study.

### *Ferrous metal ions chelating activity*

As shown in (Fig. 9), the metal chelating activities of petroleum ether fraction, EtOAc fraction, *n*-BuOH fraction, water fraction and standards were concentration-dependent. The difference between the four *D. heterophyllum* factions and the control was statistically significant ($P < 0.01$). In addition, the percentages of metal scavenging capacity of 0.5 mg/mL concentration of Petroleum ether fraction, EtOAc fraction, *n*-BuOH fraction, water fraction, α-tocopherol, BHT and BHA were found as 59.2, 78.2, 29.0, 21.8, 42.0, 58.8, 86.6 %, respectively. These results show that the ferrous ion chelating effect of EtOAc fraction was statistically similar to BHA ($P > 0.05$) but higher than BHT ($P < 0.05$) and α-tocopherol ($P < 0.05$), the petroleum ether fraction was similar to BHT ($P > 0.05$), but higher than α-tocopherol ($p < 0.05$), the *n*-BuOH and water fraction had the lowest Ferrous metal ions chelating activity.

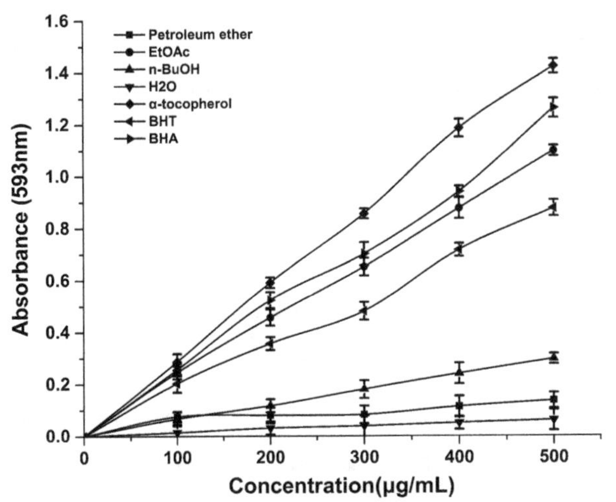


**Fig. 8** Linearity of FRAP: dose–response lines for solutions of four *D. heterophyllum* extracts. BHT, BHA and α-tocopherol were used as standard antioxidants

### *Total phenolic contents of fractions of Dracocephalum heterophyllum*

By way of the literature review (Li et al. 2012), total phenolic was of significant positive correlations with antioxidant levels. In this paper, the total phenolic contents of petroleum ether, EtOAc, *n*-BuOH and water fractions were measured by the Folin-Ciocalteu reagent. The gallic acid (100–800 μg/mL) was used to performed the calibration curves, the regression equations of it was $Y = 0.001X + 0.1693$, $R^2 = 0.9957$, and the total phenolic contents was expressed as Gallic acid mg/g of dry material. The results showed that EtOAc fraction had the highest phenolic contents of 433.7 mg gallic acid/g, *n*-BuOH and petroleum ether fraction had the phenolic contents of 110.7 and 70.7 mg/g. The water fraction had the lowest phenolic contents of 31.7 mg Gallic acid/g of dry material. This results indirectly reflect the antioxidant activities of *Dracocephalum heterophyllum* extracts.

The correlation of total phenolic contents (TPC) with DPPH assay ($R^2 = 0.9637$, $P < 0.05$), ABTS assay ($R^2 = 0.9638$, $P < 0.05$), FTC assay ($R^2 = 0.8203$, $P < 0.05$) and FRAP assay ($R^2 = 0.9991$, $P < 0.05$), respectively. The results demonstrated that the antioxidant activities of *D. heterophyllum* extracts had high correlation with total phenolic contents (TPC).

#### *The chemical constituent of Dracocephalum heterophyllum*

The main chemical constituent of petroleum ether and ethyl acetate extracts were also isolated using chromatographic techniques and identified by NMR spectroscopic methods. 5 compounds were obtained from petroleum ether extract and 4 compounds were obtained from Ethyl acetate extract. These compounds were identified as oleanolic acid (**1**) (Seebacher et al. 2003), ursolic acid (**2**) (Seebacher et al. 2003), pomolic acid (**3**) (Cheng and Cao 1992), 2α- hydroxyl ursolic acid (**4**) (Kuang et al. 1989) and Apigenin-7-O-rutinoside (**5**) (Baris et al. 2011; Wang et al. 2003) from petroleum ether extracts, rosmarinic acid (**6**) (Kuhnt et al. 1994), methyl rosmarinate (**7**) (Kohda et al. 1989), luteolin (**8**) (Geiger et al. 1995; Xu et al. 2014) and diosmetin (**9**) (Sahu et al. 1998) from ethyl acetate extracts. The compound **1–4** were triterpenes, the compound **5**, **8** and **9** were flavonoids, the compound **6** and **7** were phenolic acids (Fig. 10). The $^1H$ and $^{13}C$ NMR data are as in Additional file 1. The main chemical constituents were isolated and identified, but the anti-hepatitis, antioxidants activities of these compounds require further study.

## Conclusions

In this paper, the mouse fulminant hepatitis model induced by ConA was first used to research the anti-hepatitis activity of petroleum ether, ethyl acetate, n-butyl alcohol and water extracts of *D. heterophyllum*. The antioxidant activity was also studied by some experiment in vitro, and the bioactive substances were isolated using chromatographic techniques and identified by NMR spectroscopic methods. Our results indicated that the ethyl acetate extract of *D. heterophyllum* had the highest anti-hepatitis and antioxidants activities, followed by the petroleum ether extract. The bioactive substances may be triterpenes, flavonoids and phenolic acids, the ethyl acetate extracts of *D. heterophyllum* may be possible candidates in developing anti-hepatitis medicine.

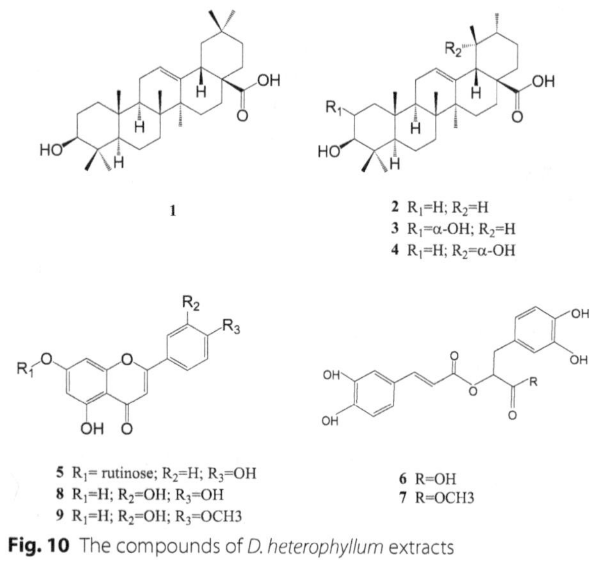


**Fig. 10** The compounds of *D. heterophyllum* extracts

#### Authors' contributions

Q-QS principal conducted the experiment and wrote the manuscripts. Associate professor Q-LW helped us to modify the manuscripts. All authors participated in drafting the manuscript. All authors read and approved the final manuscript.

#### Author details

[1] Key Laboratory of Tibetan Medicine Research, Northwest Institute of Plateau Biology, CAS, Xining, China. [2] University of Chinese Academy of Science, Beijing, China.

#### Acknowledgements

The authors gratefully appreciated the support by senior engineer Li-Juan Mei, from Northwest Institute of Plateau Biology, CAS, who helped us identifying the authenticity of *D. heterophyllum*. The authors also like to thank the grant support from Advanced technology research and development program of Qinghai Province (2014-GX-220).

#### Competing interests

The authors declare that they have no competing interests.

#### References

Baris O, Karadayi M, Yanmis D, Guvenalp Z, Bal T, Gulluce M (2011) Isolation of 3 flavonoids from *Mentha longifolia* (L.) Hudson subsp. *longifolia* and determination of their genotoxic potentials by using the *E. coli* WP2 test system. J Food Sci 76:T212–217

Benzie IFF, Strain JJ (1996) The ferric reducing ability of plasma (FRAP) as a measure of "antioxidant power": the FRAP assay. Anal Biochem 239:70–76

Blois MS (1958) Antioxidant determinations by the use of a stable free radical. Nature 181:1199–1200

Cheng DL, Cao XP (1992) Pomolic acid derivatives from the root of *Sanguisorba* officinalis. Phytochemistry 31:1317–1320

Chisari FV (1995) Hepatitis B virus transgenic mice: insights into the virus and the disease. Hepatology 22:1316–1325

Dinis TCP, Madeira VMC, Almeida LM (1994) Action of phenolic derivatives (acetaminophen, salicylate, and 5-aminosalicylate) as inhibitors of membrane lipid-peroxidation and as peroxyl radical scavengers. Arch Biochem Biophys 315:161–169

Dwyer J, Johnson C (1981) The use of concanavalin A to study the immunoregulation of human T cells. Clin Exp Immunol 46:237

Geiger H, Voigt A, Seeger T, Zinsmeister HD, Lopezsaez JA, Perezalonso MJ, Velasconegeruela A (1995) Cyclobartramiatriluteolin, a unique triflavonoid from *Bartramia Stricta*. Phytochemistry 39:465–467

Gulcin I (2006) Antioxidant activity of caffeic acid (3,4-dihydroxycinnamic acid). Toxicology 217:213–220

Hollinger FB, Liang TJ (2001) Hepatitis B virus. Fields Virol 4:2971–3036

Kikuzaki H, Nakatani N (1993) Antioxidant effects of some ginger constituents. J Food Sci 58:1407–1410

Kohda H, Takeda O, Tanaka S, Yamasaki K, Yamashita A, Kurokawa T, Ishibashi S (1989) Isolation of inhibitors of adenylate-cyclase from dan-shen, the root of *Salvia-Miltiorrhiza*. Chem Pharm Bull 37:1287–1290

Kuang HX, Kasai R, Ohtani K, Liu ZS, Yuan CS, Tanaka O (1989) Chemical constituents of pericarps of *Rosa davurica* Pall., a traditional Chinese medicine. Chem Pharm Bull 37:2232–2233

Kuhnt M, Rimpler H, Heinrich M (1994) Lignans and other compounds from the mixe Indian medicinal plant *Hyptis verticillata*. Phytochemistry 36:485–489

Li XC, Lin J, Gao YX, Han WJ, Chen DF (2012) Antioxidant activity and mechanism of *Rhizoma* Cimicifugae. Chem Cent J 6:10

Mahmood U, Kaul VK, Singh V, Lal B, Negi HR, Ahuja PS (2005) Volatile constituents of the cold desert plant *Dracocephalum heterophyllum* Benth. Flavour Frag J 20:173–175

Peng HF (1984) The effect of *Dracocephalum Heterophyllum* benth on the tolerance of acute hypoxia in the living organism. Med J Chin People's Lib Army

Pumpens P, Grens E, Nassal M (2002) Molecular epidemiology and immunology of hepatitis B virus infection-An update. Intervirology 45:218–232

Ragazzi E, Veronese G (1973) Quantitative-analysis of phenolic compounds after thin-layer chromatographic separation. J Chrom 77:369–375

Sahu NP, Achari B, Banerjee S (1998) 7,3′-Dihydroxy-4′-methoxyflavone from seeds of *Acacia Farnesiana*. Phytochemistry 49:1425–1426

Schultz U, Grgacic E, Nassal M (2004) Duck hepatitis B virus: an invaluable model system for HBV infection. Adv Virus Res 63:1–70

Seebacher W, Simic N, Weis R, Saf R, Kunert O (2003) Complete assignments of 1H and13C NMR resonances of oleanolic acid, 18α-oleanolic acid, ursolic acid and their 11-oxo derivatives. Magn Reson Chem 41:636–638

Tiegs G, Hentschel J, Wendel A (1992) A T cell-dependent experimental liver injury in mice inducible by concanavalin A. J Clin. Invest 90:196–203

Wang MF, Simon JE, Aviles IF, He K, Zheng QY, Tadmor Y (2003) Analysis of antioxidative phenolic compounds in artichoke (*Cynara scolymus* L.). J Agric Food Chem 51:601–608

Wang Y, Menne S, Baldwin BH, Tennant BC, Gerin JL, Cote PJ (2004) Kinetics of viremia and acute liver injury in relation to outcome of neonatal woodchuck hepatitis virus infection. J Med Virol 72:406–415

Wieland SF, Spangenberg HC, Thimme R, Purcell RH, Chisari FV (2004) Expansion and contraction of the hepatitis B virus transcriptional template in infected chimpanzees. P Natl Acad Sci USA 101:2129–2134

Xu H, Hu G, Dong J, Wei Q, Shao H, Lei M (2014) Antioxidative activities and active compounds of extracts from *Catalpa* plant leaves. Sci World J 2014:857982

Zhang CJ, Li HY, Yun T, Fu YH, Liu CM, Gong B, Neng BJ (2008) Chemical composition, antimicrobial and antioxidant activities of the essential oil of Tibetan herbal medicine *Dracocephalum heterophyllum* Benth. Nat Prod Res 22:1–11

Zhang CJ, Li W, Li HY, Wang YL, Yun T, Song ZP, Song Y, Zhao XW (2009) In vivo and in vitro antiviral activity of five Tibetan medicinal plant extracts against herpes simplex virus type 2 infection. Pharm Bio 47:598–607

Zoulim F (2001) Detection of hepatitis B virus resistance to antivirals. J Clin Virol 21:243–253

# 21

# Antioxidant responses and photosynthetic behaviors of *Kappaphycus alvarezii* and *Kappaphycus striatum* (Rhodophyta, Solieriaceae) during low temperature stress

Hu Li[1,2,3], Jianguo Liu[1,2*], Litao Zhang[1,2] and Tong Pang[1,2]

**Abstract**

**Background:** *Kappaphycus* are farmed in tropical countries as raw material for carrageenan, which is widely used in food industry. The sea area available for farming is one limiting factor in the production of seaweeds. Though cultivation is spreading into subtropical regions, the lower seawater temperature is an important problem encountered in subtropical regions for the farming of *Kappaphycus*. This research of physiological response to low temperature stress will be helpful for screening *Kappaphycus* strains for growth in a lower temperature environment.

**Results:** Responses of antioxidant systems and photosystem II (PSII) behaviors in *Kappaphycus alvarezii* and *Kappaphycus striatum* were evaluated during low temperature treatments (23, 20, 17 °C). Compared with the controls at 26 °C, the $H_2O_2$ concentrations increased in both species when the thalli were exposed to low temperatures (23, 20, 17 °C), but these increases were much greater in *K. striatum* than in *K. alvarezii* thalli, suggesting that *K. striatum* suffered more oxidative stress. The activities of some important antioxidant enzymes (e.g. superoxide dismutase and ascorbate peroxidase) and the hydroxyl free radical scavenging capacity were substantially higher at 23, 20 and 17 °C than at the control 26 °C in *K. alvarezii*, indicating that the antioxidant system of *K. alvarezii* enhanced its resistance to low temperature. However, no significant increases of antioxidant enzymes activities were observed at 20 and 17 °C in *K. striatum*. In addition, both the maximal efficiency of PSII photochemistry ($F_V/F_m$) and the performance index ($PI_{ABS}$) decreased significantly in *K. striatum* at 23 °C, indicating that the photosynthetic apparatus was damaged at 23 °C. In contrast, no significant decreases of either $F_V/F_m$ or $PI_{ABS}$ were observed in *K. alvarezii* at 23 °C.

**Conclusions:** It is concluded that *K. alvarezii* has greater tolerance to low temperature than *K. striatum*.

**Keywords:** Antioxidant systems, *Kappaphycus alvarezii*, *Kappaphycus striatum*, Low temperature stress, Photosystem II

## Background

*Kappaphycus alvarezii* and *Kappaphycus striatum* (Rhodophyta, Solieriaceae), two important carrageenophyte species, are used as the major commercial source of κ-carrageenan, which is widely utilized as a gelling and stabilizing agent for some food products, including frozen desserts, chocolate-flavoured milk, cottage cheese dressings and soyamilk (Doty 1973; Glenn and Doty 1990; Bixler et al. 2001; Bindu and Levine 2011). The farming of *Kappaphycus* occurs mainly in tropical southeast Asian countries (Ohno et al. 1994), such as the Philippines and Indonesia, serving as a means of livelihood for locals and as a source of revenue for the economy of these countries (Bindu 2011; Bixler and Porse 2011; Ganzon-Fortes et al. 2012). However, although the carrageenan processing technology in

*Correspondence: jgliu@qdio.ac.cn
[1] Key Laboratory of Experimental Marine Biology, National & Local Joint Engineering Laboratory of Ecological Mariculture, Institute of Oceanology, Chinese Academy of Sciences, 7 Nanhai Road, Qingdao 266071, China
Full list of author information is available at the end of the article

this industry is mature (Bixler et al. 2001), the supply of carrageenan is not adequate for the global demand. The sea area available for farming is one limiting factor in the production of seaweeds for carrageenan extraction. Currently, cultivation is spreading into subtropical regions (i.e. lower temperature regions for these seaweeds). For example, cultivation during the summer season was introduced in Shikoku Island, Southern Japan (Ohno et al. 1994). The lower seawater temperature is an important problem encountered in subtropical regions, or other low-temperature sea areas, for the farming of *Kappaphycus*.

Photosynthetic processes are exceedingly sensitive to low temperature (Allen and Ort 2001). When plants are exposed to low-temperature stress, photosynthetic enzymes may be degraded and photo-damage may occur, which would lead to decreased photosynthetic activity. The reduced photosynthesis may result in the accumulation of excess energy, which leads to the generation of numerous reactive oxygen species (ROS), such as hydrogen peroxide ($H_2O_2$) and hydroxyl free radical (·OH), which are harmful to the plant's photosynthetic apparatus. Collén et al. (1994) found that *Kappaphycus* and *Eucheuma* exposed to abiotic or biotic stresses produced $H_2O_2$. The accumulation of $H_2O_2$ within plant tissues is thought to be a crucial event in initiating adequate antioxidant responses in plants (Gechev et al. 2003). In order to gain a better understanding of the antioxidant response of *Kappaphycus* under low temperature stress, we measured the activities of antioxidant enzymes [e.g. superoxide dismutase (SOD), catalase (CAT) and ascorbate peroxidase (APX)], hydroxyl free radical (·OH) scavenging capacity and the level of malondialdehyde (MDA), an important index of oxidative lesions (Barros et al. 2003), in *K. alvarezii* and *K. striatum* thalli under low temperature stress. Moreover, since PSII is the most sensitive part of the photosynthetic apparatus, changes of PSII state may indicate changes in physiological processes in plants (Lim et al. 2014). In recent years, the chlorophyll (Chl) a fluorescence transient (O-J-I-P) has become an important tool in photosynthesis research (Strasser et al. 2004; Strauss et al. 2007), and especially in studying PSII behaviors. The changes of the PSII antenna, the reaction centers, the oxidation–reduction of the plastoquinone (PQ) pool and the electron flow limitations on the acceptor side of PSII, can be reflected by the O-J-I-P transient modulation in higher plants and algae (Stirbet and Govindjee 2012). The effect of low temperature stress on PSII was further assessed in *K. alvarezii* and *K. striatum* thalli by means of the Chl a fluorescence transient.

## Methods

### Algal sample collection and pretreatment

*Kappaphycus alvarezii* (Rhodophyta, Solieriaceae) (reddish brown, Fig. 1a) and *Kappaphycus striatum* (Rhodophyta, Solieriaceae) (green, Fig. 1b) were harvested in Li'an bay, Hainan Province, China (18°24′N, 110°03′E) (Liu et al. 2009; Pang et al. 2015). Selected healthy thalli of both species were cleaned to remove sludge, miscellaneous algal contaminants and other impurities from the surface of thalli. The algal thalli were acclimatized in a plastic tank (50 × 40 × 40 cm) with 40 L of natural seawater (about 23–24 °C, salinity 33 ‰) collected in Li'an bay (18°24′N, 110°03′E) and irradiated with light (55 μmol photons $m^{-2}$ $s^{-1}$) in a 12:12 h light:dark (L:D) cycle. They were acclimatized to their new surroundings for a day in the tank before performing the following low temperatures treatments assay.

### Low temperature treatments

Portions of thalli approximately 3 mm in diameter and 1 cm long were cut from the apices of the collected algal thalli. The samples were taken from same parts of the thallus to avoid any difference between individuals. These were all placed in plastic tanks (50 × 40 × 40 cm) with 40 L of natural seawater (about 23–24 °C, salinity 33 ‰) collected in Li'an bay (18°24′N, 110°03′E) and irradiated with a fluorescent lamp at approximately 55 μmol photons $m^{-2}$ $s^{-1}$ (Fig. 1c). Sea temperature was adjusted and maintained by an automatic temperature control unit (Guangdong Zhenhua Electrical Appliance Co., LTD). The control temperature was 26 °C, which is the optimum temperature for these two species (Lideman et al. 2013). The samples of two species were exposed to the control temperature (26 °C) and low temperatures 23, 20 and 17 °C for 2 h, respectively. The temperature treatments of *K. alvarezii* and *K. striatum* samples were shown in Fig. 1c.

### Crude extracts and antioxidant system parameters

After treatment at each temperature, algal samples of each species were immediately froze in liquid nitrogen, labeled and stored in a freezer at −80 °C for the following assay (Fariduddin et al. 2014).

The frozen samples (about 2 g) were ground in a mortar in liquid nitrogen and placed in 10-mL centrifuge tubes in ice. The algal powder was mixed with 3 mL of pre-cooled extraction buffer (50 mM phosphate buffer, pH 7.2, 0.1 mM EDTA) and homogenized using an S10 homogenizer (Ningbo Xinzhi Biotechnology Co., Ltd) in an ice-water bath (Nagarani and Kumaraguru 2012). Cell debris was removed by centrifugation at 12,000*g* for 10 min at 4 °C and the supernatant was used for the following measurements of antioxidant system parameters. Each temperature treatment had three repeats for each species.

The concentration of $H_2O_2$ in the algal thalli was assessed using a commercially available kit—Hydrogen

**Fig. 1** The photos of *K. alvarezii* and *K. striatum*. **a** *Kappaphycus alvarezii* thalli; **b** *Kappaphycus striatum* thalli; **c** A photo of temperature treatments of *K. alvarezii* and *K. striatum* samples in the tank. Algal samples were irradiated at 55 μmol photons $m^{-2}$ $s^{-1}$. The sea water temperature in the tank was adjusted and maintained by an automatic temperature-control bar

Peroxide Assay kit (Nanjing Jiancheng Bioengineering Institute). $H_2O_2$ was bound with molybdenic acid to form a complex, which was measured spectrophotometrically at 405 nm and the concentration of $H_2O_2$ was then calculated (Liu et al. 2013).

Total SOD activity in algal thalli was determined by using a Total Superoxide Dismutase Assay kit (Nanjing Jiancheng Bioengineering Institute) based on measuring the enzyme's ability to inhibit the photochemical reduction of tetrazolium blue. The developed blue color was measured spectrophotometrically at 550 nm (Das et al. 2000). CAT activity was quantified by the method of Malanga et al. (1999). The absorbance decay of $H_2O_2$ was monitored by spectrophotometry at 240 nm, with a molar extinction coefficient of $\varepsilon = 39.4\ mM^{-1}\ cm^{-1}$. APX activity was measured as described by Nakano and Asada (1981). Ascorbate consumption was monitored by spectrophotometry at 290 nm, with a molar extinction coefficient of $\varepsilon = 2.8\ mM^{-1}\ cm^{-1}$.

Hydroxyl free radical scavenging capacity (HFRSC) was determined by using a commercially available kit—Hydroxyl Free Radical Assay kit (Nanjing Jiancheng Bioengineering Institute) based on the deoxyribose degradation assay described by Halliwell et al. (2006).

To evaluate the extent of oxidative damage in algal cells under low temperature stress, the concentration of MDA, a marker of lipoperoxidation, was measured in the algal thalli by the method of Fraga et al. (1988).

#### Protein determination

Enzyme activities were expressed relative to protein concentrations, which were estimated by the method of Deng et al. (2013) using bovine serum albumin as standard. We employed the dying method of Coomassie brilliant blue G-250 to determine the protein concentration of algal samples, measuring the absorbency at the wavelength of 595 nm. Each temperature treatment had three repeats for each species.

#### Chlorophyll (Chl) a fluorescence transient measurement

Fast Chla fluorescence transients were measured with a Plant Efficiency Analyzer (Handy-PEA, Hansatech Instruments Ltd, UK). Each transient obtained from the samples was analyzed according to the JIP-test (Strasser 1978; Strasser et al. 2004; Stirbet and Govindjee 2012; Zhang et al. 2012a). Each temperature treatment had 10 repeats for each species.

The following variables were calculated directly or indirectly.

1. $\varphi_{Po}$ (or $F_V/F_m$), the maximum PSII photochemical efficiency, namely the maximum quantum yield of primary photochemistry. $F_V/F_m = \varphi_{Po} = TR_O/ABS$, where $TR_O$ and ABS denote the trapped and absorbed excitation energy fluxes, respectively.
2. $\psi_o$, the probability that a trapped exciton moved an electron into the electron transport chain beyond (plasto)quinone ($Q_A^-$). $\psi_o = ET_O/TR_O$, where $ET_O$

and $TR_O$ denote the electron transport and trapped excitation energy fluxes, respectively.

3. $\varphi_{Eo}$, the probability that an absorbed photon moved an electron into the electron transport chain further than $Q_A^-$, namely quantum yield of electron transport. $\varphi_{Eo} = ET_O/ABS$.
4. $PI_{ABS}$, the performance index on an absorption basis. $PI_{ABS} = (RC/ABS)\cdot[\varphi_{Po}/(1 - \varphi_{Po})]\cdot[\psi_o/(1 - \psi_o)]$, where RC/ABS, $\varphi_{Po}/(1 - \varphi_{Po})$ and $\psi_o/(1 - \psi_o)$ reflect the efficiency of the reaction centers, the light energy absorption efficiency and the accepting efficiency of the electron acceptor, respectively.

### Statistical analysis

Data are presented as mean ± SD for at least three replicate measurements (the number (n) of replicates is presented in each figure caption). To determine the statistical significance between experimental groups, the data were examined using a *t* test ($p < 0.05$).

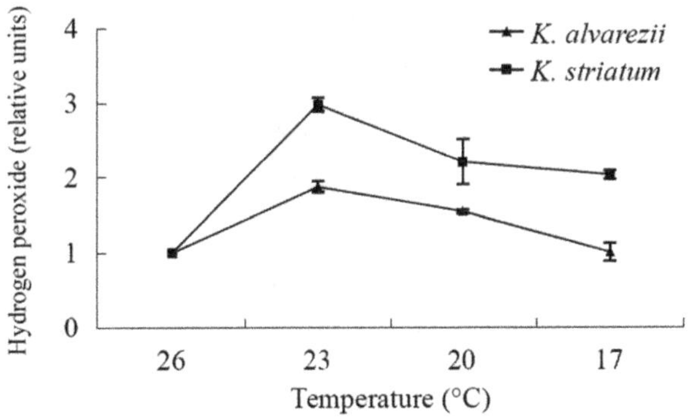


**Fig. 2** Concentrations of $H_2O_2$ in thalli of *K. alvarezii* and *K. striatum* exposed to low temperatures (23, 20 and 17 °C) for 2 h, relative to the concentration of the control of each species at 26 °C [$H_2O_2$ concentration in thalli at the control temperature (26 °C) was 17.7 ± 1.3 mmol $g^{-1}$ prot for *K. alvarezii* and 29.2 ± 2.0 mmol $g^{-1}$ prot for *K. striatum*]. Data are means ± SD of three replicates (n = 3)

## Results

### Changes in $H_2O_2$ concentrations in *Kappaphycus alvarezii* and *Kappaphycus striatum* thalli

The highest concentrations of $H_2O_2$ in both *K. alvarezii* and *K. striatum* thalli were observed at 23 °C; they decreased at 20 and 17 °C. A $H_2O_2$ concentration similar to the control (26 °C) was observed at 17 °C in *K. alvarezii* thalli but *K. striatum* thalli at 17 °C still had a higher $H_2O_2$ concentration than the control at 26 °C (Fig. 2). In other words, compared with the controls at 26 °C, the $H_2O_2$ concentration increased in both species when the thalli were exposed to low temperatures (23, 20, 17 °C), but the increase was much greater in *K. striatum* thalli than in *K. alvarezii* thalli. For example, the $H_2O_2$ concentration at 23 °C was 1.9 times as much as that of the control in *K. alvarezii* thalli, while the $H_2O_2$ concentration at 23 °C was 3 times as much as that of the control in *K. striatum* thalli. These results suggest that *K. striatum* thalli suffered more serious oxidative stress from $H_2O_2$ than *K. alvarezii* thalli.

### Antioxidant system responses of *K. alvarezii* and *K. striatum* thalli

Low temperature treatments resulted in distinct responses from the three major antioxidant enzymes (SOD, APX and CAT) in both *K. alvarezii* and *K. striatum* thalli. Compared with the control, SOD activities and APX activities increased significantly in both species when algal thalli were exposed to 23 °C (SOD activities at 23 °C were 2.1 times and 2.8 times as much as that of the controls in *K. alvarezii* thalli and *K. striatum* thalli, respectively; APX activities at 23 °C were 5.5 times and 3.1 times as much as that of the controls in *K. alvarezii* thalli and *K. striatum* thalli, respectively) (Fig. 3). At even lower temperatures (20 and 17 °C), SOD activities and APX activities of *K. alvarezii* thalli remained elevated (SOD activities were 1.8 times and 1.7 times as much as that of the control, at 20 and 17 °C, respectively; APX activities were 5.3 times and 5.9 times as much as that of the control, at 20 and 17 °C, respectively), whereas those of *K. striatum* thalli had no significant differences at either 20 or 17 °C compared with the control. Although CAT activities did not vary significantly in the two species after exposure to low temperatures (23, 20 and 17 °C) (Fig. 3), *K. alvarezii* thalli maintained higher SOD and APX activities whereas *K. striatum* thalli didn't, under the low temperatures (20 and 17 °C). In addition, compared with the control, hydroxyl free radical scavenging capacity (HFRSC) had significant increases at lowered temperatures (HFRSC were 2.9 times, 2.1 times and 2.1 times as much as the control, at 23, 20 and 17 °C, respectively) in *K. alvarezii* thalli. Hydroxyl free radical scavenging capacity (HFRSC) of *K. striatum* thalli peaked at 23 °C (HFRSC was 2.2 times as much as the control), but dropped to the same level as the control at the lower temperatures, 20 and 17 °C (Fig. 3). Taken together, these results suggest that the antioxidant system of these species increased their activities against the oxidative stress generated by exposure to a low temperature of 23 °C. The antioxidant system of *K. striatum* thalli was unable to resist the lower temperatures (20 and 17 °C) whereas that of *K. alvarezii* thalli still had resistance to the lower temperatures (20 and 17 °C). No significant differences of MDA were found in the two species during exposure to lower temperatures (23 and 20 °C) (Fig. 3). At the lowest temperature of 17 °C, a significant increase of MDA (MDA concentration at 17 °C was 2.2 times as much as

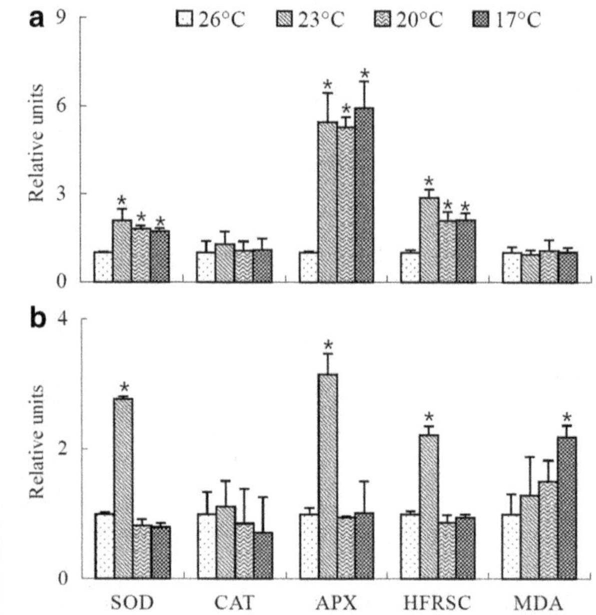


**Fig. 3** Superoxide dismutase (SOD), catalase (CAT), and ascorbate peroxidase (APX) activities and hydroxyl free radical scavenging capacity (HFRSC) and concentrations of malonaldehyde (MDA) in *Kappaphycus alvarezii* (**a**) and *Kappaphycus striatum* (**b**) thalli exposed to 26 °C (the control) and low temperatures (23, 20 and 17 °C) for 2 h, respectively. All values of the above parameters are expressed relative to those of their controls (the controls are taken as 1) and are presented in relative units. (Data are means ± SD of three replicates (n = 3). For each parameter, for each species, the treatments with a *star* (*) are significantly different from the respective control, p < 0.05)

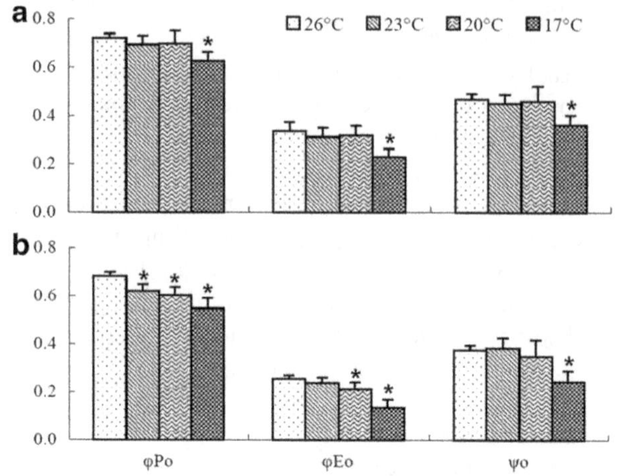


**Fig. 4** Energy flux ratios ($\varphi_{Po}$, $\varphi_{Eo}$ and $\psi_o$) of *Kappaphycus alvarezii* (**a**) and *Kappaphycus striatum* (**b**) thalli after exposure to 26 °C (the control) and low temperatures (23, 20 and 17 °C) for 2 h, respectively. (Means ± SD of ten replicates (n = 10) are presented. For each species, the treatment with a *star* (*) indicates that there is a significant difference between this temperature treatment and the control (26 °C), p < 0.05)

the control) was observed in *K. striatum* thalli, which confirmed the occurrence of oxidative damage; however, no significant increase of MDA was observed in *K. alvarezii* thalli (Fig. 3).

### Changes in photosynthetic efficiencies (or energy flux ratios)—$\varphi_{Po}$, $\psi_o$ and $\varphi_{Eo}$

Compared with the control, a significant decrease in the maximum PSII photochemical efficiency ($F_V/F_m = \varphi_{Po}$) was observed in *K. striatum* thalli at 23 °C (the $F_V/F_m$ value was 0.62 while the control was 0.68) (Fig. 4), indicating that photo-damage had occurred. However, in *K. alvarezii* thalli, there was no significant decrease of $F_V/F_m$ at that temperature (23 °C). When exposed to 17 °C, both *K. alvarezii* and *K. striatum* thalli had significant decreases in $F_V/F_m$ compared to the control ($F_V/F_m$ values were 0.63 and 0.55 at 17 °C in *K. alvarezii* and *K. striatum* thalli, respectively, whereas the respective controls were 0.72 and 0.68). Because the value of $F_V/F_m$ largely reflected the physiological activity in plants, it was inferred that physiological activity in *K. alvarezii* might have greater tolerance to low temperature stress than that in *K. striatum*.

With regard to the excitation efficiency for electron transport beyond $Q_A^-$ ($\psi_o$), both *K. alvarezii* and *K. striatum* thalli only exhibited significant decreases of $\psi_o$ when the temperature dropped to 17 °C ($\psi_o$ values were 0.36 and 0.24 in *K. alvarezii* and *K. striatum* thalli, respectively, whereas the respective controls were 0.47 and 0.37) (Fig. 4). At 23 and 20 °C, no significant decreases of $\psi_o$ were observed, indicating that the electron transport in the PSII electron transport chain was not influenced significantly in either species. Compared with the control, the quantum yield of electron transport ($\varphi_{Eo}$) in *K. striatum* thalli decreased significantly at 20 and 17 °C ($\varphi_{Eo}$ values were 0.21 and 0.13 at 20 and 17 °C, respectively, while the control was 0.26) (Fig. 4) while a significant decrease of $\varphi_{Eo}$ was only observed at 17 °C ($\varphi_{Eo}$ value was 0.23 at 17 °C while the control was 0.34) in *K. alvarezii* thalli. These results suggest that the lowered temperature of 20 °C led to a decreased quantum yield of electron transport in *K. striatum* thalli, whereas *K. alvarezii* thalli maintained a normal capacity for electron transport at that same temperature (20 °C), indicating again that *K. alvarezii* had stronger low temperature tolerance than *K. striatum* with regard to electron transport.

Since the parameters of photosynthetic efficiencies only reflect the energy cascade from light absorption to electron transport and don't indicate the specific PSII component damaged by low temperature stress, the performance index on an absorption basis ($PI_{ABS}$) and its three variables RC/ABS, $\varphi_{Po}/(1-\varphi_{Po})$, $\psi o/(1-\psi o)$, were calculated to probe the specific component of PSII possibly damaged by low temperature stress.

### Changes in the performance index—$PI_{ABS}$

$PI_{ABS}$ is more sensitive to changes of photosynthetic activity than the maximal photochemical efficiency ($F_V/F_m$). Compared with the control, there were significant decreases of $PI_{ABS}$ in *K. striatum* thalli when the temperature was below 23 °C ($PI_{ABS}$ values were 0.33, 0.32 and 0.12 at 23, 20 and 17 °C, respectively, while the control was 0.61), whereas a significant decrease of $PI_{ABS}$ was observed in *K. alvarezii* thalli only when the temperature was dropped to 17 °C ($PI_{ABS}$ value was 0.45 at 17 °C while the control was 1.71) (Fig. 5). This result suggests that the 3 °C decrement in temperature from 26 °C resulted in decreased photosynthetic activity of PSII in *K. striatum* thalli, whereas a 9 °C decrement from 26 °C was needed for a negative impact on the PSII activity in *K. alvarezii* thalli.

In order to explore responses of different parts of PSII in the two species during the low temperature stress, three variables RC/ABS, $\varphi_{Po}/(1-\varphi_{Po})$ and $\psi_o/(1-\psi_o)$, were calculated. RC/ABS, $\varphi_{Po}/(1-\varphi_{Po})$ and $\psi_o/(1-\psi_o)$ reflect the efficiency of reaction centers, the light energy absorption efficiency and the accepting efficiency of the electron acceptor, respectively. When the temperature was dropped to 23 °C, RC/ABS decreased significantly in *K. striatum* thalli (the ratio RC/ABS was 0.32 at 23 °C while the control was 0.46), whereas no significant decrease of RC/ABS was observed in *K. alvarezii* until the temperature was decreased to 17 °C (the ratio RC/ABS was 0.45 at 17 °C while the control was 0.72) (Fig. 5). This suggests that reaction centers of PSII in *K. alvarezii* thalli maintained normal efficiency at 23 and 20 °C whereas the efficiency of PSII reaction centers in *K. striatum* thalli had decreased at these temperatures. At 17 °C, PSII reaction centers of both *K. alvarezii* and *K. striatum* were damaged. However, both *K. alvarezii* and *K. striatum* thalli had significant decreases of $\varphi_{Po}/(1-\varphi_{Po})$ at 23 °C (the ratios $\varphi_{Po}/(1-\varphi_{Po})$ were 2.2 and 1.6 in *K. alvarezii* and *K. striatum* thalli, respectively, while the respective controls were 2.6 and 2.2), indicating that the low temperature reduced the efficiency of the PSII antenna in both species (Fig. 5). In addition, the accepting efficiencies of the PSII electron acceptor ($\psi_o/(1-\psi_o)$) in both *K. alvarezii* thalli and *K. striatum* thalli were impacted negatively at 17 °C (the ratios $\psi_o/(1-\psi_o)$ were 0.58 and 0.32 in *K. alvarezii* and *K. striatum* thalli, respectively, while the respective controls were 0.90 and 0.60), indicating that the acceptor sides of PSII in these two species were damaged (Fig. 5). In summary, in both species, PSII antennae were first damaged by the low temperature stress; damages to the PSII acceptor side was observed only at the lowest temperature (17 °C). In addition, the PSII reaction centers of *K. striatum* had a lower resistance to the low temperature stress than those of *K. alvarezii*.

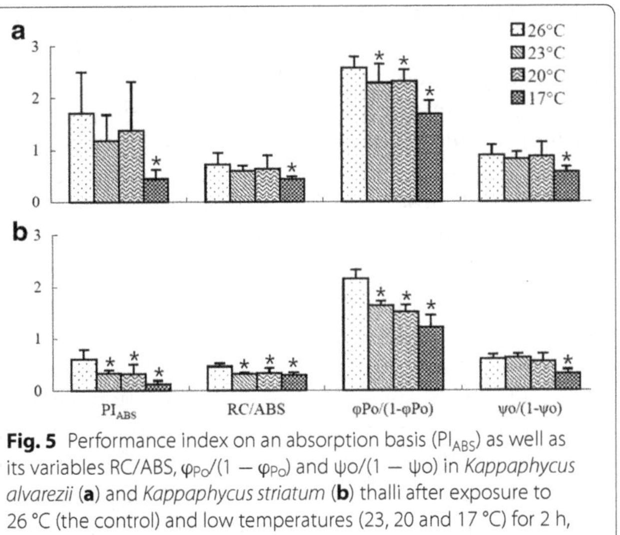


**Fig. 5** Performance index on an absorption basis ($PI_{ABS}$) as well as its variables RC/ABS, $\varphi_{Po}/(1-\varphi_{Po})$ and $\psi o/(1-\psi o)$ in *Kappaphycus alvarezii* (**a**) and *Kappaphycus striatum* (**b**) thalli after exposure to 26 °C (the control) and low temperatures (23, 20 and 17 °C) for 2 h, respectively. (Data are means ± SD of ten replicates (n = 10). For each parameter, for each species, the treatments with a star (*) are significantly different from the respective control (26 °C), $p < 0.05$)

## Discussion

PSII contains some sensitive proteins, and the state of PSII is easily affected by low temperature stress (Čajánek et al. 1998; Tang et al. 2007). After exposure to low temperature stress, physiological dysfunctions, including alteration of metabolic processes, increase in ROS and reduction of photosynthetic capacity occur in plants (Allen and Ort 2001). The harsh conditions imposed by various environmental stresses result in an increase of $H_2O_2$, which is the main ROS in *K. alvarezii* thalli (Reis et al. 2011; Ling et al. 2015). Moreover, Barros et al. (2006) found that when *K. alvarezii* thalli were subjected to two co-stressors, chilling and high light, the $H_2O_2$ level increased. In our study, increases of $H_2O_2$ levels (Fig. 2) were observed during exposure to lowered temperatures (23 and 20 °C) in both *K. alvarezii* and *K. striatum* thalli, suggesting that oxidative stress occurred, but this increases were much greater in *K. striatum* thalli than in *K. alvarezii* thalli, indicating that *K. striatum* suffered more serious oxidative stress from $H_2O_2$ than *K. alvarezii*. *Kappaphycus alvarezii* at 17 °C had a similar $H_2O_2$ level as the control at 26°. That might be due to the increase of $H_2O_2$—scavenging enzyme (such as APX) activities. Barros et al. (2003) found that *K. alvarezii* released $H_2O_2$ into the surrounding medium to avoid harmful accumulation of $H_2O_2$. But *K. striatum* at 17 °C still had a higher $H_2O_2$ level than the control at 26°. Above results indicated that *K. striatum* might suffer greater oxidative damage than *K. alvarezii* during the low temperature.

Some important antioxidant enzymes (SOD, APX and CAT) of *K. alvarezii* increased their activities when *K. alvarezii* thalli is exposed to the cold and high light stress (Barros et al. 2006). In our study, SOD and APX activities were substantially higher at 23 °C than at 26 °C (Fig. 3) in both two species, indicating the antioxidant systems of both species enhanced their antioxidant activities against the oxidative stress caused by the low temperature. However, compared with the control, no significant increases of SOD and APX activities in *K. striatum* thalli were observed at the lower temperatures of 20 and 17 °C, whereas SOD and APX activities of *K. alvarezii* thalli were still higher at these lower temperatures (Fig. 3), indicating that these two antioxidant enzymes of *K. alvarezii* thalli persisted at the lower temperatures (20 and 17 °C) but those of *K. striatum* thalli didn't. However, it is worth noting that no significant difference of CAT activity was observed during any of the low temperature treatments in either of the two species. Because CAT and APX are two important $H_2O_2$-scavenging enzymes, why doesn't CAT respond to the increase of $H_2O_2$? Barros et al. (2003) found that APX was an effective system for decomposing lower concentrations of $H_2O_2$ in *K. alvarezii*, while higher concentrations of $H_2O_2$ woud be more efficiently decomposed by CAT. Ascorbate peroxidase (APX) has a greater affinity for $H_2O_2$ than CAT. So it is surmised that concentrations of $H_2O_2$ didn't reach the level at which CAT would play a role during these low temperature treatments. Not only $H_2O_2$, but also ·OH (a powerful oxidizing agent) can cross membranes and may oxidize a variety of compounds (Nagarani and Kumaraguru 2012). Hydroxyl free radical scavenging capacity (HFRSC) was congruent with SOD and APX activities for each species (Fig. 3). A significant increase of HFRSC persisted in *K. alvarezii* thalli, whereas no significant increase of HFRSC was observed in *K. striatum* thalli at the lower temperatures (20 and 17 °C), indicating that *K. alvarezii* retained a stronger capacity for eliminating ·OH at 20 and 17 °C but *K. striatum* did not. Besides, the lipoperoxidation damage was only observed in *K. striatum* thalli (MDA, a marker of lipoperoxidation, increased significantly at 17 °C in that species) (Fig. 3). From the different antioxidant responses of these species to low temperature stress, it was inferred that *K. alvarezii* thalli might have greater low temperature resistance than *K. striatum* thalli.

Given that PSII is the most sensitive part of the photosynthetic apparatus, changes of the PSII state can indicate changes in physiological processes during low temperature stress (Lim et al. 2014). The maximum PSII photochemical efficiency decreased ($F_V/F_m$ values decreased) in both species as a consequence of low temperature stress, indicating that photo-damage had occurred such that photosynthetic activities decreased in both species.

The energy flux ratios ($\varphi_{Po}$, $\psi_o$ and $\varphi_{Eo}$), which were measured by Chl a fluorescence transients, can reflect the energy utilization from light absorption to electron transport and indicate the photosynthetic efficiencies during the energy cascade (Strasser et al. 2004). Li et al. (2014) found that the maximum quantum yield for primary photochemistry ($\varphi_{Po}$), the efficiency at which a trapped exciton moved an electron into the electron transport chain beyond $Q_A^-$ ($\psi_o$) and the quantum yield of electron transport ($\varphi_{Eo}$), all decreased sharply under temperature stress. In our results, a significant decrease of $\varphi_{Po}$ was observed in *K. striatum* thalli when the temperature dropped to 23 °C, whereas there was no significant decrease of $\varphi_{Po}$ at that temperature in *K. alvarezii* thalli (Fig. 4). On the other hand, $\varphi_{Eo}$ decreased sharply at 20 °C in *K. striatum* thalli while no significant decrease of $\varphi_{Eo}$ was observed in *K. alvarezii* thalli (Fig. 4), indicating electron transport of *K. alvarezii* PSII wasn't affected at this temperature. These results suggest that *K. alvarezii* thalli had greater low temperature tolerance than *K. striatum* thalli wtih regard to the PSII energy utilization.

$PI_{ABS}$ is more sensitive to changes in photosynthetic activity than the maximal photochemical efficiency ($F_V/F_m$) and can reflect the states of different photosynthetic components (Strasser et al. 2004). Strauss et al. (2007) found that low soil temperature decreased the value of $PI_{ABS}$ seriously in soybeans. For red seaweeds, Zhang et al. (2012b) studied the Chl fluorescence characteristics of different strains of *Porphyra yezoensis* under low temperature stress. They found that the strains Sulian and WT had higher photochemical efficiency of PSII than the strain Sutong under low temperature stress, indicating that Sulian and WT might have better tolerance to low temperature. In our study, significant decreases of $PI_{ABS}$ were observed in *K. striatum* thalli at 23, 20 and 17 °C, whereas a significant decrease of $PI_{ABS}$ was observed in *K. alvarezii* thalli only at 17 °C. It indicated that the PSII of *K. alvarezii* thalli might have greater low temperature tolerance than the PSII of *K. striatum* thalli. In order to determine the low temperature tolerance of individual photosynthetic components of PSII (reaction centers, antennae and acceptor sides of PSII) for the two species, the values of RC/ABS, $\varphi_{Po}/(1-\varphi_{Po})$, $\psi o/(1-\psi o)$, were measured (Fig. 5). From our results, PSII antennae were first damaged by the low temperature stress and damage of the PSII acceptor side was observed only at 17 °C in both species. RC/ABS decreased significantly in *K. striatum* thalli at 23 and 20 °C while no significant difference of RC/ABS was observed in *K. alvarezii* thalli at these temperatures. This suggests that efficiency of PSII reaction centres in *K. alvarezii* thalli remained at a normal level while the efficiency of PSII reaction centers in *K. striatum* thalli decreased at lower temperatures (23 and 20 °C). It was inferred that PSII reaction centers of *K. striatum* might have weaker resistance to low temperature stress than those of *K. alvarezii.*

## Conclusions

κ-Carrageenan-producing *K. alvarezii* and *K. striatum* are important commercial red seaweeds. Strains of *Kappaphycus* could be selected for temperature tolerance by using the Chl fluorescence method, which is non-destructive to the algal tissue and takes only a few secs to record (Zhang et al. 2012b). This study has shown that the excess $H_2O_2$ generated by low temperature stress can indirectly damage the PSII apparatus and these two species have somewhat different physiological responses to low temperature stress. By comparing the several important parameters of the Chl fluorescence transient, it is concluded that the PSII of *K. alvarezii* has a stronger low temperature tolerance than *K. striatum.* Given that photosynthesis has been considered to be the physiological process most sensitive to temperature damage and that PSII is the most sensitive part of the photosynthetic apparatus (Wen et al. 2005), it is inferred that *K. alvarezii* thalli might have stronger low temperature tolerance than *K. striatum* thalli as a consequence of PSII characteristics. Our results will be helpful for further studies on the low temperature tolerance of *Kappaphycus* and for screening *Kappaphycus* strains for growth in a lower temperature environment (or higher latitude regions).

#### Abbreviations

APX: ascorbate peroxidase; CAT: catalase; Chl: chlorophyll; HFRSC: hydroxyl free radical scavenging capacity; $H_2O_2$: hydrogen peroxide; MDA: malondialdehyde; OEC: oxygen-evolving complex; ·OH: hydroxyl free radical; PQ: plastoquinone; PSII: photosystem II; ROS: reactive oxygen species; SOD: superoxide dismutase.

#### Authors' contributions

HL, JGL and TP designed the research; HL performed the experiments and analyzed the data; LTZ and TP analyzed data; HL and JGL wrote the manuscript. All authors read and approved the final manuscript.

#### Author details

[1] Key Laboratory of Experimental Marine Biology, National & Local Joint Engineering Laboratory of Ecological Mariculture, Institute of Oceanology, Chinese Academy of Sciences, 7 Nanhai Road, Qingdao 266071, China. [2] Laboratory for Marine Biology and Biotechnology, Qingdao National Laboratory for Marine Science and Technology, 1 Wenhai Road, Aoshanwei Town, Jimo, Qingdao 266071, China. [3] University of Chinese Academy of Sciences, Beijing 100049, China.

#### Acknowledgements

This work was supported by the National Natural Science Foundation of China (41306154), Special Project for Marine Public Welfare Industry (201505033) and Marine Economy Innovation and Development Fund of Qingdao (2012). The authors would like to thank Zhaoliang Zheng from Lingshui Haotian Company for the provision of *Kappaphycus alvarezii* and *Kappaphycus striatum*, and thank Dr. John van der Meer (Pan-American Marine Biotechnology Association) for his assistance with proofreading.

#### Competing interests

The authors declare that they have no competing interests.

#### References

Allen DJ, Ort DR (2001) Impacts of chilling temperatures on photosynthesis in warm-climate plants. Trends Plant Sci 6:36–42

Barros MP, Granbom M, Colepicolo P, Pedersén M (2003) Temporal mismatch between induction of superoxide dismutase and ascorbate peroxidase correlates with high $H_2O_2$ concentration in seawater from clofibrate-treated red algae *Kappaphycus alvarezii*. Arch Biochem Biophys 420:161–168

Barros MP, Necchi O, Colepicolo P, Pedersén M (2006) Kinetic study of the plastoquinone pool availability correlated with $H_2O_2$ release in seawater and antioxidant responses in the red alga *Kappaphycus alvarezii* exposed to single or combined high light, chilling and chemical stresses. Biochim Biophys Acta 1757:1520–1528

Bindu MS (2011) Empowerment of coastal communities in cultivation and processing of *Kappaphycus alvarezii*—a case study at Vizhinjam village, Kerala, India. J Appl Phycol 23:157–163

Bindu MS, Levine IA (2011) The commercial red seaweed *Kappaphycus alvarezii*—an overview on farming and environment. J Appl Phycol 23:789–796

Bixler HJ, Porse H (2011) A decade of change in the seaweed hydrocolloids industry. J Appl Phycol 23:321–335

Bixler HJ, Johndro K, Falshaw R (2001) Kappa-2 carrageenan: structure and performance of commercial extracts: II. Performance in two simulated dairy applications. Food Hydrocoll 15:619–630

Čajánek M, Štroch M, Lachetová I, Kalina J, Spunda V (1998) Characterization of the photosystem II inactivation of heat-stressed barley leaves as monitored by the various parameters of chlorophyll a fluorescence and delayed fluorescence. J Photochem Photobiol B-Biol 47:39–45

Collén J, Pedersén M, Bornman CH (1994) A stress-induced oxidative burst in *Eucheuma platycladum* (Rhodophyta). Physiol Plant 92:417–422

Das K, Samanta L, Chainy G (2000) A modified spectrophotometric assay of superoxide dismutase using nitrite formation by superoxide radicals. Indian J Biochem Biophys 37:201–204

Deng Y, Wang W, Yu P, Xi Z, Xu L, Li X, He N (2013) Comparison of taurine, GABA, Glu, and Asp as scavengers of malondialdehyde in vitro and in vivo. Nanoscale Res Lett 8:1–9

Doty MS (1973) Farming the red seaweed, Eucheuma, for carrageenans. Micronsica 9:59–73

Fariduddin Q, Khan TA, Yusuf M (2014) Hydrogen peroxide mediated tolerance to copper stress in the presence of 28-homobrassinolide in Vigna radiata. Acta Physiol Plant 36:2767–2778

Fraga CG, Leibovitz BE, Tappel AL (1988) Lipid peroxidation measured as thiobarbituric acid-reactive substances in tissue slices: characterization and comparison with homogenates and microsomes. Free Radic Biol Med 4:155–161

Ganzon-Fortes ET, Trono GC Jr, Villanueva RD, Romero JB, Montaño MNE (2012) 'Endong', a rare variety of the farmed carrageenophyte *Eucheuma denticulatum* (Burman) Collins & Hervey from the Philippines. J Appl Phycol 24:1107–1111

Gechev T, Willekens H, Montagu MV, Eacute DI, Camp WIMV, Toneva V, Minkov I (2003) Different responses of tobacco antioxidant enzymes to light and chilling stress. J Plant Physiol 160:509–515

Glenn EP, Doty MS (1990) Growth of the seaweeds *Kappaphycus alvarezii, K. striatum* and *Eucheuma denticulatum* as affected by environment in Hawaii. Aquaculture 84:245–255

Halliwell B, Grootveld M, Gutteridge JMC (2006) Methods for the measurement of hydroxyl radicals in biochemical systems: deoxyribose degradation and aromatic hydroxylation. In: Glick D (ed) Methods of biochemical analysis, vol 33. Wiley, Newyork, pp 59–90

Li L, Li X, Xu X, Lin L, Zeng F (2014) Effects of high temperature on the chlorophyll a fluorescence of *Alhagi sparsifolia* at the southern Taklamakan Desert. Acta Physiol Plant 36:243–249

Lideman Nishihara GN, Noro T, Terada R (2013) Effect of temperature and light on the photosynthesis as measured by chlorophyll fluorescence of cultured Eucheuma denticulatum and Kappaphycus sp. (Sumba strain) from Indonesia. J Appl Phycol 25:399–406

Lim PE, Tan J, Phang SM, Nikmatullah A, Hong DD, Sunarpi H, Hurtado AQ (2014) Genetic diversity of *Kappaphycus* Doty and *Eucheuma* J. Agardh (Solieriaceae, Rhodophyta) in Southeast Asia. J Appl Phycol 26:1253–1272

Ling ALM, Yasir S, Matanjun P, Bakar MFA (2015) Effect of different drying techniques on the phytochemical content and antioxidant activity of *Kappaphycus alvarezii*. J Appl Phycol 27:1717–1723

Liu J, Pang T, Wang L, Li J, Lin W (2009) The reasons causing catastrophic death in tropical carrageenan producing seaweeds and their difference in resistance to illness. Oceanol Limnol Sin 40:235–241

Liu L, Liu Y, Cui J, Liu H, Liu Y, Qiao W, Sun H, Yan C (2013) Oxidative stress induces gastric submucosal arteriolar dysfunction in the elderly. World J Gastroenterol 19:9439–9446

Malanga G, Kozak RG, Puntarulo S (1999) N-Acetylcysteine-dependent protection against UV-B damage in two photosynthetic organisms. Plant Sci 141:129–137

Nagarani N, Kumaraguru AK (2012) Chemical characterization, temperature stability, and enzymatic studies on edible marine algae *Kappaphycus alvarezii* (Doty). J Aquat Food Prod Technol 21:480–492

Nakano Y, Asada K (1981) Hydrogen peroxide is scavenged by ascorbate-specific peroxidase in Spinach chloroplasts. Plant Cell Physiol 22:867–880

Ohno M, Largo DB, Ikumoto T (1994) Growth rate, carrageenan yield and gel properties of cultured kappa-carrageenan producing red alga *Kappaphycus alvarezzi* (Doty) Doty in the subtropical waters of Shikoku, Japan. J Appl Phycol 6:1–5

Pang T, Liu J, Liu Q, Li H, Li J (2015) Observations on pests and diseases affecting a eucheumatoid farm in China. J Appl Phycol 27:1975–1984

Reis MO, Necchi O, Colepicolo P, Barros MP (2011) Co-stressors chilling and high light increase photooxidative stress in diuron-treated red alga *Kappaphycus alvarezii* but with lower involvement of $H_2O_2$. Pestic Biochem Phys 99:7–15

Stirbet A et al (2012) Chlorophyll a fluorescence induction: a personal perspective of the thermal phase, the J-I-P rise. Photosynth Res 113:15–61

Strasser RJ (1978) The grouping model of plant photosynthesis. In: Akoyunoglou G (ed) Chloroplast Development. Elsevier/North Holland Biomedical Press, Amsterdam, pp 513–524

Strasser RJ, Tsimilli-Michael M, Srivastava A (2004) Analysis of the chlorophyll a fluorescence transient. In: Papageorgiou GC, Govindjee A (eds) Chlorophyll fluorescence: a signature of photosynthesis, advances in photosynthesis and respiration. Springer, Dordrecht, pp 321–362

Strauss AJ, Krüger GH, Strasser RJ, Van Heerden PD (2007) The role of low soil temperature in the inhibition of growth and PSII function during dark chilling in soybean genotypes of contrasting tolerance. Physiol Plantarum 19:89–105

Tang Y, Chen M, Xu Y, Kuang T (2007) Changes in thermostability of photosystem II and leaf lipid composition of rice mutant with deficiency of light-harvesting chlorophyll *a/b* protein complexes. J Integr Plant Biol 49:515–522

Wen X, Qiu N, Lu Q, Lu C (2005) Enhanced thermotolerance of photosystem II in salt-adapted plants of the halophyte *Artemisia anethifolia*. Planta 220:486–497

Zhang L, Zhang Z, Gao H, Meng X, Yang C, Liu J, Meng Q (2012a) The mitochondrial alternative oxidase pathway protects the photosynthetic apparatus against photodamage in *Rumex* K-1 leaves. BMC Plant Biol 12:40

Zhang T, Shen Z, Xu P, Zhu J, Lu Q, Shen Y, Wang Y, Yao C, Li J, Wang Y (2012b) Analysis of photosynthetic pigments and chlorophyll fluorescence characteristics of different strains of *Porphyra yezoensis*. J Appl Phycol 24:881–886

# PERMISSIONS

All chapters in this book were first published in BS, by Springer; hereby published with permission under the Creative Commons Attribution License or equivalent. Every chapter published in this book has been scrutinized by our experts. Their significance has been extensively debated. The topics covered herein carry significant findings which will fuel the growth of the discipline. They may even be implemented as practical applications or may be referred to as a beginning point for another development.

The contributors of this book come from diverse backgrounds, making this book a truly international effort. This book will bring forth new frontiers with its revolutionizing research information and detailed analysis of the nascent developments around the world.

We would like to thank all the contributing authors for lending their expertise to make the book truly unique. They have played a crucial role in the development of this book. Without their invaluable contributions this book wouldn't have been possible. They have made vital efforts to compile up to date information on the varied aspects of this subject to make this book a valuable addition to the collection of many professionals and students.

This book was conceptualized with the vision of imparting up-to-date information and advanced data in this field. To ensure the same, a matchless editorial board was set up. Every individual on the board went through rigorous rounds of assessment to prove their worth. After which they invested a large part of their time researching and compiling the most relevant data for our readers.

The editorial board has been involved in producing this book since its inception. They have spent rigorous hours researching and exploring the diverse topics which have resulted in the successful publishing of this book. They have passed on their knowledge of decades through this book. To expedite this challenging task, the publisher supported the team at every step. A small team of assistant editors was also appointed to further simplify the editing procedure and attain best results for the readers.

Apart from the editorial board, the designing team has also invested a significant amount of their time in understanding the subject and creating the most relevant covers. They scrutinized every image to scout for the most suitable representation of the subject and create an appropriate cover for the book.

The publishing team has been an ardent support to the editorial, designing and production team. Their endless efforts to recruit the best for this project, has resulted in the accomplishment of this book. They are a veteran in the field of academics and their pool of knowledge is as vast as their experience in printing. Their expertise and guidance has proved useful at every step. Their uncompromising quality standards have made this book an exceptional effort. Their encouragement from time to time has been an inspiration for everyone.

The publisher and the editorial board hope that this book will prove to be a valuable piece of knowledge for researchers, students, practitioners and scholars across the globe.

# LIST OF CONTRIBUTORS

**Chao Gu, Zhi-Hua Guo, Ping-Ping Hao, Guo-Ming Wang, Zi-Ming Jin and Shao-Ling Zhang**
State Key Laboratory of Crop Genetics and Germplasm Enhancement, Center of Pear Engineering Technology Research, Nanjing Agricultural University, Nanjing 210095, China

**Xu Zeng, Yuanyuan Li, Hong Ling, Sisi Liu, Mengmeng Liu, Juan Chen and Shunxing Guo**
Institute of Medicinal Plant Development, Chinese Academy of Medical Sciences & Peking Union Medical College, Beijing 100193, People's Republic of China

**Özge Çelik, Alp Ayan and Çimen Atak**
Department of Molecular Biology and Genetics, Faculty of Science and Letters, Istanbul Kultur Univesity, Ataköy, 34156 Istanbul, Turkey

**Jonathan P. Kendon, Helen Sandford, Rui Fang, Jake Bell and Viswambharan Sarasan**
Natural Capital and Plant Health, Royal Botanic Gardens Kew, Richmond, Surrey TW9 3AB, UK

**Landy Rajaovelona**
Kew Madagascar Conservation Centre, Lot II J 131 B Ambodivoanjo-Ivandry, 101 Antananarivo, Madagascar

**Edward C. Yeung**
Department of Biological Sciences, University of Calgary, Calgary, AB T2N 1N4, Canada

**Jie He, Regina M. P. Lim and Sabrina H. J. Dass**
Natural Sciences and Science Education Academic Group, National Institute of Education, Nanyang Technological University, 1 Nanyang Walk, Singapore 637 616, Singapore

**Tim W. Yam**
Singapore Botanic Gardens, National Parks Board, 1 Cluny Road, Singapore 25956, Singapore

**Hong-Chyi Jhou and Jui-Chu Yu**
The Experimental Forest, National Taiwan University, 55750 Nantou, Taiwan

**Ya-Nan Wang and Chung-I Chen**
Department of Forestry and Resource Conservation, National Taiwan University, 10617 Taipei, Taiwan

**Chung-Shien Wu**
Biodiversity Research Center, Academia Sinica, 11529 Taipei, Taiwan

**Kana Higaki, Kento Rammitsu and Yuki Ogura-Tsujita**
Faculty of Agriculture, Saga University, 1 Honjo-machi, Saga 840-8502, Japan

**Yumi Yamashita and Tomohisa Yukawa**
Tsukuba Botanical Garden, National Museum of Nature and Science, 4-1-1 Amakubo, Tsukuba, Ibaraki 305-0005, Japan

**Shih-Chang Yuan, Shih-Wen Chin, Chen-Yu Lee and Fure-Chyi Chen**
Department of Plant Industry, National Pingtung University of Science & Technology, Pingtung 91201, Taiwan

**Li Yun Yang, Shuang Long Yang and Ming Gong**
School of Life Sciences, Engineering Research Center of Sustainable Development and Utilization of Biomass Energy, Ministry of Education, Key Laboratory of Biomass Energy and Environmental Biotechnology of Yunnan Province, Yunnan Normal University, Kunming 650500, People's Republic of China

**Jun Ying Li, Jun Hong Ma, Tao Pang and Cong Ming Zou**
Yunnan Academy of Tobacco Agricultural Sciences, Kunming 650031, People's Republic of China

**Bin He**
Yunnan Tobacco Leaf Company, Kunming 650218, People's Republic of China

**Mona H. Soliman**
Biology Department, Faculty of Science, Taibah University, Al-Sharm, Yanbu El-Bahr 46429, Kingdom of Saudi Arabia
Botany and Microbiology Department, Faculty of Science, Cairo University, 12613 Giza, Egypt

**Aisha A. M. Alayafi**
Biological Sciences Department, Faculty of Science, University of Jeddah, Jeddah, Kingdom of Saudi Arabia

**Amr A. El Kelish and Abdelghafar M. Abu-Elsaoud**
Botany Department, Faculty of Science, Suez Canal University, Ismailia, Egypt

**Xu Zeng, Yuanyuan Li, Hong Ling, Juan Chen and Shunxing Guo**
Institute of Medicinal Plant Development, Chinese Academy of Medical Sciences, Beijing 100193, People's Republic of China

**Roderick Bouman**
Naturalis Biodiversity Center, Botany 2300 RA, The Netherlands
Hortus Botanicus, Leiden University, Leiden 2311 GJ, The Netherlands
Institute of Biology Leiden, Leiden University, Leiden 2300 RA, The Netherlands

**Peter van Welzen**
Naturalis Biodiversity Center, Botany 2300 RA, The Netherlands
Institute of Biology Leiden, Leiden University, Leiden 2300 RA, The Netherlands

**Sukaibin Sumail**
Herbarium, Sabah Parks, Kota Kinabalu, Sabah, Malaysia

**Antony van der Ent**
Laboratoire Sols et Environnement, Université de Lorraine, INRA, Nancy 54000, France
Centre for Mined Land Rehabilitation, Sustainable Minerals Institute, The University of Queensland, St Lucia, QLD 4072, Australia

**Guillaume Echevarria**
Laboratoire Sols et Environnement, Université de Lorraine, INRA, Nancy 54000, France

**Peter D. Erskine**
Centre for Mined Land Rehabilitation, Sustainable Minerals Institute, The University of Queensland, St Lucia, QLD 4072, Australia

**Kyu Ho Lee, Sang Woo Park, Yeon Jeong Kim and Hak Soo Seo**
Department of Plant Science, College of Agriculture and Life Sciences, Research Institute of Agriculture and Life Sciences, Seoul National University, Gwanakro 200, Gwanak-gu, Seoul 151-921, South Korea

**Yeon Jong Koo**
Department of Biological Chemistry, Chonnam National University, Gwangju 61186, South Korea

**Jong Tae Song**
School of Applied Biosciences, Kyungpook National University, Daegu 41566, South Korea

**Anne Herbst and Hendrik Schubert**
University of Rostock, Biosciences, Albert-Einstein-Straße 3, 18059 Rostock, Germany

**Xu-guang Zheng, Jun-cang Qi, Hong-shan Hui, Li-hao Lin and Feng Wang**
Department of Agronomy, College of Agriculture, Shihezi University, Shihezi, China

**Mahesh Kumar, Ajay Kumar Singh and Jagadish Rane**
School of Drought Stress Management, ICAR-National Institute of Abiotic Stress Management (NIASM), Malegaon, Baramati, Pune 413 115, India

**Susheel Kumar Raina**
School of Drought Stress Management, ICAR-National Institute of Abiotic Stress Management (NIASM), Malegaon, Baramati, Pune 413 115, India
ICAR-Central Institute of Temperate Horticulture, Old Air Field, PO Rangreth, Srinagar, J&K 190007, India

**Venkadasamy Govindasamy**
School of Drought Stress Management, ICAR-National Institute of Abiotic Stress Management (NIASM), Malegaon, Baramati, Pune 413 115, India
Division of Microbiology, ICAR-Indian Agricultural Research Institute, New Delhi 110012, India

**Ram Lal Choudhary**
School of Edaphic Stress Management, ICAR-National Institute of Abiotic Stress Management, Malegaon, Baramati 413 115, India

**Paramjit Singh Minhas**
ICAR-National Institute of Abiotic Stress Management, Malegaon, Baramati 413 115, India

**José Luis Villaseñor and Teresa Terrazas**
Departamento de Botánica, Instituto de Biología, Universidad Nacional Autónoma de México, Apartado Postal 70-367, 04510 Mexico City, Mexico

**Patricia Rivera**
Departamento de Botánica, Instituto de Biología, Universidad Nacional Autónoma de México, Apartado Postal 70-367, 04510 Mexico City, Mexico
Coordinación del Posgrado en Ciencias Biológicas, Circuito de Posgrados Ciudad Universitaria, Coyoacán, 04510 Mexico City, Mexico

**Md Nazim Uddin and Randall William Robinson**
Department of Ecology & Environmental Management, College of Engineering & Science, Victoria University, Melbourne, VIC 8001, Australia
Institute for Sustainability & Innovation, Victoria University, Melbourne, VIC 8001, Australia

**Yan-Duo Tao and Qi-Lan Wang**
Key Laboratory of Tibetan Medicine Research, Northwest Institute of Plateau Biology, CAS, Xining, China

**Hu Li**
Key Laboratory of Experimental Marine Biology, National & Local Joint Engineering Laboratory of Ecological Mariculture, Institute of Oceanology, Chinese Academy of Sciences, 7 Nanhai Road, Qingdao 266071, China
Laboratory for Marine Biology and Biotechnology, Qingdao National Laboratory for Marine Science and Technology, 1 Wenhai Road, Aoshanwei Town, Jimo, Qingdao 266071, China University of Chinese Academy of Sciences, Beijing 100049, China

**Jianguo Liu, Litao Zhang and Tong Pang**
Key Laboratory of Experimental Marine Biology, National & Local Joint Engineering Laboratory of Ecological Mariculture, Institute of Oceanology, Chinese Academy of Sciences, 7 Nanhai Road, Qingdao 266071, China
Laboratory for Marine Biology and Biotechnology, Qingdao National Laboratory for Marine Science and Technology, 1 Wenhai Road, Aoshanwei Town, Jimo, Qingdao 266071, China

**Xian Liu, Jingjing Yin, Jianqiu Yang and Yan Li**
School of Horticulture and Plant Protection of Yangzhou University, Yangzhou, Jiangsu, People's Republic of China

**Wei-Lun Chen and Jyh-Ching Chou**
Department of Natural Resources and Environmental Studies, National Dong Hwa University, Shou-feng, Hualien 97401, Taiwan

**Yen-Peng Ho**
Department of Chemistry, National Dong Hwa University, Shou-feng, Hualien 97401, Taiwan

# Index

www.ingramcontent.com/pod-product-compliance
Lightning Source LLC
LaVergne TN
LVHW080920210926
848485LV00004B/690